Marine Botany

Marine botany is an exciting and highly diverse discipline united by an ecological theme: the effects of, and adaptations to, the marine environment. With the advent in recent years of modern SCUBA and submersibles, marine botanists are now able to explore the marine environment firsthand. The development of such laboratory techniques as electrophoresis, chromatography, manometry, and chemical analyses has also enabled scientists to critically evaluate the adaptive features and acclimating abilities of marine plants.

Marine Botany, the first major text/reference in the field since 1966, brings together the important developments in our understanding of marine plant taxonomy, morphology, cytology, physiology, and biochemistry, in the unifying context of marine ecology. The material is organized in four main sections: a general introduction to marine plant life and the marine environment, highlighting the most economically important marine plants and angiosperms; a concise survey of marine algae; a discussion of environmental and ecological concepts and research methods; and a detailed examination of the principal marine plant communities. Throughout these last two sections, current investigative methods are described, providing readers with a basis for the development of marine research projects. The third section examines geological, physical, and chemical factors in the marine environment, focusing on the impact these factors have on marine plants, and pinpointing the important relationships between marine plants and their environment, interacting animals, and pollution. Practical experimental techniques are presented in sufficient detail for direct field or laboratory application.

The fourth section is the heart of the book, synthesizing material from all of the preceding chapters into a detailed ecological examination of the six major marine plant communities: Lithophytic, Coral Reef, Seagrass, Mangrove, Salt Marsh, and Phytoplankton. Together with relevant research techniques, this section includes up-to-date productivity data, physiological and structural data, zonation information, and descriptions of the morphological, anatomi-

Marine Botany

CLINTON J. DAWES
University of South Florida

A WILEY-INTERSCIENCE PUBLICATION

JOHN WILEY & SONS,   New York · Chichester · Brisbane · Toronto · Singapore

Library of Congress in Publication Data:

Dawes, Clinton J.
  Marine botany.

  "A Wiley-Interscience publication."
  Includes index.
  1. Marine flora.   2. Botany—Ecology.   I. Title.

QK931.D38   581.92   81-7527
ISBN 0-471-07844-1   AACR2

Printed in the United States of America

10 9 8 7 6 5 4 3 2 1

This book is dedicated to all marine botanists who have developed the field through research and teaching, especially E. Yale Dawson and George J. Hollenberg, my teachers, and Michael Neushul and Arthur Mathieson, who shared with me their enthusiasm when we were students. The text is also dedicated to my students and my own family, especially my wife, Kathleen, and to all who have made research and teaching in marine botany a joy.

Preface

Marine botany is an exciting and highly diverse topic united by an ecological theme: the effects of and adaptation to the marine environment. In recent years, the use of scuba and submersibles has allowed marine botanists to explore the marine environment firsthand. The development of such laboratory techniques as electrophoresis, chromatography, manometry, and chemical analyses has enabled scientists to evaluate the adaptive features and the acclimating abilities of marine plants. This text, therefore, attempts to combine the taxonomic, physiological, biochemical, and ecological aspects of marine plants under the single unifying theme of ecology.

Since the publication of E. Yale Dawson's marine botany text in 1966, the field has expanded rapidly. In 1965, I first offered a summer course using that book's title, but the need for a new text stressing ecological concepts and including some basic methodology has become increasingly apparent since then. This book is intended as a general text in marine botany and, similar to Yale Dawson's, was written primarily for upper-level undergraduate and beginning graduate students in biology. It is also intended for more advanced students and independent investigators in the field of marine biology. The book is divided into four sections: Introduction, The Algae, Ecological and Environmental Considerations, and Marine Plant Communities. My intention was to build to the final section, which examines marine plant communities and stresses ecological factors. The last two sections include methods to aid individuals or classes in field and laboratory studies. Marine fungi and bacteria are discussed in two appendices, and the approach to these two groups of organisms is functional, stressing their interaction with marine plants.

A number of marine botanists aided in the preparation of the text, though not all can be mentioned here. The following persons reviewed each of the chapters before submission: Art Mathieson (Chapters 1, 2, 3, 15), Norma Lang (Chapter 4), Ernest Truby (Chapter 5), Wayne Fagerberg (Chapter 6), Don Cheney (Chapter 7), Shirley Van Valkenburg (Chapter 8), Karen Steidinger (Chapter 9), Richard Davis, Jr. (Chapters 10, 11), Dean Martin (Chapter 12), Mike Neushul (Chapters 13, 14), Steven Murray (Chapter 15), Mark Littler (Chapter 16), Ron

Phillips (Chapter 17), Susan Bell and Robert Vadas (Chapter 18), Louis Almodover and Earl McCoy (Chapter 19), Greta Frexyell (Chapter 20), Diane Merner (Appendix A), and Warren Silver (Appendix B). Many persons furnished figures, and in some cases permission for use of a published photograph was obtained from a journal. Acknowledgments for these are given in the figure captions. Mr. Ralph Moon and Mrs. Betty Loraamm helped in the taking and printing of a number of the photographs. I especially thank four artists for producing diagrams and habit sketches: Mrs. Linda Leatherwood (Linda Baunhardt; LB), Mrs. Ana Lisa King (ALS), Miss Robin Holbrook (RRH), and Mrs. Carol Torres (C. Torres). I also wish to thank Mrs. Bonnie Diaz for typing the manuscript and my graduate students for aiding in proofing. I assume full responsibility for the text.

CLINTON J. DAWES

Tampa, Florida
September 1981

Contents

Marine Botany

PART 1

Introduction

CHAPTER 1

Marine Plants and Their Environment

This text is an ecological study of marine plants, their communities, and their environmental characteristics. Marine plants, like freshwater algae and terrestrial plants, range from small unicellular forms to large trees. The importance of marine plants to the marine environment is apparent not only in their productivity but in a variety of other ways such as prevention of substrata removal, filtration of water, and provision of a habitat for animals (Figure 1-1). It is from the primeval oceans that all life, including terrestrial plants, probably arose; thus man has an interest in the forms of plants in the marine environment.

Figure 1-1 The intertidal zone of a rocky coast at Robe, South Australia. The zone is partially exposed at low tide. Because of the heavy surf, seaweeds can extend even into the spray zone above the tides as noted by the dark coloration on the rocks. Over 300 species of algae can be found in the spray, intertidal, and subtidal regions on such a coast.

With 72% of the earch covered by oceans, it is not surprising that marine plants play such an important role in the food chains of the world. Recently man has turned to the oceans as a possible source of energy and organic matter; the marine plants are basic to such studies. Yet our knowledge of marine plants, both microscopic and macroscopic forms, is limited, and we have barely begun to utilize the vast resources present in marine communities. For too long we have considered the marine and environments secondary in importance, and we have used them as dumping grounds for terrestrial wastes.

In this first part, the general groups of marine plants, as well as the marine environment, will be reviewed. A brief summary of each group of marine plants will be included. The second chapter of this part will present data on the economics of marine plants.

MARINE PLANTS AND THE PLANT KINGDOM

Linnaeus (1707-1778) placed plants and animals in kingdoms and described a group of plants (Cryptogama) that were simple and lacked flowering organs. Since Linnaeus, many classifications of living organisms and plants have been proposed. The basic definition of a plant and, in turn, what should go into the plant kingdom are still open to question.

Classification of Organisms

The classification followed in this book is that of Whittaker (1969), who divided all living organisms into five kingdoms. Only four kingdoms are employed in this text, one prokaryotic or bacterial (Monera) and three having eukaryotic structure (Plantae, Animalia, Mycota). The term *plants* in this text includes all eukaryotic organisms (see below) that have genetically controlled photosynthetic capabilities, contain the photosynthetic pigment chlorophyll *a,* and release oxygen during photosynthesis. Some algae are colorless and have similar cytological and morphological features of specific photosynthetic forms and so are considered plants as well. Thus the classification here diverges from that of Whittaker by including many simple protistlike algae in the plant kingdom. In agreement with Whittaker, as well as Raven *et al.* (1981) and Laetsch (1979), the fungi and bacteria are placed in distinct kingdoms and in separate Appendices in this text. Figure 1-2 presents a brief outline showing suggested relationships between the major groups of organisms.

Kingdom Monera

The most basic division in living organisms is between the prokaryotic and eukaryotic organisms. It is defined in terms of cell structure and chemistry. The bacteria and blue-green algae are prokaryotic organisms, lacking membrane-

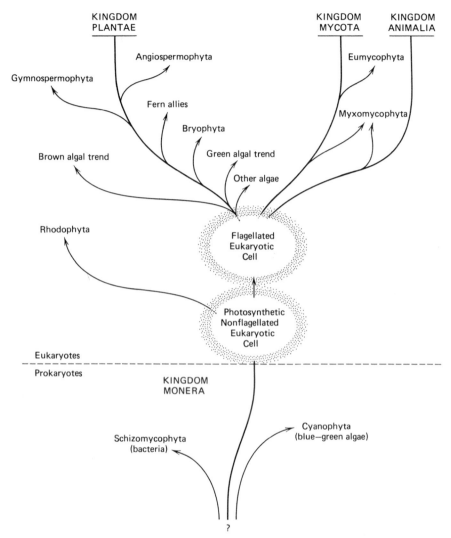

Figure 1-2 Possible relationships among the plant, fungi, and animal kingdoms can be presented as a phylogenetic tree of existing organisms. A nonflagellated and then a flagellated eukaryotic cell is proposed to have evolved only once (monophylogenetic origin), probably from a blue-green algal ancestor.

bound cellular organelles such as chloroplasts, mitochondria, nuclei, and complex flagella. The genetic material, deoxyribonucleic acid (DNA), is not organized into a large complex structure (chromosome) as in eukaryotic cells but instead is a single circular molecule entangled in the center of the cell. The DNA also lacks the basic proteins and histones, which are associated with eukaryotic DNA. In Chapter 3 (Table 3-1), the differences between eukaryotic and prokaryotic cells are summarized.

Typical sexual reproduction, as known in higher plants, is lacking within the Monera. Some genetic recombination is present in blue-green algae, and conjugation between two bacterial cells has been demonstrated in a few cases. The cell walls of prokaryotes, when present, are distinct from eukaryotic plant cell walls. The prokaryotic wall contains a series of distinct macropolysaccharides including mucopeptides, n-acetylglucosamine, and n-acetylmuramic acid. In addition, lipopolysaccharides and various amino acids such as glycine and alanine are also present within their walls.

The kingdom Monera can be separated into two divisions, the Cyanophyta, or blue-green algae, and the Schizomycophyta, or bacteria. The former have plantlike features, whereas the latter do not. Bacteria are important in a variety of mineral and nutrient cycles and in the ecology of marine plant communities. Bacteria are also involved in the decomposition and pathology of marine plants; they are dealt with in Appendix A.

The Cyanophyta, cyanobacteria, or blue-green algae (Chapter 4), are treated as algae because they possess chlorophyll *a* and release oxygen in photosynthesis through photolysis of water (splitting of water). About 50% of the roughly 5000 (?) species can occur in marine environments. Blue-green algae are prokaryotic, lack the eukaryotic cell complexity, and have no flagella (Fig. 4-1). In addition to chlorophyll *a,* these cyanobacteria have a number of unique accessory pigments including carotenoids and water-soluble biliproteins. These pigments are discussed in Chapter 3. The morphology of blue-green algae ranges from unicells to simple colonies and filaments. Blue-green algae are small (cell sizes to 30 μm diameter), with species common in all latitudes and most abundant in the intertidal zone. Since many blue-green algae have been shown to fix nitrogen, they are now recognized as an important source of nitrogen in the marine environment.

Kingdom Mycota

The fungal kingdom is briefly treated in Appendix A. They, like the bacteria, are involved in a variety of nutrient cycles and are pathogens on numerous marine plants. The fungi are not included in the plant kingdom* since they show no photosynthetic ability, although some phycologists believe the higher, more-advanced fungi may have evolved from algae. The treatment in the Appendix includes both slime molds and true fungi in the same kingdom. The emphasis in the Appendix is ecological and demonstrates the interdependence of marine fungi and plants.

Kingdom Plantae

In this text, the plant kingdom includes both unicellular photosynthetic forms that might be included as "Protista" (Whittaker, 1969) and multicellular organisms (Table 1-1). Since algae are the dominant marine plants, an entire section (Part 2)

*Most modern general botany texts now place the fungi in a kingdom separate from the plants, but they still include fungi in the text [see Raven *et al.* (1981), Laetsch (1979)].

Table 1-1   Kingdom Plantae

Division/Class	Common Name	Percentage Marine	Approximate Number of Species
Algae			
Chlorophyta	Green algae	13	7,000
Phaeophyta	Brown algae	99	1,500
Rhodophyta	Red algae	98	4,000
Chrysophyta			
Chrysophyceae	Golden-brown algae	20	650
Xanthophyceae	Yellow-green algae	15	60
Bacillariophyceae	Diatoms	50	1,000
Prymnesiophyceae		50	200
Eustigmatophyceae		?	10
Chloromadophyceae		50	50
Pyrrhophyta	Dinoflagellates	93	1,200
Euglenophyta		3	400
Cryptophyta		60	200
Nonvascular land plants			
Bryophyta	Mosses, liverworts	0	25,000
Vascular plants			
Psilophyta	Wisk ferns	0	3
Sphenophyta	Horse tails	0	15
Lycophyta	Club mosses	0	1,000
Pterophyta	Ferns	0	12,000
Gymnospermophyta	Cone plants	0	722
Angiospermophyta	Flowering plants	0.085	235,000

deals with these marine forms. Marine angiosperms (flowering plants) also occur (sea grasses, mangroves, salt marsh plants), and they form distinct marine plant communities (see Chapters 17, 18, 19).

When one compares terrestrial and oceanic vegetation, it is apparent that only a few angiosperms are truly marine (0.085% of 235,000 species). There are no representative marine forms from the mosses, liverworts, or hornworts (Bryophyta), lower vascular plants (Psilophyta, Lycophyta, Sphenophyta), ferns (Pterophyta), or gymnosperms (Gymnospermophyta).

Macroscopic algae (seaweeds) found in the marine environment primarily belong to one of three divisions, the Chlorophyta (Chapter 5, green algae), Phaeophyta (Chapter 6, brown algae), and Rhodophyta (Chapter 7, red algae). Although only about 10% of the green algae occur in marine habitats, they are particularly important in the tropics, where a variety of calcified and siphonaceous forms occur. Red algae are almost exclusively marine, as are the brown algae. Whereas the greatest variety of red algae is found in subtropical and tropical waters, brown algae are more common in cooler, temperate waters.

Microphytic algae (microscopic forms) are found in almost all groups of algae (Chapters 4, 5, 8, 9); some of the brown and red algae are simple filaments. The tiny microscopic forms are significant primary producers and form the phytoplankton community (Chapter 20). The diatoms are unicellular forms that are especially important; they are both benthic (bottom) and planktonic (free floating). The dinoflagellates, unique flagellated unicells, are found in freshwater systems and all oceans, and the blue-green algae may occur as benthic or planktonic organisms. Dinoflagellates, as well as other phytoplanktonic algae, can form blooms in the marine waters; in some cases these blooms are toxic (e.g. red tides).

Botanical nomenclature

The plant kingdom is divided into a series of *divisions* (Table 1-1) similar to the phyla of the animal kingdom. Each division can be divided into one or more *classes* that contain one or more *orders*. *Families* make up the next subgrouping and within each family the various genera and species are placed. Species are named according to the binomial system of Carolus Linnaeus as presented in his *Species Plantarum* (1753). The scientific name, which is in Latin, is a binomial; it contains a generic and a species name, as well as the author's name. Typicallly the name is a descriptive term for the organism. For example, the common green alga, called sea lettuce, would be classified as follows:

Taxonomic Unit	Name	Ending of Name
Division	Chlorophyta	-phyta
Class	Chlorophyceae	-phyceae, -cae, or -ae
Order	Ulvales	-ales
Family	Ulvaceae	-aceae
Name (genus and species)	*Ulva lactuca* Linnaeus	

Ulva is the Latin name for marsh plant, and *lactuca* is a descriptive term that means lettuce.

Note that the plant's name is italicized and the author's name is included. Also note the endings used for the various taxonomic units from division to family. In some cases, subclasses are used and the ending then is "-oideae." Following a number of algal names, are the names of two authors. For example, in the case of the green coenocytic alga *Caulerpa prolifera* (Forsskäl) Lamouroux, Forsskäl is the author of the species *prolifera,* and Lamouroux transferred that species to the present genus, *Caulerpa.*

All descriptions of new plants as well as descriptions of new taxonomic units (divisions, etc.) are governed by the rules of the International Code of Botanical Nomenclature, which can be revised every five years at the International Botanical Congress. All new genera and species must have a Latin description that is diagnostic. An example of this procedure is given in the legend of Figure 1-3.

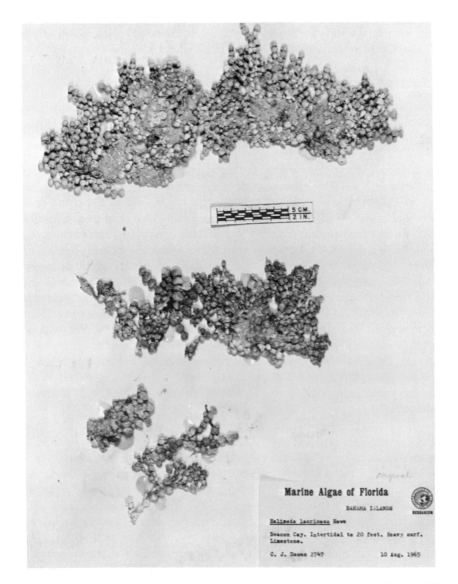

Figure 1-3 An example of a formal taxonomic description for a variety of the calcified green alga *Halimeda lacrimosa* Howe (Dawes, 1980). Whether a new genus, species, or variety is being described, a formal description must be published in Latin and usually also in the author's language. One specimen is designated the "type" or original, and its location is noted in the description. × 1.40.

Halimeda lacrimosa Howe variety *globosa* Dawes Plantae usque ad 10 cm altae, densae, plus minusve globosae; segmenta pro parte majore globosa, rare obovoidea vel pyriformia, 5–10 mm diam., cava. utriculi peripherales 35–40 μm diam., 80–100 μm longi; membrane utriculares apicales 7–9 μm crassae, post calcificatae remanentes cohaerentes. Utriculi subcorticales 100–130 μm diam., omnes ferentes 16–17 utriculi peripher-

Note that the type specimen from which the description was made is placed in a herbarium (storage site of pressed plants), and its location is given. More will be said about algal preservation and storage in Chapter 13.

Marine plant divisions

A short description of each division having members common to the marine environment is presented here. See also Chapters 3–9 and 17–19.

Chlorophyta

This division contains three classes, all eukaryotic with biflagellated stages. The flagella are of equal length and smooth (acronematic). They are grass-green because of chlorophylls *a* and *b;* reserve food is like the starch found in higher plants. The class *Chlorophyceae* is the largest. Thalli range from unicells to filaments, with parenchymatous or coenocytic (lacking cross walls) construction. Reproduction is asexual by cell division or motile or nonmotile spores, or sexual by gametes that range from isogamy (gametes look alike) to oogamy (egg and sperm). The plants occur mostly in fresh water, but there are also terrestrial and marine forms. *Prasinophyceae* is a small class of unicellular, motile or nonmotile cells or groups of cells with one, two, or four flagella. The flagella have hairs and are of equal length. Reproduction is by cell division or motile spores, with no evidence of sexual reproduction. The species are primarily found in marine environments. The class *Charophyceae* is another small group of freshwater or brackish water (low salinity to 5 ppt) green algae with true parenchymatous thalli. Sexual reproduction is oogamous with motile sperm and nonmotile eggs produced in multicellular sex organs (having sterile protective cells).

Phaeophyta

This division contains a single class, *Phaeophyceae*. These eukaryotic plants produce biflagellated zoospores and gametes. The flagella are laterally inserted, one smooth (acronematic) and one with hairs (pantonematic). Photosynthetic pigments include chlorophylls *a* and *c,* usually masked by carotenoid pigments, especially fucoxanthin, giving an olive-brown coloration. The reserve food is laminarin. Plants range from filaments to massive parenchymatous structures with a high degree of differentiation. Reproduction is by asexual motile zoospores or immotile aplanospores or by gamete fusion, which ranges from isogamy to oogamy.

---

ales. Fila littos centrales conjungentes in 3s vel 4s, rare 2s vel 6s ad nodia. Sporangia ignota.

   Plants forming dense masses to a height of 10 cm; the segments mostly spherical in shape, rarely obovoid or pyriform, 5–10 mm in diameter, hollow, the peripheral utricles 35–45 μm in diameter, 80–100 μm long, the apical walls 7–9 μm thick remaining coherent after decalcification. Subcortical utricles 100–300 μm in diameter, each bearing 16–17 peripheral utricles. Filaments of the central strand fusing at the nodes in threes and fours rarely in twos and sixes. Sporangia unknown. Type CJ Dawes No 2747 Herb. Univ. So. Fla. No 69152.

Almost all species are marine; the greatest diversity of brown algae occurs in cooler temperate waters.

Rhodophyta

The red algae are eukaryotic plants that do not produce any cells that bear flagella. The red algae have chlorophylls *a* and *d,* but these are usually masked by the water-soluble pigment phycoerythrin. A few forms are parasitic and colorless. Reserve food is a type of starch (floridean starch). Two classes can be recognized. The class *Bangiophyceae* includes mostly unicellular to simple filamentous or parenchymatous members with reproduction by cell division or by spores. Sexual reproduction has been established in only one genus of this class. Freshwater and marine forms are present.

The class *Florideophyceae* contains filamentous to complex pseudoparenchymatous plants. Reproduction is by spores (tetraspores, carpospores). The life history can have three distinct phases (triphasic). Sexual reproduction involves a type of oogamy but lacks motile sperm. The diploid zygote is retained on the female gametangial plant (haploid) and basically parasitizes the female plant while developing into a multicellular diploid carposporophyte. Subsequently, the carposporophyte produces diploid carpospores that grow into a free-living tetrasporophyte. The tetrasporophyte produces tetraspores, through meiosis, that grow into female and male gametophytes. Most members of this class are marine.

Chrysophyta

This division of eukaryotic plants can be divided into six classes. The class *Chrysophyceae* contains uniflagellated to biflagellated zooids with one hairy flagellum or two flagella of different lengths, one hairy (the longer one) and the other smooth. Chlorophylls *a* and *c* are present, but the carotenoid pigments dominate, giving a golden-brown color. Some members are colorless. The reserve food is chrysolaminarin, as it is in all the classes. Most members are unicellular and motile; a few are filamentous. Reproduction is by cell division or zoospores. Sexual reproduction, where known, is by isogamous gametic fusion. Members are primarily freshwater, but there are some marine forms.

The class *Prymnesiophyceae (Haptophyceae)* contains motile, biflagellated cells with both flagella smooth (acronematic). A specialized structure, the haptonema, may be present; it is apparently a modified flagellum. Chlorophylls *a* and *c* are present but are masked by carotenoid pigments. The reserve foods are chrysolaminarin and oil. The plants are unicellular, motile or nonmotile, and marine. Reproduction is by cell division or zoospores; sexual reproduction is unknown.

The class *Bacillariophyceae* contains the diatoms, which are golden brown because of the presence of chlorophylls *a* and *c* and the dominance of carotenoid pigments. The reserve foods are chrysolaminarin and oil. Plants are unicellular or filamentous aggregations of unicells; all of the plants have a unique cell wall composed of SiO_2 (silicon dioxide). Reproduction is by cell division and sexual reproduction is either oogamous (male gametes have a single pantonematic flagellum) or with amoebic isogametes. Plants are common as planktonic and benthic forms in marine, freshwater, and terrestrial habitats, and they are very important primary producers in such habitats.

The class *Chloromonadophyceae* (Rhaphidophyceae) can also be placed in Chrysophyta, and it contains biflagellated unicellular motile cells. The flagella are of different lengths. The presence of chlorophyll *a* and abundance of carotenoid pigments result in yellow-green cells, although some species are colorless. The reserve food is oil. Species are mostly freshwater.

Members of the class *Xanthophyceae* have biflagellated cells with flagella of different length; the longer one is hairy and the shorter one is smooth. The plants are yellow-green because of the presence of chlorophylls *a* and *c* (recently demonstrated) as well as a large quantity of carotenoid pigments. Reserve food is chrysolaminarin and oil. The plants range from unicellular to filamentous forms with some coenocytic species (lacking cross cell walls). Reproduction is by cell division, zoospores, or nonmotile spores. Sexual reproduction is oogamous in the coenocytes and isogamous in other species. But little is known about sexual reproduction in many of the species. The plants are primarily freshwater with *Vauchera* and other species occurring in salt marshes.

The class *Eustigmatophyceae* contains a small group of unicellular flagellates recently separated from the Xanthophyceae because of their distinctive ultrastructure.

Pyrrhophyta

This division includes eukaryotic, biflagellated unicellular (a few filamentous forms) algae, with one flagellum pantonematic. Two classes are usually recognized, the *Desmophyceae* and the *Dinophyceae*. The pigments consist of chlorophylls *a* and *c* and several conspicuous carotenoids. A number of species are colorless and some are parasitic. Both armored (having cellulose plates) and nonarmored forms exist. The reserve food is starch. Mitosis is distinguished by a number of features, and the chromosomes are condensed throughout the cell cycle. Sexual reproduction is poorly understood but appears to be isogamous through flagellated or amoebic gametes. Species are common in marine and freshwater habitats and may form massive blooms (red tides).

Cryptophyta

The members of this division are placed in a single class, Cryptophyceae. The plants are eukaryotic unicells that have two hairy flagella, unequal in length. The cells are a variety of colors because of the presence of chlorophyll *a*, carotenoids, and phycobilins. Trichocysts (projectile organelles) line a central reservoir. The reserve food is a type of starch. The cells have an outer firm periplast but no true cell wall. Cell division is the only known type of reproduction.

Euglenophyta

The euglenoid algae are a small group of eukaryotic grass-green plants that are unicellular or colonial. The cells are motile with two flagella usually occurring, only one emerging beyond the reservoir. The grass-green color is due to the dominance of chlorophylls *a* and *b;* however, some species lack chloroplasts. The reserve food is a type of starch called paramylum. The cells lack a cell wall and have a flexible outer periplast that can exhibit an amoebic "euglenoid" motion. Cell division is by longitudinal furrowing; sexual reproduction is unknown.

Angiospermophyta

The vascular, flowering plants are represented in the tidal marshes (salt marshes and mangrove swamps) and sea grass beds. All of these plants have vascular tissue (xylem and phloem) and are characterized by chlorophylls *a* and *b*, thus being grass green. Sexual reproduction occurs through the production of flowers and resultant seeds. Although only about 200 of the 235,000 species are present in the marine environment, the marine angiosperms can be the dominant component of many shallow-water regions. Sea grasses are monocots adapted to submerged conditions (hydrophytes) that form beds in shallow subtidal areas. They are common in subtropical and tropical regions of the world, where they are important primary producers and habitats for other organisms. The tidal marshes include two very distinctive groups of marine angiosperms that grow in environments of high water stress (xerophytes). These are the salt-marsh grasses and shrubs of temperate areas and the mangrove trees of tropical and subtropical areas. Both of these groups of angiosperms form important coastal communities, and they both contain distinctive algal floras.

THE MARINE ENVIRONMENT

About 72% of the earth's surface is covered by the ocean, which can be likened to a number of basins or shallow dishes. The lip of the basin or dish corresponds to the continental shelf and the main part is the ocean proper (the average depth is greater than 5000 m). The ocean is shallow when compared to the earth's diameter. The average depth of the ocean is equivalent to the thickness of a coat of paint on a globe 25 cm in diameter.

The distribution of marine plants is limited to the upper edge of the oceans because of light requirements for both benthic (attached) and planktonic (free-floating) species. Thus, benthic, or bottom-dwelling, forms are found in the shallower areas of the continental shelves (to about 150 m) and free-floating, or planktonic, forms are found in the upper lighted (photic) areas of the oceanic water (to about 200 m).

Vertical Distribution

As noted above, marine plants are most abundant along the coasts and in the shallower regions of the continental shelves. This is not only due to their light requirements but also due to the higher levels of nutrients washed from the continents and picked up from the shallower substrata. Three major vertical zones may

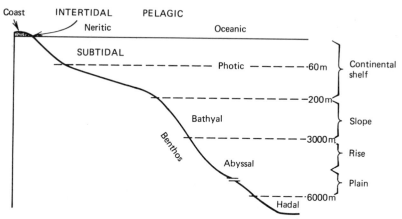

Figure 1-4 Subdivisions of the pelagic (neritic, oceanic) and benthic (intertidal, subtidal, bathyal, abyssal, hadal) zones, along with depths and relationships to the continental shelf, slope, rise, and oceanic plain.

be determined: *intertidal* (and *maritime*), *subtidal* (the submerged continental shelf), and the *basin* (continental slope and basin). The reader is referred to a number of excellent oceanographic texts (Ross, 1970; Davis, 1977) for details of these regions, as well as to Chapter 10 and Figure 1-4. As far as benthic plants are concerned, only the first two zones are directly important.

Intertidal and maritime zones

The region that is regularly covered and uncovered by the tides is called the intertidal zone. The intertidal region is adjacent to the maritime zone, which is the region above the highest tides but subject on the influence of the ocean (e.g., salt spray, wind). The intertidal zone may be a few centimeters to 10 m in height as a function of differences in tidal amplitude and wave action. If wave action is high, the upper portion of the intertidal zone directly influenced by wave action (called spray, splash, or supratidal zone) may extend a few more meters. Such physical factors as substrate, wave action, and meteorological conditions are very important in this region. An extensive discussion of these factors and the intertidal zone is presented in Chapters 11 and 13–15. In most cases, the intertidal zone can be divided into a spray zone (area directly affected by wave action but above high tide and adjacent to the maritime zone), the intertidal zone proper (area regularly exposed by tides), and an intertidal or subtidal fringe (a lower region that is submerged more often than exposed). On many rocky coasts the variety and abundance of species increases as one approaches the intertidal or subtidal fringe (mean low tide).

Subtidal zone

The area having no exposure from tidal action corresponds to the continental shelf, which usually terminates at 60–200 m at the continental slope. In the shallower subtidal areas (up to ~ 100 m), benthic algae can form extensive communities such as the giant kelp forests of the west coast of North America. At greater depths, light becomes limiting and the algal population becomes sparse and less diverse. Conditions are uniform in the mid to deep subtidal regions, and seasonal changes in the plant communities are less evident (but see Hodgson and Waaland, 1979). The diversity and abundance of shallow subtidal communities can be quite high. Just as in the subtidal fringe, the subtidal plant community can be subjected to intensive grazing pressure and competition for space.

Continental slope and ocean basin

The continental slope and ocean basin are essentially devoid of plants. However, since the deeper waters are usually nutrient rich, this area is an important source of nutrients for the shallow-water plants through vertical exchanges of water such as upwellings. This will be briefly considered in Chapter 11.

Horizontal Zonation

Ocean waters can be divided into neritic areas (coastal waters) that cover the continental shelf, and oceanic areas (open water) that are found over the ocean basin. As shown in Figure 1-4, the neritic waters cover the continental shelves, and are nutrient rich and light yellow to brown. The nutrients and color are due to the influence of the continents, runoff, and shallow water.

Oceanic water is usually nutrient poor, except in areas of vertical exchange, and is clear blue. Many areas of the oceans, because of such low productivity, are called "wet deserts."

Photic and Aphotic

The penetration of light will limit the depth to which marine plants grow. If a plant (phytoplankton cell) sinks below the light level at which it receives sufficient energy to photosynthesize, it will use up its reserve food and die. Thus, all algae have a *compensation point* regarding the needed level of light (intensity and quality) for survival. The *photic zone* is the area of sufficient light for photosynthesis (above the compensation point) and depends on water clarity. Oceanic waters are usually quite clear due to the lack of suspended material and thus have a deeper photic zone (~ 200 m) than neritic waters (~ 100 m). The *aphotic zone* is the area of insufficient light for photosynthesis.

A Summary of Factors That Influence Marine Plants

Dawson (1966) presented a list of factors that should be considered when studying a marine plant community. His list is presented in Table 1-2 with a few minor modifications. Most of these factors are dealt with in Chapters 10–12; Chapters 13 and 14 include techniques for their measurement.

Table 1-2   A List of Important Factors that Influence Marine Plant Distribution

A. Physical factors

    1. Light

        a. Intensity (varying with latitude, tidal exposure, cloud cover, shore shading, biological overshadowing)

        b. Quality (varying with water depth, transparency, tidal amplitude)

        c. Periodicity (daily; seasonal)

    2. Substrata

        a. Solidarity (bedrock, cobble, gravel, sand, mud)

        b. Texture (penetrability or suitability for attachment)

        c. Porosity (water-holding capacity)

        d. Position

            a) with regard to water availability (tidal flooding, wave wash, splash, spray, seepage, tidepool retention)

            b) with regard to wave shock or disturbance

            c) with regard to ice action or cobble scour

        e. Solubility and erosibility

        f.   Color (with regard to intertidal heat absorption, radiation, and reflection)

        g. Chemical composition

    3. Temperature

        a. Seawater temperature

            a) Annual variation

            b) Duration of maximum and minimum

            c) Diurnal variation

            d) Stratification; thermocline position with respect to tides, mixing of nutrients, etc.

        b. Air temperature during intertidal exposure

            a) Annual variation

            b) Duration of maximum and minimum

        c. Direct heat of insolation (complete exposure; tidepool exposure)

    4. Relative humidity (with respect to algae subject to exposure)

        a. Seasonal variation in conjunction with exposure

        b. Duration of minimum coincident with maximum exposure temperature

    5. Rain

        a. Seasonal extent coincident with tidal exposure

        b. Maximum duration

    6. Pressure (mainly significant with regard to effect of tidal amplitude on attached seaweeds bearing air vesicles)

B. Chemical factors

    1. Salinity

        a. Annual variation from runoff

        b. Tidal fluctuation of the halocline

        c. Maximum concentration from evaporation during exposure

Table 1-2   Continued

2. Availability of dissolved oxygen during dark-hour respiration
3. Availability of nitrogen, phosphorus, and other essential metabolic substances
4. Availability of free carbon dioxide for photosynthesis
5. pH (mainly significant in confined pools subject to marked increases during active photosynthesis)
6. Pollution
   a. Nonorganic
   b. Organic
   c. Biological

C. Dynamic factors

  1. Water movement
     a. Surf
     b. Ocean currents
     d. Maximum severity of annual storms or hurricanes
     e. Upwelling
     f. Extent of surface chop vs. calm
  2. Tidal exposure (period and amplitude)
  3. Tidal rhythm (with respect to release of reproductive bodies)
  4. Wind (with respect to coincidence with exposure)

D. Biological factors

  1. Grazing
  2. Fungal and microbial activity
  3. Competition for substrates (space)
  4. Protective cover against desiccation during exposure
  5. Light restriction by overgrowth (either by macroscopic or microscopic forms)
  6. Availability of host plants or animals for obligately epiphytic, endophytic, epizootic, endozootic, and parasitic algae

Source: Dawson (1966).

ZONATION OF MARINE PLANTS

Tidal levels have traditionally been used to separate the major zones in the marine environment (Doty, 1946, 1957). However, the intertidal zones can also be identified by characteristic populations (Lewis, 1964; Stephenson and Stephenson, 1971). A detailed discussion of these two alternative approaches is presented in Chapter 15. The purpose of this section is to introduce the terms that will be used in this text. In the author's opinion (see Chapter 13), the tides are the primary factors influencing zonation; five zones are recognized.

Figure 1-5   Maritime zones reflect not only the geological features (rocky vs. sedimentary coasts) but also the effects of wave and wind action. On the west coast of Florida, the sand dunes of low-energy beaches are stabilized by sea oats (*Uniola paniculata*) and a variety of halophytic dicot bushes. Within a few meters behind these dunes, juniper and sabal palm forests can develop because of the lack of intensive salt spray. (Photo courtesy of Lherif Loraamm, University of South Florida, Tampa.)

Maritime (Coastal) Zone

The area above the high tide mark that is essentially terrestrial but subjected to varying degrees of wave spray, mist, and salt carried by wind is included in this zone. It may be a dune region behind a sandy beach and may contain various halophytic (salt-loving) grasses such as sea oats (Figure 1-5). On rocky coasts, the area may have lichens (symbiotic fungal and algal organisms) or various spray-tolerant trees such as those found on the west coast of North America, where the Moneterey Pine has an epiphytic alga, *Trentepholia*.

Spray (Splash) Zone

The upper fringe of the intertidal zone, above the high tide mark, may be distinct if wave action is moderate to strong. The spray zone is characterized by marine plants such as lichens, blue-green algae, and marine animals such as the periwinkle *(Littorina)*. In the cooler, temperate, waters, red algae *(Porphyra, Bangia)*, green algae *(Codiolum, Urospora)*, and brown algae *(Fucus spiralis)* may occur in the spray zone (Figure 1-6). In the tropics this zone usually supports a limited blue-green algal community due to the intense sun and high temperature of the substratum.

Intertidal Zone

The area that is regularly covered and uncovered by the tidal rhythmicity is termed the intertidal region. The upper limit may be delineated by barnacles and brown algae *(Fucus, Ascophyllum)*, whereas the lower limit will be marked by an

Figure 1-6 Areas of protected coasts that lack even moderate wave action but have extensive water circulation (for example, because of tidal currents) may show an abundance of seaweeds zoned into distinct narrow bands. Thus *Porphyra* sp. (area 1, *Rhodophyta*), *Blidingia* sp. (area 2, *Chlorophyta*), *Fucus* sp. and *Ascophyllum nodosum* (area 3, *Phaeophyta*) can be seen arranged in quite distinct bands from the high tide to midtide areas on the bridge pilings at a tidal rapid region at Dover Point in the Great Bay Estuary of New Hampshire. (Photo courtesy of Donald Cheney, Northeastern University, Boston.)

upper fringe of larger brown algae, such as kelps *(Laminaria)* in color waters (Figure 1-7). In subtropical and tropical regions, the upper area will have a black band of blue-green algae of mostly crustose (growing as a crust, flat) or endophytic forms (growing into the substratum) and the lower limit will be delineated by a mixture of green algae such as *Halimeda* and *Valonia* or red algae such as *Laurencia* (Figure 1-8).

An early study of the intertidal zone of San Juan Island in Puget Sound was carried out by Muenscher (1951). He described three algal associations (Figure 1-9). The upper intertidal zone was dominated by a stiff and spiny red alga called *Endocladia,* the mid intertidal association by the large rockweed, *Fucus* and the lower intertidal association by the green leafy alga, *Ulva.* Each association was then named for these dominant species.

Subtidal Fringe

The lowest portion of the intertidal zone, which is uncovered only infrequently (for example at a period of no waves and spring tides) is designated the subtidal fringe. The marine plants growing here are only intolerant to short periods of exposure and are often tolerant to high wave action. Muenscher (1951) found that the brown kelp *Laminaria* dominated this zone on exposed coasts in Puget Sound. He also found that the *Laminaria* association contained the widest variety of algae (Figure 1-9).

Figure 1-7 The effects of heavy surf will result in a complex zonation pattern of seaweeds. On Appledore Island in the Isle of Shoals, Maine, about 20 km from Dover Point, New Hampshire (Figure 1-6) a complex vertical zonation of seaweeds is dominated by *Ascophyllum nodosum, Fucus* sp. (area A, Phaeophyta) at the higher intertidal region and *Chondrus crispus* (area B, Rhodophyta) in the lower intertidal region.

Figure 1-8 An exposed limestone intertidal region in the tropics tends to appear less populated by seaweeds than other regions because of the intense sunlight and resultant desiccation. However, as Stephenson and Stephenson (1972) noted, even a coast such as that shown here at West Summerland Key in the Florida Keys is a rich site for both marine animals and marine plants. Most of the color and shading of the limestone is due to algae growing on and in (endolithophytes) the substrate (see also Chapter 15 and Figure 15-9).

Subtidal Region

The subtidal zone extends from the lowest spring tides to the edge of the continental shelf. The greatest diversity of marine benthic algae occurs in the shallow subtidal regions, particularly if a firm substrate is available. The giant kelp forests of California (Figure 15-2) and the extensive sea grass meadows of the Gulf of Mexico (Figure 15-3) are examples of productive subtidal communities. Further information about the subtidal zone is summarized in Chapters 15, 16, and 17. The early studies of Muenscher (1951) involving the intertidal zone and subtidal fringe in Puget Sound have been extended into the subtidal region (Figure 15-12; Neushul, 1967; Hodgson and Waaland, 1979).

The use of modern diving gear such as scuba and submersibles (Figure 1-10) has allowed marine botanists to explore the deeper water subtidal floras of the world. The subtidal exploration has resulted in the identification of new species and extensions of distribution records of other algae (e.g., Eiseman, 1979).

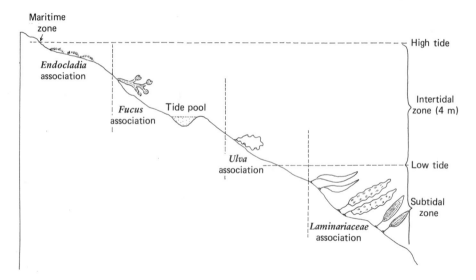

Figure 1-9 Algal associations can be used to describe intertidal areas as shown in this diagram taken from Muenscher (1951). Three algal associations were found to occur within the intertidal zone, and one shallow subtidal association was also noted on San Juan Island in Puget Sound, Washington. Each association was named for a dominant alga. *Endocladia muricata* (Rhodophyta) dominated the upper intertidal region, which also contained such algal genera as *Porphyra, Gigartina, Prionitis* (Rhodophyta), *Enteromorpha*, and *Chaetomorpha* (Chlorophyta). *Fucus distichus* ssp. *evanescens* (Phaeophyta) dominated the midintertidal region but could be replaced by *Gigartina mamillosa* if the substrate was gravel. A lower intertidal algal association was dominated by *Ulva lactuca* (Chlorophyta) but also contained a wide variety of green (5 species), brown (4 species), and red (12 species) algae. A number of kelps (brown algae), especially species of *Laminaria*, dominated the shallow-water subtidal region. See also Chapter 15 for an extension of this community into the subtidal region.

Figure 1-10 A logical extension from marine botanical studies using scuba (self contained underwater breathing apparatus) is the use of submersibles. Submarines such as the Johnson-Sea Link II extend direct observation and collection from the shallower scuba areas (up to 50 m) to the edge of the continental shelf or more. The Harbor Branch Foundation is using the Johnson-Sea Link II in studies of the benthic algal flora off the east coast of Florida. (Photo courtesy of Nat Eiseman, Harbor Branch Foundation, Fort Pierce, Florida.)

REFERENCES

Davis, R. A., Jr. 1977. *Principles of oceanography*. Addison-Wesley, Reading, Mass.

Dawes, C. J. 1980. A new variety of the calcarious coenocyte, *Halimeda lacrimosa* Howe (Chlorophyta, Siphonales). *Bull. Mar. Sci.* **30**: 142.

Dawson, E. Y. 1966. *Marine Botany, An Introduction*. Holt, Rinehart and Winston, New York.

Doty, M. S. 1946. Critical tide factors that are correlated with the vertical distribution of marine algae and other organisms along the Pacific Coast. *Ecology* **27**: 315–328.

Doty, M. S. 1957. Rocky intertidal surfaces. In Hedgpeth, J. W. (ed.). *Treatise on Marine Ecology and Paleoecology*. Memoir 67, Geological Society of America, New York, pp. 535–585.

Eiseman, N. J. 1979. Marine algae of the East Florida continental shelf. I. Some new records of Rhodophyta, including *Scinaia incrassata* sp. nov. (Nemaliales: Chaetangiaceae). *Phycologia* **18**: 355–361.

Hodgson, L. M. and J. R. Waaland. 1979. Seasonal variation in the subtidal macroalgae of Fox Island, Puget Sound, Washington. *Syesis* **12**: 107–112.

Laetsch, W. M. 1979. *Plants. Basic Concepts in Botany*. Little, Brown, Boston.

Lewis, J. R. 1964. *The Ecology of Rocky Shores*. English Universities Press, London.

Muenscher, W. L. C. 1951. A study of the algal association of San Juan Island. *Puget Sound Mar. Sta. Publ.* **1**: 59–89.

Neushul, M. 1967. Studies on subtidal marine vegetation in Western Washington. *Ecology* **48**: 83–94.

Raven, P. H., R. R. Evert, and H. Curtis. 1981. *Biology of Plants*. Worth, New York.

Ross, E. A. 1970. *Introduction to Oceanography*. Appleton-Century-Crofts, New York.

Stephenson, T. A. and A. Stephenson. 1972. *Life Between Tidemarks on Rocky Shores*. Freeman, San Francisco.

Whittaker, R. H. 1969. New concepts of kingdoms of organisms. *Science* **163**: 150–162.

CHAPTER 2

Importance of
Marine Plants

Marine plants are the primary producers of the oceans and thus form the standing crop (biomass) and determine the productivity (rate of organic production) for all communities. Several marine plants serve as habitats for economically important animals, although others act as filters and stabilizers of ocean waters and sediments (Figure 2-1). These and other ecological features of marine plant communities are described in detail in Chapters 15–20. In this chapter the economic importance of marine plants is evaluated.

Figure 2-1 The use of marine angiosperms as stabilizers of shorelines is a relatively recent event. Young seedlings of the black and white mangroves have been planted on spoil islands in Tampa Bay and after 2 yr formed erosion barriers and habitats for birds. Within about 10 yr these plantings will form an impenetrable thicket. (Photo courtesy of William Hoffman, Tampa Institute of Marine Science, Inc.)

USES OF MARINE ANGIOSPERMS

As noted earlier, there are three types of marine angiosperm communities: the subtidal seagrass beds, intertidal mangrove swamps, and salt marshes.

Seagrasses

The marine angiosperms that are referred to as "seagrasses" are not true grasses, although they are monocots. They form extensive shallow-water beds, especially in the tropics and subtropics (see Chapter 17). Seagrasses are abundant. For example, yields of 11.2 million metric tons of dry leaves of *Thalassia testudinum* (turtle grass) are estimated harvestable from the west coast of Florida (Bauersfield *et al.*, 1969). Two seagrasses found in the Gulf of California, *Ruppia maritima* (Sim W taan) and *Zostera marina* (?atan) are collected by the Seri Indians along the central western side in Sonora Mexico (Felger *et al.*, 1980). The Seri Indians toast or grind into flour the seeds of *Zostera* (Xnois) and use the leaves of both species in roofing, mats, and various balls and dolls (Felger and Moser, 1973).

McRoy and Helfferich (1980) list the following uses of seagrasses. (1) Seagrass beds have been used as a filter system for sewage effluents and as a sediment stabilizer. (2) Blades of seagrasses have low lignin and high cellulose content and hence are used in paper production. (3) The younger rhizomes of *Zostera* can be cooked and eaten directly as can the fruits from a number of seagrass species. (4) Dried seagrass blades have been used as fodder, with the best results from a mixture of seagrasses and terrestrial foods.

In addition, seagrasses can be used as fertilizers and have been thus used over the centuries. Van Breedveld (1966) found that seagrasses such as *Thalassia* (turtle grass) are particularly useful as fertilizer because of the availability of various microelements and the high levels of organic matter within their tissues. He found a doubling of yields from standard vegetable crops through the addition of dried seagrass mulch when compared to standard artificial fertilizers.

Seagrasses are also of economic interest because of their role in the stabilization of disturbed substrata, that is, after dredging or bridge or sea wall building. A number of studies have shown an increase in turbidity and a decrease in productivity at sites of dredging and filling or where death of seagrasses occurred due to diseases (see review by Phillips, 1980). Thus the development of transplantation techniques for seagrasses such as *Zostera* (Phillips, 1974) and *Thalassia* (Thorhaug, 1974) are now being actively pursued.

Mangroves

The term *mangrove* refers to an ecological group of distinct species of trees that grow in tidal marshes of subtropical and tropical regions. Mangroves play an important role in the primary production and stabilization of land as outlined in Chapter 19. Mangroves have been used for centuries as a source of lumber, tannin, dyes, fuel, and medicines. The most exploited genus, in this regard, is the red mangrove, *Rhizophora* (Morton, 1965).

More recently, mangroves are being used as stabilizers of road fills, spoil islands, and disturbed shorelines in subtropical and tropical regions (Goforth and Thomas, 1980; Hoffman and Rodgers, 1980). Plantings of the black mangrove, *Avicennia germinans*, and the white mangrove, *Laguncularia racemosa*, were established on a dredge island in Tampa Bay (Figure 2-1). After 13 months, survival was 73% and the cost of planting a hectare was $10,900, or $3.75 per transplant (Hoffman and Rodgers, 1980). In comparison, the cost of planting 1 ha with the salt marsh grass, *Spartina alterniflora*, was $4,200 or 95 cents per plug.

The bark of *Rhizophora murconata*, a species from the south central Pacific, contains about 37% tannin, which is useful in tanning heavy leather such as sole leather and in the manufacture of hot-press plywood adhesive. If treated with salts of copper or iron, the dyes extracted from the bark are chocolate brown, olive, or rust. Dyes have also been extracted from shoots (red dye) for coloring leather and other goods, and from leaves (deep-black to chestnut).

The wood of *Rhizophora* and *Avicennia* is heavy and will sink. It is a hard wood and difficult to cut when dry, but polishes well. The wood is fine grained and moderately decay resistant. It is especially valued throughout the tropics for rafters, joists, door and window framing, boat ribs, oars, pilings, and fence posts in and out of water. Golf club heads in the Bahamas are made from 14 cm^2 blocks of red mangrove wood that have been soaked for 3 yr in salt water.

A number of uses of bark extracts or extracts of the viviparous fruit are known from the tropics, where extracts are used as astringents for diarrhea and dysentery, salves for lesions, and poultices for stings from venomous animals. The fruits of *Rhizophora* are eaten when young but are bitter because of high levels of tannin. Thus they are only considered appropriate as emergency food. The leaves are used in making tea and in feeding cattle.

Salt Marsh Plants

The importance of salt marshes as primary producers and in various ecological roles is well established (Chapter 18). Direct exploitation by man seems to include only feeding cattle along temperate coasts, use as a filter system in secondary sewage treatment, and stabilization of shorelines.

USES OF MARINE ALGAE

Macroscopic (multicellular) algae are widely used by man in a number of ways. They are a direct source of food, medicine, fodder, and fertilizer, and are used as a source of salts and phycocolloids (Bonotto, 1976) and paper production (Kiran *et al.*, 1980). With the energy crisis, seaweeds are also being evaluated as sources of methane fuel. Two texts on seaweed uses are available: *Marine Algae. A Survey of Research and Utilization* (Levering, *et al.*, 1969) and *Seaweeds and*

Their Uses (Chapman, 1970) as well as a review chapter by Bonotto (1976). A brief summary of seaweed utilization is presented in Table 2-1, listing algal genera, types of utilization and areas of the world where the algae are used. Table 2-2 presents estimates of worldwide seaweed production and value in 1973. The brown and red seaweeds are the dominant algae used by man.

Marine Algae as Direct Sources of Food

Edible seaweeds have a long history, especially in China, where several algae have been used as foods since at least 1000 B.C. A number of books on seaweeds as foods, including recipes, are available. One of these is the *Book of Sea Vegetables* by Shurtleff and Aoyagi (1978); which gives Japanese, Chinese, and Hawaiian recipes. Limu (Hawaiian for seaweed) recipes are available in a small publication by Abbott and Williamson (1974), and the uses of Hawaiian seaweeds are reviewed by Abbott (1978). *The Seavegetable Book* by Madlener (1977) presents a worldwide compendium of seaweed recipes.

The Japanese and Polynesian cultures are by far the most sophisticated with regard to eating seaweeds. Some of the more common seavegetables used by the Japanese are presented in Table 2-3. In Hawaii, about 29 species of marine algae are eaten and all have common names (Abbott and Williamson, 1974; Abbott, 1978).

Chlorophycean seavegetables

The green alga *Ulva* is called *sea lettuce* in England and Scotland, *aosa* in Japanese and *palahalaha* in Hawaiian. Like its close relative *Entermorpha,* it is usually eaten fresh in salads. *Monostroma* is also cultivated in Japan and like *Ulva,* it is grown on bamboo brush called *hibi.* Species of the green algal coenocyte *Caulerpa* are eaten in Hawaii and the Phillipines, either in salads or as relishes in cooked fish and meat dishes.

Phaeophycean seavegetables

About 70 species of brown algae are harvested by man, especially species of the orders Laminariales and Fucales, which are widely used as direct food sources in eastern countries. Most are harvested from natural populations. In Japan and China, a large mariculture effort is involved in growing and selecting high-yield strains of kelps. The most important seavegetable in the Japanese diet is *kombu* (*Laminaria* spp., *Alaria*). The kelps are harvested from July to October from natural populations and in areas where large farms are maintained (Figure 2-2). The plants are collected by entangling them with hooks on long poles. After drying, *kombu* is cooked in barley water, usually with dyes to give a fresh-green color, and then dried and pressed. The pressed bundles are shredded into shavings that are dried and then packaged. Other methods are used for thicker species of *Laminaria*. *Laminaria* is cultivated by placing rocks in areas of mature kelp beds to allow the spores to settle and germinate. The procedure has been carried on by

Table 2-1 Utilization of Seaweeds

Genus	Utilization Level	Food: Cultivated	Food: Human	Food: Livestock	Industrial: Agar-Agar	Industrial: Alginates	Industrial: Paste	Industrial: Chemical Drugs	Industrial: Manure	Asia: Japan	Asia: East	Asia: West and Southeast	North America	Europe	Other (Where Used)
Green algae															
Caulerpa	x														Philippines
Chaetomorpha															South America
Codium			x					x				x			Oceania
Enteromorpha	x	x	x							x	x	x	E	x	Oceania
Monostroma	x	x	x							x	x	x		x	
Ulva	x	x	x	x					x	x	x		x	x	South America, Oceania
Brown algae															
Alaria	x	x	x	x	x					x			EW	x	
Ascophyllum		x		x									E	x	
Cladosiphon				x						x					
Cystophyllum										x		x			
Dictyota												x			
Durvillea			x												Oceania
Endarachne			x							x					Australia/New Zealand, South America

Table 2-1 Continued

Genus	Utilization Level	Food — Cultivated	Food — Human	Food — Livestock	Industrial — Agar-Agar	Industrial — Alginates	Industrial — Paste	Industrial — Chemical Drugs	Industrial — Manure	Asia — Japan	Asia — East	Asia — West and Southeast	North America	Europe	Other
Ecklonia	x	x	x			x		x	x	x	x				
Eisenia	x	x	x			x		x	x	x	x			x	
Fucus			x	x					x				W		Oceania
Analipus			x							x					
Laminaria	xx	x	x	x		x				x	x		E		
Macrocystis	x	x	x			x			x	x			W		Australia/N. Amer., South Africa, Oceania
Mesogloia										x					
Nemacystis		x	x							x					
Nereocystis	x	x	x			x			x				W		
Padina															Australia/New Zealand
Pelagophycus						x							x		
Pelvetia				x				x		x			x	x	
Petalonia			x							x					
Phyllogigas															South America
Sargassum	x		x					x	x	x	x	x	x		
Tinocladia		x	x							x					
Turbinaria												x			
Undaria	xx	x	x							x	x				

28

Red algae

Species										Region
Acanthopeltis					x					
Acanthophora					x			x	x	Oceania
Agardhiella	x							x	x	Oceania
Ahnfeltia				x	x			x	x	
Asparagopsis				x	x					
Carpopeltis	x			x	x					
Ceramium				x	x					
Chondrus	x	x								
Corallopsis	x									
Digenia										
Eucheuma	xx	x			x			E	x	
Furcellaria		x							x	
Gelidiopsis							x			
Gelidium	x	x		x	x			x	x	South America, Oceania
Gigartina	x	x		x	x			EW	x	
Gloiopeltis	x	x		x	x					
Gracilaria	x	x		x	x			x	x	Oceania
Grateloupia	x	x		x						Oceania
Griffithsia				x						
Gymnogongrus	x				x			x		Oceania
Hypnea	x	x		x	x			x		Oceania
Iridaea	x	x		x	x			W	x	
Laurencia	x				x				x	Oceania
Meristotheca	x		x	x						
Nemalion	x			x						
Pachymeniopsis	x	x			x					
Porphyra	xx	x		x	x			EW	x	South America/Australia/New Zealand
Pterocladia	x			x	x				x	Oceania

Table 2-1 Continued

| | | Used For | | | | | | | | Where Used | | | | | |
| | | Food | | Industrial | | | | | Asia | | | | | | |
Genus	Utilization Level	Cultivated	Human	Livestock	Agar-Agar	Alginates	Paste	Chemical Drugs	Manure	Japan	East	West and Southeast	North America	Europe	Other
Rhodymenia[a]	xx	x	x								x		EW	x	
Sarcodia										x		x			
Suhria															Africa
Turnerella							x			x					

Sources: Chapman, V. J. 1970. *Seaweeds and their uses*. London, Methuen. Jonston, H. W. 1966. The biological and economic importance of algae. 2. Algae as food. *Tuatara* **14**: 30–66. Neish, I. C. 1976. *Culture of Algae and Seaweeds*. FAO Technical Conference on Aquaculture, Kyoto. Saito, Y. 1976. *Seaweed Aquaculture in the Northwest-Pacific*. FAO Technical Conference on Aquaculture, Kyoto.
[a]Palmaria.

Table 2-2 Production of Seaweeds in 1973

Area	Metric Tons by Type (thousands)					U.S. Dollars by Type				
	Brown	Red	Green	Misc.	Total	Brown	Red	Green	Misc.	Total
Atlantic, N.E.										
Norway	74.2				74.2	891				891
U.K. (Scotland)	24.1				24.1	382				382
France	0.8	3.8		12.5	17.1	36	404		762	1,202
Spain	0.2	9.2			9.4	32	3,343			3,375
Atlantic, N.W.										
Canada		34.2		5.7	39.9		1,995		31	2,026
United States		1.1			1.1		64[a]			64
Atlantic, E. Cent.										
Korea Rep.[a]	107.8				107.8	12,744				12,744
Morocco[b]				1.7	1.7				85	85
Atlantic, S.E.										
South Africa	4.8	1.1			5.9	256	74			330
Atlantic, S.W.										
Argentina	1.7	21.4		1.2	24.3	37	1,155		28	1,220
Indian Ocean, W.										
Madagascar		0.3			0.3					

Table 2-2 *Continued*

Area	Metric Tons by Type (thousands)					U.S. Dollars by Type				
	Brown	Red	Green	Misc.	Total	Brown	Red	Green	Misc.	Total
Pacific, N.W.										
Japan[c]	277.7	328.1	0.7	47.7	654.2	32,970	187,312	88	2,833	223,140
Korea Rep.[a]	37.5	37.3	0.2	41.3	116.3	4,444	21,295	25	2,453	28,217
Pacific, N.E.										
United States	0.2				0.2					
Pacific, E. Cent.										
Mexico	28.2	2.1			30.3	250	150			400
Pacific, W. Cent.										
Philippines[d]							4,534[d]			
Total	557.2	438.6	0.9	110.1	1106.8	51,979	215,792	113	6,192	274,076
Price/ton						93.3	492.0	125.5	56.2	247.6

Source: Food and Agriculture Organization, 1974. *Yearbook of Fishery Statistics, Vol. 36.* FAO, United Nations Bldg. N.Y.
[a] 1972 prices.
[b] 1972.
[c] 1972 Korean prices.
[d] In 1974, Philippines produced 10,000 metric tons of *Eucheuma* with an export value of $4,534,000. Deveau, L. E. and J. R. Castle, 1976. *The Industrial Development of Farmed Marine Algae: The Case-History of Eucheuma in the Philippines and U.S.A.* FAO Technical conference on Aquaculture, Kyoto.

Table 2-3 Seaweed Utilization in Japan; Most Common Marine Plants Used as Food

Alga	Scientific Name	Harvest Preparation	Cooking Method	Use
Chlorophyta				
Miru	Codium fragile	Dried, in salt	Boiled, in soups	Garnish similar to parsley
Aosa	Ulva lactuca	Dried or used fresh	Fresh or powdered	Condiment
Awo-nori	Enteromorpha sp	Dried	Powdered	
Phaeophyta				
Kombu (Konbu)	Laminaria japonica L. angustata L. longissima L. fragilis Alaria fistulosa	Dried, shredded, powdered, or folded; sometimes canned directly	Boiled, soaked in vinegar, shredded, and sometimes candied	Used as vegetable, a relish to make tea or candy
Wakami	Undaria pinnatifida	Dried	Soaked in vinegar and soy sauce or boiled with other vegetables	Vegetable, in soup
Hijiki	Cystophyllum fusiforme	Dried	Boiled with vegetables and oil, eaten with vinegar and soy sauce	Used as vegetable
Hondawara	Sargassum enerve	Dried	Soups, sauces	As vegetable
Rhodophyta				
Nori	Porphyra lacinata P. vulgaris	Dried, roasted, or canned	Boiled, eaten directly or with other foods	Soups, and with many foods
Tengusa	Gelidium robustum	Boiled, filtered, and freeze-dried	Heated with water	Jellies, soups, saki purification
Ogo-nori	Gracilaria verrucosa	Fresh	Treated in limewater to turn green	Garnishings

Figure 2-2 The brown alga *Laminaria* sp. called *kombu* in Japanese, is used as food in Japan and China. In Shoji, Japan, bundles of *kombu* are being dried on the ground. (Photo courtesy of John West, University of California, Berkeley.)

the Japanese for over 400 years, whereas the Chinese began *Laminaria* cultivation in 1950.

The kelp *Undaria pinnatifida*, which in Japan is called *wakame*, is extensively collected, dried, and bailed in that country. The blades are used directly in soups with fish or with meat either cooked or as a pickled condiment. About 50,000 mt were harvested in 1960 in Japan alone (Levring *et al.*, 1969).

Rhodophycean seavegetables

About 200 species of red algae are harvested; of these about 90 are used as direct foods. *Porphyra* (Japanese *nori*; Hawaiian *pahe's*) is the most extensively cultivated red alga in the world. In Japan alone, the *nori* industry is valued at over $20 million (U.S. dollars) annually, with an average acre of *Porphyra* yielding $350/year. The cultivation of *Porphyra* will be discussed further in the section on marine agronomy. Because of the highly technical methods employed in cultivation, collection of natural populations of *Porphyra* has almost completely stopped.

Porphyra is eaten with rice or in hot dishes in Japan. It is prepared by washing and chopping and then drying the slurry on mats, forming the *nori* sheets, which are packaged and sold extensively. Another commonly eaten alga is *dulse* (*Rhodymenia palmata* = *Palmaria palmata*). This red alga has been used as a food in Scotland and Ireland since the 10th century. It is collected, washed, dried, and packaged. The candy-bar size packages of dulse are eaten directly or cooked as a vegetable. About 60 metric tons are harvested annually on the eastern coasts of the Canadian provinces. Irish Moss (*Chondrus crispus*), although primarily harvested for the phycocolloid extract carrageenan, is also eaten directly after drying or after cooking to make soft jellies or blancmanges. *Chondrus* is extensively harvested in the North Atlantic, especially along the eastern coasts of North America. Most of the harvesting is still carried out by hand in small boats

Figure 2-3 The red alga *Chondrus crispus* is harvested extensively for its phycocolloid, carrageenan, throughout the New England and North European coasts. In this photo, Nova Scotian fishermen in small dories are bringing *Chondrus* to a small fishing port, where it will be weighed and then dried. (Photo courtesy of FMC Corporation, Marine Colloids Division, Rockland, Maine.)

(Figure 2-3). Similar uses are known for other red algae that yield carrageenan in the Phillipines (*Eucheuma,* called agar-agar) and in Africa (*Hypnea*). In addition, agar-yielding red algae such as *Gracilaria* found in India and Hawaii are used as foods.

Seaweeds in Medicine

The Chinese and Japanese have used seaweeds to treat goiter and other glandular problems since 300 B.C. The Romans used seaweeds to heal wounds, burns, and rashes. The English used *Porphyra* to prevent scurvy on long voyages and *Chondrus* for treatment of various internal disorders.

Digenia simplex, a red alga, as well as other red algae (e.g., *Corallina officinalis*) were used as vermifuges. Kaenic acid is extracted from *Digenia* and used as a vermifuge with a price in excess of $3/lb or $6/kg. Intestinal disorders such as constipation, stomachaches, and ulcers have been treated with *Chondrus, Gracilaria, Gelidium,* and *Pterocladia*; all of these algae produce phycocolloids.

Few seaweeds are presently used directly in medicine, although some forms of phycocolloids are used in treatment of ulcers (agar, carrageenan). Phycocolloids are also used to coat pills and in the production of time-release capsules. Agar is of great importance in the culture of bacteria as demonstrated by Robert Koch in 1881.

Seaweeds as Fodder, Fertilizers, and Fuel

A number of intertidal and subtidal seaweeds have been harvested along coastal areas and used as fodder for several hundred years, especially in Europe. Algae used as direct fodder include *Palmaria* (red alga), *Alaria,* and *Laminaria* (brown algae). In England and Europe, seaweeds are collected, dried, and ground into meals (*Ascophyllum, Laminaria, Macrocystis, Phaeophyta*) and the meals are fed directly to animals or used as supplements to supply minerals.

Seaweeds such as *Sargassum* and *Ascophyllum* have also been used as green manures in coastal farming in Europe and North America. An American commercial liquid fertilizer, Agri-blend, contains *Sargassum* extracts and is used along with other fertilizers. The direct application of seaweeds in farming is a practice that extends over hundreds of years, and it is often more successful than using chemical fertilizers. The beneficial effects of seaweeds are probably due to the alginate in brown seaweeds, which improves soil structure and increases humus and the water-holding capacity of the soil. Seaweed manures tend to have more potassium than phosphorous salts; the former is better for root crops such as beets or potatoes than the latter. Seaweed manures can apparently promote the germination of certain seeds, increase resistance of plants to fungal and insect pests, and even give a certain degree of frost resistance to tomatoes (Mathieson, 1967; Ven Breedveld, 1966).

Coralline algae such as *Lithothanmium corallioides* and *Phymatolithon calcareum* are collected in the Mediterranean and the Cornish coast of England from beach deposits as well as shoals of living material. The commercial collection of the deposits, called *maerl*, off Brittany has an annual harvest in excess of 300,000 tons. *Maerl* is marketed in most west European countries, where its major use is in agriculture and horticulture, particularly in the reduction of soil acidity and as a fertilizer additive (Blunden *et al.*, 1975).

The use of seaweeds as a source of fuel (methane) is a new venture. Studies using such plants as *Macrocystis* or other large kelps are now under way. These studies are needed to develop methods to grow sufficient organic biomass that will be used to produce methane through anaerobic decomposition. Large open sea farms of *Macrocystis* are proposed for the west coast of the United States. The plants would be harvested, allowed to ferment (digest) in vats in the production of methane (see text edited by Krauss, 1977). On the east coast of the United States it has been proposed (Dr. John Ryther, personal communication) to grow the red alga *Gracilaria* in sewage-enriched tanks and to obtain methane by fermentation of the biomass. This is discussed further in the section on marine agronomy.

Industrial Uses of Seaweeds

Two major industrial uses of seaweeds are described below.

Potash, soda, and iodine

The first major industrial use of seaweeds apparently took place in the seventeenth century in France, where brown algae, especially species of *Laminaria*, were burned to obtain sodium (soda) and potassium (potash). In fact, *kelp* is a derived name originally assigned to the resulting ash. By 1800, over 20,000 metric tons of ash were produced in Scotland from drift and cut kelps. Because only 4% of the kelp's dry weight results in ash, the actual amount of harvested seaweed must have been greater than 5 million metric tons dry weight! The dominant brown algae used were kelps, *Laminaria* and *Saccorhiza*, as well as the rockweeds *Fucus* and *Ascophyllum*. The ash was important in the production of soap and potash. Around 1810 the use of algae as a source of ash declined because of the discovery that the succulent halophyte *Salicornia* had a much higher ash content. However, with the discovery that iodine could be obtained from seaweeds, the industry was again revived. By 1840 it was twice the size it was in 1800. The industry remained this size until 1870, when Chilean nitrate deposits were found, and again, the kelp industry declined. Only Japan now obtains iodine from seaweeds.

During the first world war, demands for potash and fertilizer rose so fast that the harvesting and processing of kelps, such as *Macrocystis* and *Laminaria*, once again resumed. An annual harvesting of 400,000 metric tons wet weight occurred during 1917 and 1918, and over 3,600 metric tons of potash were used for American agriculture. Acetone, a solvent used in explosives, was also obtained during this period from the breakdown of seaweeds. Although the fertilizer, potash, and acetone demands disappeared at the end of the war, the seaweed industry did not die out because of the discovery of phycocolloids.

Phycocolloid extracts

Phycocolloids are polysaccharide extracts from seaweeds, especially red and brown algae. Figure 2-4 presents a taxonomic classification of red algae that yield various phycocolloids; in addition the brown algae yield another phycocolloid, alginic acid. These compounds are also called hydrocolloids, because they can form colloidal systems, such as jellies, in water. Phycocolloids are of special importance in jelly formation since only a very low percentage (0.5–5%) is required to form firm jellies at room temperatures. The basic structure that produces the jelly strength is a double helix type of polymer that binds the water molecules (Rees, 1972).

Alginic acid

The giant kelp *Macrocystis pyrifera* contains about 2.5% algin based on wet weight. In 1970, 150,000 metric tons of *Macrocystis* were harvested by large

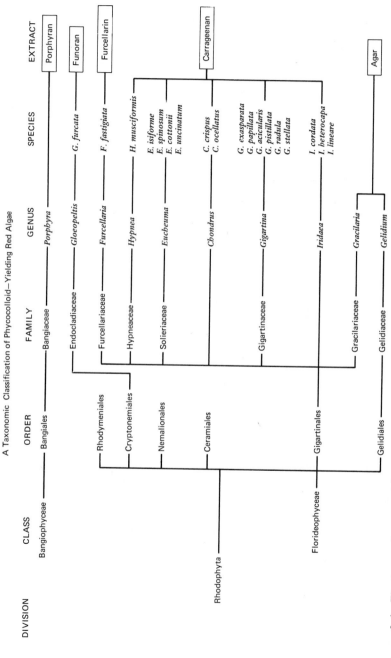

Figure 2-4 The types of extracts (phycocolloids) that are obtained from red algae. It is apparent that one order, the Gigartinales, contains the largest variety of phycocolloid-pro ducing algae. All of the genera listed in the diagram are harvested commercially for their respective product.

Figure 2-5 A view of a kelp harvester off the coast of California cutting and collecting the upper 2–3 m of the giant kelp, *Macrocystis pyrifera*. The brown alga is then processed for the phycocolloid alginate. (Photo courtesy of Kimon Bird, Harbor Branch Foundation, Ft. Pierce, Florida.)

barges along the west coast of the United States (Figure 2-5). Most (130 metric tons) of the harvest was processed for algin; the remainder was used as fodder. The mechanical harvester cuts the kelp about 1 m below the water surface. The cut weed is washed and chopped, and the algin is then extracted. Alginic acid is also obtained from other brown algae such as *Laminaria* and *Ascophyllum* in the North Atlantic and *Ecklonia* and *Sargassum* in the Pacific and Indian Oceans. The harvesting of *Laminaria* in the North Atlantic accounts for about half of the entire world production of alginic acid.

The uses of alginates are determined by their chemistry, gelling, suspending, and emulsifying properties. The most extensive uses are in the textile, paper, and paint industries, as well as in the food industries. Alginates, like the red algal extract carrageenan, are used as emulsifiers and stabilizers in dairy products (sherbets and ice cream to prevent ice crystal formation; chocolate milk, cheese, puddings, and toppings). Alginates are also used as thickeners in syrups, sauces, salad dressings (also as an emulsifier), soaps, shampoos, toothpastes, shaving creams, medicines, lipsticks, insecticides, plastics, fireproofing fabrics, and polishes. In the medical field, alginates are used to make surgical threads that dissolve and do not have to be removed. A whole-blood substitute is obtained in part from alginates, and this is used in emergency transfusions.

Silverthorne and Sorensen (1971) estimated the 1970 production of alginic acid to be about 13,000 metric tons at a value of $1.25/lb. By 1978 the value was over $1.70/lb, and the yield had increased to 15,000 metric tons. The two primary companies that produce and sell alginate in the United States are Stauffer Chemical Corporation and Kelco Chemical Corporation; both have plants in California. Kelco also has a plant in eastern Canada.

Alginic acid is a polyuronic polymer consisting of D-mannuronic acid and L-guluronic acid residues in differing amounts (Figure 2-6). Linkage is through the $(1 \rightarrow 4)$ positions, in a β form for D-mannuronic and α form for L-guluronic

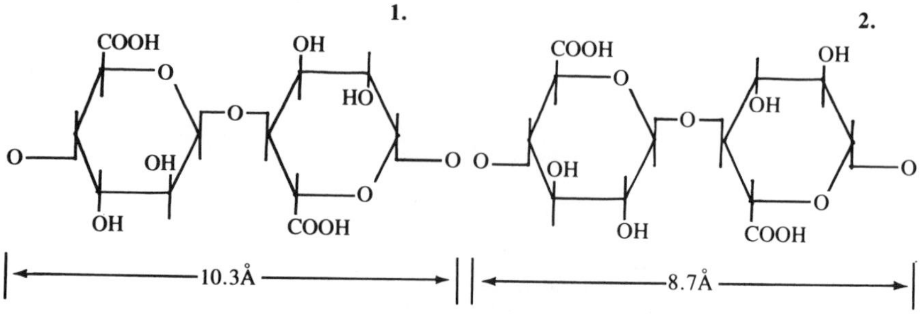

Figure 2-6 The molecular structure of alginic acid. The compound consists of two major components: (1) β-(1 → 4)-linked D-mannuronic acid (10.3 Angstrom longunit) and (2) α-(1 → 4)-linked-L-guluronic acid (8.7 Angstrom longunit). Alginic acid consists of long chains of the two components.

acid. The alginic acid content varies from 10 to 47% of the dry weight in various brown algae; it is apparently located in the cell walls. There is a wide variation in the type of alginic acid produced, even in the same species over a year's growth. This variation is now thought to be due to changes in ratios of the two residues.

Since alginic acid occurs as an insoluble salt, mostly as calcium alginate, extraction involves solubilization through base changes. The removal of impurities occurs by multiple reprecipitation and the ultimate washing of the precipitate. Free alginic acid is released by the addition of an acid (usually HCl). The acidified solution is then easily converted to sodium alginate through the addition of 3% sodium carbonate solution. It is then isolated by coagulation with ethanol or by reprecipitation as the calcium salt.

Agar

Agar is a strongly gelling hydrocolloid found in some red algae (*Gelidium, Gracilaria*). It is commercially available in powder, flakes, or strips. The name is derived from the Malayan or Ceylonese term *agar-agar*, which means jelly. The discovery of agar is told in all phycological textbooks and apparently occurred in 1658, when a Japanese innkeeper threw the excess of a cooked seaweed dish into the snow. The next morning the jelly was a papery, dry, translucent substance, which the innkeeper found he could store and convert back to the jelly by dissolving in boiling water. Agar was introduced to Europe and America from the Far East in the middle of the 1800s; it was originally used only in the manufacture of jellies and desserts.

The main use of agar as a bacteriological medium was recognized by the German wife of a doctor, who passed this information on to Robert Koch. In 1882, Dr. Koch published his now-famous note on the tubercle bacillus and the use of agar as a new solid culture medium.

Agar is useful in stomach and intestinal disorders and because of its capacity to hold water it can serve as a mild laxative. Agar is used to coat capsules, and by controlling the capsule thickness it can act as a time-release mechanism.

It is also used as a dressing for burns. Food industries utilize agar in canning as a protective gel around delicate meats and fish, as a stabilizer and protective agent in meringues, pie fillings, icings, and glazes. Other uses of agar include the sizing of fabrics, as an ingredient in waterproof paper, cloth, and glue, as a clarifying agent in the manufacture of wines, beers, and coffee, and in the preparation of special diabetic foods, where agar is used to replace starch (e.g., bread).

Silverthorne and Sorensen (1971) estimated that the worldwide annual production of agar was about 7700 metric tons at a value of $39 million in 1968. In 1977 about 10,000 metric tons were produced with a value of over $50 million. The price per pound varies from $4 to $50, due to a range of quality and purity.

Chemically, agar consists of two components, agarose and agaropectin (Figure 2-7). The main component is agarose, which is a gelling, neutral polymer having repeating units primarily of the disaccharide agarobiose (glactopyranose residues; Figure 2-7). Agaropectin is a mixture of polysaccharides containing D-galactose, 3,6-anhydro-L-galactose, D-glucuronic acid, and varying amounts of sulfate esters up to 5%. Thus there are apparently a number of types of agar based on the residues found in agaropectin.

Agar is extracted by cooking the seaweed. Initially the dried agarophytes are washed in cold water to remove excess salts. After soaking they are cooked (in the United States under 15 psi) or boiled (in Japan). A filter aid such as diatomaceous earth is added to the slurry, and the agar forced under pressure through the filter presses. The resulting sol is allowed to gel at room temperature. The gel is chopped, frozen ($-8°C$), and thawed ($10°C$). The freed liquid is removed, carrying the impurities. The resulting gel flakes may be further washed and bleached (sodium hypochlorite); ultimately they are dried to about 20% moisture and packaged.

Carrageenan

The phycocolloid carrageenan is an extract obtained from members of the red algal order Gigartinales (Figure 2-8). However, some studies suggest that car-

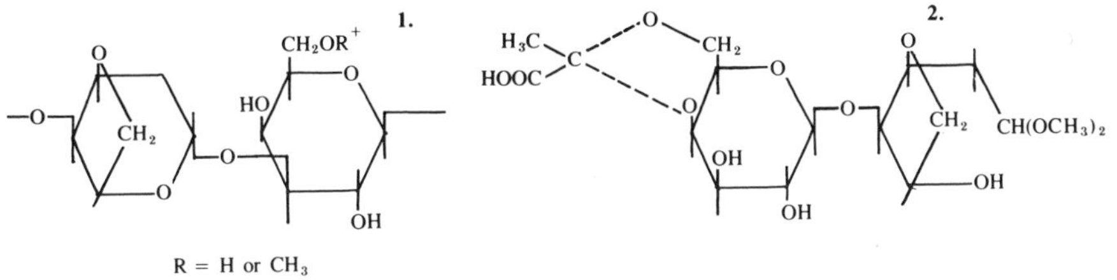

R = H or CH$_3$

Figure 2-7 The molecular structure of the two basic components of agar. This gelling hydrocolloid consists of (1) an agarobiose fraction, which is the main component, and (2) an agaropectin fraction. The latter fraction is a mixture of polysaccharides containing 3,6-anhydro-L-galactose, D-galactose and some other sugars.

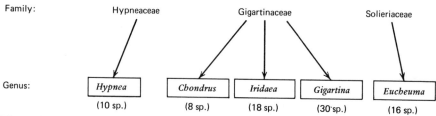

Figure 2-8 A diagram of carrageenan-yielding red algae found in the order Gigartinales. In the family Gigartinaceae, species can be found that yield κ-carrageenan in the haploid (gametophytic) phase and λ-carrageenan in the diploid (tetrasporophytic) phase. The Caribbean species *Eucheuma* only yields ι-carrageenan, while some of the Pacific species yield only κ-carrageenan.

rageenan may also be present in some members of the Cryptonemiales (*Halymenia* and *Corynomorpha*), although the chemical data is not definitive. Carrageenan is primarily extracted from *Chondrus crispus* (Irish Moss) in the North Atlantic and from *Eucheuma* spp. in the Philippines.

Carrageenan is widely used, but unlike algin, most (~ 80%) of it is used by the food industry (Table 2-4). Carrageenan is used extensively in dairy products because of its unique interaction and stabilizing effects with milk proteins. Extracts are also used in toothpaste, diet foods (the carbohydrate is not digestible by man), hand creams, soups, confections, insect sprays, water-base paints, inks, shoe stains, shampoos, cosmetics, and pharmaceutical products. Carrageenan is used in cloth sizing, printing processes, clarifying beer, and as a dental-impression compound.

The production of carrageenan almost quadrupled from approximately 2200 metric tons in 1960 to over 8640 metric tons in 1971. In 1974, over 10,000 metric tons at a cost of $44/kg were produced (Dawes, 1974). Carrageenan-yielding species of *Eucheuma* are grown extensively in the Philippines (see section on agronomy), but *Chondrus* is still harvested from natural populations in the North Atlantic.

Carrageenan is a highly sulfated (20–50% galactans) compound, whereas agar has a much lower sulfate level (less than ~5%). At present, five major forms of carrageenan are known (κ, λ, μ, γ and ι), which differ in their levels of sulfation as well as the ratios of galactose to 3,6-anhydrogalactose (Figure 2-9). The carrageenans appear to be structural components of the algal cell wall (LaClaire and Dawes, 1976). The forms are distributed between diploid and haploid phases (Dawes *et al.*, 1977a, 1977b). Some members of the Gigartinales contain κ-carrageenan in the haploid plant and λ-carrageenan in the diploid plant (*Iridaea, Chondrus, Gigartina*). Other genera of the Gigartinales only contain κ- (*Hypnea, Eucheuma denticulatum*) or ι-carrageenan (*Eucheuma isiforme, E. spinosum, E. uncinatum, Agardhiella tenera*).

Table 2-4   Uses of Carrageenan in Food Preparation

Use	Function	Approximate Amount Used
Water dessert gels (dry powders, finished gels)	Setting agent	0.70%
Dietetic jellies	Setting agent	0.50%
Pie fillings (chiffon, meringue)	Setting agent	0.50%
Syrups (chocolate, maple, etc.)	Bodying, suspension	0.20%
Fruit drink powders and frozen concentrates	Bodying and pulping effects	0.50%
Imitation coffee creams	Emulsion stabilization	<0.20%
Relishes, pizza, and barbecue sauces	Bodying	<0.50%
Buttered sauces for frozen vegetables	Cling, uniform color, mouthfeel	<0.1%
Soups	Bodying, gelling	0.2–1.0%
Toothpaste, lotions,	Bodying, emulsion stabilization	<1.0%
Suspensions (graphite, clay, etc.)	Suspending	<1.0%
Ice cream	Preventing whey separation	0.015%
Pasteurized milks (Chocolate, eggnog, fruit flavored)	Suspension, bodying	0.027%
Sterilized milks		
Chocolate, eggnog, etc.	Suspension, bodying	300 ppm
Evaporated in can	Fat stabilization	25–50 ppm
aseptic		100 ppm
900 calorie diet drinks	Suspension, bodying	250 ppm
Infant formulations (Concentrates and single strengths)	Stabilization of fat and protein	300 ppm
Puddings and pie fillings (dry powders, finished and frozen types		
Cold set without starch	Setting agent	0.5–1.0%
Cold set with starch	Setting, syneresis inhibiting	0.1–0.5%
Cooked flan or custard	Setting agent	0.3%
Cooked starch type	Anticracking, better unmolding, noncritical cooking	0.05%
Whipped products (dry, finished, frozen, aerosol)		
Creams, toppings, desserts	Fat and foam stabilization, setting	0.05–0.5%
Cold prepared milk powders		
Thickened drinks, shakes	Bodying, stabilizing overrun	0.1–0.3%

Source: Modified from a handout from Dr. Maxwell Doty, University of Hawaii, Honolulu.

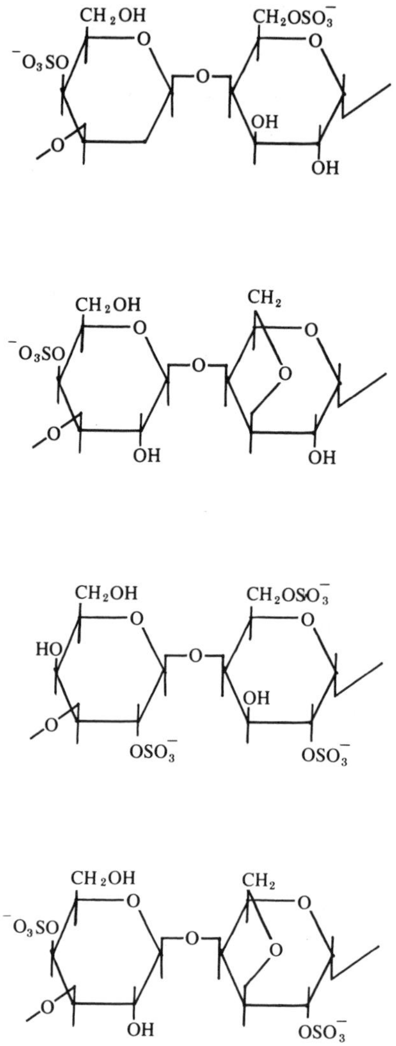

Figure 2-9 The molecular structure of four major types of carrageenan. Note the presence of 3,6-anhydrogalactose residues in both κ- and ι-carrageenan and the presence of sulfate groups in all four types.

Carrageenan is extracted by boiling the dried and ground plant material in water under slightly alkaline (pH 8.5) conditions. A filter aid such as diatomaceous earth is added to the slurry 2–5 hr after boiling, and the hot mixture is filtered under pressure. The sol is then poured into 80% isopropyl alcohol, where the carrageenan coagulates. After drying (pressing and drum drying), the sheets of carrageenan may be decolorized (with charcoal), ground, and packaged.

Other phycocolloids

Agar, algin, and carrageenan account for about 95% of the commercially available seaweed extracts. Furcellarin (Danish agar) is obtained from the red alga *Furcellaria*, another member of the Gigartinales (Fig. 2-4). Similar to agar, furcellarin when frozen will release water; however, its chemistry is similar to κ-carrageenan (D-galactose, 3,6-anhydro-D-galactose, and ester sulfate). In 1974, about 1200 metric tons were produced at a value of about $5 million. The uses of furcellarin parallel those of carrageenan, especially in dairy products.

Funoran is extracted from the red alga *Gloiopeltis furcata* (Cryptonemiales) in China and Japan. The dried plant material will completely dissolve in fresh water and no further extraction is necessary. Funoran consists of D-galactose, 3,6-anhydro-L-galactose, with a high level of ester sulfate derivatives (18%). Since it will not form gels, funoran is thought to be a sulfated agarose or agaropectin. It is used in sizing and gluing paper and textiles and in hair waving. About 700–900 metric tons were produced in 1975, with a value of over $1 million.

Porphyron is obtained by hot water extraction from the red alga *Porphyra*. It is a sulfated, linear galactan of 3-β-D-galactosyl and 4-α-L-galactosyl units, but the chain is masked by occassional D-galactosyl and L-galactosyl side units. Presently porphyron is processed in Japan and China, where it is used as a gelling agent.

Marine Agronomy (Aquaculture)

The cultivation of economically important seaweeds is in its infancy, but within the last five years, major advances and ideas have developed in ocean farming (Michanek, 1978). In a review of trends in applied phycology, Michanek divides the farming of seaweeds into six areas: general, open-sea marine farming, cottage-scale or greenhouse cultivation, wastewater treatment systems, mariculture of tropical species, and conventional cultivation schemes.

Domestication of terrestrial crop plants began 10,000 years ago, whereas the domestication of marine crop plants is only about 300 years old (Wheeler *et al.*, 1979). Four marine seaweeds are now extensively grown under agronomic conditions, the red algae *Porphyra* and *Eucheuma*, and the brown algae *Laminaria* and *Undaria*. Other algae under cultivation include the green alga *Enteromorpha* and the red algae *Gelidium* and *Gracilaria*. Even today the majority of harvested seaweeds come from natural populations, and all agonomical practices occur outside the United States (see Table 2-5).

Wheeler *et al.* (1979) have reviewed various problems facing marine agronomists, especially with relation to the cultivation of the giant kelp *Macrocystis*. He divides the adverse factors into physical (wave action, sedimentation), chemical (nutrient loading, wastewater problems) and biological (competition, grazing, encrustations, pathogens). Some of these are more important in open-ocean farming (wave action) while others become most critical in tank culture (competition, en-

Table 2-5 Cultivation of Certain Important Seaweeds

	Prophyra[a]	*Undaria*	*Laminaria*	*Eucheuma*
Location	Japan, Korea, China	Japan	Japan	Philippines, experimental in Florida
Technique	On nets held on poles parallel to water surface. Also on bamboo screens, or on buoyed floating nets	On floating ropes; 5–10 kg harvested/m of rope	On buoyed ropes, also natural growth by blasting reefs and planting stones	On nets held on poles parallel to water surface; also on lines; both on coral reefs
Source	Thalli from spores of sporophyte stage developed in tanks	Zoospores collected in tanks	Zoospores, under forced cultivation	Vegetative, pieces tied to net or line
Use	*Nori* sheets for human consumption.	Eaten dried raw or salted, *wakame*	Food, chiefly	Carrageenan
Stage used	Gametophyte	Sporophyte	Sporophyte	Vegetative

Source: Hasegawa, Yoshio. 1976. Progress of *Laminaria* cultivation in Japan. *J. J. Fish. Res. Bd. Canada.* **33**: 1002–1006. Deveau, L. E. and J. R. Castle. 1976. *The Industrial Development of Formed Marine Algae: The Case-History of Eucheuma in the Philippines and U.S.A.* FAO Technical Conference on Aquaculture, Kyoto. Saito, Y. 1976. *Seaweed Aquaculture in the Northeast Pacific.* FAO Technical Conference on Aquaculture, Kyoto.

[a]Two green algae, *Monostroma* and *Enteromorpha*, are cultivated similar to *nori* but usually at different sites.

crustation). Pathogenic problems with commercially important seaweeds are covered in the two Appendices on marine bacteria and fungi.

In a short review, Wheeler *et al.* (1979) point out that the future of marine agriculture is bright. Oriental progress in the previous 30 years has been rapid, resulting in the farming of thousands of hectars in near-shore areas. In addition to supplying food and industrial products, many seaweeds such as *Macrocystis* are being considered as sources of methane for energy production. Mathieson (1975) and Neish (1976) have presented two general reviews of seaweed agronomy. In this section farming procedures of the more important seaweeds presently under cultivation will be introduced.

Eucheuma

Two *Eucheuma* species, *E. spinosum* and *E. denticulatum*, both of which yield carrageenan, are extensively grown in the southern Philippines. Over 4000 metric tons dry weight were exported to the United States in 1977 and approximately

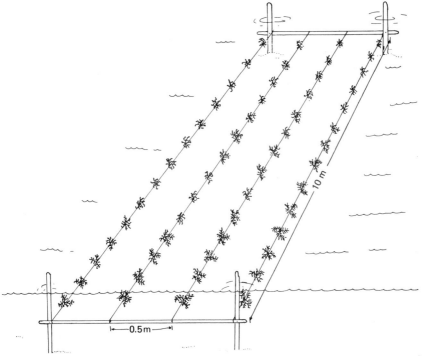

Figure 2-10 Farming of species of *Eucheuma* is done by hand in the South China Sea and the most popular method used by local fishermen is the line procedure (Ricohermoso and Deveau, 1977). Sprigs of the plant are tied on monofilament lines spaced every 0.5 m and portions of the plants are harvested throughout the year. One 10 m monofilament line can accomodate about 50 plants, yielding about 5 kg fresh weight at planting, with each line producing about 50 kg in 2–3 months. An average family farm will handle about 2,000–3,000 plants in any one month.

6000 metric tons in 1978. The farming procedure is relatively simple and has been described by Doty and Alvarez (1975). A series of 90 kg test monofilament lines are tied to wooden stakes. The cuttings, which are collected from natural populations or from selected seed stock, are tied to the 10 m nylon lines, spaced every 0.5 m (Figure 2-10). A farm usually has about 100,000 plants ha (Ricohermoso and Deveau, 1977).

Porphyra

Nori, as *Porphyra* is commonly known in the Orient, was first cultivated in Tokyo Bay in 1736 by allowing bamboo stakes to serve as attachment sites for the small blades. The discovery of the microscopic, filamentous stage that bores into oyster shells (*Conchocelis* stage) allowed a more controlled plan of cultivation. The life history is now well documented (Figure 2-11) and employed in its cultivation. *Porphyra* spores are collected on cleaned oyster shells by allowing fertile blades to float in pans over shells (Figure 2-12). Spore production occurs during

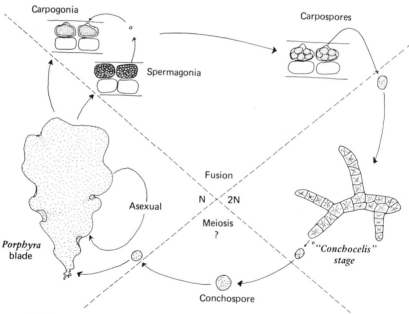

Figure 2-11 The life history of the red alga, *Porphyra* sp., is discussed in Chapter 7. There are two phases to this life history, a bladed phase that is harvested for food and a tiny filamentous or *Conchocelis* stage. The latter stage grows on shells and thus can be cultured in the laboratory (see figure 2-12). Conchospores attach to ropes in culture trays in the laboratory and these ropes are placed in the field (Figures 2-13 and 2-14). These spores germinate and grow into *Porphyra* blades, which in turn, produce what appear to be female (carpogonia) and male (spermagonia) cells. At least in one species of *Porphyra*, sexual fusion is known and the resulting diploid cell produces carpospores that germinate and grow in the filamentous "*Conchocelis*" stage.

Figure 2-12   One stage in the culture of *nori (Porphyra)* in Japan. Mature, reproducing blades of *Porphyra* are floated in tables over shells in culture labs. The resulting carpospores will settle on the shells, germinating into the filamentous *Conchocelis* stage shown here.

periods of warm water and long photoperiods. The "seeded" shells are stored over the summer (incubation of spores), and when the spores germinate, microscopic *Conchocelis* filaments are produced. Based upon previous physiological studies of light and temperature requirements, the spores can be produced at will. In October (cooler water, shorter day period), the *Conchocelis* filaments begin to produce conchospores, and the oyster shells are placed with *Porphyra* nets in trays. The conchospores are released and attach to the nets. The nets can be refrigerated and stored from October to March or April. At that time the nets are placed in the ocean (Figure 2-13) and the plants resulting from the conchospores (Figure 2-14) can be harvested through the spring and summer. Harvesting is carried out either by hand or by mechanical scrapers.

Figure 2-13   A later stage in *nori* culture. When the *Conchocelis* filaments growing in the shells reproduce in culture, the resulting conchospores are allowed to settle on *hibi* nets held in the tanks. These nets are then set out in protected estuarine waters about 0.5 m above mean low water. Note the bags containing shells are tied to the nets.

Figure 2·14 *Nori* culture. The nets can cover many hectares in protected bays such as that seen at Sendai. These are ready to harvest. (Photo Figures 2-12, 2-13, and 2-14 courtesy of John West. University of California, Berkeley.)

The successful cultivation of *nori* has resulted because of a detailed scientific knowledge of its life history. Knowledge of the importance of water temperature, photoperiod, and the life history of the plant has allowed the *nori* industry to become secure and highly productive. Over 121,000 metric tons were grown in 1978 with a value of $150 million. The recent demonstration of sexual reproduction in *Porphyra* has clarified its importance in the plant's life history (Hawkes, 1978).

Gracilaria and Gelidium

Both of these red algae are extensively harvested in Japan as a source of agar. The plants are grown vegetatively by scattering small vegetative fragments of the plants over suitable stony bottoms and allowing them to regenerate. Some rope culture (plant fragments entwined in rope filaments) of *Gracilaria* and *Gelidium* is also conducted.

Laminaria and Undaria

The two kelps *Laminaria* and *Undaria* are farmed in Japan and China. Three species of *Undaria,* or *wakame,* are consumed directly and are important edible seaweeds in Japan. Specially designed stands or blocks are established in suitable habitats for natural colonization and attachment of zoospores. Rope cultivation is also used. The zoospores of *Undaria* are released from mature blades and are allowed to settle and attach on ropes that are curled in dishes in the laboratory. After germination, the ropes are transferred to the ocean. A bamboo raft about 3.6 × 1.8 m will support sufficient rope to produce about 1 metric ton of *Undaria*. About 11,000 metric tons dry weight of these kelps were harvested in 1962.

Figure 2-15 *Kombu* culture in Japan. Mature, reproducing blades of *Laminaria* (see Chapter 6 for life history) are floated over coiled ropes in culture tables. The spores from the blades will settle on the ropes and germinate into tiny, filamentous male and female gametophytes. After fertilization, the sporophyte germinates directly on the rope.

Three species of *Laminaria*, commonly called *kombu*, are cultivated in Japan as a direct food source and for alginic acid. Both "stone plantings" and rope cultivation methods are used. Spores are germinated in the lab and the gametophytes grown in tanks (Figure 2-15). The sporophytic plants are then moved to culture beds in the ocean (Figure 2-16). Both *Undaria* and *Laminaria* are fertilized *in situ* by slow dissolving nutrient balls or with fertilizers placed in clay pots. In 1960, between 3000 and 5500 ha of coastal waters were utilized for cultivation of *Laminaria japonica* alone. In China, two new varieties of *Laminaria japonica* (Haidai) have been bred with higher yields and iodine content (Anonymous, 1976). Mutagenic X ray techniques were used in the selection of new strains.

Figure 2-16 A later stage in *kombu* culture. The ropes are tied to rafts or stakes and the *Laminaria* sporophytes are allowed to grow to maturity as seen in this culture bed of *L. angustata* in Japan. (Photos Figures 2-15 and 2-16 courtesy of John West, University of California, Berkeley.)

Green algae

Jetoegusa (Monostroma), *aonori (Entermorpha)* and *aosa (Ulva)* are edible green algae that are used in Japan. Only *Monostroma* is extensively cultivated, and it commands the highest price of any seaweed in Japan. Individual stakes or nets are placed in the ocean in the fall for the attachment of spores. Usually three harvests can be obtained from a single net or set of individual stakes.

Methane production

Probably the most unique area of seaweed farming research can be found in the proposal for large-scale industrial farming of seaweeds (Bryce, 1978) such as the giant kelp *Macrocystis pyrifera* (Wilcox, 1974; Leeper, 1976; Leone, 1979) and the red alga *Gracilaria* sp. (Ryther, 1980) in order to obtain energy. Open sea farms a hectare in size, are proposed for the cultivation of the giant kelp, and smaller tanks or basin farms have been designed for land cultivation of various red algae. The biomass resulting from photosynthesis would be digested anaerobically (fermented) and the methane gas produced would be used to generate electricity.

Two distinctive approaches have been suggested to harness the sun's energy through methane production from biomass. Ryther (1980) proposed using organic wastes as a fertilizer source to support the mariculture of *Gracilaria* in tanks or basins. Wilcox (1974) proposed the use of deep, nutrient-rich seawater, pumped to the surface, to fertilize buoyed *Macrocystis* sea farms. Ryther (1980) and Hanisak (1981) found that for every 100 g of nitrogen added to the digester in the form of *Gracilaria*, 23 g of nitrogen was recovered as solid residue and 70 g as a

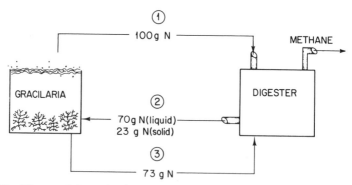

Figure 2-17   Nitrogen balance in the recycling of digester resudiues from anaerobic digestion of *Gracilaria*. (Diagram by Ryther, 1980.) In addition to methane production, almost 50% of the nitrogen can be recycled. Starting with 100 g of nitrogen in a *Gracilaria* biomass (1), 93 g of nitrogen solids and liquids are recovered after digestion (2). Of this, 73 g of nitrogen are again available in the second biomass harvest of *Gracilaria* (3).

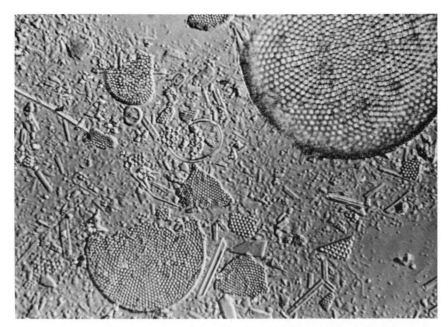

Figure 2-18 Diatomaceous earth is composed of the SiO_2 cell walls of diatoms, called frustules. This earth, which is from the Miocene, is used in a variety of polishes, insulations, and filtration systems. × 500.

liquid (Figure 2-17) and 73 g were available for a second recycle. About 0.4 liters of 60% methane/g of *Gracilaria* solids can be obtained from the process. In addition, waste materials could be used as the nutrient source. Wilcox calculated that 565,504 km^2 of coastal ocean off California could produce sufficient energy to equal all of the food (844 × 10^{15} J) and natural gas (more than 242 × 10^{17} J) presently consumed in the United States every year (Leeper, 1976)

Diatomite

The most important fossilized marine algal deposit is diatomite, which consists of diatom frustules, or cell walls, consisting of SiO_2. One large Miocene deposit, up to 300 m deep, occurs at Lompoc, California. The lightweight mass of frustules (Figure 12-18), or diatomaceous earth, can be sawed out as blocks and used as a fine abrasive in polishing compounds (toothpaste, optical powders) and for filtration (e.g., beer production, filtration of plant products). Dr. Alfred Nobel, founder of the Nobel Prizes, discovered that nitroglycerin could be safely handled when absorbed on diatomite, hence the term dynamite. Accordingly, he earned a fortune in royalties, from which the Nobel Prizes have been obtained. Diatomite is also extensively used as a lightweight insulator against sound and heat.

REFERENCES

Abbott, I. A. 1978. The uses of seaweed as food in Hawaii. *Econ. Bot.* **32**: 409–412.

Abbott, I. A. and B. Williamson. 1974. *Limu: An Ethnobotanical Study of Some Edible Hawaiian Seaweeds.* Pac. Trop. Bot. Gard. Hawaii, Kauai.

Anonymous. 1976. The breeding of new varieties of Haidai (*Laminaria japonica* Aresch.) with high production and high iodine content. *Scientia Sinica.* **19**: 243–252.

Bauersfield, P., R. R. Kifer, N. W. Duprant, and J. E. Sykes. 1969. Nutrient content of turtle grass (*Thalassia testudinum*). *Int. Seaweed Symp.* **6**: 637–645.

Blunden, G., W. W. Binns, and F. Perks. 1975. Commercial collection and utilization of Maërl. *Econ. Bot.* **29**: 140–145.

Bonotto, S. 1976. Cultivation of plants. Multicellular plants. In Kinne, O. (ed.). *Marine Ecology. Vol 3 Cultivation. Part 1.* Wiley, New York.

Bryce, A. J. 1978. *A review of the Energy from Marine Biomass Program.* Biomass Symposium, Institute of Gas Technology. General Electric Co., Philadelphia.

Chapman, V. J. 1970. *Seaweeds and their uses,* 2nd ed. Methuen, London.

Dawes, C. J. 1974. *On the mariculture of the Florida seaweed, Eucheuma isiforme.* Florida Sea Grant Prog. Rept. 5, University of Florida, Gainesville.

Dawes, C. J., N. F. Stanley, and D. J. Stancioff. 1977a. Seasonal and reproductive aspects of plant chemistry and iota carrageenan from Floridian *Eucheuma* (Rhodophyta, Gigartinales). *Bot. Mar.* **20**: 137–147.

Dawes, C. J., N. F. Stanley, and R. E. Moon. 1977b. Physiological and biochemical studies on the i-carrageenan producing red alga *Eucheuma uncinatum* Setchell and Gardner from the Gulf of California. *Bot. Mar.* **20**: 437–442.

Doty, M.S. and V. B. Alvarez. 1975. Status, problems, advances and economics of *Eucheuma* farms. *Mar. Tech. Soc. J.* **9**: 30–35.

Felger, R. and M. B. Moser, 1973. Eelgrass (*Zostera marina* L) in the Gulf of California; discovery of its nutritional value by the Seri Indians. *Science* **181**: 335–356.

Felger, R. S., E. W. Moser, and M. B. Moser. 1980 . Seagrasses in Seri Indian culture. In Phillips, R. C. and C. P. McRoy (eds.). *Handbook of Seagrass Biology. An ecosystem perspective.* Garland STPM Press, New York.

Goforth, H. W., Jr., and J. R. Thomas. 1980. *Plantings of Red Mangroves (Rhizophora mangle L.) for Stabilization of Marl Shorelines in the Florida Keys.* Technical Report 506. Naval Ocean Syst. Ctr., San Diego, Calif.

Hanisak, M. D. 1981. Recycling the residues from anaerobic digesters as a nutrient source for seaweed growth. *Bot. Mar.* **24**: 57–61.

Hawkes, M. W. 1978. Sexual reproduction in *Porphyra gardneri* (Smith *et* Hollenberg) Hawkes (Bangiales, Rhodophyta). *Phycologia* **17**: 326–350.

Hoffman, W. E. and J. A. Rodgers Jr. 1980. *A Cost/Benefit Analysis of Two Large Coastal Plantings in Tampa Bay, Florida.* Wet Lands Conference Proceedings, Tampa, Fla.

Kiran, E., I. Teksoy, K. C. Gaven, E. Guler, and H. Guner. 1980. Studies on seaweeds for paper production. *Bot. Mar.* **23**: 205–208.

Krauss, R. W. (ed.). 1977. *The Marine Plant Biomass of the Pacific Northwest Coast.* Oregon State University Press, Corvallis.

LaClaire, J. W., II and C. J. Dawes. 1976. An autoradiographic and histochemical localization of sulfated polysaccharides in *Eucheuma nudum* J. Agardh. *J. Phycol.* **12**: 368–375.

Leeper, E. M. 1976. Seaweed: Resource of 21st century? *BioSci.* **26**: 357–358.

Leone, J. E. 1979. *Marine Biomass Energy Project.* Marine Tech. Soc. Meetings, New Orleans, General Electric Report.

Levring, T., H. A. Hoppe, and O. J. Schmid. 1969. *Marine algae. A survey of research and utilization.* Cram, De Gruyter, Hamburg.

Madlener, J. C. 1977. *The Seavegetable Book.* Clarkson N. Potter, New York.

Mathieson, A. C. 1967. *Seaweed–A Growing Industry.* Pacific Search, Seattle, Wash.

Mathieson, A. C. 1975. Seaweed aquaculture. *Mar. Fish. Rev.* **37**: 1–4.

McRoy, C. P. and C. Helfferich. 1980. Applied aspects of seagrasses. In Phillips, R. C. and C. P. McRoy (eds). *Handbook of Seagrass Biology. An Ecosystem Perspective.* Garland STPM Press, New York.

Michanek, G. 1978. Trends in applied phycology. With a literature review: seaweed farming on an industrial scale. *Bot. Mar.* **21**: 467–475.

Morton, J. F. 1965. Can the red mangrove provide food, feed and fertilizer? *Econ. Bot.* **19**: 113–123.

Neish, I. C. 1976. *Culture of Algae and Seaweeds.* FAO Tech. Conf. on Aquacult, Koyoto.

Phillips, R. C. 1974. Transplantation of seagrasses with special emphasis on eelgrass, *Zostera marina* L. *Aquaculture* **4**: 161–176.

Phillips, R. C. 1980. Transplanting methods. In Phillips, R. C. and C. P. McRoy (eds.). *Handbook of Seagrass Biology. An Ecosystem Perspective.* Garland STPM Press, New York.

Rees, D. A. 1972. Polysaccharide gels. A molecular view. *Chem. and Ind.* **1972**: 630–636.

Ricohermoso, M. A. and L. E. Deveau. 1977. *Commercial Propagation of Eucheuma Spp. Clones in the South China Sea. A Discussion of Trends in Cultivation Technology and Commercial Production Pattern.* Handout at IXth Seaweed Symposium, Santa Barbara, Calif.

Ryther, J. H. 1980. *Cultivation of Macroscopic Marine Algae and Freshwater Aquatic Weeds.* Progress report, Solar Energy Research Institute. Department of Energy, Washington, D.C.

Shurtleff, W. and A. Aoyagi. 1978. *The Book of Sea Vegetables.* Autumn Press, San Francisco.

Silverthorne, W. and P. E. Sorensen. 1971. *Marine Algae as an Economic Resource.* Mar. Tech. Soc. Trans. of the 7th Ann. Mar. Sci. Conf. **7**: 523–533.

Thorhaug, A. 1974. Transplantation of the seagrass *Thalassia testudinium* Konig. *Aquaculture* **4**: 177–183.

van Breedveld, J. R. 1966. *Preliminary Study of Seagrass as a Potential Source of Fertilizer.* Fla. St. Bd. of Conserv. Mar. Lab. St. Petersburg. Sp. Sci Rep. 9.

Wheeler, W. N., M. Neushul, and J. W. Woessner. 1979. Marine agriculture: progress and problems. *Experimentia* **35**: 433–345.

Wilcox, H. A. 1974. *Proposed Program Operating Plan for an Ocean Food and Energy Farm.* Naval Undersea Center, San Diego, Calif.

PART 2

The Algae

CHAPTER 3

Introduction to the Algae

Macroscopic marine algae are commonly called *seaweeds*. However, there are more microscopic than macroscopic marine algae. The term *alga* (plural algae) is not an easy one to define. Bold and Wynne (1978) note that algae "share the more obvious characteristics with other plants while their really unique features are more subtle." Trainor (1978) states that these "simple" plants are not really simple at all.

What are algae? The technical definition presented by Trainor is clear enough: "Algae are photosynthetic, nonvascular plants that contain chlorophyll *a* and have simple reproductive structures." Probably the major distinction between algae and more complex "higher" plants lies in gamete production. In unicellular algae, the cell itself may function as a gamete, whereas in multicellular algae, specialized unicellular or multicellular structures are gametangia (gamete producers). It is important to note that *every* cell is involved in gamete production and unlike reproductive structures in "higher" plants, including mosses, there are no sterile protective cells. There is one exception to the above statement, and this will be noted in the green algal class, the Charophyceae. A second feature common to most algae is their "simple" plant body or thallus. With the exception of the larger brown algae (especially kelps and rock weeds), they have little tissue differentiation.

CHARACTERISTICS OF ALGAE

In this chapter, the major diagnostic features of the different algal divisions will be reviewed. These include cytological characteristics, photosynthetic pigments, reserve foods, cell wall structures, and composition, and flagellar types and placement. In addition, a brief introduction to algal life histories, is presented. Finally the evolution of algae is discussed briefly.

Chapters 4 through 9 summarize the various algal divisions, give a key to the major groups, and describe some of the major marine taxa. The text is primarily concerned with the marine characteristics of each algal division; thus selected

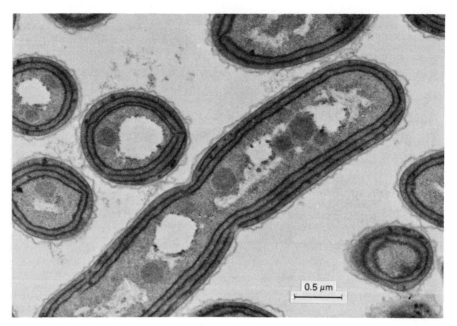

Figure 3-1 Electron micrograph of a section through the blue-green alga *Anacystis nidulans*. The bacterial, or prokaryotic, nature of the cell is evident, with a central region containing fibrils of DNA, some clear regions, and a number of peripheral photosynthetic membranes called thylakoids. The large polyhedral bodies are carboxysomes, or storage sites of a carboxylase used in photosynthetic carbon fixation. A thin, outer, irregular layer is probably a gelatinous mucopolysaccharide. × 25,900. (Courtesy of Norma Lang, University of California, Davis.)

examples are given. Since about half of the algal species are freshwater, several orders will not be covered. The reader is referred to two recent texts for a more detailed taxonomic approach to the algae (Bold and Wynne, 1978; Trainor, 1978). Detailed chemical and physiological data on algae are available in two texts edited by Lewin (1962) and Stewart (1974). A summary of algal ultrastructure (Dodge, 1973) and a review of all motile algae (phytoflagellates, Cox, 1980) have been published.

Cytological Characteristics

As pointed out in the first chapter, both the prokaryotic blue-green algae (Cyanophyta) and the eukaryotic forms of algae contain chlorophyll *a* and evolve oxygen. Bacteria and blue-green algae are prokaryotic in cell structure, as they lack membrane-bound organelles such as chloroplasts, nuclei, and mitochondria (Figure 3-1). Prokaryotic cells have smaller ribosomes (RNA, 70S*); their cell wall

*The term *70 S* refers to the rate of sedimentation in a centrifuge at a known gravity, temperature, and liquid viscosity. S is the abbreviation for a Svedburg unit.

Table 3-1   Comparison of Prokaryotic and Eukaryotic Structures

Feature	Prokaryotic	Eukaryotic
Genome	Naked molecule in cytoplasmic matrix	A true nucleus, having a nuclear envelope
Chromosome	A single, circular DNA molecule	One or more chromosomes, each consisting of numerous DNA strands and basic proteins, histones
Respiration/oxidation enzymes	On cell membrane and an extension (mesosome)	On membranes, especially present in mitochondria
Photosynthesis	Occurs on thylakoids free in cytoplasmic matrix	On thylakoids contained in chloroplasts
Microfilaments and microtubules	Absent	In all cells
Ribosomes	70S sedimentation size	80S sedimentation size in cytoplasm but smaller, 50–60S size in mitochondria and chloroplasts
Flagella	Single protein molecules composed of flagellin; external to cell membrane	Complex nine outer doublets and two single inner microtubules composed of tubulin and covered by cell membrane
Cell wall	If present, contains mucopolysaccharides, lipopolysaccharides, and special protein components	If present contains structural components, usually cellulose or other complex polysaccharide

chemistry is distinctive (see Figure 3-12) and their flagella are chemically and structurally distinct from eukaryotic forms. Table 3-1 presents a comparison of eukaryotic and prokaryotic cell characteristics.

One of the unique features of eukaryotic algae is the occurence of plastids. The photosynthetic pigments are found in the lamellae of the chloroplasts, and these lamellae are called *thylakoids*. Thylakoids may be grouped into bands of three (Phaeophyta, Chrysophyta, Pyrrhophyta), five to nine (Chlorophyta, Euglenophyta), or they may remain single (Rhodophyta). The ultrastructural views of a green (*Caulerpa*, Figure 3-2), brown (*Cutleria,* Figure 3-3) and red alga (*Eucheuma,* Figure 3-4) demonstrate the typical thylakoid banding pattern for each division. Chloroplasts can be quite unique in shape, ranging from discoid

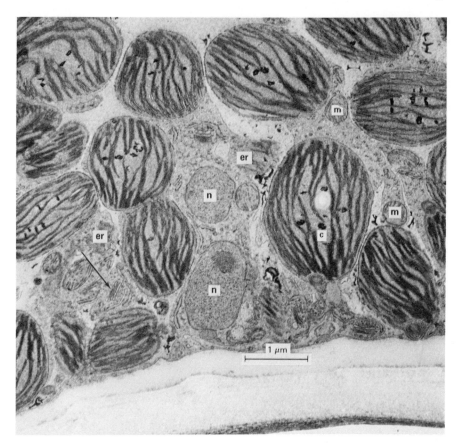

Figure 3-2 An electron micrograph of the green alga *Caulerpa verticillata* (see also Figure 5-20). Because of the large diameter of the eukaryotic cell, only a small portion of this coenocytic plant can be shown in a micrograph. Typical eukaryotic cell organelles of plants are visible including numerous chloroplasts (c) with thylakoids in bands of three to five. Two nuclei (n), one with a nucleolus, numerous mitochondria (m) a Golgi apparatus (arrow) and endoplasmic reticula (er) are evident. The cell wall is slightly separated from the protoplast because of shrinkage during preparation. × 17,280. (Courtesy of Harry Calvert, Kettering Research Laboratory, Yellow Springs, Ohio.)

(most commonly), to band-, net-, (reticulate), or cup-shaped. In some algae the plastids are parietal (along the wall), while in others they are central (axial or stellate) or radiate from the center of the cell. Plastid morphology is extensively used in the taxonomy of various algal groups.

Pyrenoids are specialized protein bodies in the chloroplasts of many algae; they are sites of starch synthesis. Pyrenoids are common to many algae, and no apparent definitive phylogenetic pattern of distribution can be suggested. Pyrenoids are granular in structure, as seen under the electron microscope, and they may be embedded (Figure 3-5) or protruding from the side of the chloroplast. The latter are called *stalked pyrenoids*. Under the light microscope they may be inconspicuous or very large; usually they appear as colorless regions in the plastid. Although amlyase, the enzyme that produces starch, has been found in the pyrenoid of green algae, the function of pyrenoids in other algae is open to question.

Mitochondria are common in all eukaryotic algae (Figure 3-5), and the cristae may be tubular or platelike. Nuclei are usually small (5–10 μm diameter, Figure 3-2), although *Acetabularia* has large nuclei (over 80 μm in diameter). Many green and red algae are multicellular and multinucleate, other green and chrysophycean algae are coenocytes: one-celled and multinucleate.

The chromosomes of algae are usually uncoiled and invisible except during mitosis or meiosis. Members of the dinoflagellate division (Pyrrhophyta) and Euglenophyta have condensed, easily viewed chromosomes during interphase.

One of the most interesting findings about algae is the unique way in which mitosis (karyokinesis, or nuclear splitting) and cell division (cytokinesis, or cell splitting) occur. Two basic forms of nuclear and cell division (Pickett-Heaps, 1975) exist, even among members of the same algal division (Chlorophyta). The two forms of karyokinesis are termed closed and open (Figure 3-6). A closed form of nuclear division is more common in algae, but it is not present in higher plants. In closed nuclear division, the nuclear membrane does not disappear during karyokinesis. Rather, microtubules penetrate each pole. At anaphase the chro-

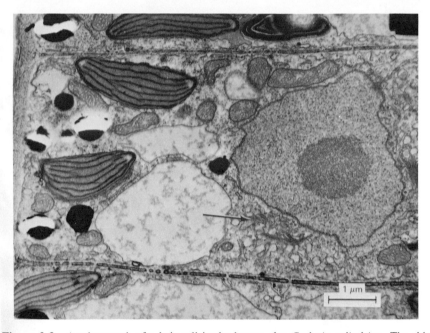

Figure 3-3 A micrograph of a hair cell in the brown alga *Cutleria cylindrica*. The chloroplasts have thylakoids grouped in bands of three. A number of mitochondria and Golgi apparati (arrow) are evident, especially near the nucleus. The newly formed cross wall (after mitosis) is perforated by plasmadesmata. × 13,000. (Courtesy of John LaClaire, University of Texas, Austin.)

mosomes are pulled apart along with the nuclear membrane. A number of variations on the theme of closed mitosis are known to occur in fungi and algae; the most extreme is within the dinoflagellates. More information about the different types of karyokinesis will be presented in the descriptions of different groups of algae. The open form of mitosis, as found in higher plants and animals, is marked by the disappearance of the nuclear membrane and the subsequent formation of spindle fibers (microtubules) that extend from each pole to the equatorial plate. The nuclear membrane reforms after the movement of the chromosomes to the two poles.

Cytokinesis in higher plants is associated with the formation of a *cell plate*, which is composed of vesicles that coalesce about the persistent spindle apparatus (Figure 3-7). A phragmoplast, which is a cluster of microtubules formed at right angles to the cell plate (Figure 3-7), produces the plate and spreads to the cell edge (wall). The cell plate becomes the first layer of the cell wall, the middle lamella. Pickett-Heaps (1975) first demonstrated the occurrence of a second type of cell division in algae involving the so-called phycoplast. The phycoplast is a layer of microtubules that lies *parallel* to the cell plate and slowly extends to the

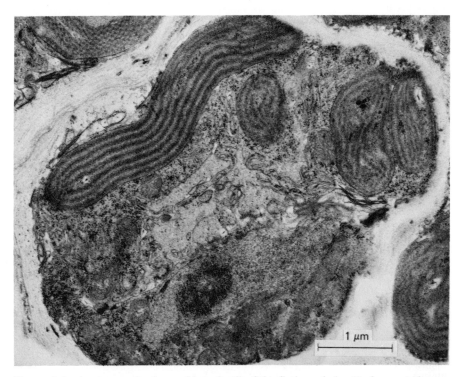

Figure 3-4 A micrograph of an epidermal cell of the fleshy red alga *Eucheuma isiforme*. The thylakoids are single, not grouped into bands in the chloroplast and show a repeating structure because of the presence of phycobilisomes (pigment granules) on the membrane. Mitochondria and a nucleus are evident in the cell. × 20,520.

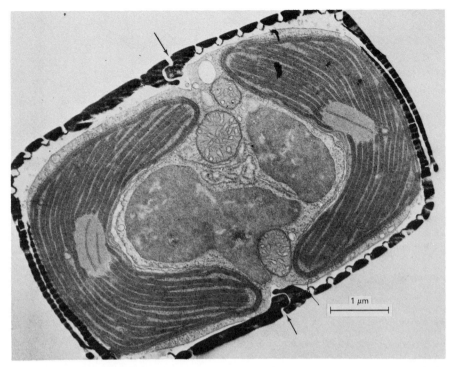

Figure 3-5 A central ultrathin section through the diatom *Navicula tripunctata*. Both the raphe and overlapping valves (arrows) of the diatom frustule are evident (see Chapter 8). The two lamellate plastids each have a central pyrenoid and thylakoids grouped into bands of three as in brown algae. Mitochondria with tubular cristae as well as lobes of a central nucleus are visible. Various pore systems are visible in the diatom frustule. × 15,600. (Courtesy of E. F. Stoermer, Great Lakes Research Station, University of Michigan.)

edge of the cell. Both forms of cytokinesis are present in the green algae, and this has been used to suggest varying phylogenetic relationships to higher plants, which only have a phragmoplast.

Vacuoles, as in all higher plants, are common structures of eukaryotic algal cells. Vacuoles in coenocytic species can be quite large. Typically, marine algae do not contain specialized crystals within their vacuoles, although viral inclusions have been found.

Photosynthetic Pigments

The grouping and separation of algae based on coloration has been used for over 100 years (Harvey, 1841). There are three major types of pigments (Table 3-2), the chlorophylls, carotenoids, and phycobilins (biliproteins). Most algal divisions are characterized by a specific set of pigments, hence the designation of green, brown, and red algae.

Table 3-2 Photosynthetic Pigments in Algae

	Cyanophyta	Rhodophyta	Chlorophyta			Phaeophyta	Chrysophyta						Cryptophyta	Pyrrhophyta	Euglenophyta
			Prasinophyceae	Chlorophyceae	Charophyceae		Chrysophyceae	Prymnesiophyceae	Xanthophyceae	Bacillariophyceae	Eustigmatophyceae	Chloromonadophyceae			
Chlorophylls															
Chlorophyll *a*	+	+	+	+	+	+	+	+	+	+	+	+	+	+	+
Chlorophyll *b*			+	+	+										+
Chlorophyll *c*							+	+	+	+			+	+	
Chlorophyll *d*		+													
Carotenoids															
α-Carotene			+									+			
β-Carotene	+	+	+	+	+	+	+	+	+	+	+	+		+	+
Flavicin	+														
Xeaxanthin	+	+	+	+	+										
Lutein		+	+	+	+										
Violaxanthin			+	+	+	+									
Myxoxanthophyll	+														
Oscillaxanthin	+														
Siphonoxanthin				+[a]											
Heteroxanthin							+	+	+	+	+				
Neoxanthin			+	+	+										
Fucoxanthin						+	+	+		+					
Diatoxanthin							+	+	+	+	+				
Diadinoxanthin							+	+	+	+	+				+
Dinoxanthin														+	
Peridinin														+	
Biliproteins															
c-Phycocyanin	+	+													
r-Phycocyanin		+													
c-Phycoerythrin	+														
r-Phycoerythrin		+													
Cryptomonad biliproteins															

Source: Stewart, 1974.

[a]Siphonoxanthin is present in siphonaceous and deep-water green algae.

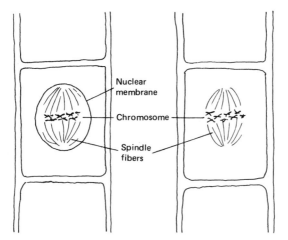

Figure 3-6 A diagram depicting open and closed mitosis (karyokinesis). The right filament cell has open karyokinesis since the nuclear membrane has broken down. In the left filament cell the nuclear membrane persists, and the spindle fibers are seen inside.

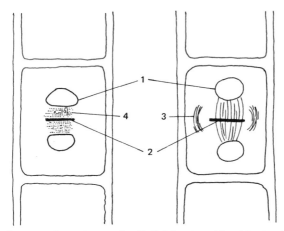

Figure 3-7 A diagram of two forms of cell division (cytokinesis). On the right, the cell shows the more typical (higher plant) form, in which a cell plate (2) and a phragmoplast (3) are developing. On the left the cell plate (2) is forming along a series of parallel microtubules called a phycoplast (4). In both cases the two daughter nuclei (1) are reforming after separation of the chromosomes.

Chlorophylls

Chlorophylls are tetrapyrroles that are arranged in a ring with a central magnesium atom and a phytol tail. Several forms of chlorophyll occur in the algae, including chlorophyll *a* which is common to all eukaryotic plants and the prokaryotic blue-green algae (Figure 3-8). There are a number of other chlorophylls, including chlorophyll *b,* which is present in all "grass green" plants (Chlorophyta, Euglenophyta, and higher plants). The distinction between the chlorophylls (*a,b,c,d*) is associated with side chains on the molecules (Figure 3-8). Chlorophylls are soluble in lipid solvents such as chloroform or acetone. Thus simple grinding of fresh plant tissues in acetone ($\sim 80\%$) is sufficient to extract the chlorophylls (see also Chapter 19).

Carotenoids

There are two classes of carotenoids, carotenes and xanthophylls. Both of the carotenoid pigments are complex hydrocarbon chains (Figure 3-9). The carotenes are unsaturated, and they transmit mostly yellow light. Xanthophylls have some degree of saturation with oxygen, and they transmit more in the orange or red wavelengths. The two carotenes, which are common to green algae and higher plants, are α- and β-carotenes (Figure 3-9). Two representative xanthophylls are also shown in Figure 3-9: zeaxanthin (green algae and higher plants) and fucoxanthin (brown algae, chrysophycean algae). Carotenoid pigments are especially soluble in ether and they can be easily separated by chromatography.

Phycobilins

Like the chlorophylls, the phycobilins are linear tetrapyrroles, but unlike chlorophylls they lack magnesium and a phytol tail. The phycobilin pigments are attached to large protein molecules, are soluble in water, and are commonly called

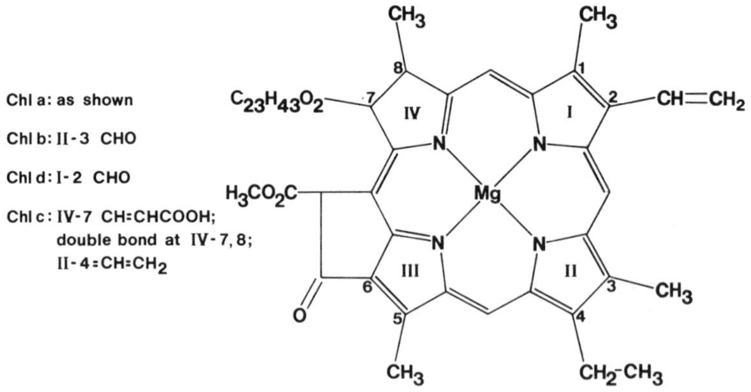

Figure 3-8 Molecular structure of chlorophyll *a* and the modifications to produce the other chlorophylls (*b, d, c*). The phytol "tail" is attached to the fourth ring.

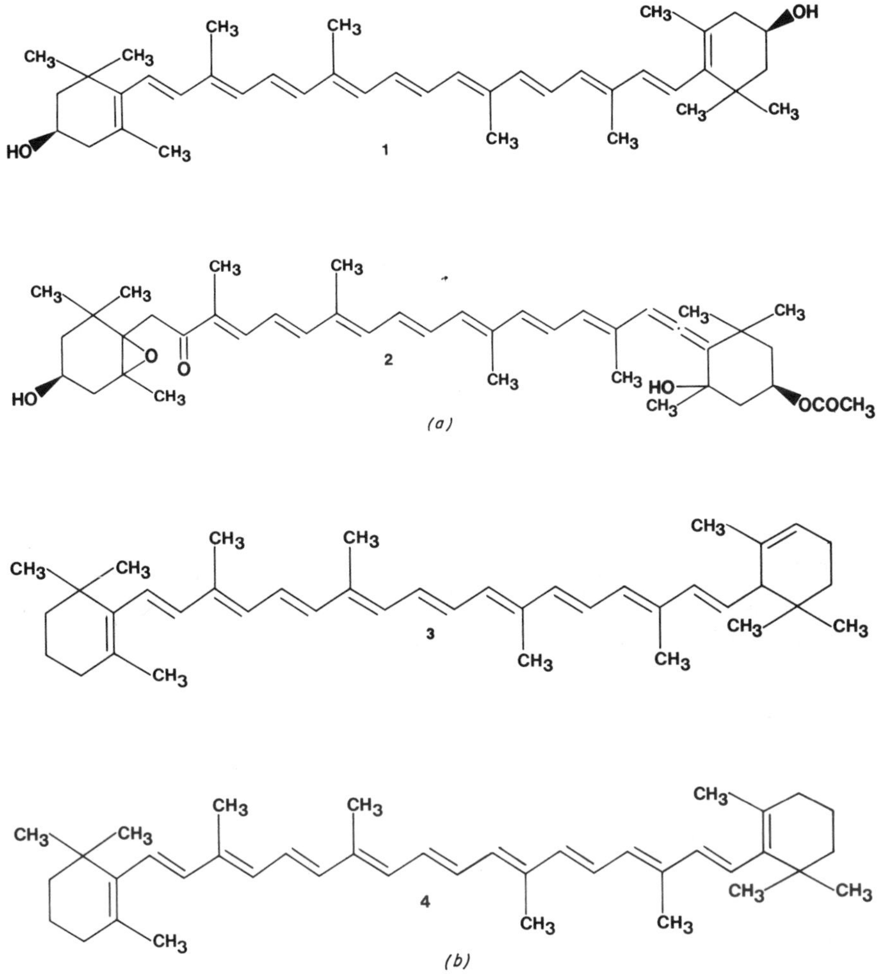

Figure 3-9 Molecular structures of four carotenoids, two carotenes and two xanthophylls. Note the presence of oxygen and hydroxyl molecules in the xanthophylls. (1) Zeaxanthin. (2) Fucoxanthin. (3) Alpha carotene. (4) Beta carotene.

biliproteins. There are two types of phycobilins (Figure 3-10), phycocyanin and phycoerythrin, which are found in blue-green, red, and the cryptomonad algae. Phycocyanin transmits blue light, whereas phycoerythrin transmits red light. Typically, deep-water red algae have higher levels of phycoerythrin, which absorbs in the blue light of the visible spectrum, than shallow-water red algae. The significance of the correlation between phycoerythrin concentration and available light has been used to explain algal distribution with depth. This concept, called chromatic adaptation, as well as methods to measure phycobilins, will be discussed in Chapter 14.

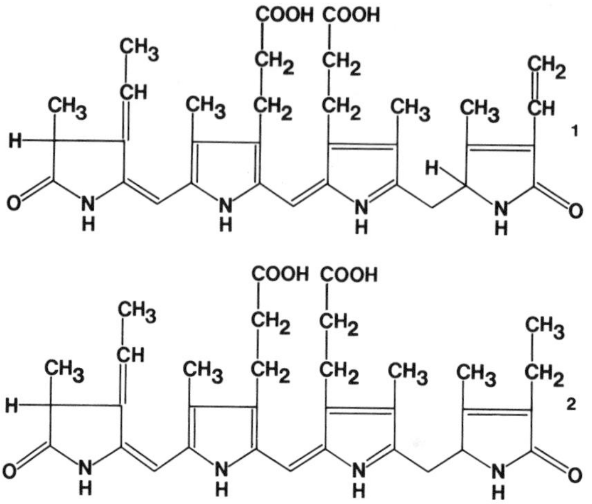

Figure 3-10 Molecular structure of the phycobilins, phycoerythrin (1) and phycocyanin (2). The differences between the two biliproteins lie in the placement of the methyl, ethyl, propionic acid residue, and vinyl side chains.

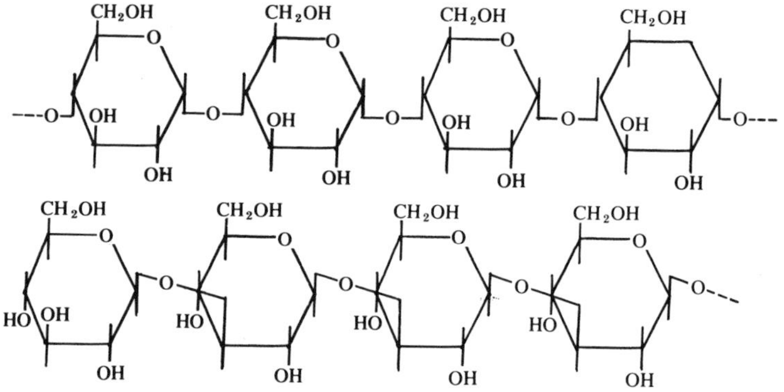

Starch (*a*)

Laminarin g-chain (*b*)

Figure 3-11 Molecular structure of green algal and higher-plant starch (*a*) and the g-chain of the brown algal reserve food, laminarin (*b*). The two types of linkage between the sugar residues can be compared; α-(1 $\rightarrow$ 4) in the starch vs. β (1 $\rightarrow$ 3) in laminarin.

Reserve Foods

As with the pigments, there are some distinctive types of photosynthetic reserves (photosynthate) in the algal divisions. In fact, recent evidence suggests that even the first formed photosynthates may be distinctive. Polymers of α-(1 $\rightarrow$ 4) glucans are characteristic of the insoluble starches of green algae (unbranched as in higher plants, Figure 3-11a) and of red and blue-green algae (branched, Floridean and Cyanophycean starches). Polymers of β-(1 $\rightarrow$ 3) glucans are characteristic of members of the brown algae (laminarin, Figure 3-11b), in Euglenophyta (paramylum), and the Chrysophyta (chrysolaminarin). Mannitol, an alcohol, apparently is found in a wide variety of algae, and it may be a common soluble reserve food throughout.

Cell Wall

Blue-green algae appear to have a typical gram-negative (see Chapter 4, Appendix B) bacterial cell wall, which includes a variety of mucopolysaccharides, mucopeptides and amino sugar polymers but few lipopolysaccharides (Figure 3-12). The basic constituent of most eukaryotic algal cell walls is cellulose (β-(1 $\rightarrow$ 4)-linked glucan), as shown in Figure 3-13. However, a number of other structural compounds can be present (Rogers and Perkins, 1968), especially in the cell walls of coenocytic green algae such as β-(1 $\rightarrow$ 3) xylan (*Bryopsis*) and β (1 $\rightarrow$ 4) mannan (*Caulerpa*).

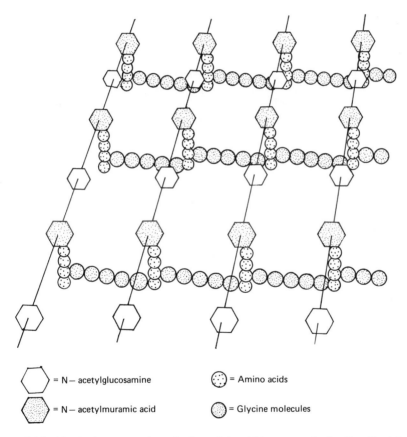

⬡ = N – acetylglucosamine   ⊙ = Amino acids

⬡ = N – acetylmuramic acid   ○ = Glycine molecules

Figure 3-12   The basic construction of a bacterial and blue-green algal cell wall might be demonstrated as consisting of four major components: n-acetylglucosamine, n-acetylmuramic acid, amino acids, and glycine molecules, as well as other components, such as lipids.

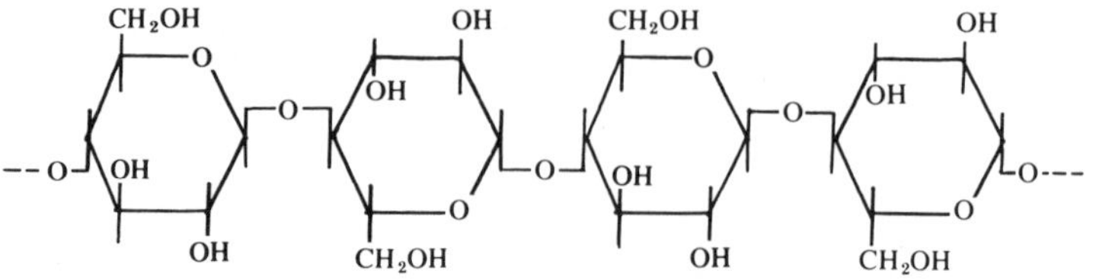

Cellulose

Figure 3-13 The molecular structure of a cellulose polymer showing the β-(1 → 4) linkage. These polymers are highly organized into crystalline structures called micelles, which, in turn, are grouped together to form microfibrils. The microfibrils are arranged in various patterns in Figures 13-14 to 13-16.

The cellulose polymer is long and is usually organized into crystalline structures called micelles. Micelles, in turn, are arranged in parallel groups to form microfibrils. The microfibrils are visible under the electron microscope (~ 18 nm diameter). The pattern of the cellulose microfibrils can be used to differentiate various algal divisions (Dawes *et al.*, 1961; Dawes, 1966). Members of the Phaeophyta and the green algal order Ulvales have cellulose microfibrils arranged in lamellae; these lamellae are placed at roughly 40–50° angles to one another (Figure 3-14). Members of the Rhodophyta have a reticulate or network pattern of microfibrils (Figure 3-15). Two marine orders of the green algae (Cladophorales and Siphonocladales) have a highly crystalline type of cellulose (no hemicelluloses or other residues) and the microfibrils are arranged in a highly ordered series of crossed lamellae as shown in Figure 3-16. Cellulose is also found in small scalelike structures that occur on flagella and the surface of the cell (organic scales, Figure 3-17). These same scales may become calcified (inorganic scales, coccoliths, Fig. 8-16). In addition, a form of α-cellulose may be deposited in plates as found in the armored forms of dinoflagellates (Figure 3-18). Other algae have specialized walls containing silicon dioxide (SiO_2) as found in the diatom cell wall (Figure 3-19). The deposition of SiO_2 is characteristic not only of diatom frustules. It also forms the skeleton of the unicellular silicoflagellates and the scales of some haptophycean algae.

Specialized cell wall polysaccharides such as agar, carrageenan (red algae), and alginic acid (brown algae) have been previously described (Chapter 2). A number of algae are calcified with either calcite or aragonite forms of calcium carbonate. Many coenocytic green algae (e.g., Dasycladales, Caulerpales) have a crystalline form of calcium carbonate called aragonite, similar to that present in echinoderms. The large Corallinaceae family of the Rhodophyta division contains calcified plants with calcite crystals. These two groups of calcified algae, along with the Coccolithophorids of the Haptophyceae, are very important contributors to the calcarious deposits in marine environments.

Flagella

Although bacteria have simple protein flagella (consisting of flagellum subunits), the blue-green algae lack flagella. In eukaryotic algae, flagellar type, number, and placement are important criteria for classification (Table 3-3). Among eukaryotic cells, the internal flagellar structure is uniform and consists of two inner microtubules and nine outer doublets of microtubules held within the cytoplasmic membrane (Figure 3-20). The basic component of the microtubules is a protein, tubulin. The surface of the flagellum can be smooth (acronematic), or have fine hairs in a single row as found in *Euglena* (stichonematic) or in two to

Table 3-3 Flagellar Characteristics (Forms,[a] Insertion, Number) *in Algae*

Cyanophyta	None
Rhodophyta	None
Phaeophyta	2, unequal, one acronematic + one pantonematic, laterally inserted
Chlorophyta	
Chlorophyceae	2 or 4, equal, acronematic, anteriorly inserted
Prasinophyceae	1, 2, or 4, equal, with scales, anteriorly inserted
Charophyceae	2, equal, with scales, anteriorly inserted
Chrysophyta	
Chrysophyceae	1, pantonematic; or 2 unequal length, 1 acronematic + 1 pantonematic; all anteriorly inserted
Prymnesiophyceae	2, equal, pantonematic, anteriorly inserted, haptonema present in some
Xanthophyceae	2, unequal, 1 acronematic + 1 pantonematic, anteriorly inserted
Bacillariophyceae	1, pantonematic, anteriorly inserted (sperm only)
Eustigmatophyceae	1, pantonematic, anteriorly inserted
Chloromonadophyceae	2, equal, 1 acronematic + 1 pantonematic, anteriorly inserted
Euglenophyta	2, unequal, only 1 emergent and stichonematic
Pyrrhophyta	2, equal, 1 each acronematic and stichonematic, anteriorly or laterally inserted
Cryptophyta	2, equal-subequal, both stichonematic

[a] Acronematic: a smooth flagellum, lacking hairs (mastigonemes).
Pantonematic: a hairy flagellum, the mastigonemes randomly arranged.
Stichonematic: a hairy flagellum, the mastigonemes in 1–2 rows.

Figure 3-14   An electron micrograph of a portion of the cell wall of *Dictyota dichotoma* (Phaeophyta). The microfibrils are arranged in parallel lamellae and a number of pit fields (sites of cytoplasmic connections between cells) are visible in the medullary cell wall. × 21,000. (Dawes *et al.,* 1961.)

three rows (pantonematic) as in members of the Phaeophyta, Chrysophyta, and Pyrrhophyta (Figure 3-21).

Scales may also be present on flagella. In addition, there is a specialized type of flagellum called a haptonema, which appears to be modified for attachment. Some chrysophycean algae have a haptonema and are, therefore, placed in the class Pyrmnesiophyceae (Haptophyceae). Table 3-3 summarizes the flagellar distribution among the algal groups. It should be noted that specialized flagella may occur on sperm (male gametes) of some brown algae.

LIFE HISTORIES

One of the unique features of many marine algae is their life histories. Drew (1955) defined a life history as "the recurring sequence of somatic and nuclear

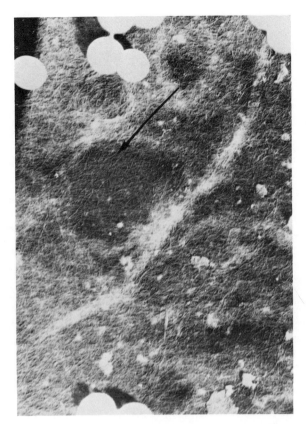

Figure 3-15 The random, reticulate pattern of microfibrils in the cell wall of *Ceramium* sp. (Rhodophyta) is evident. The thinner regions (arrow) are pit areas in the cell wall, through which plasmodesmata would connect adjacent cell prohplasts. The polystyrene balls are 0.814 μm in diameter.

phases." It is important to examine both the morphological and cytological features of an alga. Within a single division, such as the Chlorophyta, the plant one collects and identifies could be haploid, having only one set of chromosomes (*Chlamydomonas*). Another genus could have two separate plants in its life history, haploid and diploid plants (such as *Ulva*). In the Phaeophyta both diploid (*Fucus*) and haploid—diploid (*Laminaria*) life histories are known.

Although there is a fairly complex set of terms associated with the description of life histories, this text will use the simpler terms of Svedelius (1931). Thus, three types of life histories are distinguished in this text.

Haplont

A haplont is a plant with one set of chromosomes except for the diploid zygote. Meiosis is zygotic or it occurs when the zygote germinates (Figure 3–22).

Figure 3-16 (*A*) A replica of the inner wall surface. (*B*) A cross section of the wall of the green alga *Apjohnia laetevirens* (Siphonocladales), showing the crossed fibrillar pattern Alternating layers of highly crystalline cellulose microfibrils are seen in both transmission electron micrographs. In (*B*) they demonstrate a three-tiered pattern; each layer is about 65° in orientation from the previous one. The arrow in *B* indicates the outer wall layers, (*A*) × 8,400; (*B*) × 5,600. (After Dawes, 1969.)

Diplont

A diplont is a plant that has two sets of chromosomes except when the gametes are produced (Figure 3-23). Meiosis is gametic or it occurs during production of gametes.

Haplodiplont

In a haplodiplont, both a haploid and a diploid phase occur, with each phase required to complete the sexual life history (Figure 3-24, Figure 3-25). Usually both phases are free-living and the diploid phase alternates with the haploid phase. Meiosis is sporic or it occurs in the sporangia; the haploid spores germinate to produce a haploid gametophyte (i.e., a plant that produces gametangia and is haploid). The gametophyte will ultimately produce gametes through mitosis; the gametes will fuse to produce the zygote, or first diploid cell. The zygote will germinate and develop into a diploid sporophyte (i.e., a plant that produces sporangia and is diploid). The gametophyte and sporophyte may look alike and thus be designated isomorphic, as with *Ulva* (Figure 3-24), or they may look different and thus be designated heteromorphic as in the brown alga *Laminaria* (Figure 3-25).

The gametes may be identical (+ and − strains, isogamous), or one may be larger than the other (anisogamous). Many algae and most higher animals, have nonmotile eggs and motile sperm (oogamous).

EVOLUTION OF THE ALGAE

Two questions will be briefly considered in this section: the origin of the eukaryotic cell and the phylogenetic relationship between existing eukaryotic algal groups.

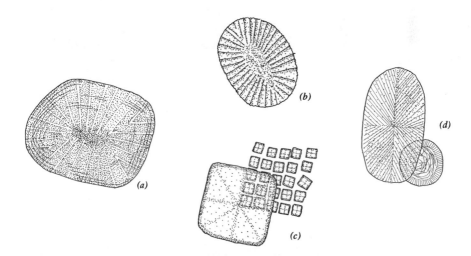

Figure 3-17 Examples of organic (cellulostic, pectic) scales found on various unicellular algae. See also Figure 8-16 for examples of inorganic scales. (*a*) *Pleurochrysis scherffelii*, a member of the Prymnesiophyceae, produces these scales internally in Golgi bodies and then transports them to the cell surface (Brown *et al.*, 1969). These scales become impregnated with inorganic substances with age (mineralization). The scale is about 1 μm in length. (*b*) *Prymnesium parvum*, for which the class Prymnesiophyceae is named, produces unmineralized plate scales that will cover the cell body (Manton and Leedale, 1963). The scale is about 0.3 μm in diameter. (*c*) *Pyramimonas parkeae* is a member of the green algal class Prymnesiophyceae and has organic body scales about 0.3 μm in size consisting of square subunits about 0.05 μm in diameter (Norris and Pearson, 1975). (*d*) *Hymenomonas carterae* is also a member of the Prymnesiophyceae and produces three types of scales, two of which are reproduced, a large eliptical scale about 2 μm in length and a smaller (0.8 μm diam.) circular scale. Both are body scales; the larger becomes mineralized with calcium carbonate to form a coccolith (Pienaar, 1969).

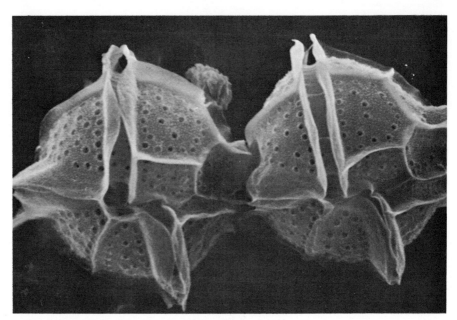

Figure 3-18 A scanning electron micrograph of two cells of *Pyrodinium bahamense,* an armored dinoflagellate, showing the thecal plates that consists of membranes and cellulose. × 39,000. (Courtesy of Karen Steidinger and Ernest Truby, Florida Department of Natural Resources Marine Laboratory, St. Petersburg.)

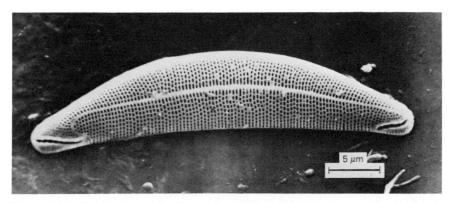

Figure 3-19 The frustule of the diatom *Pseudohimantidium pacificum* after cleaning in hydrogen peroxide and viewing with the scanning electron microscope. The valve view of the frustule shows the central (axial) striation and external groove. Unit mark equals 5 μm. (Courtesy of Robert Gibson Harbor Branch Foundation, Fort Pierce, Florida.)

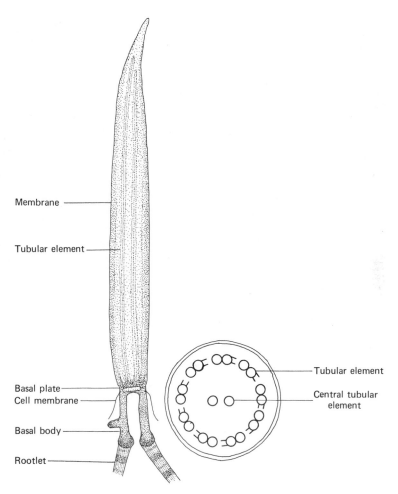

Membrane

Tubular element

Basal plate
Cell membrane

Basal body

Rootlet

Tubular element

Central tubular
element

Figure 3-20 Diagram of a typical flagellum showing the basic components. The cross section is of the flagellar portion. The rootlets can extend far back into the motile cell.

Origin of the Eukaryotic Cell

Probably the greatest difference between organisms is the distinction between prokaryotic and eukaryotic cell structure (Table 3-1). Thus far, no intermediate cell types have been found to link prokaryotic and eukaryotic cell systems. Two theories of the origin of the eukaryotic cell are available. The older one, the *Classical theory*, proposes that the organelles of eukaryotic cells (chloroplasts, mitochondria, nuclei, flagella) evolved gradually from the differentiation of a prokaryotic cell. Proponents of this theory argue that all of the present-day eukaryotic struc-

Figure 3-21 A micrograph of *Gymnodinium splendens* showing the flattened flagellum in the girdle of the cell. The scanning electron micrograph is magnified × 2,700. (Courtesy of Lana Tester, Florida Department Natural Resources Marine Laboratory, St. Petersburg).

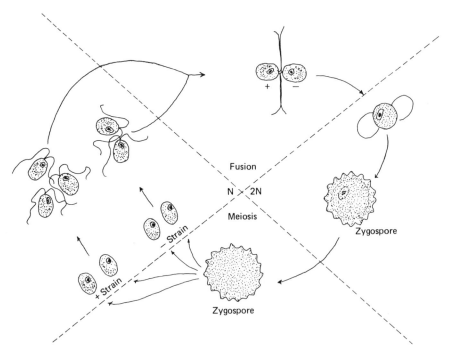

Figure 3-22 The life history of the unicellular green alga *Chlamydomonas*, a haplont. The only diploid cell is the zygote.

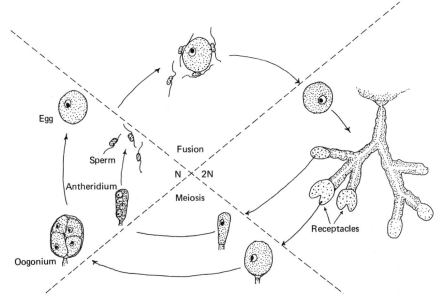

Figure 3-23 The large intertidal rock weed, *Fucus* sp., has a diplontic life history. The only haploid cells in *Fucus* are the male (sperm) and female (egg) cells.

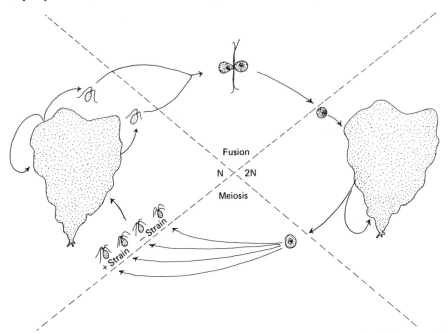

Figure 3-24 An example of a life history that has both a haploid and a diploid stage and in which the two stages look alike (isomorphic) might be the green algal genus *Ulva*. It is impossible to tell whether the green-bladed plant one collects is diploid or haploid without studying the chromosome number or following the life history.

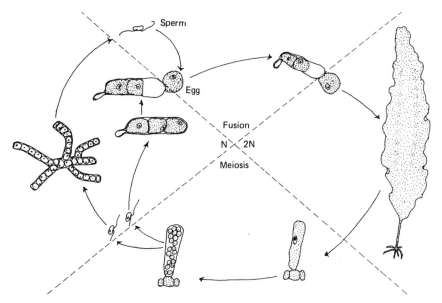

Figure 3-25 Another type of haplodiplontic life history is characteristic of the larger brown algae such as *Laminaria,* in which the diploid phase is dominant (heteromorphic). Here the plant one usually collects is diploid since the haploid or gametophytic phase is small and usually undetectable without microscopic examination. Meiosis occurs in the sporangia on the diploid (sporophytic) plant. These spores, which are haploid, germinate into small filamentous gametophytes. The gametophytes produce eggs and sperm and the resulting zygote grows into a new sporophyte.

tures are the products of gradual compartmentation of prokaryotic functions, probably in a blue-green algal ancestor (Allsopp, 1969; Jeffrey, 1971).

The *Symbiotic theory* is the more accepted theory of present-day biologists. Based on the publication of Margulis (1971), McQuade (1977), and others, the theory states that cell organelles originated by the capture of prokaryotic cells. This concept has received wide support. The symbiotic theory explains many features very well. (1) It explains the chemical differences between the inner (captured cell) and the outer (host cell) membranes that make up the cell envelope of plastids and mitochondria. (2) It also explains the different sizes of ribosomes (50–60S) characteristic of the plastids and mitochondria compared with the 80S ribosomes in the cytoplasm and nuclei in eukaryotic cells and the 70S ribosomes found in prokaryotic cells. (3) The presence of a naked DNA strand in the mitochondria and plastids supports the concept that these organelles evolved from captured prokaryotic cells. (4) Finally, a number of endosymbiotic systems exist today, and one can extrapolate back to an earlier time when the capture of some prokaryotic cells occurred. However this theory does not adequately explain the evolution of flagella, nuclear organization, and genetic control of the eukaryotic cell processes (e.g. chloroplast and mitochorndial development).

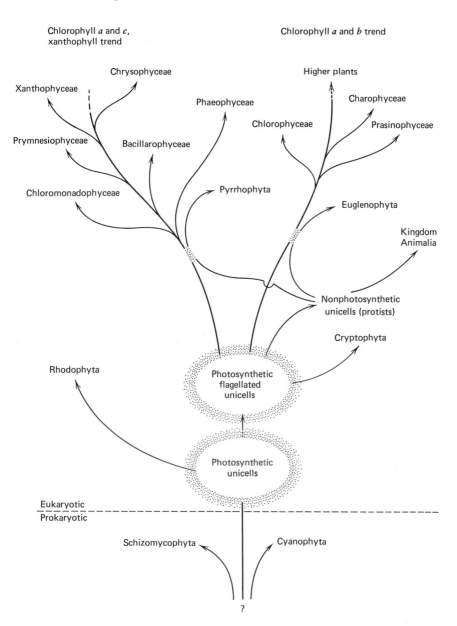

Figure 3-26 A simplified phylogenetic tree demonstrates possible relationships between the various algal divisions and classes. Note that this scheme assumes a monophyletic origin of the eukaryotic cell and of a flagellated form as well. Both the Pyrrhophyta and Euglenophyta are considered to have become "plants" through capture of other chloroplast-bearing algae.

Phylogeny of the Algae

There are probably as many schemes of algal evolution as there are publications dealing with the topic. The shcme presented here (Figure 3-26) is modified from Whittaker (1969), with additional ideas from Klein and Cronquist (1967).

The eukaryotic algal cell may have evolved symbiotically from some ancestral blue-green alga. The blue-green ancestor could have been captured and could have degenerated into a plastid in a eukaryotic protist that lacked flagella (Mc-Quade, 1977). On the other hand, the ancestral eukaryotic algal cell could have been derived from a blue-green algal cell by gradual evolution (Jeffrey, 1971). At any rate, the first eukaryotic algal cells were simple "red algae" that lacked flagella, had chloroplasts with single thylakoids, and had pigments including chlorophyll a, carotenoids, and phycobilins. From this stock, two evolutionary directions occurred. The division Rhodophyta evolved (lacking any flagellar structure) and a second ancestral stock of flagellated photosynthetic cells occurred (evolution or capture of a flagellar system). The Cryptophyta then evolved (having phycobilins) while the remaining eukaryotic algae (without phycobilins) evolved. Both the Rhodophyta and Cryptophyta can thus be considered evolutionary dead ends, or groups that were isolated early in algal evolution.

Two main lines of evolution then developed with the remaining eukaryotic algae. One line was represented by the development of chlorophylls a and c, a reserve food based on β-(1 $\rightarrow$ 3) glucan, a dominance of carotenoids, and a variety of flagellar structures and placements—i.e., the so-called brown algal trend. The second, or green algal, trend contains plants with chlorophylls a and b, carotenoids of the higher plants, and a reserve food based on an α-(1 $\rightarrow$ 4) glucan. The Euglenophyta may have evolved from a protozoan ancestor through the capture of a green algal cell. The Pyrrhophyta are also thought to have evolved from some unique protozoan ancestor that became photosynthetic through the capture of a chrysophycean alga. The symbiotic evolution of euglenoids and dinoflagellates is supported by the presence of three membranes around the plastids. In addition there are some euglenoids and dinoflagellates that have symbiotic relationships both with other algae and with animals.

The algal fossil record has representatives of most groups dating back to the Precambrian ($\sim$ 1 billion years ago, see Chapter 10). The calcified algae are best known in the fossil record, and specimens very similar to present-day forms are known since the Cambrian period (600 million years ago). Because of the scattered fossil evidence, any evolutionary scheme is based on existing specimens, and it is subject to misinterpretation. The above scheme and Figure 3-26 are presented primarily to help the reader relate one group of algae to another.

REFERENCES

Allsopp, A. 1969. Phylogenetic relationships of the procaryota and the origin of the eucaryotic cell. *New Phytol.* **68**: 591–612.

Bold, H. C. and M. J. Wynne. 1978. *Introduction to the Algae. Structure and Reproduction.* Prentice-Hall, Englewood Cliffs, N.J.

Brown, R. M., Jr., W. W. Franke, H. Kleinig, H. Falk, and P. Sitte. 1969. Cellulosic wall component produced by the Golgi apparatus of *Pleurochrysis scherffelii*. *Science* **166**: 894–896.

Cox, E. R. (ed.) 1980. *Phytoflagellates*. Elsevier/North Holland, New York.

Dawes, C. J., F. M. Scott, and E. Bowler. 1961. A light and electron microscope study of algal cell walls. I. Rhodophyta and Phaeophyta. *Amer. J. Bot.* **48**: 925–932.

Dawes, C. J. 1966. A light and electron microscope study of algal cell walls. II. Chlorophyta. *Ohio J. Sci.* **66**: 317–326.

Dawes, C. J. 1969. A study of the ultrastructure of a green alga, *Apjohnia laetevirens* Harvey with emphasis on cell wall structure. *Phycologia* **8**: 77–84.

Drew, K. M. 1955. Life histories in the algae with special reference to the chlorophyta, Phaeophyta, and Rhodophyta. *Biol. Rev.* **30**: 343–390.

Dodge, J. D. 1973. *The Fine Structure of Algal Cells*. Academic, London.

Harvey, W. H. 1841. *A manual of the British algae. Reeve, Brothers, London.*

Jeffrey, C. 1971. Thallophytes and kingdoms—a critique. *Kew Bull.* **25**: 291–299.

Klein, R. M. and A. Cronquist. 1967. A consideration of the evolutionary and taxonomic significance of some biochemical, micromorphological, and physiological characters in the thallophytes. *Quart. Rev. Biol.* **42**: 105–296.

Lewin, R. A. (ed.). 1962. *Physiology and Biochemistry of Algae*. Academic, New York.

McQuade, A. B. 1977. Origins of the nucleate organisms. *Quart. Rev. Biol.* **52**: 249–262.

Manton, I. and G. F. Leedale. 1963. Observations on the fine structure of *Prymnesium parvum* Carter. *Archiv Mikrobiol* **47**: 285–303.

Margulis, L. 1971. Symbiosis and evolution. *Sci. Amer.* **225**: 48–57.

Norris, R. E. and B. R. Pearson. 1975. Fine structure of *Pyraminonas parkeae*, sp. now (Chlorophyta, Prasionphyceae) *Arch Protistenk.* **117**: 192–213.

Pickett-Heaps, J. D. 1975. *Green Algae. Structure, Reproduction and Evolution in Selected Genera*. Sinauer, Sunderland, Mass.

Pienaar, R. N. 1969. The fine structure of *Hymenomonas* (Cricosphaera) *carterae*. II. Observations on scale and coccolith production. *J. Phycol.* **4**: 321–331.

Rogers, H. J. and H. R. Perkins. 1968. *Cell Walls and Membranes*. E. and F. N. Spon, London.

Stewart, W. D. P. (ed.). 1974. *Algal Physiology and Biochemistry*. University of California Press., Berkeley.

Svedelius, N. 1931. Nuclear phases and alternation in the Rhodophyceae. *Beih. Bot. Centralbl.* **48**: 38–59.

Trainor, F. R. 1978. *Introductory Phycology*. Wiley, New York.

Whittaker, R. J. 1969. New concepts of kingdoms and organisms. *Science* **163**: 150–162.

CHAPTER 4

Cyanophyta

The blue-green algae are placed in a single class, Cyanophyta. Blue-green algae have chlorophyll *a* and therefore possess photosystem II, releasing oxygen under normal aerobic conditions. On the other hand, this group has a prokaryotic cell structure and is thus cytologically more bacterialike than plantlike. Many bacteriologists regard members of the group as "cyanobacteria" and point out that blue-green algae can switch from the use of photosystem II and oxygen release to photosystem I alone. Thus in Padan's view (Padan, 1979) the cyanobacteria bridge the gap between photosynthetic bacteria and photosynthetic eukaryotic cells. Certainly blue-green algae or cyanobacteria play a number of important roles in the marine environment.

CYTOLOGICAL FEATURES

Pigments

Blue-green algae contain a wide variety of pigments, the more common ones include chlorophyll *a*, β-carotene, myxoxanthin, zeaxanthin, and phycobilins (*c*-phycoerythrin, *c*-phycocyanin, allophycocyanin). Usually phycocyanin is the dominant accessory pigment, causing the blue-green color. The phycobilins are found in specialized structures, the phycobilisomes, attached to the outer surface of the thylakoids.

Cytological Structure

The prokaryotic cell structure has been described in Chapters 1 and 3, and is reviewed in a number of papers (Lang, 1968). Photosynthetic lamellae (thylakoids) are concentrated in the outer pigmented cytoplasm and may encompass the entire cell (Figures 3-1, 4-1). The central region of the cell is usually less pigmented as viewed under the light microscope and contains the DNA strands. Roberts *et al.* (1977) have characterized the DNA of a colonial blue-green alga, *Merismopedia*, and found it similar to bacterial DNA. It appears there is more than one DNA strand per cell. The cytoplasm, as with most prokaryotic cells, lacks tonoplast-bound vacuoles, is viscous, and is difficult to plasmolyze. Blue-

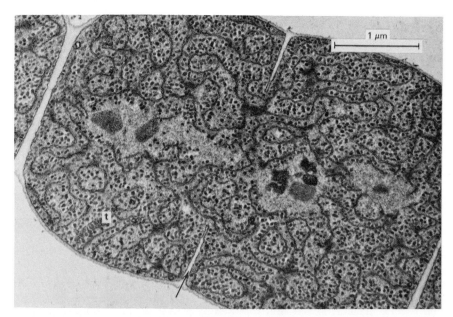

1 μm

Figure 4-1 Transmission electron micrograph of a dividing cell in a filament of the blue-green alga *Anabaena variabilis*. A developing cross wall (arrow) and the highly convoluted thylakoids (t) are visible. Some carboxysomes (polyhedral bodies in the central region) and numerous smaller granular bodies (glycogen or cyanophycean starch) are also present. The cell wall is thin and a well-defined sheath is not evident. × 23,000. (Courtesy of Norma Lang, University of California, Davis.)

green algal cytoplasm can withstand desiccation or extremes in tonicity, similar to bacterial cytoplasm. Ribosomes, lipid globules, and glycogen granules are distributed in the cytoplasm. Gas vacuoles are more common in planktonic forms. The vacuole is a group of cylindrical vesicles with thin proteinaceous membranes (unlike cell membranes). Polyhedral bodies (carboxysomes) are evident in the cytoplasm and apparently are sites of storage of ribulose bisphosphate carboxylase, which functions in photosynthetic CO_2 fixation. Microtubulerlike complexes have also been described (Bisapultra *et al.*, 1975). If these prove to be true microtubules, as found in eukaryotic cells, then the prokaryotic blue-green algal cell can be considered quite advanced over the bacterial cell.

Cell division is often by centripetal growth of a cross wall that develops from the outer wall. Microplasmodesmata occur between cells of filaments but not between cells in a colony. It is possible these are equivalent to protoplasmic connections.

Mobility

No flagella are found on the blue-green algal cells; even the simpler bacterial flagella are lacking. Many blue-green filaments show a gliding motion (up to 5

μ m/sec). An older theory held that gliding was caused by a release of gelatinous substances through pores in the side walls coupled with cellular contraction and expansion. This would occur first on one side then the other side causing a flexing or weaving motion. More recently, Halfen (1973) has shown that parallel arrays of protein microfibrils occur in the surface of the filament wall, and contraction of the microfibrils results in waves of bending of the filament. These waves are created and propagated longitudinally, displacing the filament in opposite directions against a solid substrate. Thus a forward gliding motion results.

Cell Wall Structure

A conspicuous wall and a gelatinous sheath are often present. The cell wall is composed of four layers visible under the electron microscope. The cell wall is chemically similar to the wall of a gram-negative bacterium, having a variety of simple sugars, muramic and glutamic acids, diaminopimelic acid, glucosamine, galactosamine, and alanine (see Figure 3-12). The Gram stain is a process by which the cells, mounted on a microscope slide, are first stained in crystal violet, then, after rinsing in water, in potassium iodide and iodine, then, after decolorization in 95% ethanol, in safranin, followed by a final wash. Gram-positive organisms will appear blue to black; gram-negative organisms will appear red.

The outer sheath may appear thick and obvious under the light microscope or thin and essentially invisible. The sheath consists of a variety of sugars and appears to be laid down through pores in the cell wall. The presence or absence of a sheath is used as a taxonomic feature in many classifications (Desikachary, 1959), but recently Drouet (1968) has argued that it is not a consistent feature.

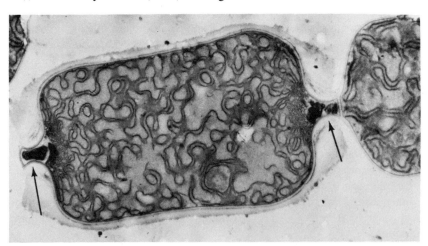

Figure 4-2 An ultrastructural view of a heterocyst of *Anabaena cylindrica* shows the polar nodules (arrows) that attach the cell to vegetative cells in the filament and contain the protein cyanophycin (Lang and Fay, 1971). Note the convoluted thylakoids in the cell and the thickened cell wall and sheath. × 13,200. (Courtesy of Norma Lang, University of California, Davis.)

Food Storage

The reserve food is called *cyanophycean starch* and is a branched carbohydrate (glycogen) which has an α-$(1 \rightarrow 4)$ linkage and is highly branched through $(1 \rightarrow 6)$ linkages. Under the light microscope the substance appears as small granules and will turn red-brown when stained with $I_2 KI$ (Gram's iodide). Chemically, it is similar to amylopectin. The large reserve granules found in specialized akinetes, resting cells, and nodules of heterocysts (see below) are composed of cyanophycin, a protein composed of arginine and aspartic acid.

MORPHOLOGY

As will be seen in the section on taxonomy, blue-green algae can be separated into three morphological groups, unicellular and colonial forms, and filaments that may or may not produce spores.

Filament Construction

A filament consists of the trichome of cells (row of cells with some portions of the cell wall shared) and its surrounding sheath. If a vegetative cell dies within the trichome, it can become a separation disc where the trichome separates. The short individual trichome segments between separation discs of a filament are called hormogonia.

The more complex forms of filamentous blue-green algae show false and true branching. False branching is simply the breaking of a trichome within the sheath. One or both ends of the broken trichomes can continue growth, which will result in the trichomes protruding through the sheath. True branching is where cell division occurs at right angles to the plane of the filament, resulting in a branch. If both a prostrate system and an erect system of branching occur, the plant is said to have heterotrichous growth.

Specialized Cells

Specialized cells include spores (exospores and endospores; akinetes) and heterocysts. Exospores and endospores are produced external to or within the parental cell respectively. Akinetes are cells with a thick envelope. The cells are usually larger in diameter and length than a vegetative cell and contain an abundance of cyanophycin (protein) granules (Figure 4-2). Heterocysts may be the same size as a vegetative cell or larger and can be distinguished by the thick cell envelope and nodules (cyanophycin composition) that connect to the adjoining cells (polar nodules, Figure 4-2). It should be noted that the walls of akinetes and heterocysts do not thicken during differentiation. Rather, a thick multilayered envelope is formed external to the original cell wall. At one time heterocysts were thought to be degenerate akinetes whose function was to act as weak links in a filament, permitting breakage. Heterocysts are now considered to be the sole sites of nitrogen fixation in filaments growing under aerobic conditions (Kulasooriya *et al.*, 1972; Rippka and Stanier, 1978). Many nonheterocystic blue-green algae can also fix nitrogen, but apparently only in cells that are under anaerobic conditions.

The internal structure of heterocysts is different from that of normal vegetative cells (Lang and Fay, 1971). In the heterocysts cells there is very little storage product, and the thylakoids are formed into a latticework, usually concentrated at the poles. The number of heterocysts increases when the level of fixed-nitrogen sources (NO_3 NO_2 NH_4) declines. The presence or absence and position of heterocysts in a filament (terminal, intercalary) have been used in taxonomy of blue-green algae. However, it is now known that under certain environmental conditions of rapid growth, few heterocysts will be produced. Also, the production of heterocysts is affected by the level of oxygen, nitrogen, light, and molybdenum (Fogg *et al.*, 1973).

REPRODUCTION

Blue-green algal reproduction include simple fragmentation (hormogonia of a filament, fragmentation of a colony), cell division resulting in two unicells (binary fission), and specialized cell production (endospores, exospores, akinetes) as described above. There is no direct evidence of sexual reproduction such as conjugation as observed in some bacteria. However, evidence of recombination has been presented through the use of induced mutations followed by retrieval of recombinant types (Orkwizewske and Kaney, 1974).

Phycoviruses (BGA viruses) have been shown to be effective in the removal of specific strains of *Lyngbya, Plectonema,* and *Phormidium* (Safferman and Morris, 1963) and may be effective in controlling naturally occurring blooms (Safferman, 1973). A large number of viruses capable of attacking many different blue-green algae have been reported (Safferman and Morris, 1977).

DISTRIBUTION AND ECOLOGY

Blue-green algae are ubiquitous. They are present in marine and freshwater environments, as well as in desert rocks and as endolithic components of the intertidal regions of the tropics (Aleem, 1980). The systematics, ecology, and biological associations of endolithic blue-green algae have been examined by Champion-Alsumard (1979), especially the genera *Hyella* and *Hormathonema.* Potts (1980) found 84 species of blue-green algae in the intertidal zone of the Gulf of Elat. Figure 4-3 demonstrates three different blue-green algae and bacterial mats that Potts found forming thick crusts in the intertidal zone of the desert coast in a mangrove swamp and open coast. The complex stratification is characteristic of both mangrove and salt lake communities as well. In these studies, species containing heterocysts were found restricted to the intertidal zone.

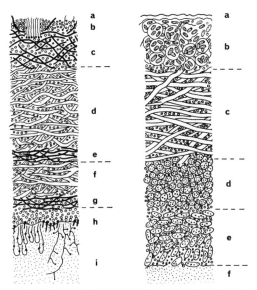

Figure 4-3 Sections through two blue-green algal mats that form in the intertidal zones (mangrove, open coast) in the Gulf of Elat (Potts, 1980). Diagram at left shows a mat taken from an intertidal zone of a mangrove swamp. Dashed lines indicate changes in color of the strata. (*a*), sediment; (*b*), *Isactis plana;* (*c*), *Lyngbya aestuarii;* (*d*), *Hydrocoleum* sp.; (*e*), dead filaments and sheaths; (*h*) sulphide zone and *Hormathonema violaceonigrum*; (*i*), beach rock with endoliths. Diagram at right shows a mat taken from the intertidal zone of an exposed coastal community. (*a*), salt crust, (*b*), *Aphanothece* sp.; (orange color); (*c*), *Microcoleus chthonoplastes*; (*d*). *Chromatium* (*e*), *Thiocystis* sp.; (*f*), black sulphide layer. (With permission of *Phycologia*.)

Some species of blue-green algae are the algal component of many lichens; other species can form extensive blooms in the marine environment. The red color of the Red Sea is said to be caused by the extensive growth of a blue-green filament, *Trichodesmium erythraeum*. This same species is extremely common in association with red tide outbreaks. In fact, the blue-green can even be responsible for red tides as reported by Aleem (1980) for the West Coast of Africa (Sierra Leone). A number of blue-green algae are toxic and in the marine environment some cause skin reactions (Hawaii: *Microcoleus lyngbyacous*) or are the source of ciguatera or fish poisoning in the tropics.

High rates of nitrogen fixation (4–12 nM C_2H_4 produced/mg chl *a* · min) have been established for a variety of marine blue-green algal communities that reduce acetylene to ethylene using the enzyme nitrogenase. Potts (1979) and Potts and Whitton (1977) demonstrated that both heterocyst- and nonheterocyst-bearing blue-green algae collected from intertidal mats in mangrove and open coastal communities of the Gulf of Elat had high nitrogenase activity. They found that nitrogen fixation was light dependent and the production of fixed nitrogen (NH_3, NO_2, NO_3) was significantly higher in the unshaded areas because of blue-green algal activity in intertidal mats. Hydrogen evolution also occurs in nitrogen-fixing blue-greens and has been a topic of interest with relation to energy production (Berchtold and Bachofen, 1979).

TAXONOMY

The blue-green algae (division Cyanophyta) and the bacteria (division Schizo-mycophyta) are placed in the kingdom Monera, since both divisions are prokary-otic in cell structure. The taxonomy of blue green algae has been the subject of numerous papers. Some bacteriologists have proposed that blue-green algae iden-tification be based only on cultured material, and all botanical classification be rejected. However, for a field phycologist such a proposal is impractical. Geitler (1932) recognized five orders, as did Desikachary (1959) and Bourrelly (1970). Smith (1950) and Papenfuss (1955) as well as Bold and Wynne (1978) grouped the blue-green algae into three orders, whereas Drouet (1968, 1973, 1978) recog-nizes only one order. Drouet combined numerous species in his revisions of the blue-green algae, and Humm and Wicks (1980) have followed Drouet in their textbook on marine blue-green algae. Drouet's concept of blue-green algal species suggests that each species has but one genotype with many ecological phenotypes (ecophenes or ecological growth forms). He assumes that the various ecophenes will revert to the same phenotype when grown under the same conditions. Unfor-tunately, a number of studies that compared growth forms under identical culture conditions do not support this concept (Stam and Holleman, 1975; Stam and Venema, 1977; Stanier *et al.*, 1971). Because of such problems, the classification followed in this text is based on that of Tilden (1910) and Geitler (1932). The system adopted recognizes three orders as outlined by Bold and Wynne (1978):

1   Producing endospores or exospores .... **Chamaesiphonales**
1   Lacking endospores or exospores 2

   2   Unicellular or colonial .... **Chroococcales**
   2   Filamentous, cells of the trichomes having a common cell wall ... **Oscillatoriales**

Chroococcales

A single family, the Chroococcaceae, is placed in this order. The plants reproduce by cell division (unicells) or fragmentation (colonies). All species are unicellular or colonial. The cells are spherical, ovoid, or cylindrical, with generic distinctions based on cell shape and planes of division. Two generic examples demonstrate the unicellular and colonial forms.

Chroococcus and Gloeocapsa

The two genera have been combined by Drouet and Daily, in part, to form the genus *Anacystis*. *Gloeocapsa zanardinii* Hauck is a common marine species (Fig-ure 4-4*a*) which is found throughout the world. The plant forms gelatinous masses on rocks, wood, and organisms. Drouet and Dailey have renamed this *Anacystis*

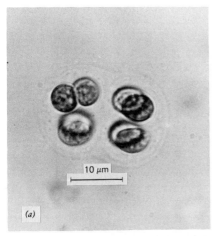
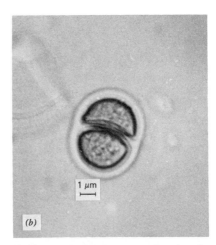

Figure 4-4 Two unicellular or colonial blue-green algae. (*a*) *Gleocapsa zanardinii* is unicellular or forms small colonies as shown in this light micrograph. (*b*) *Chroococcus dimidiatus* typically is found as a single or double cell and when divided demonstrates the typical angular cell shape of the species.

aeruginosa and include a wide variety of other species under this name as well. The cells of *G. zanardinii* are 6–12 μm in diameter and spherical after cell division. *Chroococcus dimidiatus* (Kutz.) Nagl. is a small colonial species in which the cells remain angular after division (Figure 4-4*b*). Drouet and Daily combined this species with others to form the species *Anacystis dimidiata*.

Merismopedia

The colony is a flat or curved sheet of a single layer of cells, and can be found in shallow sandy pools or in brackish water. *Merismopedia elegans* has cells that are ovoid, light green, 4–6 μm in diameter, and up to 10 μm in length. The colonies are small, with a transparent matrix (Figure 4-5).

Chamaesiphonales

Three families are recognized in this order (Bold and Wynne, 1978) and all produce endospores or exospores. Members of this order can be unicellular, filamentous, or organized as a colonial chain of cells. Drouet and Daily (1956) recognized only a single, polymorphic genus, *Entophysalis*. Two marine forms are described.

Dermocarpa

The alga is epiphytic or endophytic, initiating its growth as a single cell then producing a pad of cells. The upper pad cells of *D. olivaceae* (Reinsch) Tilden are 3–7 μm in diameter, with a distinct hyaline sheath (Figure 4-6). Members of the species often occur as epiphytes of *Cladophora* filaments. Drouet and Daily renamed the species *Entophysalis conferta*.

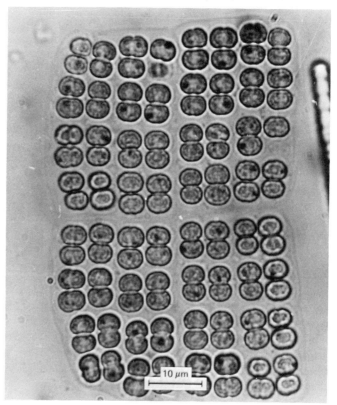

Figure 4-5 A small plate colony of *Merismopedia elegans*. Note the regular pattern of cell arrangement and the almost invisible sheath around the entire colony.

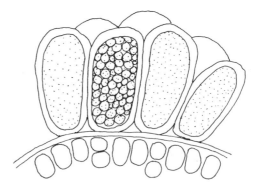

Figure 4-6 *Dermocarpa olivaceae* forms small colonies epiphytically on the surface of other algae. Asexually the cells can divide internally to produce endospores as shown in one cell. About 640 ×.

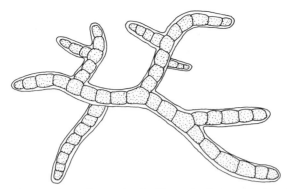

Figure 4-7 *Hyella caespitosa* forms simple filaments that penetrate calcarious matter, especially shells on beaches. As a result, the shells have a bluish-green color. The basal filaments can become more than one-cell thick (pleuriseriate). About × 190.

Hyella

The genus is characterized by branched trichomes that have normal vegetative cell division, resulting in progeny cells with common cell walls. *Hyella caespitosa* Bornet and Flanault—*Entophysalis deusta* (Menegnini) Drouet and Daily—is a branching series of trichomes that penetrate shells and limestone, especially in tropical intertidal zones. Most shells along a beach have a blue-green color, probably due to this genus. The branching is irregular (Figure 4-7, and the basal filaments may have become pluriseriate. The cells are up to 7 μm in diameter and 20 μm in length. The sheath may become stratified and is usually colorless.

Oscillatoriales

This order consists of filamentous forms of blue-green algae that do not produce specialized endospores or exospores as in the previous order. The order is usually divided up into five families, all of which have marine forms.

Oscillatoriaceae

This is the largest family of the blue-green algae, containing unbranched trichomes and lacking any specialized cells. Although a sheath is visible on the surface of most trichomes, it may be very thin so that it is not easily seen under the light microscope.

Oscillatoria

The trichomes are cylindrical and unbranched and the sheath is essentially invisible. The individual cells are usually shorter than broad, resembling discs or chips in a stack. The apical cell may be elongated or capped with a thickened end wall. Hormogonia are produced by the development of separation discs resulting from the death of an intercalary cell (Figure 4-8). Humm and Wicks (1980) list 11 species in the marine environment.

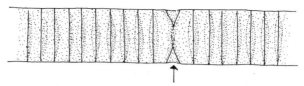

Figure 4-8 An example of the separation disc (arrow) in *Oscillatoria* sp. that results in the production of two distinct hormogonia. The genus is known for its lack of a distinct sheath and the small, almost disclike cells making up the filament. × 540.

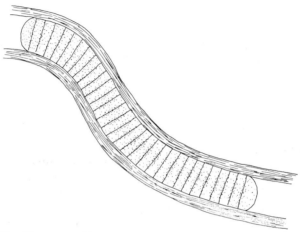

Figure 4-9 The filamentous blue-green *Lyngbya majuscula* will develop a distinctive sheath that can become stratified and contain a number of trichomes. × 140.

Lyngbya

Another example of a marine genus of this family is *Lyngbya majuscula* (Dillwyn) Harvey (Figure 4-9). The genus is characterized by a well-developed sheath, which can be pigmented. The species forms a common reddish film over the sand in quiet tropical waters.

Nostocaceae

The trichomes are unbranched but form heterocysts and akinetes. Heterocysts may be terminal or intercalary. Usually a sheath is present and the cells are spherical or ovoid in shape.

Anabaena

The plants (also spelled Anabaina) are solitary trichomes with intercalary heterocysts. Some species are planktonic. *Anabaena fertilissima* Rao has cells 4–8 μm in diameter and is slightly compressed at sites of cellular attachment (Figure 4-10). The heterocysts of *A. fertilissima* are quadrate and intercalary, whereas the akinetes, when present, are terminal and ovoid in shape.

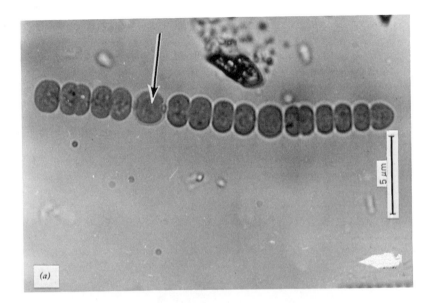

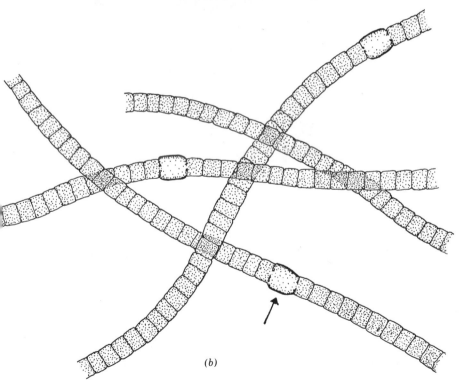

Figure 4-10 Two examples of the filamentous genus *Anabaena*. (*a*) *Anabaena fertilissima* and (*b*) *Anabaena laxa* both have heterocysts (arrows). × 200.

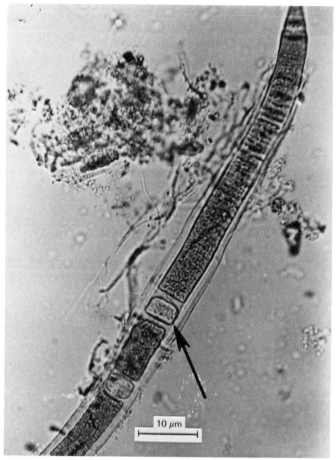

Figure 4-11 *Calothrix crustacea* is a filamentous blue-green with tapering filaments, intercalary heterocysts (arrow), and a well-developed sheath.

Rivulariaceae

The members of this family have basal or intercalary heterocysts with their trichomes tapering from base to apex or from the middle to both ends. Members of this family are especially common as gelatinous-to-firm masses or colonies on rocks of the intertidal zone.

Calothrix

The species is usually found on rocks, with the tapering filaments basally attached. The filaments have either basal heterocysts only—*Calothrix confervicola* (Roth) C. Agardh—or both basal and intercalary heterocysts (*C. crustaceae* Thuret, Figure 4-11). False branching can be common in large colonies of filaments. The sheath is usually obvious and may be confluent between trichomes.

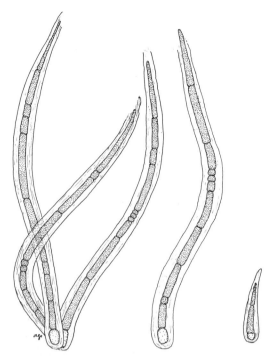

Figure 4-12 *Rivularia atra* is a filamentous blue-green with tapering filaments, basal heterocysts, and a distinct sheath around each filament. × 140.

Rivularia

The main distinction between this genus and *Calothrix* is the colonial organization. In *Rivularia,* the trichomes are organized into well-defined gelatinous colonies that may form small, distinctive spheres on intertidal rocks or algae. *Rivularia atra* Roth forms spherical, solid hard black colonies with trichomes 2–5 μm in diameter. The heterocysts are usually basal (Figure 4-12).

Scytonemataceae

The trichomes are in firm, usually pigmented, sheaths and are characterized by false branching. False branching occurs through the rupture of the sheath and outgrowth of one or both broken ends, usually near a heterocyst.

Scytonema

The filaments usually form entangled, brownish-to-bluish-green masses. The trichomes are 6–12 μm in diameter and the sheaths are lamellose and yellow-brown, as in *S. myochrous* (Dillwyn) Agardh (Figure 4-13). Heterocysts usually are quadrate or longer than their diameter. Plants can be found on rocks in the intertidal zone.

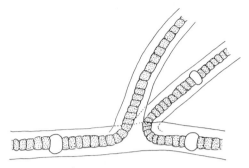

Figure 4-13 *Scytonema myochrous* shows false branching near the heterocysts. In the drawing, the false branches are paired. × 190.

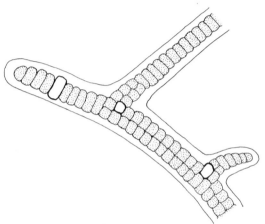

Figure 4-14 *Stigonema aerugineum* shows true branching and a pleuriseriate main axis with heterocysts adjacent to sites of branching. × 140.

Stigonemataceae

The family is considered by many phycologists to contain the most advanced blue-green algae. The species are usually filamentous with multiseriate construction, having true branching and heterotrichous construction (different types of filaments). True branching results from cell division occurring at right angles to the filament.

Stigonema

The genus is characterized by uniseriate filaments with occassional biseriate axes, especially at sites of true branching. *Stigonema aerugineum* Tilden can be found on rocky areas, especially in temporate regions of fresh water runoff or organic wastes (bird droppings). The uniseriate branches are smaller (heterotrichous) than the cells of the main axis (15 μm diameter, 6 μm long). Heterocysts are usually intercalary and about 10 μm diameter as shown for *S. aerugineum* (Figure 4-14).

Another genus placed in this family, *Mastigocoleus,* can be found in shells of molluscs (*M. testarum* Lagerheim). The genus is characterized by large terminal heterocysts, a parenchymatous prostrate system and attenuated erect branches.

REFERENCES

Aleem, A. A. 1980. Marine Cyanophyta from Sierra Leone (West Africa). *Bot. Mar.* **23**: 49–51.

Bisapultra, T., B. R. Oakley, D. C. Walker, and C. M. Shields. 1975. Microtubular complexes in blue-green algae. *Protoplasma* **86**: 19–28.

Berchtold, M. and R. Bachofen. 1979. Hydrogen formation by Cyanobacteria cultures selected for nitrogen fixation. *Arch. Microbiol.* **123**: 227–232.

Bold, H. C. and M. J. Wynne. 1978. *Introduction to the Algae. Structure and reproduction.* Prentice-Hall, Englewood Cliffs, N.J.

Bourrelly, P. 1970. *Les algues d'eau douce. Initiation à la systematique,* III. *Les algues bleues et rouges, les Eugleniens, Peridiniens et Cryptomonadines.* Ed. N. Boubee et Cie, Paris.

Champion-Alsumard, T. L. 1979. Les Cyanophycees endolithes marines. Systematique, ultrastructure, ecologie et biodestruction. *Ocean. Acta* **2**: 143–156.

Desikachary, T. V. 1959. *Cyanophyta.* Indian Council Agricultural Research, New Delhi.

Drouet, F. 1968. *Revision of the classification of the Oscillatoriaceae.* Acad. Nat. Sci. Phil., Philadelphia, Penn.

Drouet, F. 1973. *Revision of the Nostocaceae with Cylindrical Trichomes.* Nova Hedwigia, vol. 47. J. Cramer Press, Vaduz, Germany. 292 pp.

Drouet, F. 1978. *Revision of the Nostocaceae with constricted Trichomes.* Nova Hedwigia, vol. 57. J. C. Kramer Press, Vaduz, Germany.

Drouet, F. and W. A. Daily. 1956. Revision of the coccoid myxophyceae. *Butler Univ. Bot. Stud.* **12**: 1–218.

Fogg, G. E., W. D. P. Stewart, P. Fay, and A. E. Walsby. 1973. *The Blue-green Algae.* Academic, London.

Geitler, L. 1932. Cyanophyceae. In Rabenhorst, L. (ed.). *Kryptogamenflora von Deutschland,Österreich und der Schweiz,* vol. 14., Akademische Verlags Gesellschaft, Leipzig, pp. 673–1056.

Halfen, L. N. 1973. Gliding motility of *Oscillatoria:* Ultrastructural and chemical characterization of the fibrillar layer. *J. Phycol.* **9**: 248–253.

Humm, J. J. and S. Wicks, 1980. *Introduction and Guide to the Marine Blue-green Algae.* Wiley, New York.

Kulasooriya, S. A., N. J. Lang, and P. Fay. 1972. The heterocysts of blue-green algae. III. Differentiation and nitrogenase activity. *Proc. Roy. Soc. Lond.* **181**: 199–209.

Lang, N. J. 1968. The fine structure of blue-green algae. *Ann. Rev. Microbiol.* **22**: 15–46.

Lang, N. J. and P. Fay. 1971. The heterocysts of blue-green algae. II. Details of ultrastructure. *Proc. Roy. Soc. Lond.* Series B. **178**: 193–203.

Orkwizewske, K. G. and A. R. Kaney. 1974. Genetic transformation of the blue-green bacterium, *Anacystis nidulans. Archiv Mikrobiol.* **98**: 31–37.

Padan, E. 1979. Facultative anoxygenic photosynthesis in Cyanobacteria *Ann. Rev. Pl. Physiol.* **30**: 27–40.

Papenfuss, G. F. 1955. Classification of the algae. In *A Century of Progress in the Natural Sciences, 1853–1953,* Calif. Acad. Sci. San Francisco, pp. 115–224.

Potts, M. 1979. Nitrogen fixation (acetylene reduction) associated with communities of heterocystous and non-heterocystous blue-green algae in mangrove forests of Sinai. *Oecologia* **39**: 359–373.

Potts, M. 1980. Blue-green algae (Cyanophyta) in marine coastal environments of the Sinai Peninsula; distribution, zonation, stratification and taxonomic diversity. **Phycologia. 19**: 60–73.

Potts, M. and B. A. Whitton. 1977. Nitrogen fixation by blue-green algal communities in the intertidal zone of the Lagoon of Aldabra Atoll. *Oecologia* **27**: 275–283.

Rippka, R. and R. Y. Stanier. 1978. The effects of anaerobiosis on nitrogenase synthesis and heterocyst development by nostocacean cyanobacteria. *J. Gen. Microbiol.* **105**: 83–94.

Roberts, T. M., L. C. Klotz, and A. R. Loeblich III. 1977. Characterization of a blue-green algal genome. *J. Mol. Biol.* **110**: 341–361.

Safferman, R. S. 1973. Phycoviruses. In Carr, N. G. and B. A. Whitton (eds.). *Biology of Blue-green Algae,* University of California Press, Berkeley, Chapter 11.

Safferman, R. S. and M. E. Morris. 1963. Algal virus isolation. *Science* **140**: 679–680.

Safferman, R. S. and M. E. Morris. 1977. *Phycovirus Bibliography.* EPA-600/9/77–008. U.S. Environmental Protection Agency, Environmental Research Center, Cincinnati, Ohio.

Smith, G. M. 1950. *The fresh-water algae of the United States.* McGraw-Hill, New York.

Stam, W. T. and H. C. Holleman. 1975. The influence of different salinities on growth and morphological variability of a number of *Phormidium* strains (Cyanophyceae) in culture. *Acta Bot. Neerl.* **24**: 379–390.

Stam, W. T. and G. Venema. 1977. The use of DNA-DNA hybridization for determination of the relationship between some blue-green algae (Cyanophyceae). *Acta Bot. Neerl.* **26**: 327–342.

Stanier, R. Y., R. Kunisawa, M. Mandel, and G. Cohen Bazire. 1971. Purification and properties of unicellular blue-green algae (order Cyanophyceae). *Bacteriol. Rev.* **35**: 171–205.

Tilden, J. 1910. *Minnesota algae. I. The myxophyceae of North America and adjacent regions including Central America, Greenland, Bermuda, the West Indies and Hawaii. Univ. Minn. Bot. Ser. 8.,* Minneapolis.

CHAPTER 5

Chlorophyta

The green algae are common in soils and freshwater and marine environments. There are about 5500 species with about 90% found in freshwater environments. The majority of green algae are small, unicellular, or filamentous. Since green algae have the same pigments and reserve foods as higher plants, they are usually considered to be the progenitors of land plants.

CYTOLOGICAL FEATURES

Pigments

Chlorophylls *a* and *b* are present giving the green algae their typical "grass green" coloration. In addition, both α and β carotenes, lutein, and zeaxanthin are present, as in higher plants (Velicho, 1980). Siphonoxanthin, a special xanthophyll, is also present and is characteristic of the coenocytic (siphonaceous) green algae. Siphonoxanthin has also been found in deep-water species of *Ulva* (Yokohama et al., 1977), and it presumably enables deep-water species to capture light in the range of 470 nm (see Chapter 14).

Cytological Structure

The cell structure is eukaryotic and very similar to that of higher plants (Figures 3-2, 5-1). Chloroplasts usually have three to seven thylakoids grouped into bands as well as pyrenoids with peripheral starch granules. The thylakoids may be so highly organized as to resemble grana of higher plants. Pyrenoids are found in many green algae and have been shown to consist of protein and to contain starch synthetase, an enzyme involved in starch production.

The sequence of mitosis and cytokinesis in several green algae have been well studied at the ultrastructural level, and the results appear to have important taxonomic and phylogenetic implications. Although variations occur, two basic types can be described (see also Figures 3-6 and 3-7 and Chapter 3). In the first, the nuclear envelope disappears prior to the alignment of chromosomes at the metaphase plate. This is termed an open spindle and is best known in the

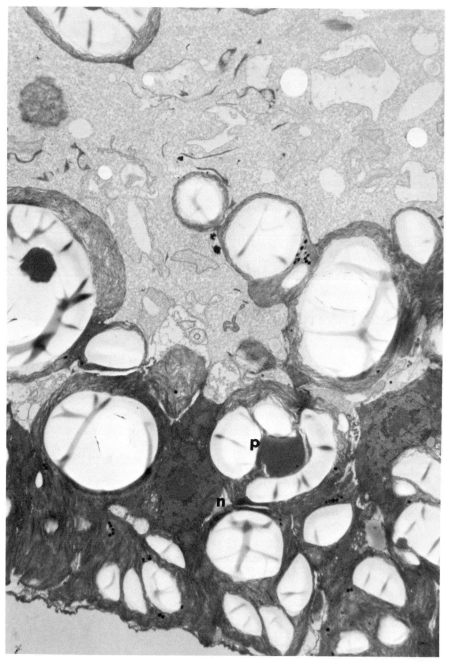

Figure 5-1 A micrograph of the green alga *Chaetomorpha aerea,* demonstrating typical green algal cell structure. Pyrenoids (*p*) with numerous starch grains are visible in the chloroplasts. Two nuclei (*n*) are also evident. × 16,000. (Courtesy of Ernest Truby, Florida Department of Natural Resources Marine Laboratory, St. Petersburg.)

Charophyceae of the Chlorophyta, and in higher plants. After this type of karyo-kinesis, the spindle remains between the widely separated nuclei, and cytokinesis procedes by furrowing of the cell membrane or by cell plate formation through a phragmoplast (a proliferation of microtubules parallel to the spindle and perpen-dicular to the plane of the future cell plate). The latter is the type found in the Charophyceae and higher plants. In the second type, the nuclear envelope remains intact, or nearly so (closed spindle), and the spindle does not persist. The nuclei lie close to each other, and cytokinesis occurs through cell membrane furrowing or cell plate formation, often in conjunction with a phycoplast (a system of microtubules at right angles to the axis of the former spindle and in the plane of cleavage).

Some phycologists propose moving all green algae with the open form of mitosis into the class Charophyceae. Then the class Chlorophyceae would include only those green algae that have a closed mitotic apparatus and a phycoplast. The Chlorophyceae are treated as the main body of green algae in this text following the presentation of Bold and Wynne (1978).

Motility

Most green algae have flagellated cells during some state of development. In many orders the flagellated cell is found in the vegetative phase, whereas in other algae only the gametes or zoospores are motile. Regardless, the flagella are usual-ly in pairs, acronematic (smooth, lacking obvious hairs), and attached to the api-cal (direction of swimming) region of the cell. The flagella are therefore alike and usually of equal length (isokont). Some members have obvious scales on the fla-gella and they are now placed in the class Prasinophyceae.

Ultrastructural studies of motile cells have shown that there are also two pat-terns of flagellar insertion and root arrangement, which also corelate with mitosis and cytokinesis. Thus algae with open persistent spindles produce motile cells with laterally inserted flagella and a root consisting of a single wide band of microtubules, whereas those with closed nonpersistent spindles produce motile cells with anteriorly inserted flagella and narrow, symmetrically arranged roots.

Cell Wall

Most of the green algae have a cellulose cell wall. However, some forms have no wall or a proteinacous type of covering. The wall may have an outer cuticle type of layer that may be protein or pectin in nature. The coenocytic members have structural carbohydrates containing mannans or xylans and thus are distinct from the other green algae. A number of the coencytes may be calcified with the ara-gonite form of calcium carbonate.

The pattern of cellulose microfibrils in the cell wall has been used to separate groups of green algae (Dawes, 1966). Two orders of green algae, both having marine members, have cell walls with highly crystalline cellulose arranged in a crossed microfibrillar pattern (Figure 3-16, Cladophorales and Siphonocladales).

Other patterns include a random matrix of microfibrils (Volvocales), microfibrils in layers but not highly organized (Ulvales), and those in which microfibrils appear to be lacking (Caulerpales).

Food Storage

The primary reserve food is true starch, which is composed of the unbranched glucose polymer amylose and a branched glucose polymer, amylopectin. Amylose contains $\alpha = (1 \rightarrow 4)$ linkages, whereas amylopectin has branches with $(1 \rightarrow 6)$ as well as $(1 \rightarrow 4)$ linkages. Oil may also be found as a reserve food and is more common in older and resting cells.

MORPHOLOGY

The majority of green algae are unicellular or colonial and may be motile or nommotile. In addition there are filamentous forms with or without branching (Ulotrichales, Cladophorales, Acrosiphoniales, Siphonocladales) and bladed forms (Ulvales, Prasiolales). A number of the green algae are multinucleate or coenocytic (multinucleate, without cross walls). If the plant body (thallus) consists of a number of coenocytic filaments that at least in part are interwoven, the plant is said to have a siphonous construction.

REPRODUCTION

Both asexual and sexual reproduction are found throughout the green algae. Asexual reproduction is by zoospores (motile) and aplanospores (nonmotile), but the latter are more common in freshwater and soil forms. Sexual reproduction, where known, is most commonly by motile gametes (isogametes, anisogametes, eggs and sperm). In some orders (Volvocales), a complete range of gametes can be found. Life histories are very diverse. The majority of green algae have haplontic life histories, but diplontic (Siphonocladales, Caulerpales, Dasycladales) and haplodiplontic types (Ulvales, Caulerpales, Ulotrichales, Cladophorales) occur as well.

DISTRIBUTION AND ECOLOGY

Since only about 10% of the roughly 5500 species of Chlorophyceae are marine, this text will be limited to a few orders. The green algae are common in the upper subtidal areas, especially in tropical and subtropical waters, where there are abundant examples of multinucleated, multicellular (Cladophorales, Siphonocladales) and siphonous or coenocytic (Dasycladales, Caulerpales) forms. Some green algae extend from the cold temperature waters of the arctic to the tropics (*Ulva, Entermorpha*). A few species are cultivated as food (Chapter 2).

TAXONOMY

Since Blackman and Tansley (1903) presented their concept of the three evolutionary trends in the green algae (Chlorophyceae), this division has had a number of major taxonomic revisions. The most recent approach has been to distinguish three distinct classes, the major group, or Chlorophyceae, and two smaller classes, the Charophyceae and the Prasinophyceae. Stewart and Mattox (1975) have proposed that the Charophyceae are actually the most highly advanced green algae cytologically and it is possible that higher plants evolved from them. They base this opinion on cytological evidence concerning flagellar apparatus and type of mitosis and cytokinesis, as well as biochemical features. Algae having open, persistent spindles and asymmetric motile cells would be placed in the Charophyceae, and those having closed, nonpersistent spindles and symmetric motile cells would make up the Chlorophyceae.

CLASS 1. PRASINOPHYCEAE

This class is characterized by unicellular, motile green algae that originally were placed in the order Volvocales of the Chlorophyceae. The flagellated forms have been reviewed by Norris (1980). A number of ultrastructural studies have been published on various members, especially the genus *Pyramimonas* (Norris and Pienaar, 1978), and cell division has been shown to involve a phycoplast in some members such as *Heteromastix* (Mattox and Stewart, 1977). Trainor (1978) and Bold and Wynne (1978) continue to place these algae in the Volvocales. The Prasinophyceae can be characterized by their (1) cell covering of one or more layers of fibrillar scales (organic); (2) flagella that are attached in a depression or groved region of the cell; (3) flagellar covering of scales and presence of thick tubular flagellar hairs; (4) complex basal body system including a transition zone, rhizoplast, and microtubular roots (Norris, 1980); (5) single, bowl-shaped plastid, central pyrenoid and surrounding starch grains; (6) mucocysts or trichocysts (not in all species); and (7) pigments, which in some cases are unique. Two examples of unicellular forms will be presented for this small class of green phytoflagellates.

Pyramimonas

The genus is characterized by a pear-shaped, quadriflagellated cell that has an apical depression from which the flagella emerge. The unicell has four lobes at the apex (Figure 5-2). Other motile, plankton representatives include the uniflagellated *Micromonas* (Manton and Parke, 1960) or biflagellated *Nephroselmis* (Butcher, 1959).

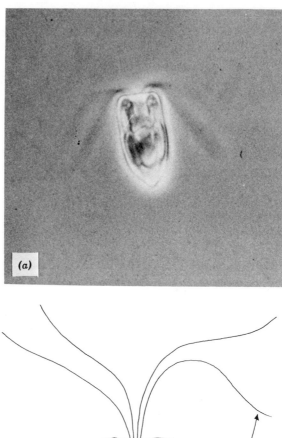

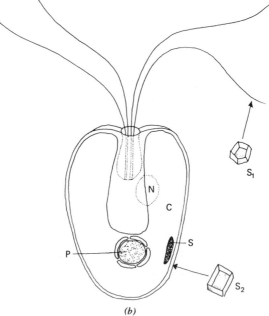

(b)

Figure 5-2 *Pyramimonas amylifera,* a unicellular green alga in the class Prasinophyceae.
(a) Light micrograph showing a living cell. (b) Diagram illustrating some of the cellular
features including two types of scales, a flagellar scale (S_1) and a body scale (S_2). The
cell has a large chloroplast (C), smaller nucleus (N), a pyrenoid with starch grains (P) and
an eyespot (S). The cells are about 5–10 μm in length. (Photo courtesy of Bill Gardiner,
University South Florida Tampa)

Prasinocladus

An example of a dendroid colonial green flagellate can be found in high intertidal pools. The plants have two to four eliptical cells that occur in the upper branches of a tubular stalk (Figure 5-3). The stalk is from the previous cell. Their motile cells are quadriflagellated, with an eyespot, chloroplast, and axial pyrenoid.

CLASS 2. CHAROPHYCEAE

The class is small, with only four genera in the North American continent. Because of their complex morphology, including an apical meristem and differentiated "leaf" and "stem" cells, some phycologists place this group in a separate division (Bold and Wynne, 1978). Some species are found in brackish water *Chara, Nitella*) and thus may be present in streams leading into salt marshes. However, only *Lamprothamnium* is known to occur in maritime habitats such as ditches, lakes, or lagoons that have been cut off from the open sea by sandbars. Bisson and Kirst (1980) have demonstrated that *Lamprothamnium* controls turgor pressure through the levels of K^+ and Cl^- in the cell vacuole. Charophytes are common in the inner Baltic Sea, where the salinity is less than 10 ppt.

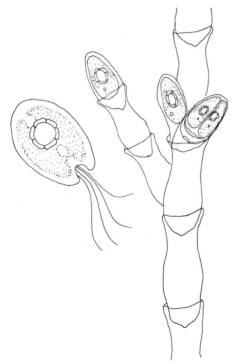

Figure 5-3 *Prasinocladus marinus* produces quadriflagellated zoospores that develop from the bottom of a depression in the apical region of the cell. The zoospores (4–5 μm diameter) will attach and produce a stalk by growing out of the old cell wall.

In addition to a highly complex apical meristem and vegetative morphology, the charophytes have sterile cells that form protective jackets around their antheridia and oogonia. This is unique with regard to all other algae and suggests adaptation to land. The life history, similar to many simpler algae, is haplontic. The plants can be large, reaching a meter or more in length, and the giant cells (up to 10 cm long) become multinucleate.

Stewart and Mattox (1975) suggested that this class should include other green algae with flagellar structures, and mitotic features similar to those found in higher plants. the authors argue that ancestors in this class evolved into the present-day land plants. A paper by Sluiman *et al* (1980) summarizes their arguments.

CLASS 3. CHLOROPHYCEAE

Originally all green algae were placed in this class. Blackman and Tansley (1903) recognized three major lines of evolution in the green algae, including a motile, colonial line (Volvocine trend), a coenocytic line (Chlorococcine trend), and a filamentous, parenchymatous line (Tetrasporine trend). Blackman and Tansley suggested that the last group was the progenitor of higher plants.

This is the dominant class of green algae. Stewart and Mattox (1975) proposed the transfer of a number of freshwater genera from this class to the Charophyceae based on cytological and biochemical data. About 10% of the species are marine; most members occur in soil or freshwater environments. Bold and Wynne (1978) recognize 15 orders, 11 of which have marine representatives. Only five orders are predominantly marine in distribution. A key to the orders containing marine species is presented in Table 5-1.

Volvocales

All members of this order are flagellated, motile, and well represented in the freshwater plankton, but relatively infrequent in marine habitats. The unicellular genera, *Chlamydomonas, Carteria,* and *Dunaliella,* occur in salt marshes and tidal pools. The fine structure of *Brachiomonas submarina* (Underbrink and Sparrow, 1968) is similar to that of *Chlamydomonas.* The cell structure has been described for a number of genera; cell division includes a closed spindle with the nuclear membrane remaining intact (Pickett-Heaps, 1973). The larger colonial forms are absent from the marine environment, although they are very common in freshwater systems. Lembi (1980) reviewed the unicellular chlorophytes and Starr (1980) has reviewed the colonial members of this order.

Tetrasporales

Members of this order occur in colonies with a gelatinous matrix. Motility is limited to brief periods or during reproduction. Two marine genera, *Pseudotetraspora* and *Palmophyllum,* are common in the tropics.

Table 5-1   A Key to the Orders Containing Marine Green Algae

1   Plants unicellular or colonial 2

1   Plants filamentous, membranous, tubular 4

   2   Cells or colonies motile during vegetative phase ....**Volvocales**
   2   Cells or colonies nonmotile during vegetative phase 3

3   Cellular organization *Chlamydomonas*-like, colonies in gelatinous matrix ....
.... **Tetrasporales**

3   Cells not *Chlamydomonas*-like, gelatinous matrix usually not present, cells may
become multinucleate ....**Chlorococcales**

   4   Cells uninucleate 5
   4   Cells large, multinucleate or coenocytic... 6

5   Plants either branched or unbranched filaments ....**Ulotrichales**

5   Plants membranous, either as a tube or sheet (monostromatic or distromatic) 10

   6   Plants multinucleate, that is, having many nuclei per cell and more than one
   cell 7
   6   Plants coenocytic, that is, having many nuclei but only one cell, a septation
   may form in gametangial production 9

7   Chloroplast discoid, life history diplontic, segregative cell division occurring ....
.... **Siphonocladales**

7   Chloroplasts reticulate or perforate and continuous, life history haplodiplontic, no
segregative cell division 8

   8   Chloroplasts, reticulate, not continuous, zoosporangia or gametangia opening
   by fissures or simple pores ....**Cladophorales**
   8   Chloroplasts parietal and perforate, continuous, zoosporangia and gametangia
   opening by operculate pores ....**Acrosiphonales**

9   Plants radially symmetrical with whorled branches, plant starting as a single
nucleus and becoming coenocytic only during phases of reproduction ....
.... **Dasycladales**

9   Plants not radially symmetrical, plants coenocytic throughout entire vegetative phase
.... **Caulerpales**

   10   Plants having lobed to cup-shaped plastids ....**Ulvales**
   10   Plants having stellate plastids ....**Prasiolales**

Source: Modified from Bold and Wynne (1978).

Pseudotetraspora

The species *P. antillarum* is common as a coating on sea grass blades and rocky
substrates in the subtidal region of tropical waters. The cells are mostly spherical
and embedded in a gelatinous, usually brownish matrix (Figure 5-4). Motile cells
are unknown for this genus, and its taxonomic position is unclear.

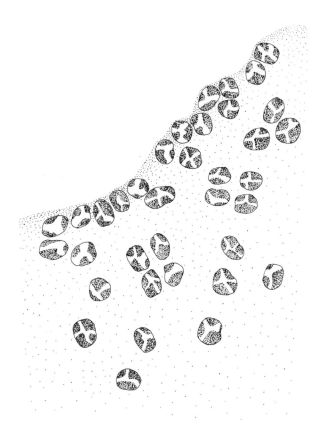

Figure 5-4 *Pseudotetraspora antillarum,* a colonial marine green alga that will attach to leaves of seagrasses or macroscopic algae. The individual cells have a single lobed chloroplast and the entire colony is enveloped in a sheath. The cells are about 5–10 μm in diameter.

Chlorococcales

The order is represented primarily by freshwater and terrestrial species, and the placement of marine species in this order varies with the authority.

Chlorella occurs in salt-marsh habitats. Other marine representatives are small endophytic plants such as *Chlorochytrium* and *Gomontia,* as well as the lithophyte and endophyte *Codiolum* (Figure 5-5). Some forms of these "genera" have been shown to be diploid stages in the life histories of other green algae. Thus the placement of these species is open to question. Wilkinson and Burrows (1972) reviewed this group, especially the shell-boring species and concluded that a number of species can be confused with *Gomontia polyrhiza*. Zoospores from cultures of *Codiolum* can germinate into several green algae. Kornmann (1978) proposed a new class, *Codiolophyceae,* to include heteromorphic plants contain-

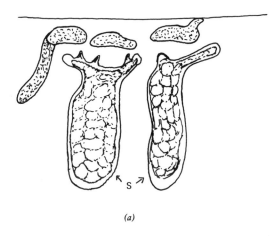

(a)

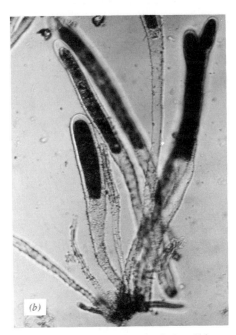

Figure 5-5 Two examples of marine members of the Chlorococcales. (*a*) *Gomontia polyrhiza* is found embedded in shells, producing a pseudofilamentous complex and sporangia (*S*). (*b*) *Codiolum petrocelidis* is a small unicellular green alga that is found embedded in the crustose red alga *Petrocelis*. Cells of both species reach 5–20 μm in diameter.

ing a *Codiolum*-like phase. *Chlorochytrium porphyrae* is an endophyte of the red alga *Porphyra perforata* which occurs in high tide zones of northern California. The spherical cells of *C. porphyrae* do not divide vegetatively. Reproduction is said to occur through the fusion of biflagellate isogametes, with the zygote invading the host to form a vegetative cell.

Ulotrichales

Plants in this order have unbranched or branched filaments, usually with a single chloroplast per cell. Most members are freshwater, but a few are present in the marine environment. Two families are presented here.

Ulotrichaceae

Members of this family are mostly unbranched filaments, and each cell has a single collar-shaped plastid.

Ulothrix

Plants are slender, unbranched filaments consisting of uninucleate cells with a parietal collar-shaped plastid. The basal cell of the filament is modified for attachment. Any cell except the holdfast cell can produce zoospores or gametes, which are first formed in the terminal cells and liberated through lateral pores (Figure 5-6). The motile cells may be quadriflagellated zoospores, biflagellated zoospores, or biflagellated gametes. Reproduction appears to be related to the photoperiod. Quadriflagellated zoospores are frequently produced. The gametes may be isogamous or anisogamous. Originally the life history was thought to be haplontic, but some species of *Ulothrix* are now known to have a *Codiolum* (zygote) stage and a heteromorphic haplodiplontic life history.

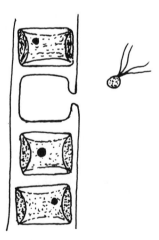

Figure 5-6 *Ulothrix* sp., a filamentous green alga that reproduces by quadriflagellated zoospores, as shown here. The chloroplast is a band, a characteristic of the family. × 450.

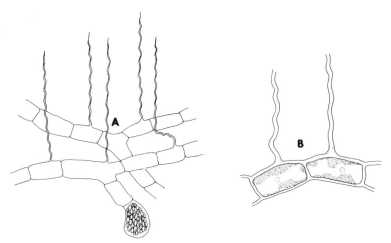

Figure 5-7 *Phaeophila dendroides*, a branching filamentous endophyte or epiphyte with cells about 20 μm in diameter, although they are quite irregular (*A*). Each cell can produce a hair, or seta (*B*) Occasionally sporangial cells can be found (*A*).

Chaetophoraceae

Members of this family are characterized by branched filaments and heterotrichous growth. Some marine members such as *Pringsheimiella scutata* are small epiphytes, which form pads of cells on *Cladophora* spp.

Phaeophila

Members of this species are common shell-boring or endophytic green algae in the North Atlantic. The plants consist of irregular cells that branch and produce erect setae, or hairs (Figure 5-7). Reproduction is by quadriflagellated zoospores produced in irregularly shaped intercalary sporangia. The zoospores produce a tube that enters the host. The cell contents then pass into the tube so that the original zoospore becomes empty and is cut off by a septum.

Ulvales

This order is characterized by parenchymatous growth, and usually the plant is a blade one to two cells thick or a tube one cell thick. In some classifications members of this order are placed in the Ulotrichales. The ultrastructure of the vegetative cells of *Enteromorpha* (McArthur and Moss, 1978) and *Ulva* (Bråten and Lovlie, 1968) as well as observations on the closed nuclear division in *Ulva* (Lovlie and Bråten, 1970) indicate the family has typical green algal cytology. Pyrenoid ultrastructure among the genera has been reviewed by Hori (1973). Life histories vary from isomorphic haplodiplontic to derived diplont or heteromorphic haplodiplontic patterns. Two families are presented here.

Monostromataceae

Members in this family are blades having only a single layer of cells (monostromatic). The initial development is from a small vesicle or tube.

Monostroma (= Ulvaria)

This plant is found in marine and estuarine waters and is easily separated from *Ulva* by its monostromatic nature; it is a more delicate and paler blade. Initial development is as a small sac that ruptures to produce the expanded blade. However, in a few species the thallus can remain saccate. Studies have shown that species of *Monostroma* are variable with regard to life histories. Some species have haplodiplontic life histories, with each phase producing spores or gametes (Dube, 1967). Other species have heteromorphic life histories in which the sporophyte is a membranous blade and the gametophyte a discoid plant. Yet in other species, the membranous blade is the gametophyte, and the zygote resembles a *Codiolum* phase. Thus the taxonomy of this genus requires more evaluation.

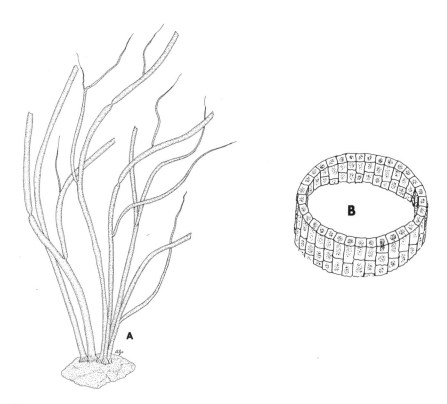

Figure 5-8 *Enteromorpha prolifera* has the characteristic tubular construction (*B*) of the genus and the species usually has numerous branches (*A*). (*A*), × 1; (*B*), × 5.

Ulvaceae

Members of this family are grass green and common in estuarine, brackish, and coastal sites. They are particularly evident in areas of pollution. Plants are parenchymatous, either forming blades (distromatic) or tubes, which may be flattened.

Enteromorpha

The characteristics now recognized for species determination in this genus are the presence or absence of a gametangial plant, the sizes of gametes and zoospores, the number of pyrenoids cell (1–10), and the size and arrangement of vegetative cells. The species are mostly tubular (Figure 5-8), simple or branched, cylindrical or compressed, sometimes only a few cells in diameter. They are attached by rhizoidal outgrowths from the basal cells. There is a single cup-shaped chloroplast in each cell. The life history is typically of an isomorphic haplodiplontic type, but some species reproduce only by zoospores, particularly under conditions of low salinity. Sporangial thalli produce 4, 8, or 16 quadriflagellated zoospores/cell. The zoospores are liberated through lateral pores in the cell wall. They settle and divide transversely to produce a uniseriate filament that develops basal attachment cells. Fertilization may be isogamous or anisogamous, and some species appear to show parthenogenetic germination of gametes.

Recent evidence suggests that the genus *Collinsiella,* a small alga formerly placed near *Prasinocladus,* is a phase in the life history of *Enteromorpha intestinalis* in Pacific Ocean material. *Collinsiella* probably represents the basal remnant of a tubular gametangial thallus from the previous season. This apparently does not occur in Japanese plants.

Ulva

Species of *Ulva* range from the arctic to the tropics in the North Atlantic and the North Pacific. The plants are distromatic blades (two cell layers thick) and attach by basal discs (Figure 5-9). The life history is similar to that of *Enteromorpha,* an isomorphic haplodiplont. The basal disc can persist and produce new fronds in the spring; thus the plant can be a pseudoperennial. Fragments of fronds also can become detached and continue to grow and reproduce, forming large free-floating masses in protected and contaminated (sewage) waters.

The morphology of *Ulva* species is very plastic, apparently due to environmental conditions. For example, the cosmopolitan species *U. lactuca* is now known to have distinct entities that are not interfertile. Rhyne (1973) has reviewed this literature for the east coast of North America.

Prasiolales

This order has relatively few genera; some are filamentous (*Rosenvingiella*) while others are foliaceous forms (*Prasiola*). The order is considered by some to be related to the Ulvales because of the leafy and parenchymatous nature of some species. However the stellate chloroplast and life history tend to argue for a separate order. Edwards (1975) found that there was a transition in morphologies be-

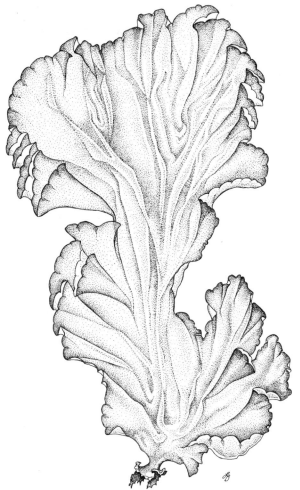

Figure 5-9 *Ulva lactuca,* a distromatic green blade alga that is common in estuarine and protected open coastal sites. × 1.

tween *Rosenvingiella polyrhiza* and *Prasiola calophylla* and that the latter species may be but an aerial ecophene of *P. stipitata.*

Prasiola

The genus *Prasiola* is represented by several species in cold and temperate waters, where the small plants are typical in high intertidal pools that are sites of bird droppings. The species produce dark-green patches and are thin, broadly ovate blades or pluriseriate filaments (Figure 5-10). The germination of a spore or zygote results in an unbranched filament, from which the membranous thallus develops. Vegetative thalli of *P. stipitata* are diploid and can form spores (mitosis) or gametes (meiosis). Spore production only occurs in the upper portion of

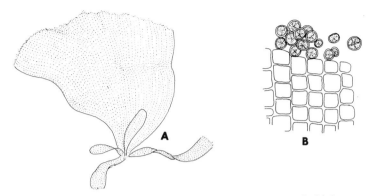

Figure 5-10 (*A*) *Prasiola stipitata*, a small-bladed plant common in higher, organic-rich tide pools. The plant will release nonmotile spores (*B*) from the blade edge. (*A*), × 4.50; (*B*) × 35.

the plant. In addition, spherical, nonmotile spores are produced and have been called "akinetes," but they differ in being thin walled. Gamete formation in *P. stipitata* usually occurs in thalli found in the older, basal portion (as in the formation of spores). Cells in the upper part or blade edge undergo meiosis and form a patchwork of male and female cells distinguished by the size of the chloroplast (females have large dark-green plastids) in each cell. At maturity, each cell produces a single anteriorly biflagellated male sperm or a nonmotile female egg, and these are discharged during a dark period. Similar life histories are known for *P. japonica* of Japan and *P. meridionalis* of Pacific North America. The life history and effects of environmental factors of *P. stipitata* are described by Friedmann (1969).

Cladophorales

This order is sometimes combined with the Siphonocladales (Chapman, 1954; Dawes, 1966) because of the highly crystalline cellulose cell wall and the multinucleate cells present in both. The distinction lies in the type of life history (mostly isomorphic haplodiplontic for the Cladophorales and diplontic for the Siphonocladales) and type of cytokinesis. Two families are presented here.

Cladophoraceae

Plants of the Cladophoraceae consist of multicellular branched or unbranched uniseriate filaments. The cells are multinucleate, and each contains a single, reticulate (netlike) chloroplast with many pyrenoids. Genera are separated based on the presence and type of branching, spore production, the type of basal cell, and the attachment rhizoids. There are problems with both generic distinction (e.g., *Rhizoclonium* vs. *Chaetomorpha*) and species separation. The general cellular ultrastructure has been described for *Chaetomorpha* (Chan *et al.*, 1978). Mitosis was found to be closed, the nuclear envelope remaining intact during chromosome alignment and separation in *Cladophora* (McDonald and Pickett-Heaps, 1976; Scott and Bullock, 1976).

Rhizoclonium

The species *R. hookeri* (Figure 5-11) consists of uniseriate, unbranched filaments about 60 μm in diameter, and a reticulate chloroplast with numerous pyrenoids. Short rhizoidal branchlets occur but not in a regular pattern. The life history is of an isomorphic haplodiplontic type. Quadriflagellated zoospores and biflagellated isogametes are produced.

Cladophora

This is probably the largest and most widespread genus of the order (Figure 5-12); the branching is lateral and attachment is by rhizoids to form a compact basal mat. Each cell is multinucleate and has a reticulate chloroplast containing numerous pyrenoids. The sporangial and gametangial plants are isomorphic (e.g., *Cladophora callicoma*; Shyam, 1980). The zoospores are quadriflagellated, and the gametes are biflagellated and isogamous. Due to the high degree of polymorphology, the taxonomy of *Cladophora* species is confusing. Van den Hoek (1979) has discussed the phytogeography of this genus in the North Atlantic, as well as the taxonomy of the genus.

Anadyomenaceae

The genera in this small family are marine tropical, bladed algae that have the characteristic filamentous construction. The two principal genera, *Microdictyon* and *Anadyomene*, have delicate, reticulate blades that result from the abundant branching of uniseriate filaments; the branching occurs in one plane. *Anadyomene* forms a delicate, netlike blade (Figure 5-13). The life histories of *Microdictyon*

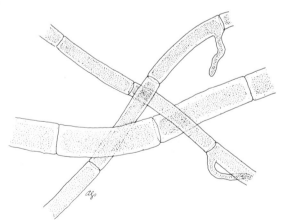

Figure 5-11 *Rhizoclonium hookeri*, a filamentous alga having occasional rhizoidal branches. The cell wall can become thickened and stratified with age similar to *Cladophora*. The filaments are 40–80 μm in diameter.

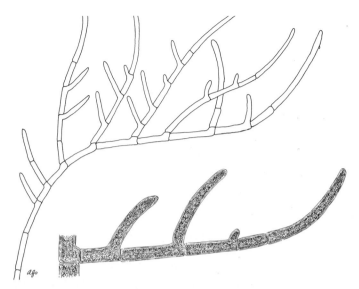

Figure 5-12 A terminal branch and detail of *Cladophora fuliginosa,* demonstrating the secund (comb-like) branching pattern. The alga is one of the largest species with regard to cell diameter (200–300 μm) and the cell wall is usually thick and stratified. The plant usually has a fungus growing in the cell walls; the relationship is analogous to that found in lichens (see Appendix A).

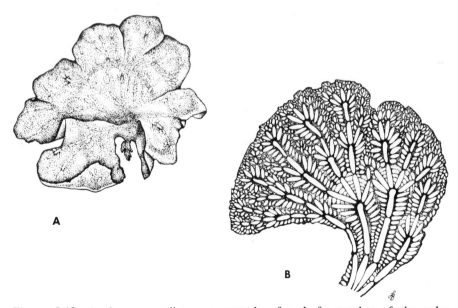

A

B

Figure 5-13 *Anadyomene stellata,* an example of a leafy member of the order Cladophorales in which the filaments are fused together (*B*) to form a blade. The blades (*A*) are about 3–6 cm in diameter and may form dense, grass-green, lettucelike clusters.

and *Anadyomene* are similar to those of the *Cladophoraceae;* they are isomorphic haplodiplontic types. Sporangial or diploid plants form quadriflagellated zoospores, whereas the morphologically similar gametangial plants produce biflagellated isogametes.

Acrosiphoniales

Members that once were included in the Cladophorales or Ulotrichales are now placed in this order because of the heteromorphic haplo-diplontic life histories and noncellulosic cell walls (Dawes, 1966). The general ultrastructure, mitosis (closed spindle), and cytokinesis have been described for *Acrosiphonia* (Hudson and Waaland, 1974).

Spongomorpha

This genus is found on the temperate and boreal coasts of the North Atlantic and North Pacific. The plant has branched filaments as well as specialized hooked spines and rhizoidlike branches that grow downward. Because of the hooked and rhizoidal branches, the plant usually becomes "ropelike" with entangled branches. The plant has uninucleated cells in contrast to a closely related genus, *Acrosiphonia*, which has multinucleated cells.

The life history is heteromorphic; the "*Spongomorpha* plants" are dioecious and produce biflagellated gametes. *Spongomorpha coalita* in California has a life history in which *Codiolum petrocelidis* is the diploid phase. A similar association has been demonstrated in the North Atlantic for the species-pairs *Spongomorpha lanosa/Chlorochytrium inclusum* and *Spongomorpha spinescens/Codiolum petrocelidis* (Kormann, 1973). Based on these life histories, Kormann (1973) has proposed a new class, the Codiolophyceae.

Urospora

The genus is worldwide, predominantly cold water and intertidal in distribution. the plants are unbranched filaments consisting of barrel-shaped cells (Figure 5-14) up to 150 μm in length and 80 μm in width. The thallus is attached by numerous rhizoids that develop from the lowermost cells. Each cell has a single reticulate or perforate chloroplast containing numerous pyrenoids. Swarmers are released laterally through a pore in the cell wall. Quadriflagellated (caudate) zoospores and biflagellated gametes are produced. *Urospora* has a heteromorphic life history involving a diploid *Codiolum* phase.

Siphonocladales

This order is entirely marine and tropical; its members are multinucleate, septate, and attached by basal rhizoids. The chloroplasts are reticulate. The major distinctions of this order are the form of cytokinesis, termed segregative cell division, and the diplontic life history. In segregative cell division, the cytoplasm divides into several protoplasmic masses of varying size, each of which becomes rounded

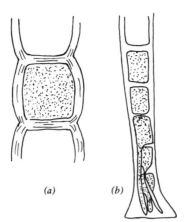

Figure 5-14 Two views of cells of the unbranched filament of *Urspora* sp. (*a*) A vegetative cell, somewhat barrel shaped with a stratified cell wall. (*b*) The filament base with a holdfast consisting of rhizoidal outgrowths developing from the lower cells. The cells are about 30–60 μ m in diameter.

Figure 5-15 *Valonia macrophysa,* an irregular cluster of dark-green cells, the largest of which will measure about 0.5–2 cm in diameter. The rhizoidal structures are from outgrowths of internal lenticular cells that have been formed through segregative cell division.

and secretes a cell wall, resulting in a number of cells. This process is especially evident in *Siphonocladus*. The taxonomic significance of segregative cell division is open to question, and it has been proposed that this order should be combined with the Cladophorales. Three families are presented.

Valoniaceae

Two genera placed in this family, *Valonia* and *Dictyosphaeria*, are pantropical in distribution. *Valonia ventricosa* is frequently used in physiological experiments because of its large size (~3 cm diameter). It is not a true coenocyte but a large multinucleate vesicle with scattered minute lenticular cells at the base. The cells, which are the products of segregative cell division, can grow out to form rhizoids. Other species of *Valonia* consist of many vesicles (Figure 5-15), whereas

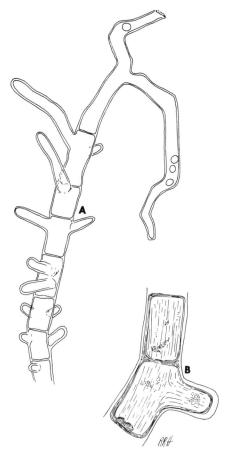

Figure 5-16 (*A*) A small portion of an irregularly branched filament of *Siphonocladus rigidus*. The filaments are 360 μm in diameter, but can reach 1 mm in diameter. (*B*) The irregular branches develop from internal cystlike cells and lenticular cells that are formed through segregative cell division.

Dictyosphaeria forms a mass of vesicles, about 2–5 mm in diameter. Swarmers thought to be gametes are liberated through pores (holes) in the wall. However, fusion has not been observed, and the occurance of a diplontic life history is uncertain. Both genera have been shown to have a closed type of mitosis (Hori and Enomoto, 1978).

Siphonocladaceae

This is a tropical family in which the species are more filamentous than the Valoniaceae. Lenticular cells are absent, but segregative cell division can result in internal spherical cells. The largest genus is *Cladophoropsis,* with some of its species extending into temperate regions. *Siphonocladus* (Figure 5-16) has irregular, club-shaped cells that ultimately become irregularly divided by transverse septa.

Boodleaceae

Two tropical genera, *Boodlea* and *Struvea*, represent this family. The plant body is reticulate, consisting of a network of uniseriate filaments. *Struvea* has a distinct blade and stalk (Figure 5-17), whereas *Boodlea* forms an amorphous, spongiose mat over the rocks. The cells of both genera are cylindrical, the netlike form resulting from anastomosis of tip cells of the filaments. The plants' life histories are unknown, but swarmers have been reported in both genera.

Caulerpales

The order contains coenocytes that may be siphonous (interwoven filaments). The ultrastructure of this order has been extensively studied, especially the genera *Caulerpa* (Dawes and Rhamstine, 1967), *Bryopsis* (Burr and West, 1970), and *Penicillus* (Turner and Friedmann, 1974). Genera of the Caulerpales have numerous discoid-to-fusiform chloroplasts that contain the typical photosynthetic pigment siphonoxanthin. The cell wall is noncellulosic; the structural component is mannan or xylan.

Bryopsidaceae

This family includes genera that have a heteromorphic haplo-diplont life history (Bold and Wynne, 1978). Rietema (1975) has reviewed the various types of life histories and has confirmed the heteromorphic nature of the genera placed in this family.

Derbesia

Members of the genus are filamentous coenocytes producing mats of subdichotomously branched siphons with broad, pluglike septa that are produced by annular growth. The species are worldwide and can be found on rocky, open coasts and estuarine muds of mangrove swamps. Sporangia are cut off by septa and produce zoospores having whorls of flagella (stephanokonts) at the anterior

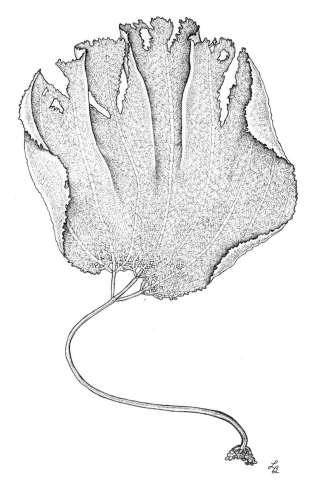

Figure 5-17 *Struvea pulcherrima* is an elegant, bright-green, stalked blade found in deeper tropical (~ 30–50 m) waters. The blade is composed of anastomosizing filaments; the stalk is the main filament. Plants will reach up to 20 cm in width. The stalk is attached by rhizoidal outgrowths at its base.

pole (Figure 5-18). The zoospore of *D. marina* develops into a large vesicle anchored by a perennial rhizoidal system. The vesicle was named and placed in a separate genus, *Halicystis*. In *Halicystis ovalis*, the gametangial thalli are dioecious, producing biflagellated anisogametes from reproductive areas in the vesicles. In a like manner, *Derbesia tenuissima* is related to *Halicystis parvula*. Furthermore, the zoospores of *Derbesia neglecta* (western Mediterranean) develop into *Bryopsis halymeniae*. All three "species-pairs" are examples of heteromorphic life histories. However, in some parts of the world both "partners" are not found, suggesting that each member can reproduce asexually as well.

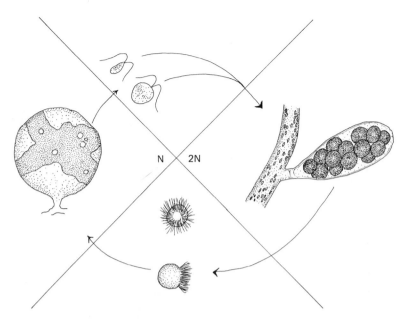

Figure 5-18 The heteromorphic life history of *Derbesia/Halicystis,* in which a portion of the coenocytic (2N) *Derbesia* filament bears a zoosporangia (~30 μm diameter). The multiflagellated zoospores released are haploid because of meiosis, and these grow into the spherical *"Halicystis"* gametophyte (N) that is also coenocytic (~ 1 mm diameter). The gametophyte will then produce anisogametes, which upon fusion form a zygote that germinates into a new *Derbesia* plant. Neumann (1974) has shown a number of variations in the life history of *Derbesia* including an alternation with *Bryopsis*.

Bryopsis

This genus has a distinct main axis with radially or pinnately arranged branchlets (Figure 5-19). The lateral branchlet may cut off from the axis by a transverse septum and produce biflagellated anisogametes. Earlier it was believed that the plants of *Bryopsis* were the only morphological phase in the life history, thus thought to be a typical diplontic life history. Now the coenocytic genus is considered to contain algae that have a heteromorphic life history as well; the diploid phase is *Derbesia* and the haploid phase *Halicystis* or *Bryopsis*. In other species of *Bryopsis* (*B. plumosa, B. hypnoides, B. monoica*), the zygote develops into a small siphonous thallus (previously unnamed diploid phase) that eventually gives rise to multiflagellated zoospores similar to those formed by *Derbesia*. In North Sea populations of *Bryopsis plumosa*, the small siphonous thallus gives rise directly (without zoospore formation) to the pinnate, gametangial thallus characteristic of Bryopsis. In this last case, meiosis may not occur so that the heteromorphic life history may be lacking. Kermarrec (1980) has followed the life history of *Bryopsis plumosa* and *B. hypnoides,* and he determined that meiosis is

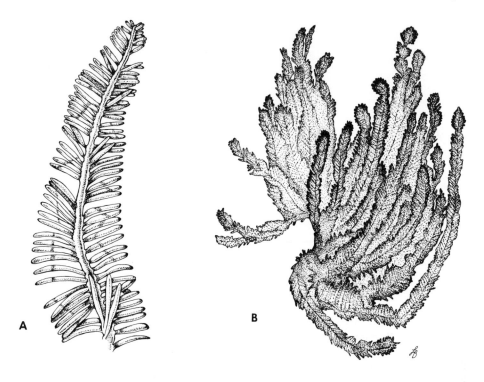

A   B

Figure 5-19 *Bryopsis pennata* has featherlike coenocytic fronds (*A*) and is common in tidal pools and the intertidal fringe of rocky coasts. The plants are typically a dark-green color, and are about 2–5 cm tall (*B*).

gametic and the two morphological phases of the life histories for these species were both diploid. The life histories for *Derbesia* and *Bryopsis* are reviewed by Neumann (1974) and Tietema (1975).

Caulerpaceae

The family is monotypic with coenocytic species of *Caulerpa* distributed throughout the tropical marine waters as well as the temperate waters of Australia.

Caulerpa

The uniqueness of the true coenocytic species lies in the size of many of the plants, some species reaching a meter or more in length. The plants have erect blades (assimilators), horizontal stems, and colorless rhizoids which penetrate the substrate (*Caulerpa cupressoides*, book cover; *C. verticillata,* Figure 5-20). Although there are no cross walls within the siphon or tubular structure, there is an elaborate network of internal trabeculae, or wall struts (Figure 5-20). Species are usually separated based on the morphology of the erect photosynthetic portion, or *assimilators*.

Figure 5-20 *Caulerpa verticillata,* one of the more delicate species of the highly polymorphic and coenocytic genus (see also Figure 3-2). Whereas other species will have blades 10–15 cm in length, this species is usually about 1 cm tall and the branchlets about 0.8 mm in diameter arranged in whorls around the main axis. The rhizome will have a diameter of about 80 μm. Typical of the genus, trabeculae (wall struts) are present throughout the cell interior (*W*). Rhizoids usually occur at the site of assimilator development.

Because they are continuous siphons, these plants are susceptible to wounding. When cut, the blade or rhizome will exude a yellow, sticky mass that hardens to a plug (wound plug) within a few minutes. The same type of wound plug occurs if a dull razor blade is pressed against any part of the plant, even though no cut results. Unlike *Bryopsis*, which produces a proteinaceous wound plug

(Burr and West, 1971), *Caulerpa* produces a carbohydrate wound plug (Dawes and Goddard, 1978).

Reproduction in the species seems to be primarily through fragmentation, although production of anisogametes has been described (Lohr and Dawes, 1974). The gametes are formed in unmodified areas of the plant without separation by cross walls and are liberated through superficial papillae in gelatinous extrusions. The life history appears to be diplontic, although meiosis has not been observed during gamete formation.

Figure 5-21 *Halimeda discoidea,* a segmented and calcified member of the Codiaceae. The geniculae (*G*) are noncalcified regions between segments, and the individual siphons can be teased out in these regions. In deeper waters, the basal, flattened segments can reach a diameter of ∼3 cm. The plants are found attached in soft substrates as well as rocky ones because of the presence of an extensive rhizoidal system. This species is usually less calcified than others and so is more easily decalcified to demonstrate the utricles (see Figure 5-23*B*) that form the surface of the segments.

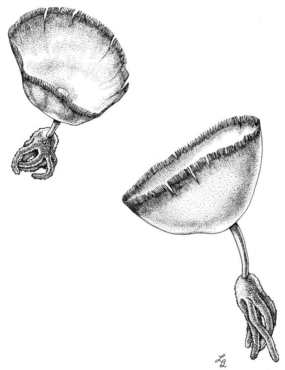

Figure 5-22   Udotea is a member of the Codiaceae with species that form stalked blades or cups as shown here for *U. cyathiformis*. The blade or cup portion of the plant is composed of parallel siphons, whereas the stalk will also have a corticating layer of branchlets. This species reaches lengths of 5–12 cm and is more typical of deep waters in tropical regions. Note the massive rhizoidal base with adherent sand grains permitting attachment in soft substrates.

Codiaceae

The coenocytic filaments are usually woven together to form siphonous plants in this family. The morphology varies from thinly branched plants of *Chlorodesmis* to the dense and calcified thalli of *Halimeda* (Figure 5-21) and *Udotea* (Figure 5-22). Other forms of siphonous green algae include *Penicillus* and *Codium* (Figure 5-23). The calcified forms deposit an aragonite form of calcium carbonate. Reproduction and life histories are poorly known, but where reported the gametes are anisogamous, and the life history is diplontic.

Codium

Codium is a worldwide genus, common throughout the temperate and subtropical regions of the world. The plants consist of coenocytic filaments (siphons) that form the spongy macroscopic plant body (Figure 5-23a). If the siphons fail to aggregate, they may be misidentified as *Derbesia*. The thallus can be prostrate

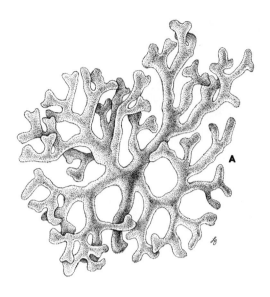

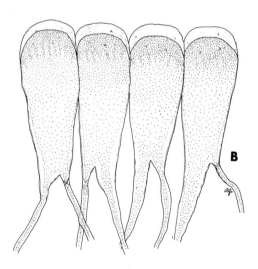

Figure 5-23 *Codium*, as represented here by *C. taylori*, is an uncalcified member of the Codiaceae (*A*). The soft, cylindrical branches may be flattened at the dichotomies. The construction is of siphons with utricles (*B*) that form the cortex of the plant. The plant will reach 10 cm in length. The utricles are 125–300 μm in diameter and 2–5 diameters long, and have thickened end walls in *C. taylori*.

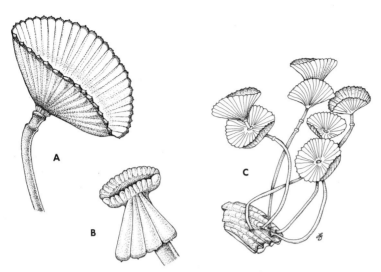

Figure 5-24 *Acetabularia crenulata* has a slender, calcified stipe about 7 cm tall and a concave terminal disk (*C*). The disk is 6–18 mm in diameter and composed of 30 to 80 moderately calcified rays (*A*). A crown of projections called the *corona superior* develops in the center and above the rays (*B*) and a similar set of projections called the *corona inferior* forms below the rays (*A*). The corona superior when young may have a series of colorless hairs, but these are lost when the rays form. The species is usually found attached to shell fragments as shown in (*C*).

and encrusting or erect. The medullary filaments are colorless and produce a cortex of inflated utricles of distinctive shape and end-wall thickening (Figure 5-23*b*). The shape, terminal-wall thickness and size of the utricles are important taxonomically. The cytoplasm contains numerous nuclei and small discoid chloroplasts lacking pyrenoids. The life history is diplontic. Gametangia are borne laterally on the utricles. The saclike gametangia are separated from the utricles by an annular ingrowth of the cell wall. The gametangia produce anisogametes.

Dasycladales

This order contains calcified members with fossil records of about 10 genera dating back to the Ordovician. The present forms are all tropical or subtropical. All have whorled branching and produce cysts containing haploid secondary nuclei from which the anisogametes are released. The life history is diplontic. The family Dasycladaceae includes all living genera that have radial symmetry with an erect main axis and one to many whorls of lateral branches. The thallus initially has a large primary nucleus, but through nuclear fragmentation (amitosis) the coenocytic condition develops prior to reproduction. Meiosis may occur during or just after fragmentation. Koop (1979) suggests that meiosis occurs in the primary nucleus. The genus *Acetabularia* (Figure 5-24) has been used extensively in stud-

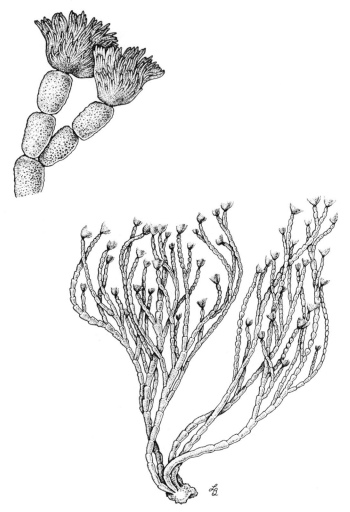

Figure 5-25 *Cymopolia barbata,* an interesting tropical member of the Dasycladales in that the segments are heavily calcified so the typical whorls of branchlets are masked. Only the terminal whorls, where calcification is lacking, are visible. Members of the species will reach 20 cm in length, although smaller sizes are more typical.

ies of nuclear control of cell morphology (Gibor, 1966). Other common genera include *Cymopolia* (Figure 5-25), which is heavily calcified and has numerous whorls of branchlets, and *Batophora*. Little is known about reproduction and cytology for most genera.

Acetabularia

Members of this genus have an erect calcified stalk and a rhizoidal base. From the stalk apex, whorls of deciduous hairlike appendages are formed and shed, leaving persistent scars. The stalk may die back to the perennial rhizoidal base and regenerate two to three times before becoming reproductive. At reproduction, the nucleus divides (see above) through fragmentation, meiosis, and mitosis. The daughter nuclei move into a whorl of developing gametangial rays (Figure 5-24). Uninucleate cysts, which may be calcified, are formed within each gametangium. The cysts are liberated and may act as a resting stage for 2–3 mo. before the nucleus undergoes mitosis and produces biflagellate isogametes.

REFERENCES

Bisson, M. A. and G. O. Kirst. 1980. *Lamprothamnium*, a curyhaline charophyte. I. Osmotic relations and membrane potential at steady state. *J. Exp. Bot.* **31**: 1223–1235.

Blackman, F. F. and A. G. Tansley. 1903. A revision of the classification of the green algae. *New Phytol.* **1**: 17–224.

Bold, H. C. and M. H. Wynne. 1978. *Introduction to the Algae. Structure and Reproduction.* Prentice-Hall, Englewood Cliffs, N.J.

Bråten, T. and A. Lovlie. 1968. On the ultrastructure of vegetative and sporulating cells of the multicellular green alga *Ulva mutabilis Foyn. Nytt. Mag. Bot.* **15**: 209–219.

Burr, F. A. and J. A. West. 1970. Light and electron microscope observations on the vegetative and reproductive structures of *Bryopsis hypnoides. Phycologia* **9**: 17–37.

Burr, F. A. and J. A. West. 1971. Protein bodies in *Bryopsis hypnoides,* their relationship to the wound healing and branch septum development. *J. Ultrast. Res.* **35**: 476–498.

Butcher, R. W. 1959. *An Introductory Account of the Smaller Algae of British Coastal Waters. Part I: Introduction and Chlorophyceae.* Ministry of Agriculture, Fisheries and Food. HMS Stationary Office, London.

Chan, K-Y, S. L. L. Wong, and M. H. Wong. 1978. Observation on *Chaetomorpha brachygona* Harv. (Chlorophyta, Cladophorales). 1. Ultrastructure of the vegetative cells. *Phycologia* **17**: 419–429.

Chapman, V. J. 1954. The siphonocladales. *Bull. Torrey Bot. Club* **81**: 76–82.

Dawes, C. J. 1966. A light and electron microscope study of algal cell walls. II. Chlorophyta. *Ohio J. Sci.* **66**: 317–326.

Dawes, C. J. and R. H. Goddard. 1978. Chemical composition of the wound plug and entire plants for species of the coenocytic green alga, *Caulerpa. J. Exp. Mar. Biol. Ecol.* **35**: 259–263.

Dawes, C. J. and E. Rhamstine. 1967. An ultrastructural study of the giant green algal coenocyte *Caulerpa prolifera. J. Phycol.* **3**: 117–127.

Dube, M. A. 1967. On the life history of *Monostroma fuscum* (Postles *et* Ruprecht) Wittrock. *J. Phycol.* **3**: 64–73.

Edwards, P. 1975. Evidence for a relationship between the genera *Rosenvingiella* and *Prasiola* (Chlorophyta). *Br. Phycol. J.* **10**: 291–297.

Friedmann, I. 1969. Geographic and environmental factors controlling life history and morphology in *Prasiola stipitata* Suhr. *Osterr. Bot. Z.* **116**: 203–225.

Gibor, A. 1966. *Acetabularia*: a useful giant cell. *Sci. Amer.* **215**: 118–124.

Hori, T. 1973. Comparative studies of pyrenoid ultrastructure in algae of the *Monostroma* complex. *J. Phycol.* **9**: 190–199.

Hori, T. and S. Enomoto. 1978. Developmental cytology of *Dictyosphaeria cavernosa*. II. Nuclear division during zooid formation. *Bot. Mar.* **21**: 477–481.

Hudson, P. R. and J. R. Waaland. 1974. Ultrastructure of mitosis and cytokinesis in the multinucleate green alga *Acrosophonia*. *J. Cell. Biol.* **62**: 274–294.

Kermarrac, A. 1980. Sur la place de la meiose dans le cycle de deux Chlorophycees marines: *Bryopsis plumosa* (Huds.) C. Ag. et *Bryopsis hypnoides* Lamouroux (Codiales). Cahs. Biol. Mat. **21**: 443–446.

Koop, H. U. 1979. The life cycle of *Acetabularia* (Dasycladales, Chlorophyceae): A compilation of evidence for meiosis in the primary nucleus. *Protoplasma* **100**: 353–366.

Kornmann, P. 1973. Codiolophyceae, a new class of Chlorophyta. *Helgo. Wiss. Meeresunt.* **25**: 1–13.

Lembi, C. A. 1980. Unicellular chlorophytes. *In* Cox, E. R. (ed.) *Phytoflagellates*, Elsevier/North Holland, New York, pp. 5–60.

Lohr, C. A. and C. J. Dawes. 1974. Light and electron microscope studies on the gametes of the green alga *Caulerpa* (Chlorophyta, Siphonales). *Fla. Scientist* **37**: 45–49.

Lovlie, A. and T. Bråten. 1970. On mitosis in the multicellular alga *Ulva mutabilis* Foyn. *J. Cell Sci.* **6**: 109–129.

Manton, I. and M. Parke. 1960. Further observations on small green flagellates with special reference to possible relative of *Chromulina pusilla* Butcher. *J. Mar. Biol. Assoc. U.K.* **39**: 275–298.

Mattox, K. R. and K. D. Stewart. 1977. Cell division in the scaly green flagellate *Heteromastix angulata* and its bearing on the origin of the Chlorophyceae. *Amer. J. Bot.* **64**: 931–945.

McArthur, D. M. and B. L. Moss. 1978. Ultrastructural studies of vegetative cells, mitosis and cell division in *Enteromorpha intestinalis* (L.) Link. *Br. Phycol. J.* **13**: 255–267.

McDonald, K. L. and J. D. Pickett-Heaps. 1976. Ultrastructure and differentiation in *Cladophora glomerata*. I. Cell division. *Amer. J. Bot.* **63**: 592–601.

Neumann, K. 1974. Zur Entwicklungsgeschichte und Systematik der Siphonalen Grunalgen *Derbesia* und *Bryopsis*. *Bot. Mar.* **17**: 176–185.

Norris, R. E. 1980. Prasinophytes. In Cox, E. R. (ed.). *Phytoflagellates*, Elsevier/North Holland, New York.

Norris, R. E. and R. N. Pienaar. 1978. Comparative fine-structural studies on five marine species of *Pyramimonas* (Chlorophyta, Prasinophyceae). *Phycologia* **17**: 41–51.

Pickett-Heaps, J. D. 1973. Cell division in *Tetraspora*. *Ann. Bot.* **37**: 1017–1025.

Rietema, H. 1975. Comparative investigations on the life-histories and reproduction of some species in the siphoneous green algal genera *Bryopsis* and *Derbesia*. Ph.D. dissertation, Rijksuniversiteit, Groningen.

Rhyne, C. F. 1973. Field and experimental studies on the systematics and ecology of *Ulva curvata* and *Ulva rotunda*. Univ. N. C. Sea Grant Publ. UNC-SG-73-09.

Scott, J. L. and K. W.Bullock. 1976. Ultrastructure of cell division in *Cladophora*. Pregametangial cell division in the haploid generation of *Cladophora flexuosa*. *Can. J. Bot.* **54**: 1546–1560.

Shyam, R. 1980. On the life cycle, cytology and taxonomy of *Cladophora callicoma* from India. *Amer. J. Bot.* **67**: 619–624.

Sluiman, H. J., D. R. Roberts, K. D. Stewart, and K. R. Mattox. 1980. Comparative cytology and taxonomy of the *Ulvaphyceae*. I. The zoospore of *Ulothrix zonata* (Chlorophyta). *J. Phycol.* **16**: 537–545.

Starr, R. E. 1980. Colonial Chlorophytes. In Cox, E. R. (ed.). *Phytoflagellates*, Elsevier/North Holland, New York, pp. 147–164.

Stewart, K. D. and K. R. Mattox. 1975. Comparative cytology, evolution and classification of the green algae with some consideration of the origin of other organisms with chlorophylls *a* and *b*. *Bot. Rev.* **41**: 104–135.

Trainor, F. R. 1978. *Introductory Phycology*. Wiley, New York.

Turner, J. B. and E. I. Friedmann. 1974. Fine structure of capitular filaments in the coenocytic green alga *Penicillus*. *J. Phycol.* **10**: 125–134.

Underbrink, A. G. and A. H. Sparrow, 1968. The fine structure of the alga *Brachiomonas submarina*. *Bot. Gaz.* **129**: 259–266.

Van den Hoek, C. 1979. The phytography of *Cladophora* (Chlorophyceae) in the northern Atlantic Ocean, in comparison to that of other benthic algal species. Helgo. Wiss. Meeresunt. **32**: 374–393.

Velichko, I. M. 1980. The pigment complex of green filamentous algae. *Hydrobiol. J.* **16**: 35–43.

Wilkinson, M. and E. M. Burrows. 1972. An experimental taxonomic study of the algae confused under the name *Gomontia polyrhiza*. *J. Mar. Biol. Assoc. U. K.* **52**: 49–57.

Yokohama, Y., A. Kageyama, T. Ikawa, and S. Shimura. 1977. A carotenoid characteristic of chlorophycean seaweeds living in deep coastal waters. *Bot. Mar.* **20**: 433–436.

CHAPTER 6

Phaeophyta

The 1500 species of brown algae are almost exclusively marine, having only three freshwater species. The largest seaweeds are brown algae, and they occur primarily in temperate and cold water regions, where they often dominate the subtidal and intertidal zones. The brown algae are also the most complex in morphological and anatomical development. Many species are harvested as direct foods and for the extract alginic acid (Chapter 2).

CYTOLOGICAL FEATURES

Pigments

The brown algae contain both chlorophylls *a* and *c* along with α and β carotene and an abundance of xanthophylls, especially fucoxanthin (also flavoxanthin and violaxanthin). The xanthophylls give the distinctive brown coloration to this group of seaweeds.

Cytological Structure

The vegetative eukaryotic cell shows a number of common features among brown algae (Figure 6-1); also Figure 3-3). All have uninucleate cells and chloroplasts with thylakoids in bands of three with a girdling lamella just inside the chloroplast. The relationship between the outer chloroplast envelope membrane and the endoplasmic reticulum which can be continuous with the outer nuclear envelope, has been described by Bouck (1965). He also showed that the Golgi apparatus is clearly associated with the nuclear envelope. The chloroplasts may be large and few per cell (laminate or stellate) or small and numerous (discoid). Pyrenoids are present in various members throughout the division (Hori, 1972). Numerous vesicles containing polyphenolic by-products of photosynthesis may be present. These are called physodes or fucosan granules (Ragen, 1976) and are especially common in intertidal species such as *Fucus*. Physodes appear to have a number of functions such as excluding excessive sunlight to protect chloroplasts, acting as

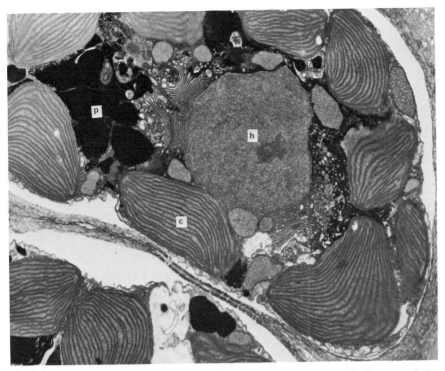

Figure 6-1 An epidermal cell of *Sargassum filipendula* as seen with the transmission electron microscope demonstrating the typical brown algal cytology. A central nucleus (n) surrounded by Golgi bodies and chloroplasts (*c*) is seen in the cell. In addition, physodes (p), densely staining bodies, are visible toward the inside of the cell. Plasmodesmata traverse the wall separating the two epidermal cells. × 11,250. (Courtesy of Wayne Fagerberg, Southern Methodist University, Dallas, Texas.)

antifoulants (Sieburth and Conover, 1965), and forming the wound plug during traumatic excision (Fagerberg and Dawes, 1976).

Motility

All of the brown algae thus far studied produce a motile (heterokont) cell, either zoospores or gametes or both (Figure 6-2). These reproductive cells are flagellated, having a pair of laterally inserted flagella; the longer posterior flagellum is smooth (acronematic) and the anterior flagellum is hairy (pantonematic). The hairs (mastigonemes) are arranged in two bilateral rows along the flagellum. Their structure has been compared for six species and appears uniform (Loiseaux and West, 1970). The motile cells are usually pyriform or kidney-bean shaped. Some differences are known with regard to motile cell morphology, but these can usually be intrepreted to be derived forms (e.g., Dictyotales, Fucales).

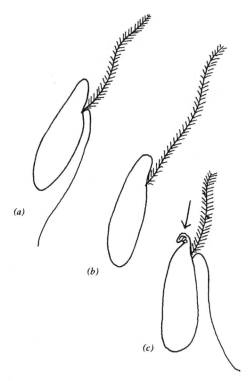

Figure 6-2 Three types of motile cells found in the Phaeophyta. (*a*) The typical motile cell common to most brown algae. (*b*) The modified sperm cell characteristic of the Dictyotales. (*c*) The specialized sperm cell of the Fucales, the arrow pointing to the proboscis. Note that the pantonematic flagellum is shorter than the acronematic flagellum in (*C*). (After Bold and Wynne, 1978.)

Cell Wall Structure

In all of the brown algae examined, the structural component is cellulose, although it may account for only a small portion of the wall by weight (Dawes *et al.*, 1961). The cellulose microfibrils are arranged in alternating lamellae of parallel microfibrils. These lamellae appear to alternate with layers of amorphous material that may be alginic acid as well as other polysaccharides. Plasmodesmata are common and usually are found to penetrate the cell wall in specific regions, producing well-defined pit fields seen in both light and electron micrographs (see Figure 3-14).

Food Storage

The reserve food is a $\beta = (1 \to 3) =$ linked glucan that has $\beta = (1 \to 6)$ side chain likages and is called laminarin (Figure 3-11). Mannitol, an alcohol, is also present, in abundance in some cases.

MORPHOLOGY

The simplest forms are filamentous branching plants such as *Ectocarpus*. Filamentous construction is considered the basic type of growth pattern in brown algae from which parenchymatous plants have been derived. The orders that are considered to contain advanced species of brown algae are characterized by highly differentiated forms such as kelps (Laminariales) and rockweeds (Fucales). In these orders the plant is differentiated into a well-developed attachment (hapteron, holdfast), stalk (stipe, stem), blades (lamina, leaves), apical (Fucales) or intercalary (Laminariales) meristems, and flotation devices (bladders, floats).

In addition, a high degree of tissue differentiation can be found in these same

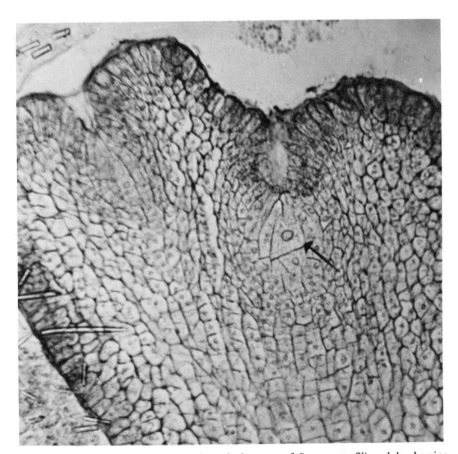

Figure 6-3 A longitudinal section through the apex of *Sargassum filipendula* showing the apical cell (arrow) and surrounding meristematic cells. A branch "bud" has begun to develop on the left side of the main axis and contains the same type of apical meristem, but the plane of section does not reveal the central cell. Note the pit at the tip of both the main axis and the branch, a feature typical of the Fucales. × 75.25 (Courtesy of Wayne Fagerberg, Southern Methodist University, Dallas, Texas.)

orders. A well-differentiated apical meristem with distinctive meristematic zones (Figure 6-3) is characteristic of the Fucales (Fagerberg and Dawes, 1976). Both orders have a high degree of tissue differentiation. The mature stem or leaf of a typical fucalean alga (*Sargassum*) will include an epidermal layer, an outer photosynthetic cortex, an inner storage cortex, and a central conducting medulla. Translocation of photosynthates in the medulla has been demonstrated for a number of kelps (Nicholson and Briggs, 1972; Buggeln, 1976; Buggeln and Lucken, 1979).

The type of apical growth will greatly influence the morphology and anatomy of the alga and thus the location of cell division is used as a taxonomic character. The main site of cell division can be apical, and distinct apical cells may be evident (apical growth); or the meristem can occur at the base of a hair, resulting in an occasional division that continues hair growth (trichothallic growth). The meristem can be in a specific site within the plant (intercalary growth called a meristoderm) or distributed throughout the plant (diffuse growth).

REPRODUCTION

Asexual reproduction in the simpler forms of brown algae is commonly through zoospore production or fragmentation. Sexual reproduction is well known for many of the brown algae, and except for the Fucales, the typical life history is haplodiplontic (see Figure 3-25). The sporophytic generation reproduces the gametophytic generation through zoospores that are the result of meiosis (but see Fucales). Some orders are characterized by isomorphic sporophytes and gametophytes, while others have heteromorphic, dominant sporophytes. Members of the order Fucales have a true diplontic life history (Fig. 3-23). All forms of gametes can be found in the Phaeophyta; reproduction thus includes isogamy (Chordariales, Dictyosiphonales, Ectocarpales), anisogamy (Cutlariales, Ralfsiales, Sphacelariales) and oogomy or eggs and sperm (Desmarestiales, Dictyotales, Fucales, Laminariales, Sporochnales, Tilopterdiales).

Two main types of sporangia are found in brown algae although there are a number of modifications that will be noted for specific groups. Multicellular, or plurilocular, sporangia are found on both haploid (gametophytes) and diploid (sporophytes) plants. The plurilocular sporangia produced by gametophytes will release gametes through mitosis. In many genera, especially in the simpler species, the gametes can also parthenogenetically grow into new gametophytes, allowing a method of asexual reproduction. If plurilocular sporangia are present on the sporophyte, the resulting diploid zoospores (through mitosis) will germinate into more sporophytes.

Unilocular sporangia are unicellular, the site of meiosis, and usually only found on the diploid sporophytes (but see Müller, 1972). The unilocular sporangia produce through meiosis four or more haploid zoospores (except in the Dictyotales, where nonmotile aplanospores are produced) that germinate into gametophytes (usually dioecious). Whereas the diploid plant can have both plurilocular and unilocular sporangia, a haploid plant can produce only plurilocular sporangia. The reproductive cycles are influenced by photoperiod, temperature, nutrients, and salinity (e.g., Florida Sargassum; Prince and O'Neal, 1979). Some examples of the importance of physical and chemical factors in reproduction are presented in Chapter 14.

DISTRIBUTION AND ECOLOGY

Most of the larger brown algae are cold-water species, forming large kelp beds along the coasts of the North Atlantic (*Laminaria, Alaria*) and the North Pacific (*Nereocystis, Macrocystis*). The order Fucales is especially well represented in the cooler waters of the southern hemisphere (Southern Australia, Tasmania, New Zealand, Africa), with some genera found worldwide in the intertidal zones of temperate waters. A wide variety of brown algae also occur in subtropical and tropical waters; members of the Dictyotales (*Dictyota, Padina*) and Fucales (*Sargassum, Cystoseira*) are most common. The importance of the larger benthic plants in the marine environment is evident, as they provide habitats (kelp forests, rockweed beds) and food for herbivores (see Chapter 15).

TAXONOMY

Traditional classification divided the division Phaeophyta into three classes or subclasses based on life history, plant construction, and growth pattern (see Papenfuss, 1951 for explanation). Recent studies discount the importance of one or more of these criteria, and the classes have been dropped (Trainor, 1978; Bold and Wynne, 1978). The filamentous forms are considered to be the most primitive types because of their isomorphic alternation of sporophytic and gametophytic generations and a high degree of plasticity in their life histories because of asexual reproduction. The more advanced brown algae, the kelps and rockweeds, lack asexual spore production and exhibit reduction of the gametophytic generation, as well as oogamous sexual reproduction. A key to all 12 orders of brown algae is given in Table 6-1.

Table 6-1   A Key to the Orders of the Phaeophyta

1	Life cycle diplontic, lack a gametophytic phase.................................**Fucales**	
1	Life cycle haplodiplontic, having both a gametophytic and a sporophytic phase... **2**	
	2	Plants filamentous or pseudoparenchymatous in construction.................... **3**
	2	Plants parenchymatous in construction, at least in one phase of the life history **7**
3	Life history isomorphic to slightly heteromorphic....................................... **4**	
3	Life history heteromorphic having a small gametophyte and dominant and larger	

Table 6-1   Continued

sporophyte **5**

   4 Spores germinating to produce a loosely filamentous development, more than one plastid per cell, pyrenoids present .... **Ectocarpales**

   4 Spores germinated to produce a discoid-type of development, a single plastid per cell, pyrenoids absent... **Ralfsiales**

5 Sexual reproduction having eggs and sperm (oogamous)............................. **6**

5 Sexual reproduction isogamous, gametes look alike ....**Chordariales**

   6 Growth trichothallic a single filament terminating each axis (uniaxial organization)... **Desmarestiales**

   6 Growth trichothallic, a conspicuous tuft of filaments terminating each axis and a meristematic zone found at the base of the filaments................ **Sporochnales**

7 Isomorphic life history (except the derived form of *Cutleria*), both phases parenchymatous in construction.. **8**

7 Heteromorphic life history, only one phase of parenchymatous construction, the other phase being filamentous or pseudoparenchymatous **11**

   8 Trichothallic growth.. **9**

   8 Growth by apical cell(s)... **10**

9 Filamentous construction, uniseriate apical regions, multiseriate basal regions, forming quadrinucleate monospores on diploid plants..........................**Tilopteridales**

9 Nonfilamentous construction, forming only unilocular sporangia on the diploid plant ....**Cutleriales**

   10 Plants erect and flattened, 4–8 nonmotile spores formed per unilocular sporangium, oogamous sexual reproduction...................................... **Dictyotales**

   10 Plants erect and terete, numerous motile spores formed per unilocular sporangium, anisogamous sexual reproduction .... **Sphacelariales**

11 Vegetative cells with one platelike chloroplast and conspicuous pyrenoid, larger plant bearing plurilocular sporangia only....................................... **Scytosiphonales**

11 Vegetative cells with many chloroplasts per cell, with or without pyrenoids, larger plant having unilocular sporangia... **12**

   12 Growth apical or diffuse, isogamous sexual reproduction .....**Dictyosiphonales**

   12 Growth intercalary with a defined meristem, oogamous sexual reproduction.... ....**Laminariales**

Source: Adapted from Bold and Wynne (1978).

Ectocarpales

This order is usually considered to be the simplest and most primitive because its members are uniseriately branched filaments with a heterotrichous organization. Growth is usually diffuse rather than from a distinct apical or intercalary meristem. Most members exhibit an isomorphic alternation of a gametophytic and a sporophytic generation. However, a few species have been found to have a slightly larger sporophyte, and this can be construed as a heteromorphic haplodiplontic life history. All are placed in the single family.

Ectocarpaceae

The species of this family are found throughout the world from cold water to the tropics. Species are separated by the shapes of the chloroplasts and plurilocular sporangia. For example, *Pilayella* has intercalary sporangia; both plurilocular and unilocular sporangia occur among vegetative cells of the filament. Members of other genera have terminal sporangia. The chloroplasts are ribbon-shaped (*Ectocarpus*), multilobed, radiate (*Bachelotia*), or discoid (*Giffordia*). An excellent ultrastructural study was carried out on *Pilayella* by Markey and Wilce (1976) demonstrating the reproductive stages in the life cycle.

Ectocarpus

These plants are simple branched filaments that grow by diffuse or trichothallic growth. The chloroplasts are large laminate forms with few per cell. The genus has been used as the example of a primitive brown alga in most botany texts because it has an isomorphic haplodiplontic life history; the life history is plastic, the plurilocular sporangia are found on both the haploid and diploid plants and are capable of reproducing the same phase (asexual reproduction). (Figure 6-4). The zygote produces a diploid plant that bears both plurilocular and unilocular sporangia. The plurilocular sporangia produce swarmers that germinate only into more diploid plants. The unilocular sporangia produce multiples of four haploid zoospores through meiosis, and these germinate to produce the gametophytes. The gametes are isogamous or anisogamous and are produced from the plurilocular sporangia of gametophytes.

Sporophytes can be smaller or more robust or even turn out to be a distinctly different species (heteromorphic life histories). Müller (1972) has shown that *E. siliculosus* can have a very complex life history with diploid or haploid "gametophytes" and haploid, diploid, or tetraploid "sporophytes." Müller (1979) has also demonstrated that gametes of *E. siliculosus* plants collected throughout the North Atlantic will fuse, and this supports the taxonomic criteria of the species.

Ralfsiales

Most of the genera found in these families are crustose and all members start out as a discoid type of structure when a spore germinates. Some members have a single large stellate chloroplast in each cell. Gametes are anisogamous. Of the three families placed in this order, only the Ralfsiaceae will be discussed here.

Ralfsiaceae

Members of this family are discoid pads ranging from cold to warm temperature waters. The plants can be easily overlooked since they appear as brown-to-golden blotches or spots on the rocks. Removal usually requires careful scraping of the rocks.

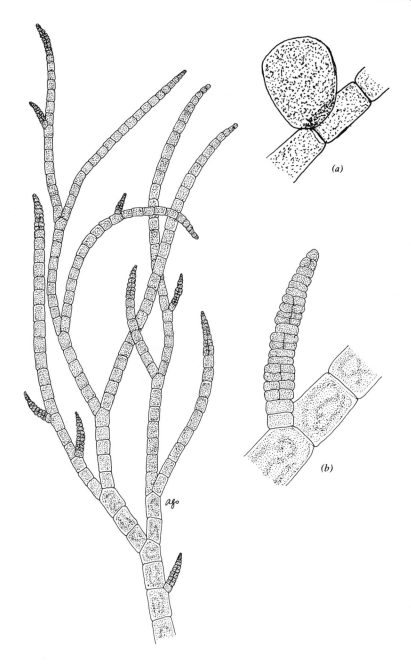

Figure 6-4 A branch of *Ectocarpus siliculosus* bearing immature plurilocular sporangia. (*a*) A unilocular sporangium. (*b*) A mature plurilocular sporangium. The conical plurilocular sporangia are 50–60 μm in length, the unilocular sporangia are about 30–60 μm in diameter. Cells of the main axis measure 40–60 μm in diameter.

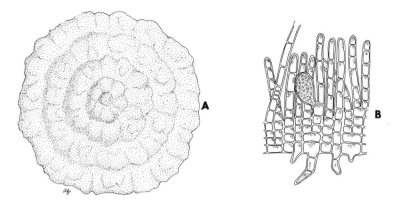

Figure 6-5 *Ralfsia*, a crustose genus that is common on rocks of colder waters in the North Atlantic and Pacific (*A*). The species shown here, *R. pacifica*, forms small (2–5 cm) dark-to-golden-brown crusts and consists of compactly arranged filaments. The cross section (*B*) reveals a unilocular sporangium that has developed from the side of an erect filament. The unilocular sporangia are spherical in *R. pacifica*, whereas in other species they can be elongate, up to 60 μm long.

Ralfsia

The crustose thallus of *R. pacifica* forms patches on the rocks and is closely adherent to the substratum (Figure 6-5). The plant consists of a basal layer of radiating filaments, each cell of which produces an erect filament. The erect filaments form a tightly compacted crust. Both sporangial and gametangial plants are found, suggesting a typical isomorphic life history, but gametic fusion has not been confirmed even though motile cells are released. Sporangial plants bear lateral unilocular sporangia. The plurilocular sporangia of the gametophyte are modified vegetative cells of the filaments and may be uniseriate or multiseriate. Culture studies indicate that at least some species of *''Ralfsia''* are the alternate life history phases of other advanced brown algae found in the Scytosiphonales.

Sphacelariales

Members of this order may have either derived filamentous or parenchymatous construction, depending on whether one is studying the smaller forms (*Sphacelaria*) or the larger, heavily corticated species (*Cladostephus*). The plants have a large apical cell and the branching is radial. The lenticular chloroplasts lack pyrenoids. The fine structure of *Sphacelaria* has been described in a number of papers (Galatis *et al.*, 1977). The life history is isomorphic and sexual reproduction will vary from isogamy (*Cladostephus*), anisogamy (*Sphacelaria*), to oogamy (*Halopteris*).

Sphacelaria

Species of this genus are common throughout the North Altantic, including the Caribbean. The plants are basically filamentous but become multiseriate in older

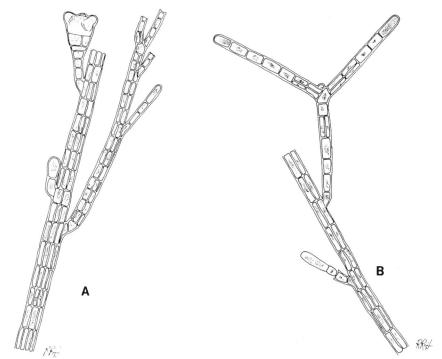

Figure 6-6   Two examples of propagule for the genus *Sphacelaria* (*A*) The compact propagule characteristic of *S. tribuloides* (× 175). (*B*) The triradiate propagule produced by *S.
fucigera* (× 175). Both species, as with the genus, have large distinctive apical cells, seen
in the lateral branches, and are multiseriate in mature regions of the axes.

portions by longitudional cell divisions. The prominent apical cell and the presence of specialized reproductive branchlets, called propagule, distinguish this genus (Figure 6-6). The propagule are short branch systems that can detach and
develop new plants (compare Figure 6-6*a* and 6-6*b*). The slightly larger sporophytes produce unilocular sporangia that are the site of meiosis. Female gametophytes are larger than male gametophytes and produce larger plurilocular
sporangia (gametangia), which in turn produce larger gametes (anisogamy). According to Colijn and van den Hoek (1971) temperature and photoperiod are critical in the control of sporangial development as well as propagulae production.

Tilopteridales

This small and poorly known order contains a few genera of filamentous-to-mul-
tiseriate brown algae. Growth is trichothallic; the cell divisions are intercalary.
Monosporangia occur on both gametophytes (having one large nucleus) and sporophytes (four nuclei), the latter being the site of meiosis. At least in *Haplospora
globosa* (Figure 6-7), the gametophyte produces eggs and sperm and the life history is isomorphic. *Tilopteris mertensii* is found in the North Atlantic along the

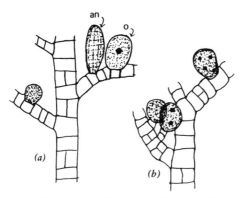

Figure 6-7 *Haplosphora globosa*, a member of the Tilopteridales, resembles *Sphacelaria* in that it is multiseriate in older sections of the plant. However, the gametophytes (*a*) bear oogonia (*o*, arrow) and antheridia (*an*, arrow); the latter plurilocular sporangia produce sperm. Mature axes will be about 20 μm in diameter, the unilocular sporangia (*b*) about 40 μm in diameter.

New England coast and apparently reproduces only asexually through monospores, but the plants do bear both monosporangia and plurilocular sporangia (South and Hill, 1971).

Cutleriales

Only two genera are included in this small order, *Cutleria* and *Zanardinia*. Both are characterized by trichothallic growth, pseudoparenchymatous structure, and anisogamy. A third possible genus, *Microzonia,* is of questionable taxonomy with only sporangial plants known. *Cutleria* has an alternation of heteromorphic phases, whereas *Zanardinia* has an alternation of isomorphic phases.

Cutleria

The gametophyte of *Cutleria* is an erect, highly divided strap-shaped thallus that develops from a perennating base, whereas the sporophyte is a prostrate, crustose plant (Figure 6-8). The gametophytes produce two sizes of plurilocular organs and anisogametes. Unilocular sporangia occur on the surface of the sporophyte. A light and electron microscope study on the female plant of *Cutleria* revealed typical brown alga cytology; mitosis is carried out by means of a closed spindle (La-Claire and West, 1978).

Dictyotales

This order contains one family in which the algae have blades with parenchymatous construction and distinct apical cells. The group is pantropical in distribution, containing about 20 genera. The life histories in the Dictyotaceae are isomorphic haplodiplontic types. Sexual reproduction is oogamous with uniflagellated sperm. Unilocular sporangia produces nonmotile spores through meiosis, the

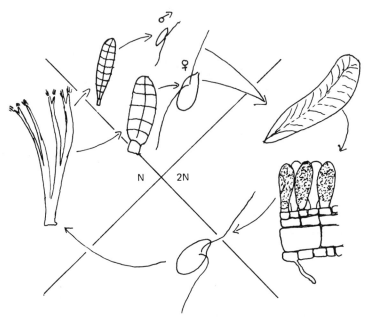

Figure 6-8 The life history of *Cutleria multifida*. The diploid or sporophytic stage is a small (1–5 cm diameter) crustose brown-to-black plant originally named *Aglaozonia*, which produces unilocular sporangia on the upper surface. The haploid zoospores resulting from meiosis in the unilocular sporangia give rise to the erect, bladed gametophyte (10–15 cm tall), which produces two types of plurilocular sporangia, these produce anisogametes. Growth is trichothallic; thus a conspicuous tuft of hair is evident, especially at the apical region of the gametophyte. The sporophyte has a marginal meristem more similar to members of the Dictyotales.

only such case in the Phaeophyta. The genera are distinguished by the number of apical cells per branch, the number of medullary cells seen in cross section, and the number of spores produced per sporangia. Examples include the fan-shaped blade of *Padina*, which has a curled margin, a marginal row of apical cells, and may have some superficial deposition of calcium carbonate (limestone). Another example is *Zonaria*, a common genus of the west coast of the United States, having a marginal row of apical cells and forming fan-to-strap-shaped blades. The cell structure of *Padina vickersiae* is typical of brown algae; no major cytological distinctions are evident between the vegetative cells of the gametophytes and sporophytes of *Padina* (Fagerberg and Dawes, 1973). The nuclear envelope does not disappear during mitosis in *Dictyota*.

Dictyota

The plant is flat, dichotomously branched, and has a single apical cell at the tip of each branch (Figure 6-9). The cortex has only one layer of chloroplast-containing cells on each surface covering the single layer of larger, quadrate medullary cells. The plant is attached by a cluster of rhizoids at the base. Diploid sporo-

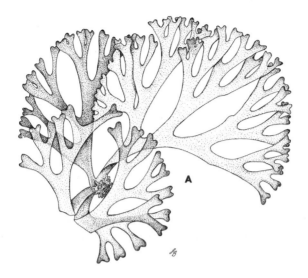

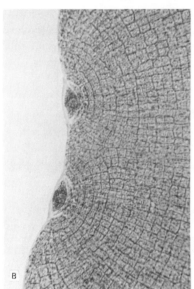

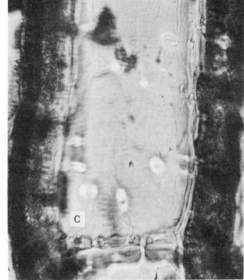

Figure 6-9 *Dictyota dichotoma,* probably the most widespread species of the genus, being found throughout tropical and subtropical waters of the world. The dichotomously branched, flattened thallus (*A*) grows by means of a single apical cell at the tip of each branch (*B*). The thallus is parenchymatous, having a single layer of large, rectangular medullary cells and a covering layer of epidermal cells on both surfaces (*C*). Pits and pit fields are evident in the cross section (*C*). The plant will reach 2–3 dm in length but is usually smaller. (*B*) × 360. (*C*) × 215. (Dawes *et al.*, 1961.)

phytes bear sporangia in irregular patches over the surface of the blade; in the sporangia four spores are produced through meiosis. Sterile hairs, or paraphyses, accompany each sporangial series. The gametophytes of most species are dioecious and bear either antheridia (plurilocular sporangia) or oogonia that are modified plurilocular sporangia in which only one egg is produced. The sperm are uniflagellate, lacking chloroplasts; the eggs are nonmotile with numerous plastids. The sperm, although uniflagellated, has two basal bodies, suggesting that the posterior flagellum has been evolutionarily lost. Williams (1905) found that reproduction is periodic, depending on the tidal amplitude, but more recently Foster *et al.* (1972) could not confirm this.

Chordariales

This is a large and diverse order with about four families. The members are characterized by heteromorphic life histories, having small gametophytes and large sporophytes. The plant is of filamentous to pseudoparenchymatous construction, often slimy because of its mucilaginous structure and loose construction. Isogamy is common in the few cases where sexual reproduction is known. The zygote may produce a microdiploid thallus, the *plethysmothallus,* which can either asexually reproduce itself through plurilocular sporangia or develop directly into the typical macroscopic sporophyte. Ultrastructural studies carried out on members of this order, such as *Eudesme* (Cole, 1969), show typical brown algal cytology.

Myrionemataceae

This family consists of small epiphytic algal discs with heterotrichous erect filaments. Each cell of the basal layer will produce an erect filament or a reproductive organ. Life histories with typical heteromorphic haplodiplonts as well as modifications are known.

Myrionema

The plant is epiphytic on other algae and seagrasses, forming a monostromatic basal layer from which vertical filaments develop (Figure 6-10). Plurilocular and unilocular sporangia are found on separate plants. Although microscopic, the unilocular sporangia-bearing plants (sporophytes) are larger than the plurilocular-bearing plants (gametophytes); hence the life history can be considered to be heteromorphic. Not all "species" produce plurilocular sporangia, suggesting mitosis and not meiosis may occur in the unilocular sporangia.

Elachistaceae

The pincushion-shaped thalli in this family are the result of a compact pseudoparenchyma, so that the heterotrichous nature of the plant is not always evident.

Elachista

The small spheroid-to-cushion-shaped plants of this genus occur as epiphytes on larger brown algae such as *Fucus* and *Ascophyllum*. The basal filaments produce

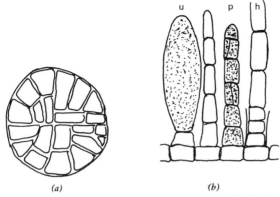

Figure 6-10 A young sporophyte of *Myrionema strangulans* is disclike (*a*) and usually epiphytic on other algae or seagrasses. This monostromatic layer will produce erect hairs (h; 50–100 μm long) and sporangia [(u), unilocular; (p), plurilocular) on separate plants but shown together in (*b*). The discs can attain a diameter of about 1–2 cm.

rhizoidlike branches that penetrate the host as well as erect filaments that have intercalary growth. Unilocular and plurilocular sporangia have been reported for *E. fucicola*, but the life history is not well understood. It appears that there may be an asexual type of heteromorphic life history with the unilocular sporangia producing zoospores *without* meiosis. Smaller thalli or microthalli (plethysmothalli) result from the germination of the zoospores and produce plurilocular sporangia, which in turn produce zoospores that germinate into the macrothalli of *E. stellaris* (Wanders *et al.*, 1972).

Leathesiaceae

The thalli of members of this family are soft, globular forms with abundant mucilage. Anatomically there is a colorless medulla and smaller chloroplast-bearing cortex and epidermis. The construction is pseudoparenchymatous.

Leathesia

This genus is common throughout the cooler waters and at one time was thought by Linneaus to be a jellylike fungus. The construction is convoluted, brainlike and hollow (Figure 6-11). Plurilocular sporangia and unilocular sporangia are found in the cortex of the same plant. Thus the plurilocular sporangia permit asexual reproduction of the sporophyte. The unilocular sporangia produce zoospores that germinate into a filamentous stage, which in turn produce plurilocular sporangia (gametangia). The resulting anisogametes apparently fuse, with the zygote growing into the sporophyte.

Chordariaceae

Chordaria and *Cladosiphon* are typical of the family and order in that they are gelatinous and filamentous to pseudoparenchymatous. The sporophyte is domi-

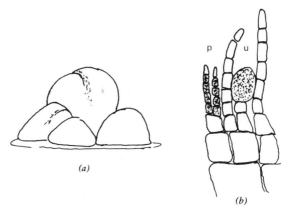

Figure 6-11 *Leathesia difformis*, a spongy, amorphous mass that is usually hollow (*a*). (*b*) When reproductive, the sporophyte produces both plurilocular (*p*) and unilocular (*u*) sporangia as well as sterile hairs. The plants form irregular growths over other algae and rocks, with diameters of about 5–10 cm.

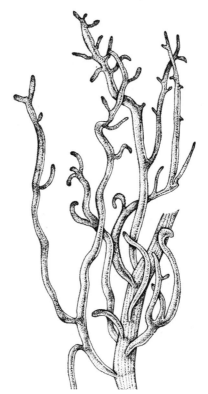

Figure 6-12 *Cladosiphon occidentalis*, an epiphytic brown alga common on seagrasses in warm, temperate, and tropical waters. The golden-to-dark-brown plant is gelatinous with irregular branching, 1–3 dm tall. Construction is by filaments embedded in mucilage. Both plurilocular and unilocular sporangia are found on the sporophyte.

nant and may have both types of sporangia. The gametophyte is a small filamentous free-living plant.

Chordaria

The highly branched sporophyte is a common epiphyte of larger algae as well as lithophytic in the North Atlantic during most of the mid- to-late summer. The terete axis is pseudoparenchymatous, having a medulla of compact filaments and a cortex of short radiating filaments. The entire structure is embedded in a thick mucilage, and the brown-to-black sporophyte is slippery but firm. Kornmann (1962) found that *C. flagelliformis* has a heteromorphic but asexual life history. The unispores produced by the "macrothallus" germinate into small filamentous plants that bear only plurilocular sporangia. The plurispores form the macrothallus again. No meiosis or fusion has been observed.

Cladosiphon

Unlike *Chordaria*, members of this genus have soft sporophytes that are completely filamentous (Figure 6-12). Sporophytes occur along the Atlantic coast as far south as Central America in the spring. The soft, easily squashed thallus demonstrates the typical construction of this order and family. Unilocular and plurilocular sporangia can be found on the same plant. The life history is not known.

Sporochnales

This is a small order with about six genera placed in one or two families; two are present in the Gulf of Mexico. The sporophyte is macroscopic and has a trichothallic growth pattern. An intercalary meristem occurs at the apex of each branch and at the base of each hair, resulting in a small tuft of hairs at the apex and solid parenchymatous tissue in the body.

Sporochnus

The life history (Figure 6-13) is of a heteromorphic haplodiplontic type; the microthalloid gametophytes produce eggs and sperm, either monoecious or dioecious. Apparently *Neria*, which occurs in the Gulf of Mexico along with *Sporochnus*, has the same type of life history. No plurilocular sporangia are evident on the macroscopic sporophytes of the plats in these two genera, although unilocular sporangia are common on mature plants, suggesting an obligate alternation of gametophyte and sporophyte generations, providing that sexual reproduction does occur.

Desmarestiales

This order is small, with one or two families, but has a worldwide distribution in the cold temperate waters. The most common genus, *Desmarestia*, may even be the dominant plant in marine flora of the Antarctic. All of the members of the order are characterized by a macroscopic sporophyte and trichothallic growth.

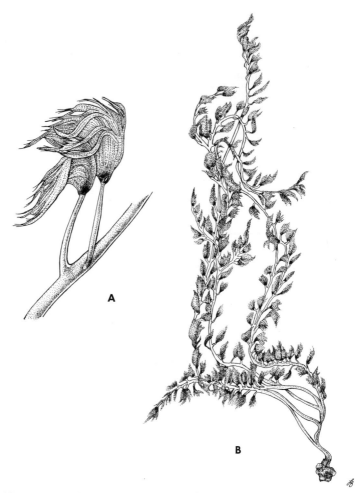

Figure 6-13 (*B*) *Sporochnus pedunculatus*, a wiry sporophyte that will reach 3 dm in length. The trichothallic growth is evident at the apices of the branchlets (*A*) where tufts of filaments are formed. Unilocular sporangia also are found at these sites.

However, the trichothallic growth is less evident because the hairs resulting from meristematic activity at the apex of branches fall off as the plant matures. The gametophyte is a microscopic filament.

Desmarestia

This genus occurs in cool-to-cold temperate waters of both hemispheres and shows a wide variety of forms (polymorphic). Species found in the Pacific Northwest will range from highly bladed forms to dissected, wiry types (Figure 6-14). Members of the genus can be divided into two groups; sporophytes with opposite branching have monoecious gametophytes, whereas sporophytes with alternate branching have dioecious gametophytes. The life history is heteromorphic with

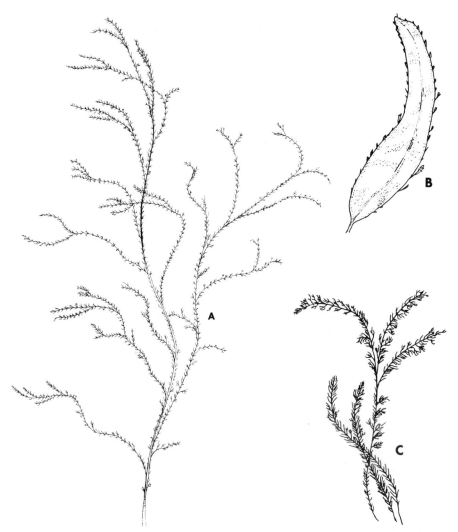

Figure 6-14 The polymorphic genus *Desmarestia* has a wide variety of species. Some species are wiry, such as *D. aceulata* (*A*) and *D. viridis* (branchlets, *C*), whereas others, such as *D. ligulata*, are more foliose (a blade, *B*).

the gametophytes producing sperm and eggs. One feature of this genus is the presence of free sulfuric acid (0.44*N*) in the vacuole (Meeuse, 1956). The low pH (0.8–1.8) can result in rapid detoriation after collection and leakage of the acid into collecting buckets with subsequent destruction of other algae.

Dictyosiphonales

Four families are placed in this large order and are characterized by a heteromorphic haplodiplontic life history where the sporophyte is the macroscopic plant with apical or diffuse growth. The gametes, where known, are isogamous. The members are considered primitive when compared to other brown algae that have heteromorphic life histories. This is because of the production of plurilocular sporangia and asexual reproduction by the sporophyte. A unique intermediate stage in sporophyte development, the plethysmothallus, can also be present and is similar to that described in the Chordariales. Three families will be presented here.

Striariaceae

Stictyosiphon

The terete, highly branched sporophyte (Figure 6-15*a*) has four large medullary cells when young but can become hollow with age. Cortical cells develop into

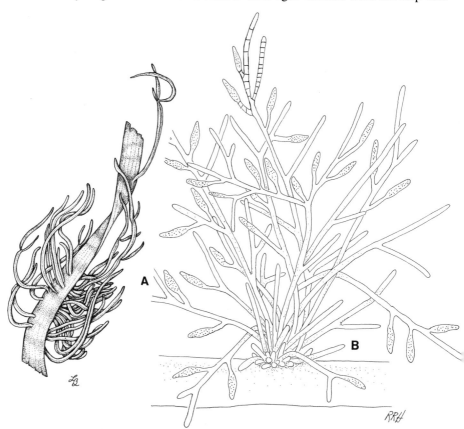

Figure 6-15 *Hummi onusta,* a combination of two previously existing species of brown algae. The sporophyte is *Stictyosiphon subsimplex* (*A*), which is also as an epiphyte on a seagrass blade. The gametophyte is *Myriotrichia onusta,* a small filamentous epiphyte (*B*) found on the same grass blades. The sporophyte has four large medullary cells when young and becomes hollow and bears unilocular sporangia when mature. The gametophyte is a branching uniseriate filamentous pad bearing plurilocular sporangia. (*A*) × 1; (*B*) × 4.50; filaments 7–8 μm diameter.

either unilocular or plurilocular sporangia and sterile hairs are common over the surface of the sporophyte. Fiore (1977) demonstrated through culturing that *Stictyosiphon subsimplex* is the diploid phase of a heteromorphic life history, and *Myriotrichia onusta* (Figure 6-15*b*) is the gametophyte. He combined these two algae and renamed them *Hummi onusta*. The Mediterranean species *S. adraiaticus* also has a heteromorphic life history.

Punctariaceae

Members of this family may have saccate or flattened thalli arising from a discoid base. *Soranthera*, a saccate member of this family has a typical heteromorphic haplodiplontic life history.

Punctaria

Members of this genus are foliose; the plants are small and epiphytic on other algae or attached to rocks. Blade thickness varies according to species (2–4 up to 4–7 cells). Plurilocular and unilocular sporangia develop from epidermal cells along with sterile hairs (paraphyses). Filamentous stages have been found in culture and may be the result of either unilocular or plurilocular spores from the sporophyte. A heteromorphic haplodiplontic life history was first proposed (Sauvagean 1917), but more recent work suggests that unilocular spores may not always be the products of meiosis, the regeneration of the sporophyte is asexual, that is, directly from the unilocular spores (see Bold and Wynne, 1978).

Dictyosiphonaceae

This family has unilocular sporangia that are sunken in the plant surface. The sporophthic thalli are cylindrical or saccate but always parenchymatous.

Dictyosiphon

This genus has a solid parenchymatous construction with a distinct apical cell. *Dictyosiphon foeniculaceus*, common in the North Atlantic, may occur as an epiphyte on *Chordaria* and reach lengths of 1m or more. Plants may become hollow when mature (Figure 6-16). Only unilocular sporangia are present on the sporophyte, and these are sunken on the surface of the plant. The life history is hetermorphic with small filamentous gametophyte-bearing plurilocular sporangia, which produce isogametes.

Scytosiphonales

The members of this order were previously placed in the Dictyosiphonales but have recently been separated because of the presence of a pyrenoid in each cell and the occurrence of only plurilocular sporangia on the macroscopic plants. It also now appears that a *Ralfsia*-like crustose plant (sporophyte?) may be the alternate phase in their life histories. No evidence of a sexual cycle is available, and perhaps the two phases are only asexual or morphological forms of the same plant. Cole (1970) found typical brown algal cytology and a few distinctive ul-

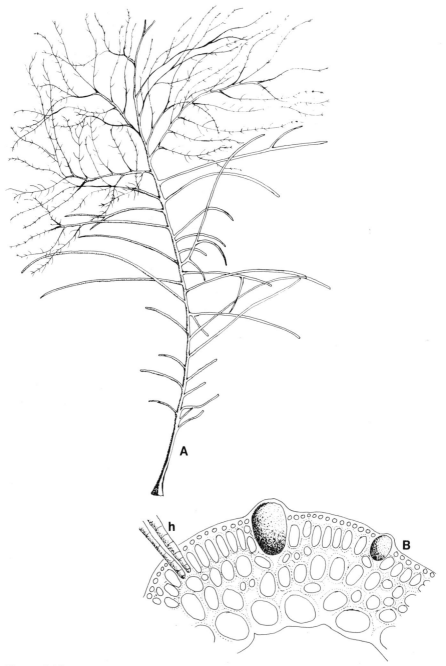

Figure 6-16 (A) *Dictyosiphon foeniculaceus*, a common epiphyte of many larger algae in the North Atlantic that can reach lengths of about 1 m. The plant is parenchymatous and solid when young but becomes hollow when older (B) and bears unilocular sporangia and hairs (h) in the outer cortex. (A) × 0.50; (B) × 88.

trastructural features for *Colpomenia, Petalonia,* and *Scytosiphon.* A single plastid with a central pyrenoid occurs in each cortical and sporangial cell, in contrast to other brown algae. *Colpomenia* and *Hydroclathrus* have convoluted hollow thalli and grow on intertidal rocks in warmer waters of the North Atlantic. Another genus that may belong to this order is *Rosenvingea*; a tubular, branching plant common from deeper waters of the Gulf of Mexico and the Caribbean. *Colpomenia, Hydroclathrus,* and *Rosenvingea* are hollow parenchymatous plants that bear only plurilocular sporangia. Another genus, *Petalonia,* consists of small blades with parenchymatous medullae about two-to-three cells thick. Its small epidermal cells can develop into plurilocular sporangia, and the resulting zoospores germinate into small crustose thalli or blades. Development depends on the temperature and photoperiod (Wynne, 1969).

Scytosiphon

This plant is an unbranched tubular axis with several layers of parenchymatous cells. The tube is usually constricted, but this appears to be phenotypically variable (Figure 6-17*a*). If Tatewaki (1966) is correct, *Scytosiphon* is the gametophyte, bearing only plurilocular sporangia (Figure 6-17*b*), and a *Ralfsia*-like stage is the sporophyte, bearing the unilocular sporangia. However, Kristiansen and Pedersen (1979) found only asexual reproduction in the Danish populations of *Scytosiphon lomentaria.* Swarmers from the tubular, erect plants germinated directly into prostrate systems ranging from filamentous to *Ralfsia*-like crusts.

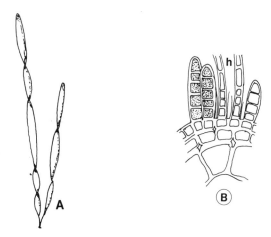

Figure 6-17 *(A) Scytosiphon lomentaria,* a tubular, unbranched plant that is hollow when mature and will grow to 20 cm in length. *(B)* As with members of other genera placed in this order (Scytosiphonales) only plurilocular sporangia are found on the plant along with sterile hairs *(h).* The plant is parenchymatous in construction. (A) × 0.50; (B) × 132.

Laminariales

Plants placed in this order are the largest and most complex forms of the brown algae. They are found primarily in the subtidal region along almost every coast in cold-water zones. The plants, called kelps, are obligate heteromorphic haplodiplonts with oogamous sexual reproduction. The gametophyte is always a small filamentous plant, whereas the sporophyte is large and parenchymatous. Members of this order have typical brown algal cytology both in their sporophytes and gametophytes, as well as their zoospores (Gherardini and North, 1972).

Most of the kelps are perennials; their sporophytes can be divided into at least three parts: holdfast, stipe, and blade(s). Growth in length occurs through an intercalary meristem that is found at the junction of the blade and stipe. If the plant is multibladed this is usually due to splits extending down into intercalary meristem. Tissue differentiation is complex in this order with an epidermis, outer and inner cortex, and central medulla. Trumpet hyphae and sievelike elements are common in the medulla (Figure 6-18). A number of ultrastructural studies have been conducted on these medullary sieve elements and trumpet hyphae (Schmitz and Srivastava, 1976). Distinct physiological roles for each tissue are evident; photosynthesis occurs in the epidermis and outer cortex and food storage in the inner cortex. The medullary cells are the sites of translocation in the kelps (Parker and Hiber, 1965; Van Went and Tammes, 1972; Schmitz and Srivastava, 1980). The order can be divided into four families based on morphological differentiation, number of blades, and location of the sori (clusters of unilocular sporangia).

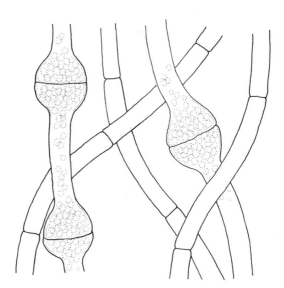

Figure 6-18 Trumphet hyphae of *Laminaria* spp., showing the typical swollen cross walls. These are sometimes called sieve cells. They measure about 20 μm in diameter at the sieve plate.

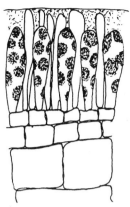

Figure 6-19 A reproductive area on the surface of a sporophytic blade of *Laminaria* contains the unilocular sporangia that will produce the zoospores.

Chordaceae

Chorda

There is no distinction between the stipe and blade in members of this monotypic family and genus. Rather, the entire plant is but one long "cord" reaching up to 10 m in length (Figure 6-19). *Chorda filum* has dioecious gametophytes, whereas *C. tomentosa* has monoecious gametophytes. The sporophyte is an annual or pseudoperennial and the holdfast is a small disc.

Laminariaceae

Plants placed in this family (*Laminaria, Agarum, Costaria*) usually have only one blade, or at least any splits in the blade do not penetrate into the intercalary meristem found at the blade base. The unilocular sporangia are found on sori on the surface of all blades and not restricted to specialized "sporophylls."

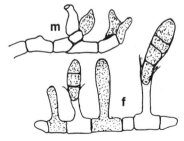

Figure 6-20 The zoospores of *Laminaria* will grow into microscopic male (*m*) and female (*f*) gametophytes. The cells measure about 20 μm in diameter. Each antheridium releases a single biflagellated sperm, whereas each oogonium retains its single egg. The female gametophyte is shown with developing oogonia and developing sporophytes.

Laminaria

This genus is widespread and can become the dominant plant of the subtidal fringe and subtidal zones of many cold-water shores. *Laminaria* has a holdfast with one or more stipes (see Figure 3-25). Most species are perennial and the blade tip, the oldest portion, often will erode away. Unilocular sporangia occur on the blades during the late summer. Gametophytes have been found both in the field and grown from zoospores in culture. They are small monoecious filaments, the female gametophyte being much more long-lived (Figure 6-20). Species are distinguished by blade shape, the solid or hollow nature of the stipe, and presence or absence of mucilage ducts. Many species of *Laminaria* are grown for food (*kombu*) especially in China (Chapter 2; Figures 2-15, 2-16).

Lessoniaceae

This family contains the largest members of the kelps and is characterized by having longitudinal splits extending into its intercalary meristem. Thus, a number of blades are produced, each with a meristem. Some species have specialized sporophylls (*Macrocystis*), while in others the unilocular sporangia occur on vegetative blades (*Nereocystis*). Other genera placed in this family include the erect intertidal sea palm *Postelsia* (Figure 15-2), which is common in the more northern waters of California to Alaska, and *Pelagophycus*, a deep-water kelp.

Macrocystis

The giant kelp, *Macrocystis*, dominates the shallow subtidal waters of the west coast of North and South America, forming kelp forests (Figure 15-3A). The life history is a typical heteromorphic haplodiplontic type (Neushul, 1963). The holdfast of *Macrocystis* is massive and the plant can reach a length of more than 100 m (Figure 6-21). The plant consists of a number of intertwined stipes arising from

Figure 6-21 The giant kelp, *Macrycystis pyrifera*, produces a massive holdfast as shown on this specimen being examined by Mike Neushul. A number of stipes are produced and become intertangled to form the main axis. Sporophylls are produced from the lower portion of the plant.

the holdfast, each stipe having a series of *Laminaria*-like blades. Growth is from a terminal blade that undergoes asymmetric fission to produce lateral blades. Splitting, as found in the family, occurs in the meristematic region by gelatinization of the inner tissue. Specialized sporophylls are found at the base of each stipe. The plant is extensively harvested in southern California and in Peru for the cell wall phycocolloid alginic acid (Chapter 2).

Alariaceae

This family is characterized by a *Laminaria*-like blade and the presence of specialized spore-bearing blades, sporophylls. Splits do not extend into the intercalary meristem, and even if there are multiple blades, these are limited in size. Genera include *Alaria, Egregia,* and *Eisena. In Egregia,* the primary blade is replaced by the stipe, which forms a compressed structure that produces lateral blades and small floats (pneumatocysts).

Alaria

This genus has a conspicuous midrib and a biennial blade that will become shred ded because of wave action (Figure 6-22). Sporophylls are present at the base o the stipe. There are 14 species in the Pacific Northwest that Widdowson (1972 found difficult to separate because environmental conditions can produce varia tions in distinguishing features. The species are restricted to the 20°C isotherm, s

Figure 6-22 *Alaria esculenta,* a good example of the family Alariaceae with its main blade and basal sporophylls. The plants, as seen in this lower tide pool at the Isle of Shoals, have a midrib that may be the only portion remaining in wave-beaten coasts. (Photo courtesy of Don Cheney, Northeastern University, Boston.)

the plants are most common in the cold, arctic areas, especially in sites exposed to heavy surf. Translocation has been demonstrated in *A. esculenta* by Buggeln and Lucken (1979).

Fucales

In some taxonomic reviews, this order is placed in a distinct class or subclass because of the diplontic life history (see Figure 3-23). Meiosis precedes gameto-genesis in producing the eggs and sperm. the order contains widely diverse genera that are mostly cold-water and intertidal in distribution and appear to have a center of distribution in Australia and New Zealand. *Sargassum,* however, is a large genus that is pantropical and temperate. The growth occurs through an api-cal cell in a pit at the tip of a branch in the sporophyte, although when the plant is young, trichothallic growth is evident. The initial three-sided apical cell re-mains in *Sargassum* but changes to a four-sided cell in *Fucus*, resulting in terete and flattened thalli, respectively.

Members of the order Fucales have typical cell organelles and have been used extensively in studies of cell development and tissue differentiation, because of their well-developed apical meristem (Fagerberg and Dawes, 1976, 1977). The epidermis, outer and inner cortex, and medulla all develop from the distinct apical

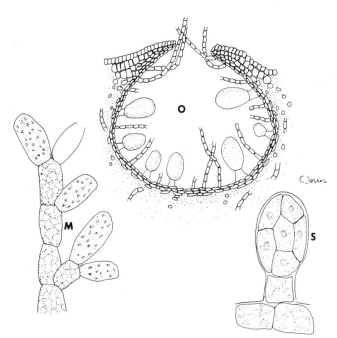

Figure 6-23 *Fucus vesiculosus.* (*o*) A conceptacle-bearing oogonia each with 8 eggs; (*s*) a single oogonium with eggs and (*m*) an antheridial branch. Meiosis occurs in the produc-tion of the eggs and sperm in *Fucus;* thus its life history is diplontic. The conceptacals, or reproductive pits, are found on specialized swollen branch tips of the plant.

meristem, which has a large apical cell. In *Fucus,* mitosis is the open type, and during cytokinesis a cleavage furrow forms without any microtubules being evident (Brawley *et al.,* 1977).

Reproduction occurs in specialized and usually swollen branches called receptacles. The sex organs are found in cavities or pits called conceptacles (Figure 6-23). The number of eggs produced per oogonia vary according to genus, but all oogonia have eight nuclei because of one mitotic division after meiosis. The number of resulting eggs is eight for *Fucus,* four for *Ascophyllum,* two for *Pelvetia,* and one for *Hesperophycus* and *Cystoseira.*

Because of sperm structure, Manton (1964) has suggested that *Fucus* is the most advanced fucoid in the order. The biflagellated sperm found in the family *Fucaceae* have a longer posteriorly directed flagellum unlike that of other brown algae. Egg and sperm release is linked to the tidal cycle. Exposure at low tide results in frond shrinkage and extrusion of mucilage, carrying out eggs and sperm masses. The incoming tide causes the outer walls of the eggs to rupture because the innermost layer (endochite) imbibes water. When the endochite becoming gelatinous, the freed sperm can penetrate the eggs. Nizamuddin (1962) reviewed the Fucales and recognized eight families. The three most common families of the Northern Hemisphere are presented here.

Figure 6-24 *Ascophyllum nodosum* and the ecad *scorpioides* can be seen in the same photograph of the intertidal zone in Great Bay, New Hampshire. The "normal plants" are visible at the base of the boulders; the ecad is seen on the mud in front. (Photo courtesy of Don Cheney, Northeastern University, Boston.)

Fucaceae

Members of this family have a flattened morphology and a four-sided apical cell. Vesicles or floats may be present in some of the species. Members include *Fucus, Pelvetia, Ascophyllum* (North Atlantic) and *Hesperophycus* (Pacific). Species of *Ascophyllum* have linear, compressed axes without a distinct midrib. Two free-living ecads of *A. nodosum* occur in the high and low salt marshes of New England, *mackaii* and *scorpioides* (Figure 6-24). Regeneration of these ecads is by fragmentation, and they are considered to be "end points" of *A. nodosum*. Attached plants of *A. nodosum* are harvested from the intertidal zones of western Europe and used as fertilizer, either directly or after digestion.

Figure 6-25 (*A*) A sterile branch tip of *Sargassum pterpleuron* with a close-up of the leaf with a raised midrib (costate). (*B*) A fertile branch of *Turbinaria turbinata* and a close-up of two leaves with fertile receptacles in their axes. The receptacles bear the conceptacales where meiosis occurs as in *Fucus* (Figure 6-23). (*A*) × 2.75; (*B*) × 3.50.

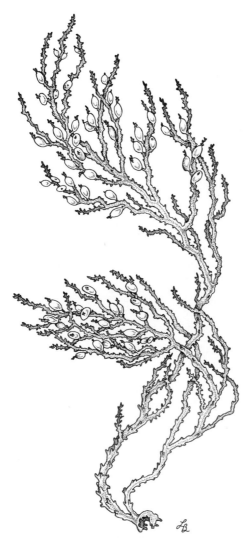

Figure 6-26 *Cystoseira myrica* as a young plant will bear minute leaves and floats as shown here. As the plant matures, the stipe may flatten and receptacular branches will form. × 1.75.

Fucus

This genus has a high degree of morphological variability and most species have a flattened, dichotomously branched form (Figure 3-23). Species are common throughout the intertidal zones of the North Atlantic and North Pacific. *Fucus* has been used extensively in studies on polarity, embryo development (Hurd, 1920; Whitaker, 1940; Nakazawa, 1957; Brawley *et al.*, 1977) and cell wall formation and chemistry (Stevens and Quatrano, 1978).

Sargassaceae

There are a number of subtropical and tropical genera such as *Sargassum* (Figure 6-25*a*) and *Turbinaria* (Figure 6-25*b*) in this small family. The family is characterized by a radial organization because of the three-sided apical cell. Floats or air bladders are common, and lateral branches occur in the axes of subtending leaves.

Sargassum

This genus contains plants that show a high degree of morphological differentiation, having a holdfast that is perennial (Prince and O'Neal, 1979), a stipe, leaves, lateral branches and air bladders (Figure 6-25). This differentiation extends to the anatomical features as well, and ultrastructural features can be quantitatively correlated with physiological aspects of various tissues (Fagerberg *et al.*, 1979). Two sterile species, *Sargassum natans* and *S. fluitans*, are found as free-floating species in the Sargasso Sea. *Sargassum muticum*, a Japanese species, has now invaded not only the west coast of North America but also southern England. In *Sargassum*, the receptacles are modified branches found in the upper portions of the plant (Figure 6-25).

Cystoseiraceae

Cystoseira

Although similar to the Sargassaceae, this family does not have branches that arise in the axis of leaves. The genus has a number of species along the west coast of the United States and one species in the Caribbean (Figure 6-26) but is most prominent in the Southern Hemisphere. The older portion of the stipe is flattened, and only in the younger portion can floats, leaves, and receptacles be found.

REFERENCES

Bold, H. C. and M. J. Wynne. 1978. *Introduction to the Algae. Structure and Reproduction.* Prentice-Hall, Englewood Cliffs, N.J.

Bouck, G. B. 1965. Fine structure and organelle associations in brown algae. *J. Cell Biol.* **26**: 523–537.

Brawley, S. H., R. S. Quatrano, and R. Wetherbee. 1977. Fine-structural studies of the gametes and embryo of *Fucus vesiculosus* L. (Phaeophyta) III. Cytokinesis and the multicellular embryo. *J. Cell Sci.* **24**: 275–294.

Buggeln, R. G. 1976. The rate of translocation in *Alaria esculenta* (Laminariales, Phaeophyceae). *J. Phycol.* **12**: 439–442.

Buggeln, R. G. and S. Lucken. 1979. Kinetic characteristics of photoassimilate translocation in *Alaria esculenta* (Laminariales, Phaeophyceae). *Planta* **147**: 241–245.

Cole, K. 1969. The cytology of *Eudesme virescens* (Carm.) J. Ag. II. Ultrastructure of cortical cells. *Phycologia* **8**: 101–108.

Cole, K. 1970. Ultrastructural characteristics in some species in the order Scytosiphonales. *Phycologia* **9**: 275–283.

Colijn, F. and C. van den Hoek. 1971. The life-history of *Sphacelaria furcigera* Kutz.

(Phaeophyceae) II. The influence of daylength and temperature on sexual and vegetative reproduction. *Nova Hedwigia* **21**: 899–922.

Dawes, C. J., F. M. Scott, and E. Bowler. 1961. A light- and electron-microscopic survey of algal cell walls. I Phaeophyta and Rhodophyta. *Amer. J. Bot.* **48**: 925–934.

Fagerberg, W. R. and C. J. Dawes. 1973. An electron microscopic study of the sporophytic and gametophytic plants of *Padina vickersiae* Hoyt. *J. Phycol.* **9**: 199–204.

Fagerberg, W. R. and C. J. Dawes. 1976. Studies on *Sargassum*. I. A light microscopic examination of the wound regeneration process in mature stipes of *S. filipendula*. *Amer. J. Bot.* **63**: 110–119.

Fagerberg, W. R. and C. J. Dawes. 1977. Studies on *Sargassum*. II. Quantitative ultrastructural changes in differentiated stipe cells during wound regeneration and regrowth. *Protoplasma* **92**: 211–227.

Fagerberg, W. R., R. Moon, and E. Truby. 1979. Studies on *Sargassum*. III. A quantitative ultrastructural and correlated physiological study of the blade and stipe organs of *S. filipendula*. *Protoplasma* **99**: 247–261.

Fiore, J. 1977. Life history and taxonomy of *Stictyosiphon subsimplex* Holden (Phaeophyta, Dictyosiphonales) and *Farlowiella onusta* (Kutzing) Kornmann in Kuckuck (Phaeophyta, Ectocarpales). *Phycologia* **16**: 313–320.

Foster, M. S., M. Neushul, and E. Y. Chi. 1972. Growth and reproduction of *Dictyota binghamiae* J. G. Agardh. *Bot. Mar.* **15**: 96–101.

Galatis, B., C. Katsaros, and K. Mitrakos. 1977. Fine structure of vegetative cells of *Sphacelaria tribuloides* Menegh. (Phaeophyceae, Sphacelariales) with special reference to some unusual proliferations of the plasmalemma. *Phycologia* **16**: 139–151.

Gherardini, G. L. and W. J. North. 1972. Electron microscopic studies of *Macrocystis pyrifera* zoospores, gametophytes and early sporophytes. Proc. Seventh Internat. Seaweed Symp. Wiley, New York.

Hori, T. 1972. Further survey of the pyrenoid distribution in Japanese brown algae. *Bot. Mag. Tokyo* **85**: 125–134.

Hurd, A. M. 1920. Effect of unilateral monochromatic light and group orientation to the polarity of germinating *Fucus* spores. *Bot. Gaz.* **70**: 25–50.

Kornmann, P. 1962. Die Entwicklung von *Chordaria flagelliformis*. *Helgo. Wiss. Meeresunt.* **8**: 265–279.

Kristiansen, A. and P. M. Pedersen. 1979. Studies on life history and seasonal variation of *Scytosiphon lomentaria* (Fucophyceae, Scytosiphonales) in Denmark. *Bot. Tidkr.* **74**: 31–56.

LaClaire, J. W., II, and J. A. West. 1978. Light- and electron-microscopic studies of growth and reproduction in *Cutleria* (Phaeophyta) I. Gametogenesis in the female plant of *C. hancockii*. *Protoplasma* **97**: 93–110.

Loiseaux, S. and J. A. West. 1970. Brown algal mastigonemes: Comparative ultrastructure. *Trans. Amer. Microsc. Soc.* **89**: 524–532.

Manton, I. 1964. A contribution towards understanding of 'the primitive fucoid'. *New Phytol.* **63**: 244–254.

Markey, D. R. and R. T. Wilce. 1975. The ultrastructure of reproduction in the brown alga *Pylaiella littoralis*. I. Mitosis and cytokinesis in the plurilocular gametangia. *Protoplasma* **85**: 219–241.

Meeuse, B. J. D. 1956. Free sulfuric acid in the brown alga, *Desmarestia*. *Biochim. Biophys. Acta.* **19**: 372–374.

Miller, D. G. 1972. Studies on reproduction in *Ectocarpus siliculosus*. *Soc. Bot. Fr. Memoires* **1972**: 87–98.

Müller, D. G. 1979. Genetic affinity of *Ectocarpus siliculosus* (Dillw.) Lyngb. from the Mediterranean, North Atlantic and Australia. *Phycologia* **18**: 312–318.

Nakazawa, S. 1957. Developmental mechanics of Fucaceous algae. VI. A unified theory on the polarity determination in *Coccophora*, *Fucus*, and *Sargassum* eggs. *Tohoku Univ. Rep. Ser. 4. (Biology)* **23**: 119–130.

Neushul, M. 1963. Studies on the giant kelp, *Macrocystis*. II. Reproduction. *Amer. J. Bot.* **50**: 354–359.

Nicholson, N. L. and W. R. Briggs. 1972. Translocation of photosynthate in the brown alga *Nereocystis*. *Amer. J. Bot.* **59**: 97–106.

Nizamuddin, M. 1962. Classification and the distribution of the Fucales. *Bot. Mar.* **4**: 191–203.

Papenfuss, G. F. 1951. Problems in the classification of the marine algae. *Svensk Bot. Tidskr.* **45**: 4–11.

Parker, B. C. and J. Huber. 1965. Translocation in *Macrocystis*. II. Fine structure of the sieve tubes. *J. Phycol.* **1**: 172–179.

Prince, J. S. and S. W. O'Neal. 1979. The ecology of *Sargassum peteropleuron* Grunon (Phaeophyceae, Fucales) in the waters off South Florida. I. Growth, reproduction, and population structure. *Phycologia* **18**: 109–114.

Ragen, M. A. 1976. Physodes and the phenolic compounds of brown algae. Composition and significance of physodes "in vivo." *Bot. Mar.* **14**: 145–154.

Sauvageau, C. 1917. Sur un nouveau type d'Alternance des Generations chez les algues brunes (*Dictyosiphon foenicalaceus*) *Compt. Rend. de l' Acad. Sci. Paris* **161**: 796–799.

Schmitz, K. and L. M. Srivastava. 1976. The fine structure of sieve elements of *Nereocystis lutkeana*. *Amer. J. Bot.* **63**: 679–693.

Schmitz, K. and L. M. Srivastava. 1980. Long distance transport in *Macrocystis integrifolia*. III Movement of THO. *Plant Physiol.* **66**: 66–69.

Sieburth, J. M. and J. T. Conover. 1965. *Sargassum* tannin, an antibiotic which retards fouling. *Nature* **208**: 52–53.

South, G. R. and R. D. Hill. 1971. Studies on marine algae of Newfoundland. II. On the occurrence of *Tilopteris mertensii*. *Can. J. Bot.* **49**: 211–213.

Stevens, P. T. and R. S. Quatrano. 1978. Cell wall assembly in *Fucus* zygotes. II. Cellulose synthesis and deposition is controlled at the post-translational level. *Develop. Biol.* **62**: 518–525.

Tatewaki, M. 1966. Formation of a crustaceous sporophyte with unilocular sporangia in *Scytosiphon lomentaria*. *Phycologia* **6**: 62–66.

Trainor, F. R. 1978. *Introductory Phycology.* Wiley, New York.

Van Went, J. L. and P. M. L. Tammes. 1972. Experimental fluid flow through plasmodesmata of *Laminaria digitata*. *Acta Bot. Neerl.* **21**: 321–326.

Wanders, J. B. W., C. van den Hoek, and E. N. Schillern-van Nes. 1972. Observations on the lifehistory of *Elachista stellaris* (Phaeophyceae) in culture. *Netherlands J. Sea Res.* **5**: 458–491.

Whitaker, D. M. 1940. Physical Factors of growth. *Growth* Suppl. 79–90.

Whittaker, D. M. 1931. Some observations on the eggs of *Fucus* and upon their mutual influences in the determination of the developmental axis. *Biol. Bull.* **61**: 294–308.

Widdowson, T. B. 1972. A taxonomic revision of the genus *Alaria* Greville. *Syesis* **4**: 11–49.

Williams, J. L. 1905. Studies in the Dictyotaceae. III. The periodicity of the sexual cells in *Dictyota dichotoma*. *Ann. Bot.* **19**: 531–560.

Wynne, M. J. 1969. Life history and systematic studies of some Pacific North American Phaeophyceae (brown algae). *Univ. Calif. Pub. Bot.* **50**: 1–88.

CHAPTER 7

Rhodophyta

The red algae consist of approximately 4000 species and comprise the largest proportion of macroscopic seaweeds. Most members are marine with only about 3 freshwater; these are found primarily in streams, although a few unicellular forms occur in soil. The marine forms occur in a great variety of habitats, from the intertidal zone to deep water. A number of red algae are economically important both as direct human food and as a source of various phycocolloidal extracts (Chapter 2). Red algae range in morphology from simple filaments to more massive and relatively complex thalli.

CYTOLOGICAL FEATURES

Pigments

The pigments include chlorophyll a and perhaps chlorophyll d, α and β carotene, lutein, zeaxanthin, and the phycobiliproteins, r-phycocyanin and r-phycoerythrin. The biliproteins are present on structures called phycobilisomes. They are 35 mm in diameter and are found on the surface of the thylakoids. Phycoerythrin is often the dominant pigment and the cause for the red coloration of red algae.

Cytological Structure

Although strictly eukaryotic in cell structure, the red algae are unique because they lack flagella or any part of the flagellar system including the centriole. The chloroplasts have thylakoids that occur singly, that is, not in bands as in the case of the green or brown algae (Figure 7-1; also see Figure 3-4). A girdle or boundary thylakoid is found encircling the plastid in more advanced red algae. Pyrenoids are common in the chloroplasts of the simpler red algae but are not associated with the common storage product floridean starch. Nuclei, mitochondria, and dictyosomes are typical of the eukaryotic cells. A complex involving a Golgi body, mithochondria, and floridean starch granule has been reported in red algae involved in active growth and wall synthesis (Aghajanian, 1979). Mitosis is closed, and the nuclear membrane remains throughout karyokinesis (McDonald, 1972).

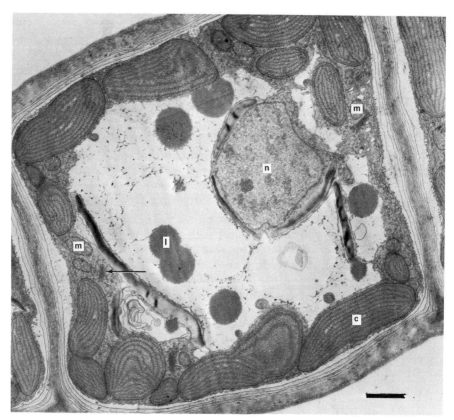

Figure 7-1 The vegetative cell from a filament of the marine alga *Antithamnion defectum,* showing typical red algal cytology. Floridean starch grains surround a central nucleus (n). The numerous discoid chloroplasts (c) have thylakoids that are not grouped in bands. Golgi bodies (arrow) and mitochondria (m) are evident in the peripheral cytoplasm and lipid bodies (l) are seen in the vacuole. Unit mark is 1 μm. (Courtesy of David Young, University Southern California, Los Angeles.)

Motility

Because the red algae lack flagella or any indication of their presence, some investigators believe the red algae evolved directly from a blue-green algal ancestor or from a nonflagellated eukaryotic ancestor (see Figure 3-26). Sexual reproduction is well known in red algae and in all cases spermatia are passively carried to the female gametophyte. All spores are nonmotile as well.

Cell Wall Structure

In general, the cell wall usually contains a small amount of the structural component cellulose. The cellulose microfibrils are not organized into lamellae and occur in a random pattern (Figure 3-15). Much of the wall is gelatinous or amorphous and contains a variety of sulfated galactan polymers, some of which are economically important (e.g., agar, carrageenan, funoran, furcellarin). Because a large percentage of the cell wall of red algae is amorphous material, the wall usually appears transparent under the light microscope.

A number of red algae are calcified; most of these are in the family Corallinaceae. The form of calcium carbonate is calcite and is therefore distinct from that found in the cell walls of calcified green algae, which contain aragonite. Many of the calcified red algae are important reef builders on coral reefs (see Chapter 16), as well as contributors to the calcareous sand of the tropics.

Food Storage

The primary reserve food is floridean starch, an insoluble and refractive $\alpha = (1 \rightarrow 4) =$ linked glucan with $\beta = (1 \rightarrow 6)$ side chains. This compound is similar to cyanophycean starch and glycogen in animals. Floridean starch granules occur outside of the chloroplast, free in the cytoplasm, and are visible as refractive bodies under the light microscope. As with cyanophycean starch, floridean starch reacts with Gram's iodide to produce a brownish color rather than the deep blue color produced by green algal starch. Lee (1974) has used starch grain type and chloroplast structure as phylogenetic indicators in the more primitive red algae.

MORPHOLOGY

The two major taxonomic groups (classes) of red algae are relatively distinct morphologically. The Bangiophyceae are generally simpler forms, ranging from unicellular to simple filaments and blades, one-to-two cells thick. Filamentous and bladed forms of this more "primitive" group will have a holdfast of some sort, and most have a diffuse type of cell division (mitosis occurring throughout the plant).

The majority of red algae, however, are placed in the second class, Florideophyceae. This class contains a wide variety of forms, ranging from simple filaments to highly developed parenchymatous-structured thalli. The basic construction throughout the class is filamentous, and growth is usually apical with one or more well-defined apical cells. In a few families (e.g., Corallinaceae), apical growth is supplemented or replaced by diffuse intercalary growth.

None of the red algae attain the morphological or anatomical complexity found in the larger brown algae (kelps, rockweeds). Anatomically, the more complex red algae have an epidermis, cortex, and a central medulla (Figure 7-31b). Red algae are often small, but may reach lengths of over 1 m. Morphologically they vary in complexity from simple filaments or blades to more complex forms with a stem, holdfast, bladders, and/or blades (e.g., *Botryocladia*).

One particularly important cytological feature found in all the advanced red algae, and many of the more primitive forms, is the pit connection. Pit connec-

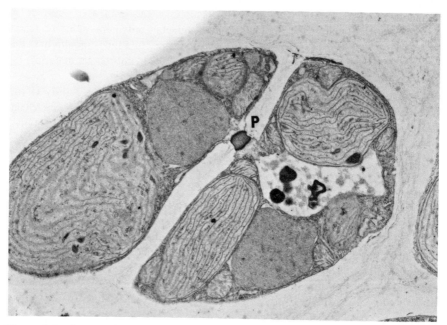

Figure 7-2 A transmission electron micrograph of two epidermal cells of *Hypnea musciformis,* showing typical red algal cytology and a pit connection (*P*). Each cell has a number of chloroplasts, mitochondria, and a single nucleus. Because these are young epidermal cells, the vacuoles are small. The cell walls appear somewhat structureless except for the pit connection. × 11,100.

tions are easily seen under the light microscope and appear as a lens-shaped plugs in the cell wall between two cells. Studies with the electron microscope (Figures 7-1, 7-2) have demonstrated (Myers *et al.*, 1959) that pit connections may be of two types, the open form (immature stage, without plug) and closed form (mature stage, with a dense plug). The pit connection, when mature, consists of a number of layers of wall material (Ramus, 1969). Pueschel (1980) has found the plug core contains protein while the plug cap, in part, contains polysaccharides. Pit connections, because they are produced during septa formation between daughter cells, can be used to trace cell orgin.

One of the unique assemblages of red algal species are the various red algal parasites. Parasitic red algae occur in four of the five major orders of the class Florideophyceae (Evans *et al.*, 1978) and are particularly common in the orders Gigartinales, Cryptonemiales, and Ceramiales. The parasitic red species are designated adelphoparasites if they are taxonomically closely related to their hosts (in the same family) or alloparasites if unrelated to their hosts (Fan, 1961). Evans *et al.* (1978) list 50 species of adelphoparasites and 12 species of alloparasites. They note that in most cases little has been done to determine the exact symbiotic relationships except for the initial identification. Evolution of adelphoparasites may have occurred through mutation of spores of a parent plant, resulting in total or partial loss of the autotrophic habit. For example, tetraspores of *Agardhiella tenera* sometimes germinate directly on the parent plant, producing an epiphyte that is dwarfed and pigmented. On the other hand, alloparasites may have evolved through obligate epiphytism. The obligate condition may result due to a dependence on some host substance (Fan, 1961). Evans *et al.* (1978) review the characteristics of parasitic red algae as well as some of the cytological studies of their relationships. Light and electron microscopic as well as physiological studies have shown the presence of secondary pit connections between host and parasite cells and the transfer of nutrients (Goff, 1979; Krugens and West, 1973).

REPRODUCTION

Asexual reproduction is common in the simpler red algae and may be by cell division, spores, or simple fragmentation. Sexual reproduction has been well documented for a variety of species in the class Florideophyceae but has been found in only a few species of the class Bangiophyceae. Most of the advanced red algae thus far studied appear to have a life histories as described here. The life history includes free-living male and female gametophytes (haploid) and two diploid phases, one free-living (tetrasporophyte) and another (carposporophyte) that is parasitic on the female gametophyte. Thus the life history is termed triphasic and can be described as "diplo-diplohaplont."

In the Florideophyceae, sexual reproduction is a form of oogamy with the "sperm" (spermatium) nonmotile and the "egg" (carpogonium) retained within or on the female gametophyte. A typical triphasic life history of a red alga is presented for *Eucheuma* in Figure 7-3. Starting with the free-living diploid plant, the tetrasporophyte contains specialized cells (tetraspore mother cells) that undergo meiosis and produce haploid tetraspores. The pattern of division within the Florideophyceae tetrasporangia varies among orders and includes several types (e.g., cruciate, zonate, tetrahedral, Figure 7-10). Upon their release and germination, the tetraspores develop into free-living male and female (haploid) gametophytes, which in the case of *Eucheuma* appear identical to the tetrasporophyte (isomorphic alteration of generations). The mature male gametophyte produces spermatangial cells that give rise to spermatia. In *Eucheuma,* these spermatangial cells are modified epidermal cells that bud off uninucleate spermatia (Figure 7-3). In many filamentous forms, spermatangial branches may be formed that produce dense clusters of spermatia. The female gametophyte produces specialized cells called carpogonia, which are produced from either normal vegetative cells or special groups of cells called carpogonial branches (Figure 7-3). The carpogonium has an elongated hairlike extension called a trichogyne. The development of the carpogonial branch and carpogonium are important taxonomic features in the Floridiophyceae.

Fertilization takes place after a spermatium has attached to the trichogyne of the carpogonium, and its nucleus has migrated down to the carpogonium proper and fused with the egg nucleus. The cell (zygote) begins a series of mitotic divi-

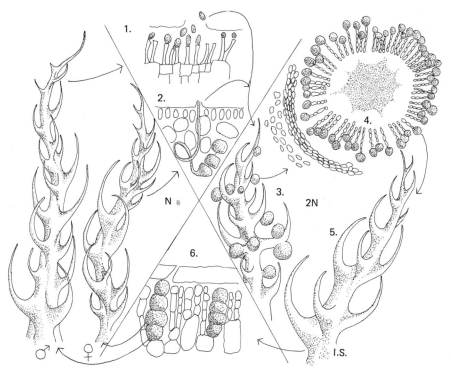

Figure 7-3 The triphasic life history of *Eucheuma isiforme*. On the diploid (2 N) side, numbers 1, 2, 3 & 6 represent haploid tissue while numbers 4–5 represent diploid tissue. The female gametophyte (haploid, #3) is shown because it bears spherical cystocarps in which a diploid fusion cell and resulting carpospores (carposporophyte) are found (4). The carpospores give rise to the diploid tetrasporophyte (5) that bears tetraspores (6) that are the result of meiosis. The tetraspores germinate to produce male and female gametophytes (haploid) that are morphologically identical to the sporophyte. The male gametophyte bears spermatia (#1) and the female carpogonial branches carpogonia (2). Upon fusion of a spermatium with a carpogonium, the carposporophyte (4) develops on the female gametophyte (3).

sions that result in the production of a small mass of diploid cells firmly attached to the female (haploid) gametophyte. The group of diploid cells is called the carposporophyte and is the first diploid phase. The carposporophyte can be thought of as a small parasite on the female gametophyte, taking nourishment from the plant by fusion with surrounding nutrient-enriched cells. Diploid nuclei (in *Eucheuma* and many advanced red algae) may also be transferred to other receptive cells (auxiliary cells) throughout the female gametophyte through specialized transfer tubes (ooblast filaments). These auxiliary cells may be close to the carpogonium (procarp) or separated (nonprocarp). The result is that from one fertilization, a number of carposporophytes can develop. Each carposporophyte produces many diploid carpospores that may subsequently grow into new tetrasporophytes. These carposporophytes may be enclosed by filaments produced by the female gametophyte and this enclosure is called a pericarp. The combined pericarp (haploid) and carposporphyte (diploid) is called a cystocarp.

Two questions are often asked by students studying red algae: (1) Why is there such an elaborate triphasic life history with a parasitic carposporophyte? (2) What is the role of such an elaborate fertilization process, which may include specialized nutrient cells (nurse cells), auxiliary cells, and ooblast filaments? The best explanation may be that they are adaptive mechanisms, compensating for the lack of motile cells. Thus the more spores (carpospores) produced from a single fertilization, the better the chances of the progeny surviving (Searles, 1980). Once fertilization has occurred, it is selectively advantageous to produce as many diploid spores (carpospores) as possible. Consequently, natural selection would appear to favor the elaboration of the carposporophyte. The diploidization (e.g., transfer of a diploid nucleus) of auxiliary cells allows the red algae to take advantage of a single fertilization and produce many carposporophytes, resulting in tens to hundreds of thousands of carpospores. Although this explanation is only a hypothesis, it does appear that the red algae have been very successful despite the potential problem of nonflagellated spores or sperm.

The triphasic life history can be modified in many advanced red algae, and may include a heteromorphic alternation of tetrasporic and gametophyte plants. For example, *Gigartina papillata* and *Petrocelis middendorffii* are the gametophytic and tetrasporic phases, respectively in the same life history (Polanshek and West, 1977). There are also forms of red algae in which a complete loss of sexuality appears to have occurred, and apomictically produced carpospores or tetraspores simply reproduce the same organism (e.g., some strains of *Gigartina stellata*). Finally, it should be noted that for many red algae, the life history is either poorly or incompletely understood.

DISTRIBUTION AND ECOLOGY

Red algae can be found both intertidally and subtidally throughout the world, and are most abundant in the tropics. Because of the accessory photosynthetic pigment r-phycoerythrin, red algae can grow at greater depths than most other algae and usually dominate deep water floras. At least a partial explanation for their dominance in deep water is the correlation between the absorption peak of r-phycoerythrin and the blue-green light available at greater depths. This is the basis of the theory of chromatic adaptation that is discussed further in Chapter 14.

As pointed out in the section of the cell wall structure, a number of red algae are calcified and play a significant role in reef building, especially in forming the algal ridge of atolls. The 40 genera and 600 species of calcified red algae range from the cold waters of the arctic to the tropical atolls of British Honduras (Belize) in the Caribbean. Marl—the calcified skeleton of the red alga *Corallina*—is used in place of lime to reduce soil acidity. It is collected on the southern coasts of England and France.

Red algae are eaten by a variety of fish and invertebrates and probably are an important food source in the food webs of benthic communities. Red algae are important to man (see Chapter 2) as a direct food (*Palmaria, Porphyra, Eucheuma*) as well as a source of phycocolloid extracts such as agar (*Gelldium, Pterocladia, Gracilaria*), carrageenan (*Chondrus, Hypnea, Eucheuma*), furcellarin (*Fucellaria*) and funoran (*Porphyra*).

TAXONOMY

Two classes are recognized in this treatment of the Rhodophyta, the Bangiophyceae and Floridiophyceae (Pappenfuss, 1966; Bold and Wynne, 1978). The two classes can be separated by a number of vegetative and reproductive features. A key to all the orders is given in Table 7-1; it was modified from Bold and Wynne (1978). Modern taxonomic approaches have been summarized in a text on brown and red algae by Irvine and Price (1978).

Bangiophyceae

In this class, plastids are simple and few and pit connections are not easily visible. Pyrenoids are common. Morphology is simple. Plants range from unicellular and filamentous forms to simple blades. Life histories appear to be asexual with the exception of *Porphyra* and perhaps *Bangia*. (Cole and Conway, 1980).

Florideophyceae

Members of this class have numerous discoid plastids, and pit connections are usually easily visible under the light microscope. Pyrenoids are uncommon. Morphology ranges from simple filaments to complex parenchymatous structures, but always basically filamentous. Sexual, triphasic life histories are generally the rule.

It should be noted that the features used to separate the two classes are not without overlap, especially between the more advanced forms of the Bangiophyceae (e.g., order Bangiales, *Porphyra*) and the more primitive forms of the Florideophyceae (e.g., order Nemalionales, *Nemalion, Acrochaetium*). Furthermore, because the mode of reproduction is not well understood, in many cases ordinal placement of many genera in the Florideophyceae is difficult and uncertain.

Little is known about the phylogeny of red algae. Dixon (1973) aptly summarized the problems when he stated that "most speculation on the evolutionary origin and relationships of the red algae and of the various classes and groups within that division can be dismissed as mere flights of phylogenetic fancy." Even such a basic question as whether the major class of red algae, the Florideophyceae, was derived from the Bangiophyceae is not easy to answer, although there is evidence to support this from the presence of cellulose in the cell wall. Perhaps each of these classes arose from some common ancestor.

Table 7-1   A Key to the Orders of Marine Red Algae

1   Pit connections absent or at least not visible under the light microscope. Cell division intercalary, plants simple, mostly sexual reproduction.................................   2

1   Pit connections present and visible under the light microscope. Cell division apical, sexual reproduction is common...   4

   2   Unicellular forms, may form simple aggregations of cells or colonies..............
       .... **Porphyridiales**
   2   Plants multicellular of various morphologies (blades, filaments)................   3

3   Plants filamentous, asexual reproduction only, by monospores and not by specialized cell division... **Goniotrichales**

3   Plants filamentous, blade-like or parenchymatous, asexual reproduction by monospores that are formed in narrow filaments and produced from vegetative cells.... **Bangiales**

   4   Carposporophyte developed directly from fertilized carpogonium, auxiliary cells absent.. **Nemalionales**
   4   Carposporophyte either developing from a supporting cell with which the fertilized carpogonium fuses or from an auxiliary cell.......................................   5

5   Carposporophyte developing from supporting cell with which the fertilized carpogonium fuses, no auxiliary cell.. **Gelidiales**

5   Auxiliary cells present and carposporophytes developing from them................   6

   6   Auxiliary cell that receives zygote nucleus produced after fertilization and produced from the supporting cell... **Ceramiales**
   6   Auxiliary cell present prior to fertilization, source of production variable.....   7

7   The auxiliary cell a part of a vegetative filament, in the normal pattern of branching, auxiliary cell intercalary... **Gigartinales**

7   The auxiliary cell not intercalary in a vegetative filament.............................   8

   8   The auxiliary cell the terminal cell of a two-celled filament produced from the supporting cell of the carpogonial branch............................. **Rhodymeniales**
   8   Auxiliary cell found on a specialized filament either produced from the supporting cell or not (procarp and nonprocarp)................................. **Cryptonemiales**

Source: After Bold and Wynne (1978).

CLASS 1. BANGIOPHYCEAE

This class contains relatively few species and appears to be the more primitive group in the red algae. Essentially, the plants are simple in construction ranging from unicells to pseudo or true filaments and thin blades. The cells are uninucleate and usually contain a single, large, lobed-stellate plastid and a single pyrenoid. Much confusion is centered around the function (sexual or asexual) of various "spores" produced by members of this class. Sexual reproduction has been confirmed in only a few cases and life histories are generally not known. Pit connections may be present; they are smaller than those found in the Florideophyceae, and not easily viewed under the light microscope. Ultrastructural studies have been carried out on several of members of this class including *Erythrotrichia* (McBride and Cole, 1972), *Porphyridium* (Gantt *et al.*, 1968; Neushul, 1970) and *Smithora* (McBride and Cole, 1969). Three orders are recognized in this text as having marine members.

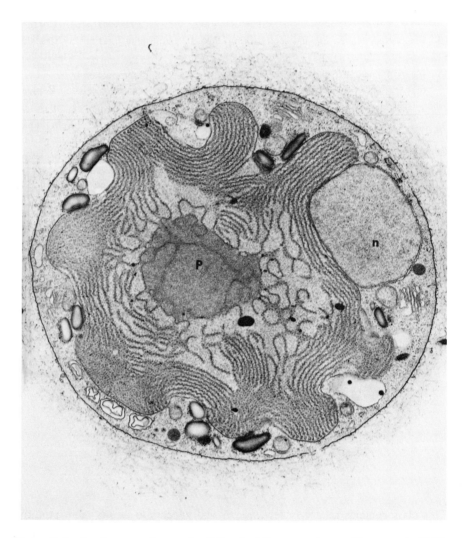

Figure 7-4 An electron micrograph of *Porphyridium aerugineum* (Gantt *et al.* 1968) showing the central lobed chloroplast with a pyrenoid (*p*). A nucleus (*n*), numerous Floridean starch grains and mitochondria are visible around the cell periphery. (Courtesy of Beth Gantt, Radiation Biology Laboratory, Smithsonian Institute, Washington, and with permission of the *Journal of Phycology*.)

Porphyridiales

The first order is characterized by plants that are unicellular or irregular cellular masses. Sexual reproduction is unknown.

Porphyridium

Probably the best-known member of the order, this unicellular genus (Figure 7-4) occurs on the surface of damp soil (e.g., greenhouses) and as plankton. The cells are globose, with a single large stellate chloroplast and a nucleus located between the arms of the chloroplast. A central pyrenoid is usually present in the chloroplast. Reproduction is by cell division.

Goniotrichales

This order is characterized by branched or unbranched filaments and intercalary cell divisions. Most plants occur as minute epiphytes on larger algae. Reproduction is asexual by monospores; sexual reproduction is unknown.

Goniotrichum

The genus *Goniotrichum* is a common microscopic epiphyte on many macroscopic algae such as *Cladophora* and *Chaetomorpha*. The plant consists of branched filaments embedded in a gelatinous matrix. The filaments are usually one cell thick when young (Figure 7-5). Each cell has a typical stellate chloroplast.

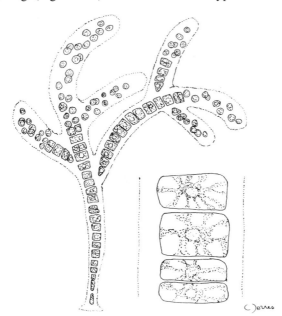

Figure 7-5 Diagrams of *Goniotrichum alsidii,* showing a portion of the branching plant and the irregular arrangement of the cells in the gelatinous sheath. Details of two cells (~5 μm diameter) with stellate plastids are shown on the right.

Monospores are produced by the rounding up of a vegetative cell; they are released after breakdown of the gelatinous filamentous wall.

Bangiales

This order is the largest of the class, and can be divided into three families. The plants are multicellular and appear as filaments or blades. There is an apical cell in the early stages of development but ultimately cell division is diffuse. A variety of spores are produced by members of this order and, in at least one species of *Porphyra*, sexual reproduction has been confirmed (Hawkes, 1978). Two families will be presented here, a third family, the Boldiaceae, is a monotypic, freshwater family and is omitted in this discussion.

Family Erythropeltidaceae

Members of this family have creeping or erect filaments or blades that develop from a basal cushionlike pad of cells (Figure 7-6).

Erythrotrichia

The genus contains unbranched, filamentous epiphytes on other algae (Figure 7-6). The plants are differentiated into a prostrate basal portion (holdfast cells) and one or more erect filaments, thus demonstrating heterotrichous growth.

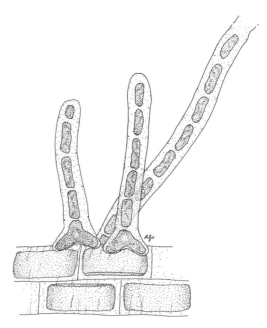

Figure 7-6 Basal portion of two young plants of *Erythrotrichia carnea* with a more mature filament attached to an *Enteromorpha* branch. Although not visible, the chloroplast in each cell is stellate. Cells are about 10 μm in diameter.

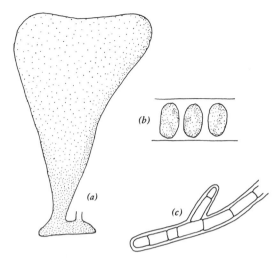

Figure 7-7 A habit sketch (*a*) and cross section (*b*) of *Smithora naiadum* found growing on the seagrasses *Phyllospadix* and *Zostera*. The red-to-deep-purple blades are monostromatic and were removed from the genus *Porphyra* by Hollenberg (1959). Richardson and Dixon (1969) found a *Conchocelis* phase (*c*) in the life history. The blades are monostromatic (*b*) and will reach 50 mm in length and 20–30 μm in thickness.

Monospores are common and are produced when oblique cell divisions give rise to unequally sized daughter cells, the smaller becoming the spore. Although there are reports of sexual reproduction in this genus (i.e., spermatia formation, fusion of filament cells), no convincing evidence has yet been published.

Smithora

Members of the genus *Smithora* have membranous, monostromatic blades much like those of *Porphyra*. The plant occurs as an obligate epiphyte on the seagrasses *Phyllospadix* and *Zostera* along the Pacific Coast (Figure 7-7). It has a perennial cushion of cells from which the blades arise.

Hollenberg discovered that these small "*Porphyra*-like" epiphytes were distinct from *Porphyra*. He described their asexual mode of reproduction as deciduous areas on the blade that contain monosporangia. These monosporangial sori becomes gelatinous and break off (Hollenberg, 1959). Cells associated with sexual reproduction (i.e., spermatia, carpospores) have been noted in the literature, but to date no firm evidence for sexual reproduction is available for *Smithora*. However, Richardson and Dixon (1969) reported that *Conchocelis* filaments arise *in situ* from vegetative blade cells of *Smithora* (Figure 7-7*c*).

Family Bangiaceae

This family has three genera, of which *Bangia* and *Porphyra* are the most diverse. Both genera appear to have heteromorphic life histories (Cole and Conway, 1980), but in only one species of *Porphyra* has sexual reproduction been confirmed (Hawkes, 1978). In both genera, the life histories are highly influenced by physical factors (see below). The family is characterized by a rhizoidal holdfast (Figure 7-8)

Bangia

This genus is an unbranched, uniseriate or multiseriate, filamentous alga, which is common in rocky intertidal zones (Figure 7-8). *Bangia atropurpurea* is apparently worldwide in distribution, found in marine and freshwater habitats. The plant begins development as a uniseriate filament, gradually becoming multiseriate and producing a massive rhizoidal holdfast (Figure 7-8*b*). The life history appears to be haplo-diplontic, that is, having two different phases (Nichols and Veith, 1978), the *Bangia* phase and a small filamentous perennating phase (see Cole and Conway, 1980, for details). Reproduction by both phases is asexual monospores. Lin *et al.* (1977) have compared the ultrastructure of these two phases and found that they differ in plastid shape, position, organization of thylakoids, pyrenoid structure, and presence of pit connections. Because the chromosome number is the same in both phases, it appears that sexual reproduction is lacking. Also, spores from each phase do not necessarily produce the alternate phase. The production of monospores by the *Bangia* stage appears to be controlled by photoperiod and temperature (Richardson and Dixon, 1968). For example, when monospores germinated under a short photoperiod (less than 12 hr light), the sporelings were bipolar and developed into *Bangia*. If the monospores were germinated under long photoperiods (more than 12 hr light), the sporelings were unipolar and grew into the filamentous *Conchocelis* stage.

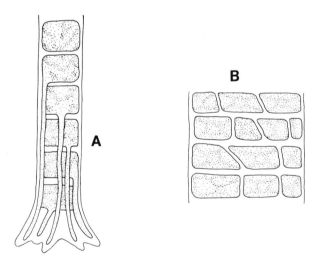

Figure 7-8 (*A*) Bangia fuscopurpurea, a filamentous red alga with a holdfast of rhizoids. (*B*) The main axis will become multiseriate when mature. The uniseriate basal portion will measure 40–60 μm in diameter, whereas the multiseriate upper region will reach 150 μm in diameter.

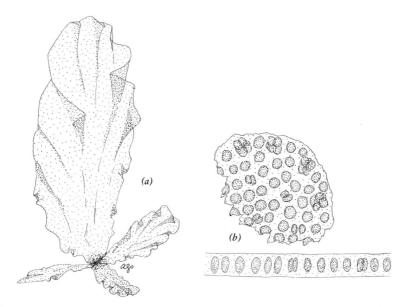

Figure 7-9 *Porphyra leucosticta,* a monstromatic (*b*) bladed (*a*) plant found attached to rocks and other algae in the intertidal regions in New England. The blade reaches 10–15 cm in length and 25–50 μm in thickness. The color of the plant ranges from a rose red to a deep purple, depending upon the amount of phycorytherin in the chloroplast.

Porphyra

Porphyra is an important seaweed economically, eaten directly as food in Japan (where it is called *nori*) and China (Chapter 2). The plant consists of a monostromatic or distromatic blade, depending on the species, and a rhizoidal holdfast (Figures 7-9*a*, 7-9*b*). There are a variety of species, several of which occur on the rocks of the upper intertidal zone. As with *Bangia, Porphrya* species have been found to have heteromorphic life histories. Dr. Kathleen Drew (1949) was the first to demonstrate that *Porphyra* and *Conchocelis* were the haploid and diploid phases, respectively, of the same life history. In addition to making ultrastructural observations, Hawkes (1978) demonstrated that sexual reproduction occurs in *P. gardneri.* The life history of *Porphyra* is presented in Chapter 2 (Figure 2-11). A single blade of *Porphyra* may produce both spermatia and large mammilated carpogonial cells. At fertilization, the resulting diploid nucleus divides mitotically, producing diploid carpospores that subsequently grow into the small filamentous *Conchocelis* phase. Because sexual reproduction has not been demonstrated for most species of *Porphyra* it may be wise to use the older terms α- and β-spores, instead of carpospores and spermatia for species other than *P. gardneri.* Spermatangial areas/(β-sporangia) on *Porphyra* blades appear pale yellow and can easily be distinguished from carpogonial areas (α-sporangia), which are heavily pigmented. Asexual reproduction occurs in *Porphyra* by the production of monospores and in the *Conchocelis* stage by the production of conchospores. Although meiosis has not been observed directly, the conchospores germinate and grow into haploid *Porphyra* plants, and thus meiosis is inferred. Photoperiod and temperature play a role in the maturation of the spores in both phases.

CLASS 2. FLORIDEOPHYCEAE

This class contains most of the red algal species, with plants ranging from simple filaments to massive parenchymatous blades. The basic construction of all members is filamentous, so that even the most complex, parenchymatous thalli can be considered pseudoparenchymatous. Development occurs when apical cells give rise to filaments, which in turn combine to form a more complex plant body. Although cell division is generally apical, a few cases of intercalary cell divisions occur, especially in the families Corallinaceae and Delesseriaceae.

Pit connections may be of two types in the Florideophyceae, primary and secondary. Primary pit connections are the result of incomplete closing of a cross wall between daughter cells and the subsequent formation of a spindle-shaped plug. In the case of secondary pit connections, two nonrelated cells may develop a connection after coming in contact with each other and produce a similar appearing plug.

A triphasic life history has been demonstrated for a great many members of this class and is the most common type of life history. However, only a few life histories have been studied sufficiently to determine where meiosis occurs. Most members of the Florideophyceae are dioecious with spermatangia and carpogonia on separate plants. A variety of tetrasporangia and other types of sporangia (e.g., monosporangia, bisporangia, and polysporangia) are produced by different genera within the class (Figure 7-10).

Six orders have traditionally been recognized in this class (Dixon, 1973). The

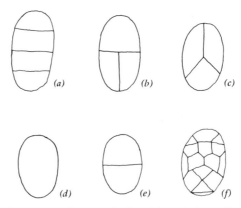

Figure 7-10 The various types of sporangia found in red algae. Tetrasporangia may be zonately divided (*a*), cruciately divided (*b*), or tetrahedrally divided (*c*). Specialized sporangia in the red algae include monosporangia (*d*), bisporangia (*e*), and polysporangia (*f*).

major distinctions between each class are based on sexual features, especially the nature of the carpogonial branch, presence of an auxiliary cell, and association of these two. It should be noted that distinctive morphological forms may be found in the same order. Thus, ordinal placement of many members of this class is uncertain and many phycologists find the taxonomic basis of the Florideophyceae difficult to use. Even so, no better taxonomic scheme has been proposed.

Nemalionales

This order is considered by most phycologists to contain the most primitive algae in the class. Some texts (Bold and Wynne, 1978) now include the Gelidiales within the Nemaliionales, because of the lack of an auxiliary cell in both orders. However, because of the heteromorphic life histories found in many members of this order (see Figure 7-13) and the distinctive plant body of the Gelidiales, the two orders will be treated separately in this text. Certainly the wide variety of life histories in the Nemalionales suggests a diverse, even artifical assemblage. Four families are included in this account; a fifth family, Batrachospermaceae, is restricted to fresh water.

Acrochaetiaceae

Members of this family are generally small, branched filaments that often occur as epiphytes or endophytes. Some of the "species" in the family are stages in the life histories of other red algae. The two best known genera are *Rhodochorton* and *Acrochaetium*. However, Woelkerling (1973) has combined most species of these two genera into the genus *Audouinella*. West has published a series of studies on various species of *Rhodochorton* (e.g., West, 1979). Two of the species he studied only produced tetrasporangia in culture (*R. concrescens, R. membranaceum*), and a third species (*R. purpueum*) had a typical triphasic life cycle that was controlled by temperature and photoperiod.

Audouinella (Acrochaetium)

Plants of this genus consist of uniseriate, branching filaments (Figure 7-11), which include lithophytes, epiphytes, epizoics, endophytes, and endozoics. The common asexual reproductive structures found in most species are monosporangia, with the monospores apparently reproducing the same plant. West (1968) found a typical triphasic life history in cultures of *Acrochaetium pectinatum*. Tetrasporangia are produced by some species, whereas spermatia, carpogonia, and carpospores have been described for others.

Helminthocladiaceae

Members of the family Nemalionaceae are included here. Where sexual reproduction is known, the algae show heteromorphic triphasic life histories, with the gametophytes erect plants of multiaxial construction and tetrasporophytes small filamentous plants. Most gametophytes are gelatinous at first, but become firmer

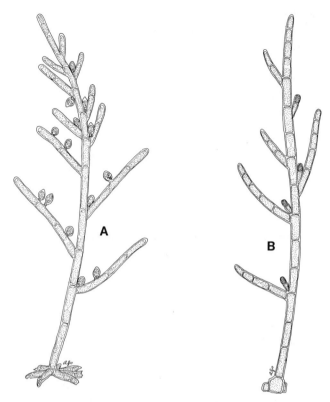

Figure 7-11 Two examples of the genus *Acrochaetium*. (*A*) *A. globosum*, having a filamentous base. (*B*) *A. sargassi*, having a globular basal cell. Both examples bear
monosporangia. The cells of the lower and main axes measure about 5–6 μm in diameter.

with maturity or even calcified (e.g., *Liagora*). The tetrasporangial plants are
small and filamentous and in some cases closely resemble "species" of
Acrochaetium.

Nemalion

The gametophytes of *Nemalion* and *Helminthocladia* (Figure 7-12) are soft, gelatinous, threadlike thalli composed of intertwined, branching filaments. There is
little differentiation between the cortex and the inner medullary area. Carpogonial
branches are three to four cells in length (Figure 7-13) and bear a very long,
projecting trichogyne. Spermatangial branches are modified cortical cells and
form clusters of spermatia (Figure 7-13*a*). After fertilization, the zygote nucleus
in the carpogonium divides, producing an upper and a lower cell. The
carposporophyte develops from the upper cell and produces carpospores without
any specialized surrounding layer of gametophytic filaments (pericarp), as is
found in the more-advanced orders of the Florideophyceae. Culture studies of
Nemalion have demonstrated that the *Nemalion* phase alternates with an *Ac-*

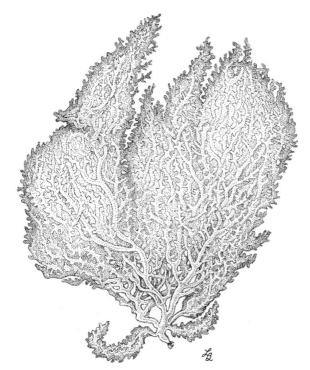

Figure 7-12 An example from the family Nemalionaceae, the very gelatinous species *Helminthocladia clavadosii,* a deep-water tropical member of the family. The plant habit (× 1) depicts a soft, gelatinous plant that is extensively and progressively branched.

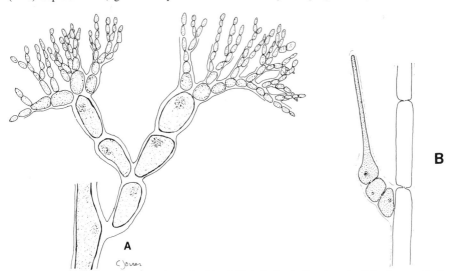

Figure 7-13 The carpogonial branch (*B*) of *Helminthocladia clavadosii* is three celled. The plants will reach 4–5 dm in length; the carpogonial branch cells are 20 μm in diameter. The spermatangial branch (*A*) produces clusters (2–3 μm) of spermatia at each tip.

rochaetium-like plant that produces tetrasporangia and monosporangia (Umezaki, 1972). Meiosis has not been observed.

Bonnemaisoniaceae

This was the first family of the order in which a heteromorphic life history was established. The gametophytes are erect and obvious, whereas the tetrasporophytes are small and filamentous. One advancement in this family is the existence of a well-developed pericap; that is, the carposporophyte is surrounded by gametophyte filaments.

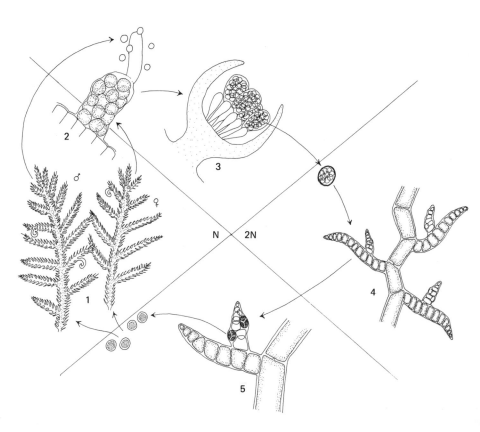

Figure 7-14 An example of a heteromorphic life history in the family Bonnemaisoniaceae, taken from *Bonnemaisonia geniculata* as described by Shevlin and Polanshek (1978). Homothallic male and female gametophytes (*1*) produce the respective spermatial and carpogonial branches (*2*). After fertilization, a carposporophyte (*3*) is formed that produces large carpospores containing numerous starch grains. These carpospores germinate to produce a yet-unidentified filamentous sporophyte that produces tetraspores (*5*) through meiosis. The tetraspores grow into gametophytes.

Bonnemaisonia

The species *B. hamifera* is named after the hooklike, hamate branches that result in entanglement with other algae (Figure 7-13). The plant appears to have been introduced from Japan. By 1900 it was also known for the south coast of England and now is known throughout northern Europe and the northeast coast of North America. It is also present in southern California and Baja California (Mexico).

The gametophyte is erect, usually deep red, and grows to be up to 15 cm in length. The plant's internal construction is uniaxial, which may be obscured by the heavy cortication surrounding the central siphon. The male and female reproductive structures are similar to those of *Nemalion* with the exception that a pericarp (sterile gametophyte filaments) develops around the carposporophyte. The diploid plant is a uniseriate filament which is branched and bears a "gland" cell (colorless refractive type of vesiculate cell) at the upper end of each major axial cell. The diploid plant was originally thought to belong to a different genus, *Trailliella*, until its life history was completed in culture. The *Trailliella*-stage produces cruciately divided tetraspores.

A heteromorphic life history has also been reported for *B. geniculata* and an unknown filamentous phase (Shevlin and Polanshek, 1978) and is diagrammed in Figure 7-14. Another species, *Asparagopsis taxiformis*, has been found to be gametophytic and to alternate with a filamentous, diploid phase called *Falkenbergia intricata*.

Chaetangiaceae

The gametophytes in this family may reach 10 cm or more in height and have an erect bushy habit. The cortical cells form a continuous surface and the construction is multiaxial. A pericarp is formed around the carposporophyte. Most genera placed in this family are tropical, such as *Galaxura* (Figure 7-15) and *Scinaia*. Both of these genera have heteromorphic life histories; *Galaxura* has two free-living and strongly dimorphic gametophyte and tetrasporophyte stages, and *Scinaia* has a small *Acrochaetium*-like tetrasporophytic phase (Svedelius, 1944; van den Hoek and Cortel-Breeman, 1970). A taxonomic review of *Galaxura* and associated problems with species pairing (gametophytic and tetrasporophytic phases of the same life cycle) has been presented by Papenfuss and Chiang (1969).

Gelidiales

This order is small, consisting of one or two families (Gelidiaceae, Wurdemanniaceae). The members lack an auxiliary cell and all life histories thus far studied have an isomorphic alternation of generations. The genera are usually wiry and yield the economically important phycocolloid agar. Some are also grown as food. Fan and Papenfuss (1959) have described four species that are parasitic (adelphoparasites) on various members of the Gelidiales.

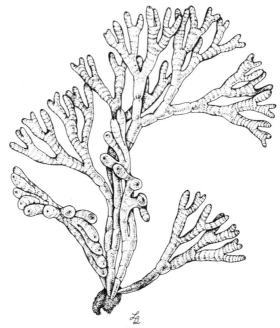

Figure 7-15 *Galaxura obtusata*, a tropical member of the Chaetangiaceae, is calcified. The plant will reach 10–20 cm in length and is rose-red or white if heavily calcified. The annulation is due to production of hairs that project through the cortex. × 1.25.

Gelidiaceae

The tetrasporangia found in this family are cruciately divided. The carpogonia are sessile on intercalary cells of third-order branches (Figure 7-16*a*). After fertilization, a multinucleate fusion cell is formed by fusion with surrounding, nutrient-rich cells (nurse cells) in the female gametophyte. Irregular-shaped "gonimoblast" filaments develop from this fusion cell and bud off carpospores. A greater number of tetrasporophytic than gametophytic plants have been reported from field populations.

Examples of genera placed in this family are *Pterocladia, Gelidiopsis, Gelidiella,* and *Gelidium.*

Gelidium

This genus has the largest number of species in the order and contains species ranging from minute (1–2 mm) plants (*G. pusillum*) to larger forms such as *G. robustum* (15–30 cm tall). The plants have single apical cells and are uniaxial in construction; they can form dense, wiry tufts (Figure 7-17*b*). The uniaxial nature is usually obscured by a dense, central medulla that contains large thin-walled cells and small, thick-walled cells (rhizines). The latter cells appear lens-shaped in cross section (Figure 7-16). The rhizines are internal rhizoids that are produced below the apical meristem and grow through the medullary tissue. The relative

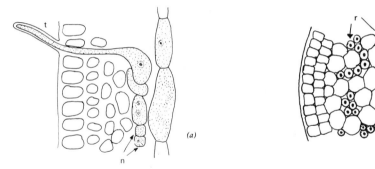

Figure 7-16 (*a*) A carpogonial branch of *Gelidium* with a projecting trichogyne (*t*) and two basal nurse cells (*n*) attached to the supporting cell (*s*). (*b*) A cross section through a mature stem of *Gelidium* showing the thick-walled rhizomes (*r*) that occur in the outer region of the medulla. The epidermal cells in both drawings are 4–10 μm in diameter.

position of the rhizines and number of apical cells are diagonistic characters for generic separation. The plant axes may be terete or flattened, depending on the species (Figure 7-17).

Cryptonemiales

Cryptonemiales is a large order with eight families and approximately 100 genera. The order shows a wide variety of morphologies and internal constructions. The members include gelatinous, fleshy, cartilaginous, and even calcified forms. The main feature of the order is that the auxiliary cell is recognizable before fertilization and is formed from a special filament of the female gametophyte. The auxiliary cell will receive the zygotic nucleus and be the site of the carposporophyte. The auxiliary cell filament is identical to the carpogonial filament with cells of a denser cytoplasmic content than the vegetative cells (Fig. 7-18*a*, 7-18*b*). The auxiliary cell filament and the carpogonial filament may develop from a common cell or arise at a distance from one another. Dixon *et al.* (1972) have reviewed life history studies on various members of this order. It is apparent that a number of species demonstrate heteromorphic life histories, at least in culture (e.g., *Halymenia floresia*, *Acrosymphyton purpuriferum*, *Thuretellopsis peggiona*, *Pikea californica*).

Choreocolaceae

This family contains a number of reduced parasites on members of the Rhodomelaceae (*Choreocolax* on *Polysiphonia*; *Harveyella* on *Rhodomela* and *Odonthalia*). Because the parasites are not found in the same order as the host, but are unrelated, they are called alloparasites. A thorough study of the alloparasite *Harveyella mirabiles* and its infection on *Odonthalia flocossa* has demonstrated a transfer of nutrients through secondary pit connections (Goff, 1979).

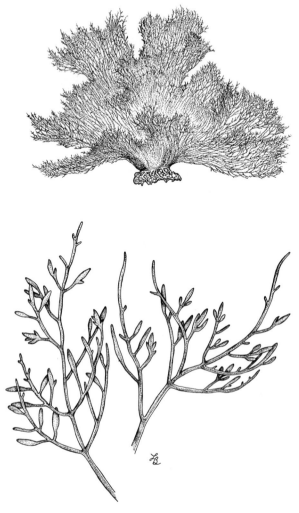

Figure 7-17 The habit sketch and detail of individual branches of a smaller species of *Gelidium, G. crinale,* showing the terete branches with flattened apices. A dense tuft, as shown above, will reach 5 cm in height and the individual branches will be 0.1–0.2 mm in diameter.

Corallinaceae

This family is the largest of the order. The members are calcified with a calcite form of calcium carbonate. The family can be divided into the crustose, nonarticulated subfamily Melobesioideae (Figure 7-20) and the articulated or "jointed" subfamily Corallinoideae (Figure 7-21). Excellent reviews of the crustose and articulated corallinae have been published by Littler (1972) and Johansen (1974). The ultrastructure of various species is presently being investigated (Borowitzka

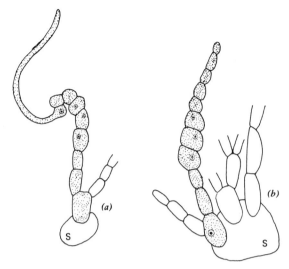

Figure 7-18   The carpogonial (*a*) and auxiliary cell (*b*) branches as found in *Dudresneya*. The supporting cell (*s*) in both branches is a normal vegetative cell, whereas the branches are specialized ones. The terminal carpogonial cell with trichogyne is visible in (*a*).

and Vesk, 1979; Giraud and Cabioch, 1979). Members of both subfamilies are widely distributed in all seas. In the tropical waters, coralline algae form extensive algal ridges of atolls; in colder waters, extensive beds of coralline algae occur in subtidal zones and form marl on the shore.

Members of the Corallinaceae have extensive fossil records because of their calcification. Genera can be identified as far back as the Jurassic, with their most extensive distribution occurring from the Cretaceous onward. Atolls may contain fossil beds of coralline algae of considerable thickness.

The Corallinaceae also differ from other red algae in possessing reproductive structures formed in specialized sunken cavities called conceptacles, which are open to the exterior by one or more pores (Figure 7-19*a*, 7-19*b*). The gametophytes are dioecious and bear distinctive spermatangial and carpogonial conceptacles. The carpogonial branch is two celled, with its basal cell serving as the auxiliary cell. Large numbers of carpogonial branches are formed in a carpogonial conceptacle. After fertilization and transfer of the zygote nucleus to the auxiliary cell, all the auxiliary cells within a conceptacle participate in producing a single large fusion cell from which gonimoblast filaments arise and produce carpospores. Throughout the entire family, the tetrasporangia show a zonate arrangement of tetraspores.

Lithothamnion

The genus has a pantropical distribution, an epilithic crustose habit and many species (Figure 7-20*a*). Cells of the prostrate filaments give rise first to "cap cells" and then to cells that function as the sites of meristematic activity, giving rise by their divisions to the upright filaments. Thus a type of intercalary cell

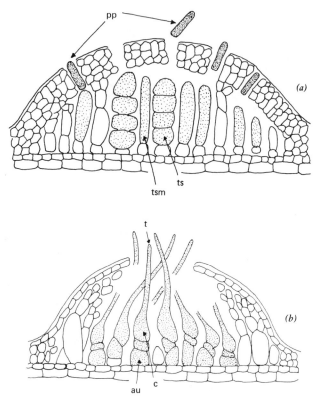

Figure 7-19 Two examples of conceptacles as found in the family Corallinaceae. (*a*) The tetrasporangial conceptacle of *Lithothamnion*, which has multiple pores. The pore plugs (*pp*) are lost in the formation of the pores. Both tetrasporangial mother cells (*tsm*) and tetraspores (*ts*) are shown. (*b*) A procarpial situation with the supporting cell of a carpogonial branch being the auxiliary cell (*au*). The carpogonium (*c*) bears a trichogyne (*t*).

division occurs in the Corallinaceae. The conceptacles may be embedded on the surface and in this genus all have a number of holes or ostioles (Figure 7-19*a*).

Porolithon

This crusotose coralline alga is pantropical in distribution and a major constituent of algal reefs. It is like *Lithothamnion* in having a crustose structure but differs in that rows of "hererocysts" or large cells occur scattered through the tissue (Figure 7-20*b*). These heterocysts are different from those of blue-green algae and represent the remains of basal hair cells. The conceptacles have a single pore, while in other genera the conceptacles are multipored.

Jania and Amphiroa

These genera are examples of articulated coralline algae (Figure 7-21). The erect fronds arise from a crustose base that may be extensive and persist for a considerable time. The erect "fronds" contain segments or articulations and joints

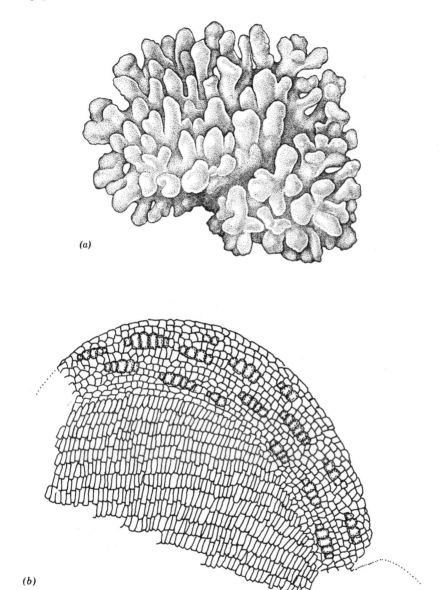

(a)

(b)

Figure 7-20   (*a*) *Lithothamnion incertum*, an example of a nonarticulated, massive coralline (× 3.80). (*b*) A cross section of a tip of a single protuberance of another nonarticulated coralline, *Porolithon,* revealing a two-layered anatomy, one layer of prostrate cell rows
(hypothallus) and the other of erect cell rows (perithallus) (× 100). Scattered heterocysts
or large cells are characteristic of *Porolithon* and are found in the perithallus.

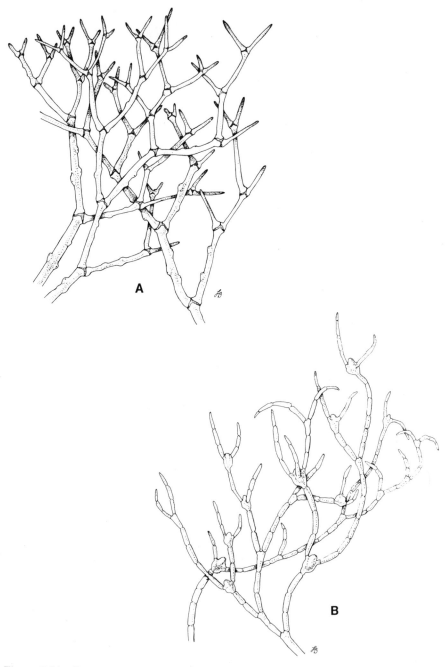

Figure 7-21 Two examples of articulated coralline algae. *Amphiroa fragilissima* (*A*) and *Jania capillacea* (*B*) both have genicula where calcification does not occur. Thus the plants are flexible. Both diagrams show conceptacles being borne on the branchlets; in the case of *Jania*, the conceptacles are found in the axes of branches. × 4.50 (A) and (B).

(*Amphiroa*, Figure 7-21*a*; *Jania*, Figure 7-22*b*). The "articulations," or intergenicula, represent the calcified portions of the thallus, whereas the joints, or genicula, are not calcified and remain somewhat flexible. Construction is multiaxial, with filaments of limited growth giving rise to the calcified intergenicula, and the multiaxial core forming the genicula.

Dumontiaceae

The main characteristic of this family is that the fertilized carpogonium fuses with an intermediate cell in the carpogonial branch from which ooblast (connecting) filaments develop. The gametophytes of this family consist of erect fronds arising from prostrate discoid bases. Most members are uniaxial in construction, although *Dilsea* is multiaxial. Gametangial and tetrasporangial plants may be isomorphic as in the case of *Cryptonemia, Dumontia* (Figure 7-22), *Dudresnaya*, and *Dilsea*; or

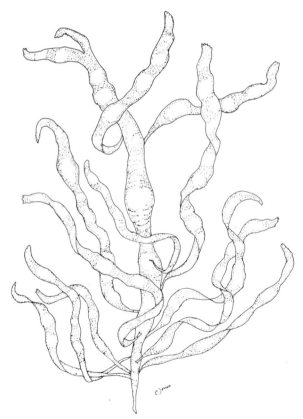

Figure 7-22 *Dumontia incrassata*, an irregularly branched plant that becomes inflated when mature and will harbor the green algal endophyte *Ulonema rhizophorum* in the older, torn regions of the branches. The plants can become quite large, to 6 dm in length, but more commonly are 1–2 dm tall. When young, the plants are terete, solid, and rose-red.

they may be heteromorphic, with a crustose or discoid sporophyte (e.g., *Pikea* or *Acrosymphyton*). The tetrasporongial stages of some plants are still unknown (e.g., *Farlowia*).

Dumontia

One species, *D. incrassata* occurs throughout the North Atlantic (Figure 7-22). The life history is isomorphic; both tetrasporophytes and gametophytes are terete and irregularly branched. The plants develop new uniaxial fronds from a discoid base during the spring. As the plant matures, it becomes hollow and irregularly inflated; the single apical cell is usually lost. The tetrasporophyte produces cruciately divided tetraspores in the inner cortex.

Endocladiaceae

Members of this family develop from a single apical cell and have a prominent axial filament. The carpogonial branches are two celled and develop from the same supporting cells as the auxiliary cell. Members are similar in many respects to the Dumontiaceae. Genera placed in this family include *Endocladia* and *Gloiopeltis*.

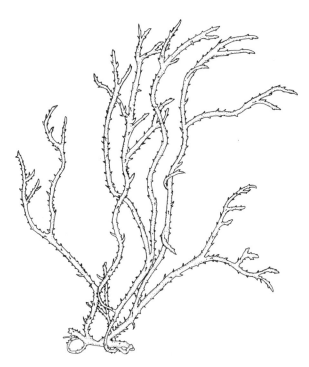

Figure 7-23 *Endocladia muricata,* a short, erect plant, usually dark in color and intertidal in distribution on the west coast of the United States. The wiry plants will reach 4 cm in height and have spines 0.5 mm long.

Endocladia

This genus is particularly common on the west coast of North America. *Endocladia muricata* occurs from Alaska to Baja California, Mexico (Figure 7-23). It is a common plant in the upper intertidal zone. The plants are small, wiry, highly branched, and develop from a small discoid base. The life history is isomorphic. The apical cell is evident and the uniaxial organization is obvious in cross section. The carpogonial branch and auxiliary cell filament develop from a common supporting cell, and the carposporophyte is enclosed by an urn-shaped pericarp. Tetrasporangia are irregularly tetrahedral and are produced at the surface of the tetrasporophyte.

Cryptonemiaceae

Members of this family are erect and arise from discoid bases. The plants are of multiaxial construction and usually consist of highly divided blades. The family is characterized as having auxiliary cells that develop at a distance from the carpogonial branches. The carposporophytes are immersed, and there is no pericarp. Life histories in this family are isomorphic. A large number of genera occur in the tropics (e.g., *Halymenia, Corynomorpha, Cryptonemia, Grateloupia*). An example of a temperate Pacific genus would be *Prionitis*.

Halymenia

Although most of the *Halymenia* species are bladed, the polymorphic genus also has some terete, bushy species, as does the related genus *Grateloupia*. The species may be gelatinous, soft and fleshy, or even membranous blades (Figure 7-24). Construction is multiaxial with a medulla of slender filaments, usually embedded in a jellylike substance. Many of the species have specialized medullary cells called *ganglia* that interconnect through filaments throughout the plant (Figure 7-24*b*). Carpogonial branches consist of two cells, and auxiliary cells are intercalary. Unlike the rest of the family, *Halymenia* has a pericarp of slender filaments, but as with other genera, the carposporophyte is immersed, producing swellings on the blade. The tetrasporophyte produces tetrahedral tetrasporangia (Figure 7-24*b*).

Gloiosiphoniaceae

The gametophytes in this family are soft and bushy, and resemble members of the Dumontiaceae. The plants are terete and erect, and arise from crustose discoid bases. Construction is uniaxial, but it can be masked by rhizoidal growth and cortical development. Where it is known, the tetrasporophyte is a crustose, discoid plant. Carpogonial and auxiliary cell branches arise from the same supporting cell.

Gloiosiphonia

The widely distributed species *G. capillaris* occurs in both the North Atlantic and North Pacific (Figure 7-25). The gametophytes are only found in spring and early

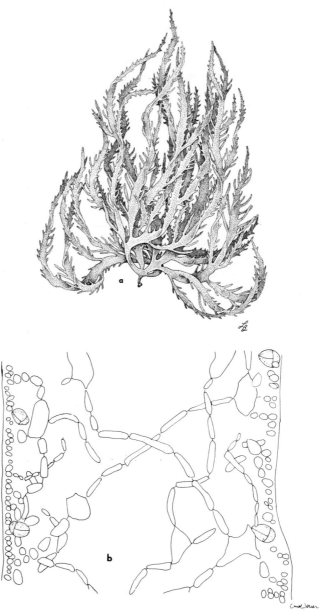

Figure 7-24 *Halymenia floresia*, an example of a large genus found in tropical and temperate waters. (*a*) The habit sketch reveals a soft and gelatinous bladed plant; it is rose-red and will attain lengths of 4 dm. The plant is pinnately divided, although other species are large single blades or even terete forms. The medulla of many species contains stellate "ganglia"-like cells (*b*) as shown in *H. hollenbergii*, a west coast species of North America. These cells measure about 50 μm in diameter. Note also the tetrahedrally divided tetrasporangia in (*b*).

Figure 7-25 *Gloiosiphonia capillaris*, a species of the west coast of North America that will reach 15 cm in height, but more commonly the rose-colored plant is about 7–9 cm tall. The plant is gelatinous to slightly firm, semiterete to flattened, and occurs in the low intertidal region and tide pools during the spring.

summer and consist of irregularly cylindrical thalli with spindle-shaped branches and discoid bases. Gametophytes are monoecious; the spermatangia are produced in superficial patches, whereas the carpogonial and auxiliary cell branches are produced from the same basal or supporting cell. Carposporophytes are immersed and lack a pericarp. Culture studies have shown that the carpospore produces a prostrate, crustose tetrasporophyte. The construction of the tetrasporophyte is by erect filaments up to seven cells high embedded in a gelatinous matrix. Tetrasporongia show all patterns of division (zonate, cruciate, and tetrahedral).

Kallymeniaceae

In this family, construction is multiaxial, and the plants range from simple or divided blades to strap-shaped forms. All develop from small discoid bases, and where known have isomorphic life histories. The internal anatomy of most genera consists of a thin cortex and a thick medulla of filaments. Gametophytes are dioecious. Spermatangia are scattered in superficial patches. Carpogonial and auxiliary cell arrangements show a wide range of development and include a transition from *nonprocarpial* to *procarpial* systems. The term *procarp* refers to a regular, close association between the carpogonial and auxiliary cells. The carposporophytes are immersed and large with a well-developed pericarp. The cystocarps frequently have beaked ostioles. The tetrasporophytes produce cruciately divided tetraspores in the cortex of the blade in special proliferations. Examples of genera included in this family are *Kallymenia* and *Callophyllis* (Figure 7-26), which occur on both coasts, and *Pugetia* (*Callophyllis firma*, North Pacific). Another common North Atlantic species, *Euthoria cristata*, has recently been reassigned to *Callophyllis* as *C. cristata*.

Callophyllis

Pacific coast species are mostly subtidal and include *C. obtusifolia*, *C. crenulata*, and seven other species. The plants are bladed with a wide variety of forms; all, however, lack ribs or veins (Figure 7-26*a*). Internal construction is multiaxial, with a thin cortex and a medulla consisting of large and small cells (Figure 7-26*b*). The small cells are filaments that develop from cortical cells. The carpogonial branches develop from a large, lobed, supporting cell. The supporting cell functions as the auxiliary cell. After receiving the zygote nucleus, the auxiliary cell fuses with surrounding cells. The carposporophyte is surrounded by a well-developed pericarp. Tetrasporophytes have cruciately arranged tetraspores that are scattered over the surface, or they are restricted to special proliferations. They are formed from cells of the inner cortex.

Peyssonneliaceae (Squamariaceae)

This family has members that are prostrate or crustose and lightly calcified. In some species only tetrasporophytes are known; thus these may represent stages in the life history of genera in which only gametophytes are presently known. In other cases an isomorphic life history is known. Genera placed in this family include *Peyssonnelia, Cruoriopsis, Rhodophysema,* and some times *Hildenbrandia*.

Peyssonnelia

Species of this genus are distributed worldwide. The plants are crustose and are attached by ventral rhizoids. The internal construction of the crusts consists of a basal cell layer from which vertical rows of cells are produced. In some species both gametophytic and sporophytic phases are present, whereas in others only tetrasporophytes are known. The carpogonial branches and auxiliary branches are four celled. Tetrasporongia occur in shallow pits and are cruciately divided.

Gigartinales

Gigartinales is one of the largest orders, with about 20 families and over 150 genera (Bold and Wynne, 1978), which range from crusts and blades to fleshy plants. Both isomorphic (*Eucheuma*) and heteromorphic (*Gigartina*) life histories

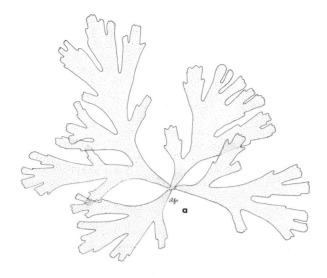

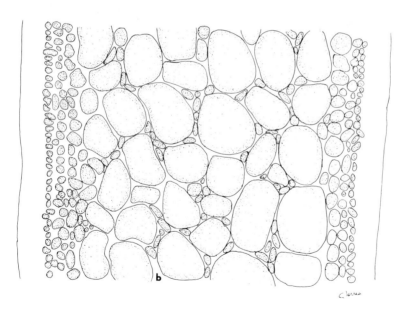

Figure 7-26 The genus *Callophyllis* is a large, polymorphic one with representatives on both coasts of North America. (*a*) *Callophyllis flabellata* is a common subtidal species of the west coast that will reach about 12 cm in length. (*b*) In cross section, the parenchymatous construction is evident. A characteristic of the genus is the uniseriate filaments that develop from epidermal cells and grow among the medullary cells. The larger cells can reach 40 μm in diameter.

are known in the order. The single most unifying characteristic of the order is that the auxiliary cell is an ordinary intercalary vegetative cell. Auxiliary cells may be in close proximity (procarpial) or distant from the carpogonium (nonprocarpial).

Many of the members of this order are economically important (see Chapter 2 and Figure 2-8), producing various phycocolloids, such as carrageenan (e.g., *Chondrus, Eucheuma*). Thus a number of biochemical and ecological studies have been carried out on these plants (Dawes, 1979; Doty, 1979; Harvey and McLachlan, 1973; see also Chapter 2). Several parasitic genera are also contained in this order such as the adelphoparasites *Hypneocolax* on *Hypnea* and *Gracilariophila* on *Gracilaria* (Fan, 1961). For brevity only six families are presented here.

Cruoriaceae

These plants are crustose, composed of prostrate radiating filaments from which erect filaments arise. The erect filaments may be embedded in mucilage. The auxiliary cell is remote from the carpogonium, and the ooblast (connecting) filament is quite long. Many species lack a gametophytic phase and have only a tetrasporophytic stage. A number of "species" appear to be the tetrasporophytic stages in heteromorphic life histories of other red algae. For example, a *Cruoriopsis*-like plant is involved in the life history of *Gloiosiphonia*, a member of the Cryptonemiales. *Petrocelis middendorffii* represents the tetrasporangial phase of a species of *Gigartina*. Plants of *Cruoria rosea* are the tetrasporangial phases in the life histories of *Halarachnion ligulatum* and *Turnerella pennyi*. Thus the circumspection and validity of this family is a matter of controversy.

Solieriaceae

Plants in this family may be cylindrical or foliaceous, and simple or branched to a variable degree. Internal construction is multiaxial. The life histories are isomorphic with nonprocarpial carposporophytes. The tetrasporangia have zonately arranged tetraspores, whereas the cystocarps are immersed or emergent, usually with an obvious ostiole. All of the genera placed in this family apparently are carrageenophytes and include *Agardhiella, Neoagardhiella, Eucheuma, Opuntiella*, and *Solieria*.

Agardhiella

Wynne and Taylor (1973) proposed that *A. tenera* be moved to *Solieria tenera* and *A. ramosissima* be placed in a new genus, *Neoagardhiella*. In a description of marine algae of the west coast of Florida, Dawes (1974) presented the arguments for these taxonomic changes. Because the taxonomic status of *Agardhiella* still seems to be in question (Kraft and Wynne, 1979), the name is retained here.

Agardhiella baileyi (Neoagardhiella baileyi) is found on the Pacific and Atlantic shores of North America and is characterized by a simple basal attachment and thin-walled medullary filaments. *Agardhiella tenera (Solieria tenera)* (Figure 7-27) is a warm-water plant, with thick-walled, continuous longitudinal medullar

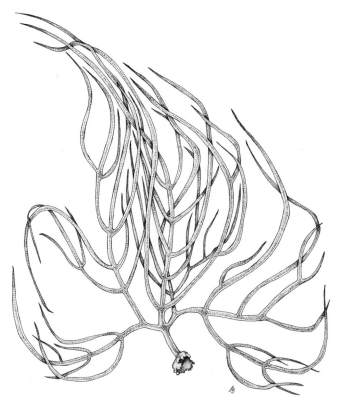

Figure 7-27 *Agardhiella tenera (Soleria tenera),* a terete, bushy plant that can reach 3 dm in length. It is attached by a fibrous base. The deep-to rose-red plant is found on both the eastern and western coasts of Florida and yields ι-carrageenan. The branches range from 1–4 mm in diameter and taper from base to tip.

filaments and a loose fibrous basal attachment. *Agardhiella ramosissima (Neoagardhiella ramosissima)* is a compressed-to-flattened species; the other two are terete. Cystocarps are embedded and the carpogonial branches are three celled.

Eucheuma

This genus is pantropical and is an important source of carrageenan. It is extensively farmed in the Philippines (see Chapter 2). Ultrastructural studies have been carried out to evaluate carrageenan localization (LaClaire and Dawes, 1976). The species are polymorphic, ranging from finely terete to coarsely and irregularly branched. *Eucheuma isiforme* is an example of the latter type (Figure 7-28*a*). The genus has a large fusion cell (Figure 7-3) and the cystocarps are usually found on specialized projections or papillae (Figure 7-3). In construction, the plant consists of a thick cortex and medullary filaments that may be densely packed (Figure 7-28*b*), randomly dispersed, or lacking.

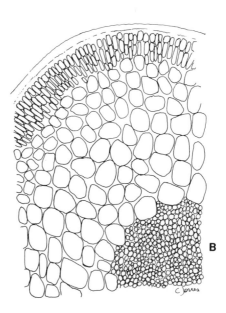

Figure 7-28 The genus *Eucheuma,* a pantropical carrageenophyte that is well represented in both Pacific and Caribbean waters. (*A*) *Eucheuma isiforme* is an erect (to 5 dm tall),
bushy plant of the Caribbean that is firm to cartilaginous and ranges from deep red to light
yellow. The branches are beset with opposite or whorled spines. (*B*) In cross section, the
medulla is composed of densely packed filaments.

Hypneaceae

This family has few genera; these are mostly tropical and contain plants with single apical cells. A uniaxial construction is evident in young branches but is usually obscured in older plants. The plants have procarpial arrangements. Life histories are isomorphic and tetrasporangia are zonately divided. The cystocarp is spherical and is formed on a short, terminal axis.

Hypnea

This genus produces the phycocolloid κ-carrageenan and is common in tropical and subtropical waters of the Atlantic and Pacific (Figure 7-29). The plants are bushy or tufted, with cylindrical axes 1–2 mm in diameter. The tips of many axes are elongate and may be characteristically hooked as in the hamate branches of *H. musciformis* (Figure 7-29).

Figure 7-29 *Hypnea*, a worldwide genus, is harvested for k-carrageenan. The most common species is *H. musciformis*, characterized by hemate or hooked branch tips that entangle with other seaweeds. The main branches are about 1–2 mm in diameter, and the plants can reach 10–30 cm in length.

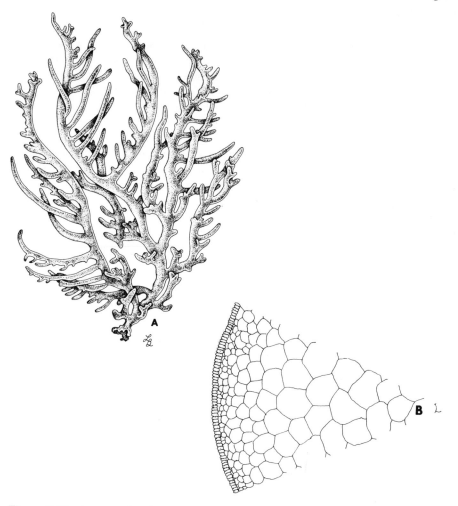

Figure 7-30 An example of a coarse species of the worldwide genus *Gracilaria* is *G. debilis*, which is common throughout the Caribbean. The plants may be 2 dm in height and have a cervicorn or secund branching pattern (*A*). The plants range from deep red (deep-water plants) to a dull brown or yellow (shallow-water plants). (*B*) The parenchymatous construction is visible in cross section.

Gracilariaceae

This family is characterized by multiaxial construction and a parenchymatous anatomy (Figure 7-30*b*). Tetrasporangia are cruciately divided and occur scattered over the surface of the thallus. Male and female gametophytes are found, and in one subgeneral group the spermatangial branches occur in pits. The mature cysto-carp protrudes from the thallus proper, with an obvious ostiole. The largest genus is *Gracilaria* with over 100 species ranging from cool to tropical waters.

Gracilaria

This genus is highly polymorphic with species ranging from terete highly branched forms to compressed or flattened blades. An example of a more coarse form is *G. debilis* (Figure 7-30a). A number of species are harvested for the phycocolloid agar. Where known, the life history is isomorphic, although in many cases reproductive plants are not found. In fact, vegetative reproduction by fragmentation appears to be very common among species in the genus. Tetrasporangia are cruciately divided. Gametophytes are dioecious, and spermatangia may occur in pits. The cystocarps are prominent, having a well-developed pericarps and evident ostioles.

Phyllophoraceae

Members are multiaxial with a large-celled pseudoparenchymatous medulla. Plants range from terete forms to blades or crusts. The medulla is large celled and the tetrasporangia occur in nemathecia (raised over the surface). A variety of life histories are present, including both isomorphic and heteromorphic forms. In addition, cases have been reported of gametophytic plants bearing carposporophytes that produce cruciately divided carpospores, and no free living tetrasporophyte. Such a pattern suggests that meiosis is zygotic. Genera having varied life histories include *Gymnogongrus*, *Ahnfeltia*, and *Phyllophora*.

Gymnogongrus

The gametophytes of this genus are dichotomously branched, occurring on both coasts of North America (Figure 7-31). Two distinctive life histories are known.

Figure 7-31 Although many of the members of the Phyllophoraceae are bladed or foliose, *Gymnogongrus griffithsiae* is terete and grows in tufts 2–5 cm high. The cystocarpic plant shown here is repeatedly branched in the upper region and the tips can be flattened.

Figure 7-32   Species of *Gigartina* on the west coast of the United States are highly varied in morphology, but bladed forms predominate. The species *G. acicularis,* found on the east coast of Florida, is a small (to 8 cm tall) terete plant attached by a fibrous base.

In some species (e.g., *G. platyllus*), the carpospores are cruciately divided (carpotetrasporangia), suggesting meiosis. Tetrasporophytes are known for only some species and may be similar to gametophytes (isomorphic) or distinct and occur as prostrate crusts (heteromorphic).

Ahnfeltia and Phyllophora

Ahnfeltia plicata is found in both the North Atlantic and on the west coast of North America and is a source of agar. Gametophytes are erect, wiry, black-appearing plants found in intertidal and subtidal zones. Monosporangia produced by the gametophytes repeat the phase. In another species, *A. gigartinoides*, it appears that carpospores produce a crustose plant resembling *Petrocelis*. At least in one species of *Phyllophora*, a leafy subtidal species called *P. truncata*, the tetrasporophytic phase grows from the haploid gametophyte, and thus can be considered parasitic on the haploid plant.

Gigartinaceae

This family contains erect, terete (Figure 7-32) or foliose plants that have perennial basal discs. Construction is multiaxial with a filamentous medulla. The cells of the cortex are often arranged in perpendicular rows. Although most life histories are isomorphic, examples of heteromorphic life histories have been reported. John West and his students (West *et al.*, 1978) have found that a *Petrocelis*-like crustose plant is the tetrasporophyte in the life cycles of a number of species of *Gigartina*, including *G. agardhii, G. stellata*, and *G. papillata*. Gametophytes in this family are dioecious. Where known, spermatangia are produced superficially. Tetrasporangia are cruciately divided. Four genera are recognized in this family: *Chondrus, Gigartina, Irideae,* and *Rhodoglossum*; all are carrageenophytes. Recent revisions would include *Iridaea* and *Rhodoglossum* in the genus *Gigartina*.

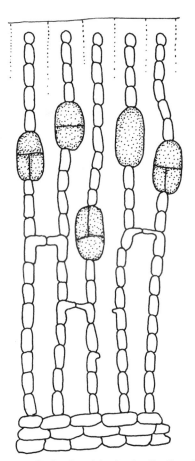

Figure 7-33 The genus *Petrocelis* is placed in the family Crouriaceae and is a crustose, tetrasporangial plant consisting of a hypothallus and erect filaments embedded in a gelatinous matrix (perithallus). The species shown here, *P. middendorffii*, apparently is the tetrasporangial phase of *Gigartina papillata*. Other species, *P. cruenta* and *P. anastomosans*, are apparently also tetrasporophytic phases of *G. stellata* and *Ahnfeltia gigartinoides*, respectively. Thus, the genus is taxonomically confused. Note the connections between the perithallial cells. Cells are 5–10 μm in length, tetrasporongia 10–20μm.

Gigartina

These erect plants are multiaxial and some of the bladed forms may reach 1 m in breadth and more that 2 m in length, while other species are small, terete forms as shown in Figure 7-32. Life histories are isomorphic in some species, while in others the tetrasporophyte is lacking or crustose. The heteromorphic species are placed in the subgenus *Mastocarpus*. The crustose plants resemble *Petrocelis* (Figure 7-33). A good example of this genus is *G. stellata*, which occurs throughout the North Atlantic. It is an important source of carrageenan. The plants have holdfasts that continue to produce new fronds through a perennating holdfast. The gametophyte is a narrow stripelike plant with an expanded blade that is dichotomously branched and about 20 cm tall. The spermatangia are formed on short papillae, and the mature cystocarps occur as conspicuous swellings borne on longer papillae. Tetrasporophytes are absent or may be a crustose phase formerly referred to as *Petrocelis*.

Chondrus

Chondrus crispus (Figure 7-34) is a widespread species of the North Atlantic and is a highly important source of κ-carrageenan (gametophytes) and λ-carrageenan

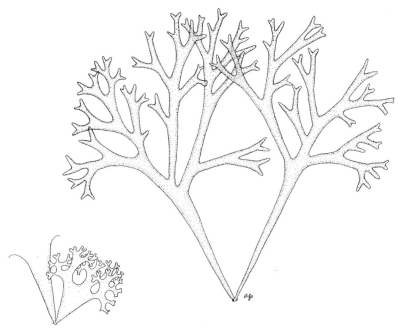

Figure 7-34 *Chondrus crispus,* harvested extensively along the northeastern coasts of North America, is a highly polymorphic species. The blades develop from disclike holdfasts and reach 8–15 cm in length. They may be broad and flat or narrow and compressed. Two examples are shown here. Reproductive structures (not shown) are evident as raised areas on the blades.

(sporophytes). Its biology, ecology, physiology, and biochemistry have been summarized (Harvey and McLachlan, 1973). The plants are erect fronds (20 cm) arising in tufts from a perennating discoid base. The upper portion is a fanlike blade, the form being extremely variable. The species has a typical triphasic red algal life history.

Rhodymeniales

About 12 genera are placed in three families in this order. The plants are multiaxial, many have a hollow construction, and the typical forms are blades or terete branches. Most life histories are isomorphic, but anomalous life cycles are also known (Palmariaceae). The order is characterized by a procarp consisting of carpogonial branches that contain three or four cells and by auxiliary cell branches containing two cells, formed before fertilization (Figure 7-37b). Tetrasporangia are cruciately or tetrahedrally divided.

Rhodymeniaceae

Plants in this family have cruciately divided tetrasporangia and typical triphasic, isomorphic types of life histories. The plants may be flat or cylindrical and include genera that have solid (*Rhodymenia*, Figure 7-35; *Fauchea, Leptofauchea*) or hollow (*Chrysymenia*, Figure 7-36, and *Botryocladia*) internal construction with no longitudinal filaments.

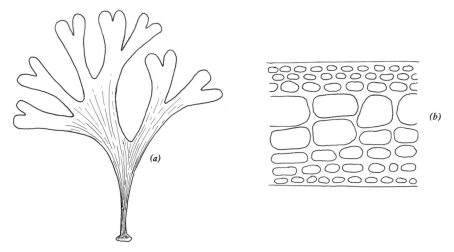

Figure 7-35 *Rhodymenia pseudopalmata*, a flat, dichotomously branched blade having a fibrous to discoid holdfast and reaching 4 cm in height (*a*). In cross section (*b*) the plant is parenchymatous and of two layers: internal oblong medullary cells and an outer layer of regularly arranged cortical cells.

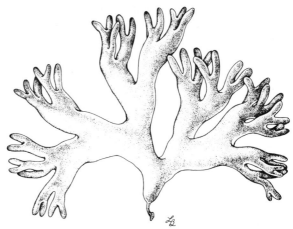

Figure 7-36 *Chrysymenia halymenioides,* a hollow member of the Rhodymeniaceae. It lacks internal filaments, a feature of the other family, Champiaceae. The species is saccate and dichotomously branched, reaching 7–10 cm in length. It is found in deeper waters throughout the Caribbean.

Rhodymenia

This genus is typified by a flattened morphology, at least in the upper portion of the blades (Figure 7-35*a*). About 50 species are contained in this genus. These are primarily found in the Northern Hemisphere and are especially common in deeper (30–50 m) water. The plants are parenchymatous, with large medullary cells and smaller cortical cells arranged in rows (Figure 7-35*b*). Gametophytes are dioecious, with spermatangia scattered in superficial terminal clusters. The supporting cell bears a three- to four-celled carpogonial branch and a two-celled auxiliary cell filament. The carposporophyte forms outwardly with a domed pericarp. Tetrasporangia are found in superficial terminal clusters.

Champiaceae

Plants in this family are at least in part hollow with cavities lined by prominent longitudinal filaments that are absent in the hollow members of the Rhodymeniaceae. Tetrasporangia are tetrahedrally divided but are not produced in all species; polysporangia may be produced instead. Common genera include *Champia* (Figure 7-37), *Gastroclonium,* and *Lomentaria*.

Champia

These plants are common in subtropical and tropical seas. They are branched, tubular plants usually constricted at intervals and having internal cellular septa at the constrictions (Figure 7-37*a*). Plants occur as tufts, to 5 cm tall. The life history is isomorphic, with dioecious gametophytes and a procarp association (Figure 7-37*b*). Tetrasporangia are found in the cortex in small pits and are tetrahedrally divided.

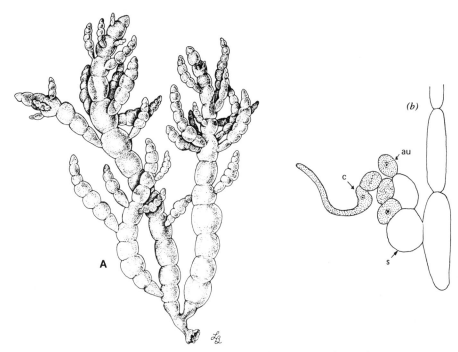

Figure 7-37   The genus *Champia* is typical of the family Champiaceae in having hollow segments with internal longitudinal filaments. The species shown here, (*A*), *C. salicornioides*, can have segments up to 4 mm in diameter, and the plant can reach 12 cm in length. (*b*) The carpogonial branch has a procarpial organization with the auxiliary cell (*au*) attached to the same supporting cell (*s*) as the carpogonium (*c*).

Palmariaceae

This newly described family presently contains three genera (Guiry, 1974). The species were removed from the Rhodymeniaceae because of the absence of carpo-gonial plants as well as the presence of stalk cells on the tetrasporangia. Howev-er, van der Meer and Todd (1980) found that tetraspores from diploid *Palmaria palmata* plants (Figure 7-38*a*) germinated into robust male and small, short-lived female haploid plants. After fertilization of the carpogonia (no carpogonial branches nor auxilliary cells were observed), the diploid tetrasporangial plants developed directly on the gametophytes from the zygotes. Thus there appears to be no carposporophyte. The development of tetrasporangia (Figure 7-38*b*) in-volves the enlargement and periclinal division of a cortical cell to give two prod-ucts of markedly different size. Pueschel (1979) has described the tetraspore ultrastructure in *P. palmata*. The outermost cell enlarges to form the cruciately divided tetrasporangium, whereas the innermost one forms the stalk cell. The three genera placed in this family are *Palmaria*, *Halosaccion*, and *Leptosacus*.

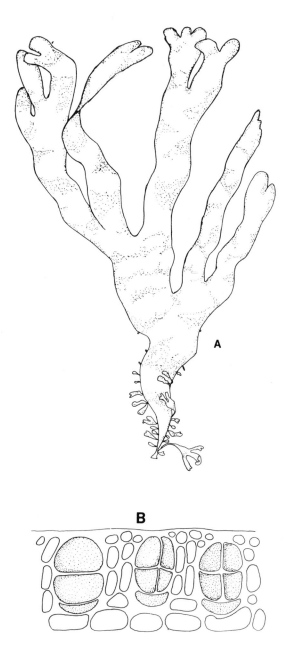

Figure 7-38 *Palmaria palmata* (*Rhodymenia palmata*), a strap-shaped blade that when young is dichotomously branched. When mature it may reach lengths of 5 dm (*A*). Guiry (1974) renamed this species because of the apparent lack of female gametophytes and a stalk cell present in tetrasporangial development as seen in cross section (*B*). The tetrasporangia occur in sori and are 30 × 50 μm, lying near the blade surface (*B*).

Palmaria

The plant that was originally called *Rhodymenia palmata* (Figure 7-38*a*) is dried and commonly eaten as dulse, especially in the Canadian maritime provinces. Although primarily found in the North Atlantic, a variety, *mollis*, is found on the west coast of the United States. The plant consists of a blade that can become highly divided over a period of years. Primary blades reaching 1.8 m in length are common. Internal construction is multiaxial, with a parenchymatous inner cortex containing large, spherical cells and an outer cortex of small cells. Spermatangial plants are common, but no female gametophytes were found until the study of van der Meer and Chen (1979). The tetrasporangia have stalk cells typical of the family.

Ceramiales

Ceramiales is the largest and most clearly defined order in the red algae, having four families, about 240 genera, and over 1000 species. The major characteristic that delimits members of this order is the production of the auxiliary cell *after* fertilization and directly from the supporting cell of the carpogonial filament. In polysiphonous forms, the supporting cell is usually a pericentral cell (Figure 7-34*b*). The carpogonial branch is four celled. A detailed account of the reproduction is given for *Polysiphonia* in Figures 7-43 and 7-44. Although the reproductive features are well defined, morphologies vary greatly, ranging from filaments (uniseriate or polysiphonous) to delicate blades and massive parenchymatous thalli. Most of the known life histories in this order are of the typical triphasic type.

Ceramiaceae

Plants in this family are terete and uniseriate, although some may be extensively corticated (*Spyridia*). A distinct pericarp is lacking; however, involucral filaments may be losely arranged around the carposporophyte. The genera may completely lack cortication (*Antithamnion, Griffithsia*), be corticated only in bands (*Ceramium, Crouania*), or be extensively corticated (*Spyridia, Ptilota*). *Griffithsia* is unique because of the large size of the multinucleated cells of some species (*G. globulifera*), so the plant has been used in experimental studies on cell repair or regeneration (Waaland and Cleland, 1974). Prince (1979) recently demonstrated that the life history of *Crouania pleonospora* is typically triphasic isomorphic (*Polysiphonia*-type).

Antithamnion

The taxonomy of species placed in this genus has been under review, and a number of new genera have been established (*Hollenbergia, Antithamnionella*). The plants are small, uniseriate, delicate algae completely lacking cortication (Figure 7-39). Branching is opposite with refractive gland cells that may have a secretory function (Young and West, 1979) present on the upper side of the determinate (of definite length) branches (Figure 7-39). The lateral branches are deter-

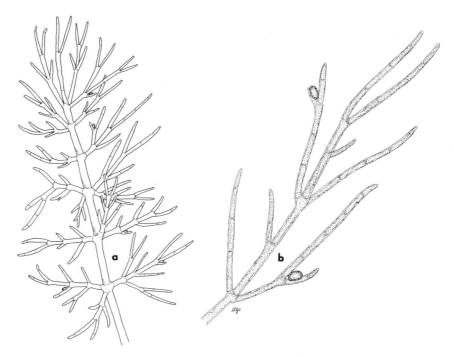

Figure 7-39 (*a*) *Antithamnion cruciatum*, a uniseriate, tufted plant, the cells of the main axis reaching 90 μm in diameter. (*b*) Lateral branches bear "gland" or "refractive" cells in the axis of the branchlets (Young and West, 1979). The plants are common epiphytes on coarse algae and rocks, and species are found throughout the world in both temperate and tropical waters. See also Figure 7-1.

minate so that the form of the thallus is very well defined. A typical triphasic life history has been recorded for some, but not all, species (see Bold and Wynne, 1978). Gametophytes and tetrasporophytes, if present, are isomorphic. The male gametophyte bears clusters of spermatangia on the adaxial sides of determinate branches. The female gametophyte bears a four-celled carpogonial branch on the lowermost cell of a determinate branch with the trichogyne directed apically. The auxiliary cell is produced from the supporting cell after fertilization. The carposporophyte is relatively well exposed, being surrounded by a few involucral filaments. In some species, the tetrasporophyte appears to be the most common phase found in nature, producing tetrahedrally or cruciately divided tetrasporangia.

Delesseiraceae

This family has about 75 genera and consists of plants that are mostly bladed, often very delicate and membraneous (i.e., monostromatic to multistromatic). Some of the most beautiful species of red algae belong to this family because of the veinlike polysiphonous structures in the blade medians. Two types of growth patterns can be found: distinct apical growth and marginal meristematic growth. The latter type of growth can become intercalary, resulting in blades with proliferations. Two subfamilies are distinguished, based upon the positions of the procarps and cystocarps. In the Nitophylloideae, procarps and resulting cystocarps occur scattered over the entire thallus surface, while in the Delesserioideae, procarps and cystocarps are only found along the midrib. Genera characteristic of this family include *Platysiphonia, Membranoptera, Delesseria, Caloglossa, Grinnellia, Phycodrys,* and *Polyneura.*

Caloglossa

This genus is probably the most widespread of all the genera in this family, occurring in mangrove swamps and salt marshes and ranging from tropical to temperate tidal marshes. Thus this genus has been used in physiological studies (see Chapter 14). The plants are dorsiventrally flattened, having narrow, dichotomously branching blades that are monostromatic except for the polysiphonous "veins" (Figure 7-40). Attachment is through rhizoids on the ventral surface. The primary branches are exogenous in origin (apical cell), whereas the lateral branches originate endogenously (from intercalary meristem of midrib). Carpogonial branches

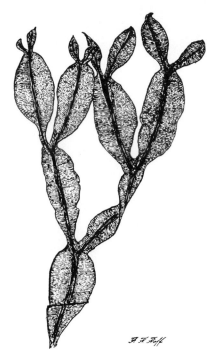

Figure 7-40 *Caloglossa leprieurii,* a small (1–3 cm long) bladed alga that is rose to purple. The plant is monostromatic except for the polysiphonous "veins." Segments are 4–6 mm in length. Members of the species are common as epiphytes on the bases of salt marsh and mangrove plants along the subtropical and temperate coasts of eastern North America.

are four celled, developing from a supporting cell that is also one of the pericentral cells. Tetrasporangia are cruciately divided.

Dasyaceae

This is a small family whose members demonstrate sympodial growth. That is, the apical cell is constantly replaced by a lateral cell so that the growing point is frequently shifted. Typically there are five pericentral cells. The axis is covered

Figure 7-41 The habit and detail of a branchlet of *Dasya caraibica*, revealing the delicate monosiphonous ultimate branches. The bushy plants can reach 20 cm in length and the monosiphonous branchlets 24 μm in diameter. As with other members of the genus, *D. caraibica* has a polysiphonous main axis and the uniseriate branchlets dehisce along the older regions of the axes.

by short lateral pigmented filaments. The pericarp is well developed in this family, and two groups of sterile cells are present. Tetrasporangia are born on specialized branches called stichidia. Some typical genera include *Dasya, Dasyopsis,* and *Heterosiphonia.*

Dasya

This plant is a bushy (generic meaning) red alga (*D. caraibicea*, Figure 7-41). The most common species in the North Atlantic is *D. baillouviana* (*D. pedicellata*), which occurs from the Caribbean into the North Atlantic and more recently to Northern Europe. Some of the species may reach 50 cm in length. Gametophytes produce clusters of spermatangial (yellowish-red) branches. Female gametophytes have a four-celled carpogonial branch born on a pericentral (support) cell, along with a group of sterile cells. The cystocarp becomes stalked and is vase shaped. The tetrasporophytes bear numerous stichidia, each having serially maturing and tetrahedrally divided tetraporangia.

Rhodomelaceae

This family is the largest of the order, having more than 100 widely differing genera. The polysiphonous construction is obvious in some genera and obscured in others. Typically, there are five pericentral cells that are cut off in an alternating sequence, the last formed opposite the first one. The apical cell is prominent. In some members extensive divisions of the pericentral cells occur and a massive even hollow, thallus is formed. Exogenous filamentous branches are formed outwardly from the apical cell, and these trichoblasts bear the sexual organs. In some species, these exogenous branches may be formed in all thalli and persist, whereas in others, they are ephemeral and dioecious. A small basal scar cell persists after the trichoblast falls off; it is cut off from the apical cell before the pericentral cells. The cells of the trichoblasts are usually colorless (but see *Brongniartiella*). The life histories are isomorphic and the gametophytes are dioecious. Members of this family are common to most parts of the world and include *Bryothamnion, Chondria, Digenia, Bostrychia, Laurencia, Acanthophora, Wrightiella* (Figure 7-42), and *Polysiphonia* (Figure 7-43, 7-44). A number of genera are adelphoparasites on other members of this family: *Levingiella* on *Pterosiphonia* and *Janczewskia* on *Laurencia*. Studies of parasitic species in this family have been carried out by Krugens and West (1973).

Polysiphonia

The genus has over 150 species and is found worldwide. The plants are usually radially organized and well branched (Figure 7-43a). Growth is by a well-defined apical cell that cuts off proximal cells that in turn cut off the pericentral cells and a central siphon or axial cell. Species may have rhizomes with rhizoids (Figure 7-44c) or holdfasts. The number of pericentral cells of this genus range from four to twenty. The pericentral cells are connected by primary pit connections to the central siphon and to each other by secondary pit connections. Trichoblasts arise exogenously and the basal *scar cell* can be found after they fall off. Trichoblasts

Figure 7-42 A member of the Rhodomelaceae, *Wrightiella tumanowiczi*, has a polysiphonous main axis that is beset with monosiphonous and pigmented branchlets as shown in the detailed inset. The plants can become quite large, reaching 75 cm in length. The four pericentral cells of the major axes can be hidden because of cortication. The ultimate monosiphonous branchlets are about 20–30 μm in diameter.

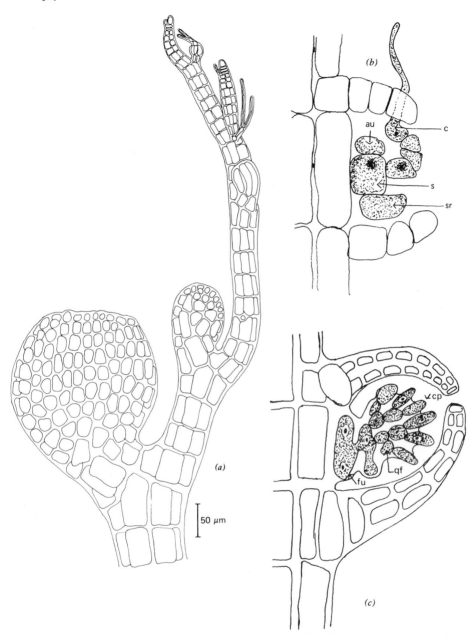

Figure 7-43 The reproductive structures found on the female gametophyte of *Polysiphonia denudata*, demonstrating the reproductive structures of the order Ceramiales. (*a*) Young, immature carpogonial branches are found as well as mature cystocarps. (*b*) A mature and recently fertilized carpogonium (*C*) is shown with the newly formed auxiliary cell (*au*) as well as the supporting cell (*s*) and sterile cell (*sr*). (*c*) A young carposporophyte with developing sterile filaments with carpospores (*cp*), gonimoblast filaments (*gf*) and the original fusion cell (*fu*).

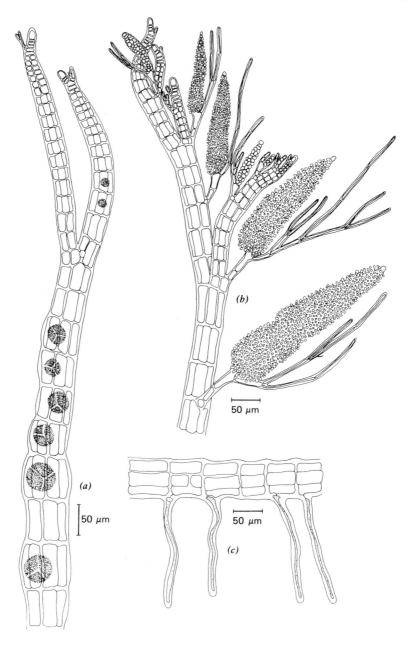

Figure 7-44 (*a*), A branch of *P. denudata* bearing internal tetrasporangia. (*b*) A branch of a male gametophyte showing the spermatangial branches. (*c*) A short rhizome segment with rhizoids. All unit marks equal 50 μm. (Figures 7-43*a*, 7-44*a*, 7-44*b*, and 7-44*c* courtesy of Don Kaprun, University of South Carolina, Wilmington.)

are formed in a regular spiral sequence. Endogenous branches are less common and originate from the axial cell after the pericentral cells have been formed. In most species, the branches develop from cells cut off from the basal cell of a trichoblast (Figure 7-43a). The branch may be formed in the axis of a trichoblast or it may replace a trichoblast. The scar cells, or the basal cells of trichoblasts remaining after the trichoblast is lost, may serve as the site of formation of lateral axes in other species. The male gametophytes produce specialized spermatogonial branches that are equivalent to the trichoblasts (Figure 7-44b).

Carpogonial branches are formed on young trichoblasts near the apex, one from each trichoblast (Figure 7-43a). A trichoblast that is to produce a carpogonial branch becomes polysiphonous, and the last pericentral cell to be formed functions as the fertile cell from which the carpogonial branch develops. The last pericentral cell then divides unequally to produce a large carpogonial branch initial and a smaller supporting cell. The carpogonial branch is a four-celled branch and the supporting cell cuts off one basal sterile cell and two flanking sterile cells. After fertilization, the carpogonium increases in size and the trichogyne is cut off and disappears. The fusion nucleus divides, as do the sterile cells. The basal sterile cell divides once and each of the lateral cells twice. The supporting cell (pericentral cell) enlarges and the auxiliary cell is produced, underneath the carpogonium (Figure 7-43b). Through simple fusion or by a connecting ooblast filament, the zygote nucleus is transferred to the auxiliary cell. The carpogonium and the cells of the carpogonial branch will degenerate. The auxiliary cell produces on its outer side a gonimoblast mother cell from which a branched gonimoblast develops with the terminal cells converting into carposporangia (Figure 7-43c). The surrounding sterile cells produce the ostiolate pericarp that encloses the carposporophyte (Figure 7-43a). The tetrasporophyte bears the tetrahedral tetrasporangia on fertile tiers (Figure 7-44a).

REFERENCES

Aghajanian, J. G. 1979. A starch grain-mitochondrion-dictyosome association in *Batrachospermum* (Rhodophyta). *J. Phycol.* **15**: 230–232.

Bold, H. C. and M. J. Wynne. 1978. *Introduction to the Algae*. Prentice-Hall, Englewood Cliffs, N.J.

Borowitzka, M. A. and M. Vesk. 1979. Ultrastructure of the Corallinaceae (Rhodophyta) II. Vegetative cells of *Lithothrix aspergillum*. *J. Phycol.* **15**: 146–153.

Cole, K. and E. Conway. 1980. Studies in the Bangiaceae: reproductive modes. *Bot. Mar.* **23**: 545–553.

Dawes, C. J. 1974. *Marine Algae of the West Coast of Florida*. University of Miami Press, Coral Gables, Fla.

Dawes, C. J. 1979. Physiological and biochemical comparisons of *Eucheuma* spp. (Florideophyceae) yielding *iota*-carrageenan. In Jensen, A. and J. R. Stein (eds.). *Proc. Ninth Intern. Seaweed Symp.*, Science Press, Princeton, N.J., pp. 199–207.

Dixon, P. S. 1973. *Biology of the Rhodophyta*. Hafner Press, New York.

Dixon, P. S., S. N. Murray, W. N. Richardson, and J. L. Scott. 1972. Life history studies in genera of the Cryptonemiales. *Soc. Bot. Fr. Memoires* **1972**: 323–332.

Doty, M. S. 1979. Status of marine agronomy, with special reference to the tropics. In Jensen, A. and J. R. Stein (eds.). *Proc. IX Intern. Seaweed Symp.*, Science Press, Princeton, N.J., pp. 35–58.

Drew, K. M. 1949. *Conchocelis*-phase in the life history of *Porphyra umbilicas* (L) Kutz. *Nature* **164**: 748.

Evans, L. V., J. A. Callow, and M. E. Callow, 1978. Parasitic red algae: an appraisal. In Irvine, D. E. G. and J. H. Price (eds.). *Modern Approaches to the Taxonomy of Red and Brown Algae*. Academic, New York, pp. 87–110.

Fan, K-C. 1961. Studies on *Hypneocloax*, with a discussion on the origin of parasitic red algae. *Nova Hedwigia* **3**: 119–128.

Fan, K-C. and G. F. Papenfuss. 1959. Red algal parasites occurring on members of the Gelidiales. *Madrono* **15**: 33–38.

Gantt, E., M. R. Edwards, and S. F. Conti. 1968. Ultrastructure of *Porphyridium aerugineum* a blue-green colored rhodophytan. *J. Phycol.* **4**: 65–71.

Giraud, G. and J. Cabioch. 1979. Ultrastructure and elaboration of calcified cell-walls in the coralline algae (Rhodophyta, Cryptonemiales). *Biol. Cellulaire* **36**: 81–86.

Goff, L. J. 1979. The biology of *Harveyella mirabilis* (Cryptonemiales, Rhodophyceae). VII. Structure and proposed function of host-penetrating cells. *J. Phycol.* **15**: 87–100.

Gordon-Mills, E. M., J. Tas, and E. L. McCandless. 1978. Carrageenans in the cell walls of *Chondrus crispus* Stack. (Rhodophyceae, Gigartinales) III. Metachromasia and the topooptical reaction. *Phycologia* **17**: 95–104.

Guiry, M. D. 1974. A preliminary consideration of the taxonomic position of *Palmaria palmata* (Linnaeus) Stackhouse = *Rhodmenia palmata* (Linnaeus) Greville. *J. Mar. Biol. Assoc. U.K.* **54**: 509–528.

Harvey, M. J. and J. McLachlan. 1973. *Chondrus crispus*. *Proc. Nova Scotian Inst. Sci.* **27** (suppl.). Halifax, Nova Scotia.

Hawkes, M. W. 1978. Sexual reproduction in *Porphyra gardneri* (Smith et Hollenberg) Hawkes. (Bangiales, Rhodophyta). *Phycologia* **17**: 326–350.

Hollenberg, G. J. 1959. *Smithora*, an interesting new algal genus in the Erythropeltidaceae. *Pac. Natur.* **1**: 3–11.

Irvine, D. E. G. and J. H. Price (eds.). 1978. *Modern approaches to the Taxonomy of Red and Brown Algae*. Academic, New York.

Johansen, H. W. 1974. Articulated coralline algae. *Oceanogr. Mar. Biol. Ann. Rev.* **12**: 77–127.

Kraft, G. T. and M. J. Wynne. 1979. An earlier name for the Atlantic North American red alga *Neoagardhiella baileyi* (Solieriaceae, Gigartinales). *Phycologia* **18**: 325–329.

Krugens, P. and J. A. West. 1973. The ultrastructure of an alloparasitic red alga *Choreocolax polysiphoniae*. *Phycologia* **12**: 175–186.

LaClaire, J. W., II and C. J. Dawes. 1976. An autoradiographic and histochemical localization of sulfated polysaccharides in *Eucheuma nudum* (Rhodophyta). *J. Phycol.* **12**: 368–375.

Lee, R. E. 1974. Chloroplast structure and starch grain production as phylogenetic indicators in the lower Rhodophyceae. *Br. Phycol. J.* **9**: 291–295.

Lin, H-P., J. R. Swafford, and M. R. Sommerfeld. 1977. Comparative ultrastructure of alternating vegetative phases of *Bangia fuscopurpurea* (Bangiophycidae, Rhodophyta). *Bot. Mar.* **20**: 339–344.

Littler, M. M. 1972. The crustose Corallinaceae. *Oceanogr. Mar. Biol. Ann. Rev.* **10**: 311–347.

McBride, D. L. and K. Cole. 1969. Ultrastructural characteristics of the vegetative cell of *Smithora naiadum* (Rhodophyta). *Phycologia* **8**: 177–186.

McBride, D. L. and K. M. Cole. 1972. Fine structural studies on several Pacific coast representatives of the Erythropeltidaceae (Rhodophyta). *Proc. Seventh Intern. Seaweed Symp.* Wiley, New York, pp. 159–164.

McDonald, K. 1972. The ultrastructure of mitosis in the marine red alga *Membranoptera platyphylla*. *J. Phycol.* **8**: 156–166.

Myers, A., R. D. Preston, and G. W. Ripley. 1959. An electron microscope investigation into the structure of the floridean pit. *Ann. Bot. N.S.* **23**: 257–260.

Neushul, M. 1970. A freeze-etching study of the red alga *Porphyridium*. *Amer. J. Bot.* **57**: 1231–1239.

Nichols, H. W. and G. M. Veith. 1978. Development of *Bangia atropurpurea*. *Phytomorphology* **28**: 322–328.

Papenfuss, G. F. 1966. A review of the present system of classification of the Florideophycidae. *Phycologia* **5**: 247–155.

Papenfuss, G. F. and Y-M. Chiang. 1969. Remarks on the taxonomy of *Galaxura* (Nemalionales, Chaetangiaceae). *Proc. Sixth Intern. Seaweed Symp. Pergamon, New York. pp. 303–314.*

Polanshek, A. R. 1974. Hybridization studies of *Gigartina agardhii* and the gametophytes of *Petrocelis franciscana*. *J. Phycol.* **10** (suppl.): 3 (abst.).

Polanshek, A. R. and J. A. West. 1977. Culture and hybridization studies on *Gigartina papillata* (Rhodophyta). *J. Phycol.* **13**: 141–149.

Prince, J. S. 1979. The life cycle of *Crouania pleonospora* Taylor (Rhodophyta, Ceramiales) in culture. *Phycologia* **18**: 247–250.

Pueschel, C. M. 1979. Ultrastructure of tetrasporogenesis in *Palmaria palmata* (Rhodophyta). *J. Phycol.* **15**: 409–424.

Pueschel, C. M. 1980. A reappraisal of the cytochemical properties of rhodophycean pit plugs. *Phycologia* **19**: 210–217.

Ramus, J. 1969. Pit connection formation in the red alga *Pseudogloiophloea*. *J. Phycol.* **5**: 57–63.

Richardson, W. N. and P. S. Dixon. 1968. Life history of *Bangia fuscopurpurea*. (Dillw.) Lyngb. in culture. *Nature* **218**: 496–497.

Richardson, W. N. and P. S. Dixon. 1969. The Conchocelis phase of *Smithora naiadum* (Anders.) Hollenb. *Br. Phycol. J.* **4**: 181–183.

Shevlin, D. E. and A. R. Polanshek. 1978. Life history of *Bonnemaisonia geniculata* (Rhodophyta): a laboratory and field study. *J. Phycol.* **14**: 282–289.

Svedelius, B. 1944. *Galaxura*, a diplobiontic Floridean genus within the order Nemalionales. *Farlowi* **1**: 495–499.

Umezaki, I. 1972. The life histories of some Nemaliales whose tetrasporophytes were unknown. In Abbot, I. A. and M. Kurogi (eds.). *Contributions to the Systematics of Benthic Marine Algae of the North Pacific, Jap. Soc. Phycol.*, Kobe, pp. 231–242.

van den Hoek, C. and A. M. Cortel-Breeman. 1970. Life-history studies on Rhodophyceae III. *Sciania complanata* (Collins) Cotton. *Acta Bot. Neerl.* **19**: 457–467.

van der Meer, J. P. and L. C-M. Chen. 1979. Evidence for sexual reproduction in the red alga *Palmaria palmata* and *Halosaccion ramentaceum*. *Can. J. Bot.* **57**: 2452–2459.

van der Meer, J. P. and E. R. Todd. 1980. The life history of *Palmaria palmata* in culture. A new type for the Rhodophyta. *Can. J. Bot.* **58**: 1250–1256.

Waaland, S. D. and R. Cleland. 1974. Cell repair through cell fusion in the red alga *Griffithsia pacifica*. *Protoplasma* **79**: 185–196.

West, J. A. 1968. Morphology and reproduction of the red alga *Acrochaetium pectinatum* in culture. *J. Phycol.* **4**: 89–99.

West, J. A. 1979. The life history of *Rhodochorton membranaceum*, an endozoic red alga. *Bot. Mar.* **22**: 111–115.

West, J. A., A. R. Polanshek, and D. E. Shevlin. 1978. Field and culture studies on *Gigartina agardhii* (Rhodophyta). *J. Phycol.* **14**: 416–426.

Woelkerling, W. J. 1973. The morphology and systematics of the *Audouinella* complex (Acrochaetiaceae, Rhodophyta) in northeastern United States. *Rhodora* **75**: 529–621.

Wynne, M. J. and W. R. Taylor. 1973. The status of *Agardhiella tenera* and *Agardhiella* baileyi (Rhodophyta, Gigartinales). *Hydrobiologia* **43**: 93–107.

Young, D. N. and J. A. West. 1979. Fine structure and histochemistry of vesicle cells of the red alga *Antithamnion defectum* (Ceramiaceae). *J. Phycol.* **15**: 49–57.

CHAPTER 8

Chrysophyta

For the most part, the plants in this division are small, freshwater species, and little is known about their life histories. The classification used in this text recognizes the six classes presented by Bold and Wynne (1978).

CYTOLOGY

Pigments

Chlorophylls a and c are present in at least some members of all classes. The distinctive gold-to-yellow-green coloration of most members is due to the dominance of carotenoids. Fucoxanthin occurs in three classes, and β-carotene in all six classes. Specific xanthophylls (Table 3-2) are characteristic of certain classes (e.g., violaxanthin in the Eustigmatophyceae).

Cytological Structure

Cell structure is eukaryotic with typical organelles. The chloroplast contains thylakoids grouped into bands of three as well as a girdle thylakoid that is found inside the plastid envelope and encircling the thylakoids (Figure 8-1). Pyrenoids and stigmas (eyespots) may be present, and in the Eustigmatophyceae are a distinctive characteristic. Contractile vacuoles are common in the motile unicellular freshwater species but not in marine forms.

Motility

The number and types of flagella (Table 3-3) are used to separate the classes and orders. Flagellar characteristics are given for each class.

Cell Wall

Members of the various classes include naked (rhizopodial) forms, as well as those with scales (Figure 3-18), siliceous walls (frustules of diatoms, Figure 3-19; statospores), and typical cellulose cell walls.

Food Storage

The common reserve food is a β-$(1 \rightarrow 3)$-linked glucan (chrysolaminarin, leucosin) that is similar in structure to the laminarin of the brown algae. Oil is also common as a reserve food, especially in the diatoms.

MORPHOLOGY

The most common forms are unicellular, although a number of filamentous, coenocytic, and parenchymatous members are present. Because of the wide diversity of morphologies, many phycologists have proposed that the classes, especially the Chrysophyceae and Xanthophyceae, show parallel evolution with the Chlorophyceae.

ECOLOGY

Most of the classes are predominantly freshwater species with a dominance of unicellular planktonic forms. Several genera can form toxic blooms in the marine environment (e.g., *Phaeocystis*). Many of the planktonic species, especially the diatoms, are important primary producers of open oceanic waters and in the shallow-water locales.

REPRODUCTION

Only a few life histories are known, and these are mostly haplontic, except for the diatoms whose life histories are diplontic. Gametes are usually isogamous, but anisogamy and oogamy are also present in some classes. Asexual reproduction is relatively common and in at least two classes a specialized spore called a statospore may be produced. The statospore is a cell that is formed within a vegetative cell (endoplasmic formation). The cell wall consists of two halves of unlike sizes (Chrysophyceae) or similar sizes and overlapping halves (Xanthophyceae). Usually the statospore wall is composed of silicon dioxide (SiO_2).

TAXONOMY

Specifics of each class will be given and only orders containing predominantly marine genera will be presented. A key to the six classes is presented in Table 8-1.

Table 8-1   A Key to the Classes of Algae in the Division Chrysophyta

1	Vegetative cells surrounded by an ornamented cell wall consisting of two halves and containing SiO_2 ....**Bacillariophyceae**
2	Vegetative cells not covered by a SiO_2 cell wall.. **2**
	2 The presence of a haptonema (a modified flagellum) as well as scales covering the unicell common.. **Prymnesiophyceae**
	2 A haptonema lacking, scales lacking or inconspicuous **3**
3	Unicells having an eyespot independent from the chloroplast, dictyosomes lacking and girdle lamellae not present in the chloroplast........................ **Eustigmatophyceae**
3	If an eyespot is present it is contained in the chloroplast, girdle lamellae present in the chloroplast and dictyosomes present.. **4**
	4 Trichocysts (discharge bodies) and kinetochores present in the grove of the unicell .... **Chloromonadophyceae**
	4 Trichocysts and kinetochores absent.. **5**
5	Fucoxanthin the dominant xanthophyll, the cell a golden brown color ....**Chrysophyceae**
5	Fucoxanthin lacking, cells a yellow-green color ....**Xanthophyceae**

Source: After Bold and Wynne (1978)

CLASS 1. CHRYSOPHYCEAE

This class contains a diverse group of "golden algae" ranging from phytoplankton to benthic, intertidal species. Morphologically, the species are unicellular (amoeboid, flagellated, or coccoid), colonial or palmelloid, filamentous, or parenchymatous. By far the majority are flagellated unicells, and most other forms have a flagellated or "monad" stage. Pienaar (1980) has reviewed the flagellated chrysophytes.

Both chlorophylls *a* and *c* are present as well as diatoxanthin and the dominant carotenoid fucoxanthin. Chloroplasts are single to numerous per cell, and are usually parietal. The chloroplasts contain thylakoids in patterns typical of the division (Figure 8-1). Food reserves include chrysolaminarin, found in vesicles and fat droplets, which are distributed in the cytoplasm. The presence and number of flagella are used for taxonomic purposes. If two flagella are present they are unequal in length, the longer one being pantonematic (hairy) and the shorter being acronematic (smooth) and having a basal flagellar swelling. Amoeboid cells are common and a number of flagellated species can become amoeboid as well.

If present, the cell wall may contain cellulose or have scales of organic (cellulose) or inorganic (silicon dioxide, calcium carbonate) composition.

Reproduction is usually by cell division of either motile or nonmotile cells. Specialized cysts, called statospores, are flask-shaped siliceous structures that function as resting stages during adverse environmental conditions. The cyst, or statospore, wall is of two parts, the main flask-shaped portion and a small plug that fills the pore, or opening.

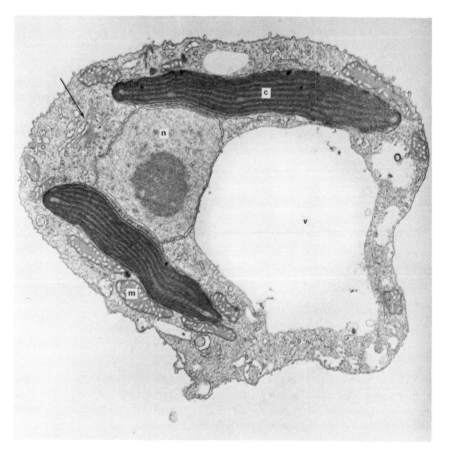

Figure 8-1 A near-median section of *Ochromonas danica* (Slankis and Gibbs, 1972), showing the two lobes of the single chloroplast (c), a central nucleus (n) with an associated Golgi body (arrow), and vacuole, and peripheral mitochondria (m). The vacuole (u) contains leucosin and occupies the posterior half of the cell. A girdle lamella or thylakoid surrounds the other thylakoids within the chloroplast lobes. × 14,680. (With permission of the *Journal of Phycology.*)

Sexual reproduction has rarely been described, but where known the gametes are isogamous and the life history is haplontic. In some cases the "statospore" described for a species was actually the zygote resulting from gametic fusion.

Taxonomically, 12 orders can be grouped into three subclasses using the presence or absence and the number and type of flagella. Selected genera found in five orders having a number of marine representatives will be presented.

Ochromonadales

Members of this order are biflagellated and unicellular. The genus *Ochromonas* is a typical example. Different species of *Ochromonas* may be auxotrophic (require

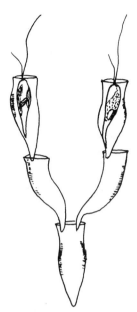

Figure 8-2   A small colony of *Dinobryon* sp. in which two living cells are seen in the top two loricas. The cells have unequal flagella and are attached at the inside of the rim. The cells have a diameter of about 6–8 μm.

vitamins), heterotrophic (phageotrophic, saprophytic), or autotrophic (photosynthetic). This genus has been the subject of extensive biochemical and ultrastructural investigations. Each cell has two flagella of markedly unequal length, and two chloroplasts (Figure 8-2). Gibbs (1962) in a thorough study of *Ochromonas danica* found that the nuclear envelope surrounded the chloroplast, a trait that appears to be characteristic of the order. Slankis and Gibbs (1972) described a closed form of mitosis in the same species; the nuclear membrane remains intact until chromosome movement at anaphase. A "rhizoplast" type of structure produces the microtubules involved in spindle formation. A number of multicellular members of the Chrysophyceae reproduce asexually by the formation of *Ochromonas*-like motile cells. Thus the actual number of species may be open to question.

Another member, *Dinobryon,* can be found in salt marshes. *Dinobryon* is a multicellular aggregation of cells held within loricas (Figure 8-2). Each cell within the colony is attached at its base to an urn-shaped lorica. Thus a *Dinobryon* colony consists of a number of cells attached to each other at the lorica rims. The lorica consists of cellulose and protein (Herth and Zugenmaier, 1979).

Chromulinales

The order Chromulinales is usually placed in the subclass Uniflagellatae. The type genus, *Chromulina* (Figure 8-3) is quite similar to *Ochromonas* except that it

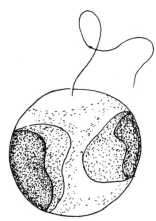

Figure 8-3 A single cell of *Chromulina* sp. with two chloroplasts. The spherical, uniflagellated cell is 10–30 μm in diameter.

has a single, pantonematic flagellum. However, through use of the electron microscope it has become evident that there is usually a vestigial, smooth flagellum that is not visible under the light microscope. With this finding some phycologists believe the uniflagellated condition is a more recent, derived one. The cells are naked (no cell wall) with one-two chloroplasts. Reproduction is by cell division or cyst formation. Other genera in this order include the amoeboid genus *Chrysoamoeba*, which has a single plastid and nucleus, as well as long rhizopodia radiating from the central protoplasm.

Another member of this order is *Pseudopedinella,* placed in the family Pedinellaceae. Although uniflagellated, there is a second, nonemergent flagellum as well as vestigial anterior tentacles and a trailing rhizopodium (Ostroff and Van Valkenburg, 1978). The rhizopodium is without supporting microtubules, and it appears to arise from a large posterior vesicle close to the Golgi apparatus. Ostroff and Van Valkenburg found that it has a stalked pyrenoid similar to that of diatoms but unlike pyrenoids of all other chrysophycean members.

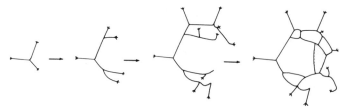

Figure 8-4 Skeletal development of *Dictyocha fibula* according to Van Valkenberg and Norris (1970). The star-shaped silica skeleton is composed of tubular elements that are formed by dichotomous outgrowths of a vacuole, and fusion is through anastomosizing of the tubular elements. (With permission of the *Journal of Phycology.*)

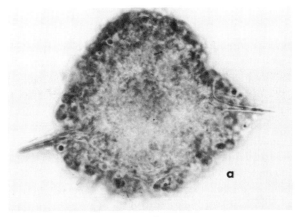

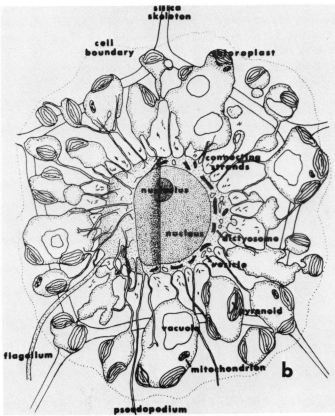

Figure 8-5 (*a*) A living cell of the silicoflagellate *Dictyocha fibula*. The cell is about 40 μm in diameter and is naked with two projecting spines from the internal silica skeleton. (*b*) A diagram of the various components of the cell from the study of Van Valkenberg (1971b). Note the highly globular cytoplasm forming the external portion of the protoplast, which has a "sunburst form." Note also the central nuclear region [(*b*) is with permission of the *Journal of Phycology*].

Dictyochales

This order, which is commonly known as Silicoflagellatae, is also placed in the subclass Uniflagellatae, and the motile members have been reviewed by Van Valkenburg (1980). The order is characterized by marine, uniflagellated single cells, the protoplasm of which is draped over a star-shaped silica skeleton (Figure 8-4). The plastids are golden brown and discoid, and the cells are uninucleate. Amoeboid pseudopodia are present. Because of their siliceous skeleton, the fossil record is quite good, indicating that silicoflagellates have existed since the early Cretaceous. The fossils are frequently found in diatomaceous earth that contains larger, more elaborate skeletons than those present today.

Probably the best studied species of silicoflagellates is *Dictyocha fibula* (Van Valkenburg and Norris, 1970; Figure 8-4). The complexity of the skeleton of this species (Van Valkenburg, 1971a) and the protoplast (Van Valkenburg, 1971b) is most impressive. Figure 8-5*a* is a photograph of a living cell, showing a part of the silica skeleton. Figure 8-5*B* is a diagrammatic view of the cell ultrastructure. Van Valkenburg and Norris (1970) have found that the skeleton is not as useful in taxonomic identifications as thought because of polymorphism.

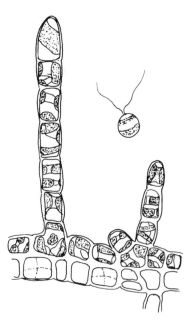

Figure 8-6 *Rhamnochrysis aestuarinae*, an example of a marine filamentous chrysophyte with lobed or laminate chloroplasts. The basal cells are 16–19 × 14–16 μm; these can become multiseriate. The erect branch cells are about 12 × 5–6 μm. The zoospores are 3–4 μm in diameter and are biflagellated. (After Wilce and Markey, 1974).

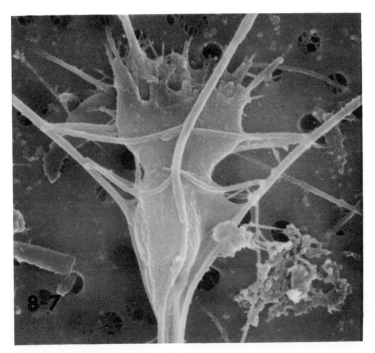

Figure 8-7 A choanoflagellate, *Calliacantha simplex,* collected in the North Pacific is seen with the scanning electron microscope. Note the skeletal system and protoplasmic collar. (Courtesy of Beatrice Booth, University of Washington, Seattle.) × 7,000.

Phaeothamnionales

This order has filamentous (simple, branched, or thalloid member) plants which produce biflagellated zoospores similar to those of *Ochromonas*. Thus they are placed in the subclass Biflagellatae. *Thallochrysis* is a branching filamentous form, whereas *Chrysomeris* is a parenchymatous marine genus. The filamentous genus *Apistonema*, which has both freshwater and marine species, may be the alternate phase of some coccolithophoran alga (Prymnesiophyceae). Wilce and Markey (1974) described another filamentous-to-parenchymatous marine chrysophyte called *Rhamnochrysis* (Figure 8-6). The species occurs in the upper intertidal zone (*Calothrix* zone) of New England salt marshes, and appears to tolerate only medium salinities (~ 20 ppt). Each cell contains two band-shaped chloroplasts, a single nucleus and usually a number of spherical, refractive chrysolaminarin bodies. As the plant ages, it becomes multiseriate in culture and releases *Ochramonas*-like zoospores.

Craspedomonadales

Some taxonomic treatments separate this order from the Chrysophyceae, but most include it as a separate order. The "choanoflagellates," as they are commonly

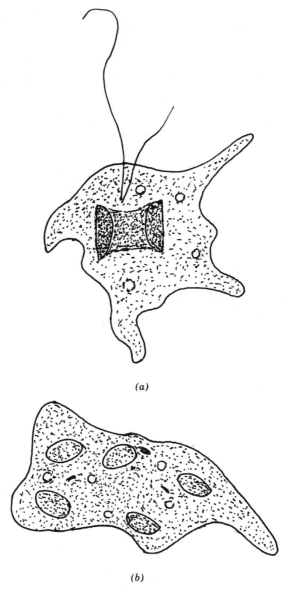

(a)

(b)

Figure 8-8 Two examples of rhizopidial/amoeboid cells. (a) *Chloromeson* sp. is a biflagellated rhizopodial cell with a single plastid. The cells are about 5–10 μm in diameter. (b) *Rhizochloris mirabilis* is an amoeboid cell with a number of discoid chloroplasts. Cells will reach 10–15 μm in length.

called, are unique marine nanoplanktonic algae. They are so named because they bear a cytoplasmic collar near the flagellar insertion point (Figure 8-7). This group appears to be most closely related to the Chromulinales.

CLASS 2. XANTHOPHYCEAE (TRIBOPHYCEAE)

The yellow-green algae contain a number of carotenoids, including β-carotene, diadinoxanthin, and vaucheriaxanthin in addition to chlorophylls a and c. Fucoxanthin is apparently absent (Hibberd, 1980a). The reserve food is again chrysolaminarin, although oil can also be found in the resting cells. Flagellated cells have pairs of unequal anterior flagella, the longer one being pantonematic and the shorter acronematic with a basal flagellar swelling. Hibberd (1980a) has reviewed the flagellated forms of this class. Morphological types vary from unicellular or colonial motile forms, to immobile coccoid or palmelloid forms, to filamentous forms. There are a few coenocytic types as well, so that members of this class, like the Chrysophyceae, are thought to show parallel evolution with the Chlorophyceae.

Reproduction is most commonly asexual, either by the production of non-motile spores (aplanospores) or zoospores. Statospores are commonly observed, especially in brackish and freshwater species. The statospores are internally formed cysts having two overlapping silica walls of equal size, unlike those of the Chrysophyceae. Upon germination, the statospore produces either a zoospore or an amoeboid cell. Although sexual reproduction is poorly known for this class, all three types of gamete production are found (isogamy, anisogamy, oogamy).

Six orders can be recognized for this small class. Only two will be presented here because few members are marine.

Rhizochloridales

The order is sometimes called Chloramoebales, because its members are amoeboid or have rhizopodial development. Members of another order, Heterochloridales, are included here as well because both orders display some degree of amoeboid movement. The former order was separated by having a high degree of metaboly (amoeboid movement), whereas members of the latter order had a lower degree of metaboly. *Chloromeson* (Figure 8-8) has a firm cell wall for part of its life history and would have been placed in Heterochloridales. *Rhizochloris* is an example of a truly amoeboid cell and would have been placed in Rhizochloridales (Figure 8-8*b*). Both can be found in muddy salt marsh areas. *Olisthodiscus luteus* is another naked unicell that is biflagellated, but it has no pseudopodia. Cattolico *et al.* (1976) have shown that it has a number of discoid chloroplasts surrounding a central nucleus with typical cell organelles (Figure 8-9). There is some question about the placement of this brackish-water genus, but for the present it would appear to be a xanthophycean alga.

Vaucheriales

Although there are other orders of filamentous and coccoid xanthophytes, the most unique group of marine members are the coenocytic forms (Vaucheriales; old name Heterosiphonales). The most widespread and best studied genus of this

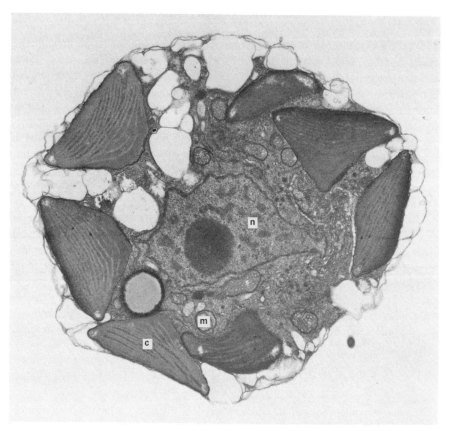

Figure 8-9 *Olisthodiscus luteus*, an elliptical cell that contains several discoid chloroplasts (c) and has two unequal-length flagella, the longer being pantonematic (Cattolico *et al.*, 1976). The electron micrograph reveals the central nucleus (n) and surrounding mitochondria (m) and chloroplasts. × 9,400. (Courtesy of R. A. Cattolico, University of Washington, Seattle.)

order is *Vaucheria*. Because it is olive-green, *Vaucheria* was placed in the order Caulerpales of the class Chlorophycae until it was observed that the reserve food was oil, not starch, and that the pigments corresponded to those of the Xanthophyceae. *Vaucheria* is a common inhabitant of brackish-water environments such as mud flats of tidal marshes.

Vaucheria is a highly branched coenocytic tube (Figure 8-10) ranging from 6 to 40 μm in diameter and attached to the mud by colorless, basally branched "rhizoids." When the plant reproduces, septa are formed separating the reproductive organs from the vegetative cytoplasm. Ott and Brown (1978) have described the ultrastructure and development of *Vaucheria*. Asexual reproduction is by fragmentation or large zoospores, which have numerous pairs of flagella (multiflagellated) of unequal length. Sexual reproduction is oogamous with the formation of

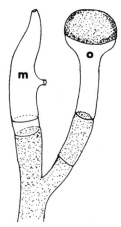

Figure 8-10 An antheridium (*m*) and an oogonium (*o*) on a terminal filament of the coenocyte *Vaucheria bermudensis*. The antheridium is empty and has two pores from which the sperm were released. The oogonium has one fertilized egg (oospore). Filament diameter is 18–50 μm.

eggs and sperm (biflagellated, unequal flagellar lengths). The oogonia are large and bulbous, and are separated by a septation from the vegetative filament, while the antheridia in dioecious species are usually septate hooklike branches adjacent to the oogonia (Figure 8-10). Numerous sperm are released from the antheridia and swim to the beak of the bulbous oogonium. A single egg is found within the oogonium, and upon fertilization a thick-walled zygote is formed. Meiosis is zygotic and the life history is haplontic. The siphon that develops from the zygote is haploid.

CLASS 3. PRYMNESIOPHYCEAE (HAPTOPHYCEAE)

The members of this class were originally placed in the class Chrysophyceae. Hibberd (1976) has reviewed the class and its flagellated members (Hibberd, 1980b), especially the distinctive ultrastructural features of its motile cells. The two most evident reasons for its separation are the presence of a *haptonema* and/or coccoliths. The haptonema is a specialized projection that arises near the pair of flagella. Ultrastructurally it consists of three concentric membranes surrounding six tubules, so it is distinct from a flagellum (the plasmalemma surrounding nine outer doublets and two inner single tubules, see Figure 3-20). The haptonema can function as an attachment organ, and some are able to coil or straighten rapidly. The second feature is the presence of scales that are either organic (cellulose), or calcified (coccoliths). These scales are formed in Golgi vesicles and become deposited on the outside of the cell. Calcified coccoliths are evident in the fossil record, especially in Cretaceous deposits such as those found in the White Cliffs of Dover. The calcified coccoliths were well known from deep-sea deposits, but until 1940, they were thought to be of inorganic origin and not the products of nannoplankton.

The haptophycean algae were originally placed in the Chrysophyceae because they had the same pigments (including fucoxanthin) and food reserves (chryso-laminarin). Most of the species are motile unicells, usually having two ac-ronematic flagella of equal or occasionally unequal length. A flagellar swelling or an eyespot are not common in the flagellated members (Hibberd, 1980b).

The cell walls are composed of cellulose or a mucopolysaccharide matrix in sessile species, whereas phytoflagellates have complex and delicate organic scales (Figure 3-17) that can become mineralized (Figure 8-16). The organic matrix of the scales is apparently composed of cellulose, the scales are organized in one-to-many layers around the cell and are contained in a gelatinous outer matrix. Coc-colithophorids also have scales of calcium carbonate (Figure 8-16). Asexual re-production is by binary cell division or by the release of motile or nonmotile cells. However, there is much plasticity in the mode and type of asexual repro-duction. Depending on the age of a culture, a number of morphological forms can occur. Little is known about sexual reproduction, and recently both haplontic as well as isomorphic and heteromorphic haplodiplontic life cycles have been demonstrated. Much confusion exists as to the classification and placement of genera within this class. Three orders are described here.

Isochrysidales

Members of this order have reduced haptonemas or lack them entirely; their mo-tile cells have two acronematic flagella. Scale production is variable, ranging from no scales to organic or calcified ones. *Isochrysis* (Figure 8-11) has a highly reduced haptonema or none at all. It has organic (cellulose) scales as well as some calcareous elements in a thick mucilage.

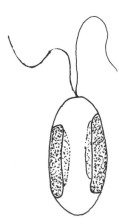

Figure 8-11 *Isochrysis galbana*, a biflagellated member of the phytoplankton that has two laminate chloroplasts per cell. The cell reaches 8 μm in diameter and bears organic scales (see Figure 3-17).

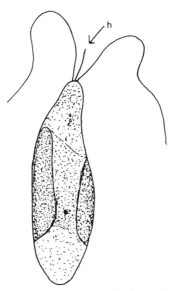

Figure 8-12 *Prymnesium saltans,* a phytoflagellate (10–12 μm in diameter) that is common in tidal marshes and bears a single, laminate chloroplast, two equal length flagella and a haptonema (*h*). The cell bears simple organic scales (see Figure 3-17). Many of the species can form blooms in estuaries, releasing an extracellular toxin that causes fish kills.

Prymnesiales

The members of this order are mostly motile unicells with variable lengthened haptonemas and equal flagellae. Several genera, including *Prymnesium* (Figure 8-12) and *Chrysochromulina* (Figure 8-13) have been extensively studied.

Each cell has two plastids, and organic scales (Figure 3-17) are common among members of this order. *Prymnesium* is responsible for fish kills in brackish water. One member, *Phaeocystis,* produces acrylic acid, which can kill penguins feeding on fish where this alga is abundant. The acrylic acid apparently kills the intestinal flora of the penguins.

Coccolithophoridales

This order is probably an artificial grouping of genera possessing calcified scales. Most species are marine and can account for a significant portion of the primary productivity of the nannoplankton (Chapter 20). Some taxonomic problems have arisen because a few neritic species have heteromorphic, haplodiplontic life histories including two distinctive genera. For example, the palmelloid or filamentous mass of cells called *Apistonema* is the haploid phase in a life history with *Hymenomonas* (Figures 8-14, 8-15) the diploid phase. Both phases have the same type of calcified scales and reproduce by means of zoospores. An important taxonomic feature of the order is the presence of calcified (inorganic) scales, or coccoliths, which contain calcite crystals deposited in an organic matrix (Pienaar, 1969). The

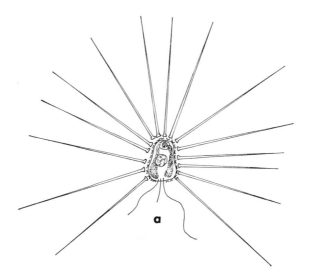

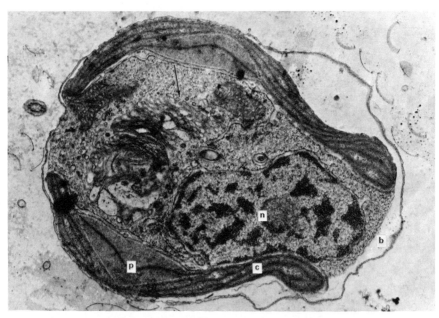

Figure 8-13 (a) *Chrysochromulina spinifera,* bearing two types of scales, long coccolith spines and small convex organic ones. The cells are 8–10 μm in length and have two lobed chloroplasts along with two unequal-length flagella and a haptonema (central strand in drawing). (b) The cell structure is typical of a chrysophyte with a lobed chloroplast (c) and pyrenoid (p). A central nucleus (n) and Golgi body (arrow) are also visible. × 2,860. (After Pienaar and Norris, 1979; (b) courtesy of R. Pienaar, University of Natal, Durban, South Africa.)

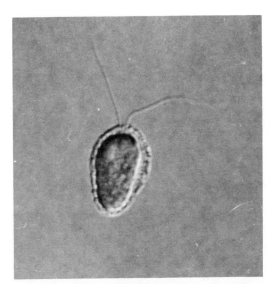

Figure 8-14 *Hymenomonas carterae* bears three forms of coccoliths on the motile cell. A living cell has a pair of flagella, is oval in shape, and is about 5 μm in diameter. (Courtesy of R. Pienaar, University of Natal, Durban, South Africa).

scales or coccoliths are found in the fossil record as far back as the lower Jurassic of the Mesozoic period. Figure 8-16 demonstrates several coccolith morphologies. These coccoliths, which can be quite elaborate, have an organic substructure that is calcified within the Golgi body vesicle before transport to the cell surface (Paasche, 1968; Pienaar, 1969).

CLASS 4. CHLOROMONADOPHYCEAE (RHAPHIDOPHYCEAE)

This class is small, consisting mostly of freshwater species, although Loeblich and Fine (1977) suggest that marine chloromonads are quite common in the neritic environment. The phytoflagellates are frequently flattened dorsiventrally. They lack cell walls and an eyespot but usually have mucocysts (trichocysts, ejectile bodies), which are refractive under the light microscope. Two flagella arise near the cell apex, usually from a shallow pit; one extends anteriorly and is hairy (tubular flagellar hairs) whereas the other lacks hairs (acronematic) and trails posteriorly (Heywood, 1980). Kinetochores, organelles uncommon to algae, are also present in members of this class. The Golgi apparatus is found in the anterior part of the cell and is associated with the nucleus. Mitosis is closed, with polar fenestrations occurring. Contractile vacuoles are found in the cell anterior and appear to be limited to freshwater species. The chloroplasts are numerous and discoid with thylakoids in bands of three and the typical girdle band of thylakoids. Pigments are typical of the division except for the lack of fucoxanthin. The main reserve food appears to be oil rather than chrysolaminarin (Heywood, 1980).

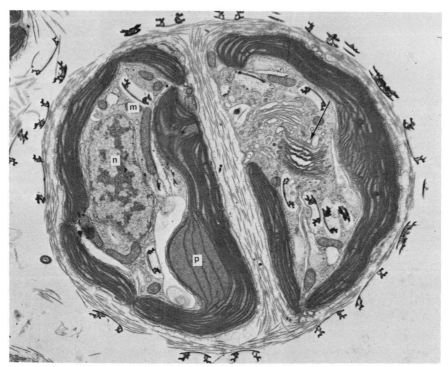

Figure 8-15 Under the electron microscope, typical cell structure is visible in two coccoid cells of *Hymenomonas carterae*, including a posterior nucleus (n), Golgi bodies (arrow), mitochondria (m) and chloroplasts (c) with a pyrenoid (p). Coccoliths can be seen on the surface of the gelatinous sheath. × 2,300. (Courtesy of R. Pienaar, University of Natal, Durban, South Africa).

Hornellia (*Chattonella*) is responsible for extensive fish kills off the Malabar coast of India. The biflagellated cell has a plastic periplast and a shallow infolding associated with an apical furrow (Figure 8-17). Reproduction is by logitudional cell division. Fusion has been reported, but the process of sexual reproduction is poorly known. Ejectile bodies (trichocysts, mucocysts) are located along the cell periphery.

Chattonella subsalsa is a small, motile unicell that has typical chloromonad features (Loeblich and Fine, 1977). Mucocysts are present in the posterior region of the cell. The golden chloroplasts are surrounded by four membranes; the outer two are part of the endoplasmic reticulum and the inner two are the chloroplast envelope membranes.

CLASS 5. EUSTIGMATOPHYCEAE

This small class was first proposed by Hibberd and Leedale (1970), who separated 12 genera and 15 species of coccoid xanthophycean species based on cellular

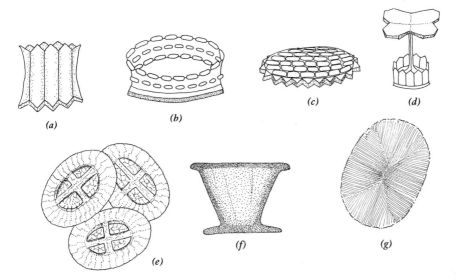

Figure 8-16 Examples of inorganic scales, coccoliths, found on members of the Prymnesiophyceae. See also examples of organic scales in Figure 3-17. (*a*) *Hymenomonas globosa* (Gayral and Fresnel, 1976). The crown-shaped coccolith is about 1.5 μm in width and is hollow. (*b*) *Cricosphaera roscoffensis* (Gayral and Fresnel, 1976). These coccoliths reach 2.5 μm in diameter and are hollow rings with various beadlike thickenings. (*c*) *Pappomonas flabellifera* (Manton and Oates, 1975). The coccoliths are about 1.5 μm in length and consist of a mozaic of quadrangular bands of CaCO$_3$ that are formed on an organic matrix. (*d*) *Papposphaera lepida* (Manton and Oates, 1975). The coccoliths are calcified appendages that radiate out from a small base. Note the stalk, cap of four plates, and the complex base. Coccolith length is about 1 μm. (*e*) *Coccolithus neohelis* (West, 1969). The coccoliths are 2.5 μm in length and are solid elliptical plates. (*f*) *Ochrosphaera verrucosa* (West, 1971). The coccoliths are truncated cones, about 0.75 μm tall. (*g*) The organic matrix for a coccolith (see also Figure 3-17) is shown after removal of the CaCO$_3$.

morphology and ultrastructure (Figure 8-18). In a later publication, Hibberd and Leedale (1972) presented the diagnostic features of this class as follows:

1 Zoospores are elongate to flask shaped and not distinctly bilaterally symmetrical.

2 Nuclei are pyriform, with the anterior end elongated toward the basal body and not structurally associated with the chloroplast.

3 The zoospore has a single, long chloroplast, and pyrenoids are absent in the zoospore but can be present in the vegetative (coccoid) cell. The pyrenoid is large and polyhedral and attached to the inner face of the chloroplast by a narrow connection.

4 The chloroplast is surrounded by a layer of endoplasmic reticulum that is not continuous with the nucleus.

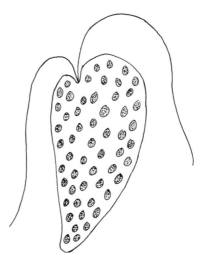

Figure 8-17 *Hornellia marina* is a chloromonad with numerous discoid chloroplasts. The phytoflagellate is biflagellated, measures about 12–15 μm in length, and has trichocysts along the cell periphery.

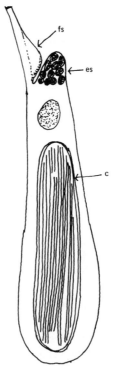

Figure 8-18 A diagram of a typical eustigmatophycean motile cell after Hibberd and Leedale (1972). (*fs*) Flagellar swelling that is associated with the eyespot (*es*). (*c*) Chloroplast that has an endoplasmic reticular envelope and is separate from the nucleus. The thylakoids form parallel bands in the chloroplast.

5   The zoospore has a single, anteriorly directed flagellum—with bilateral hairs and a swelling at its proximal end—pressed against the cell at the site of the large eyespot.

6   The eyespot is an irregular aggregate of electron-dense droplets at the extreme anterior end of the zoospore and remote from the chloroplast.

7   Golgi bodies are absent in the zoospores.

8   The vegetative cell is coccoid with a single, bowl-shaped to lobed chloroplast. The cell wall is in one piece (not overlapping halves).

Only chlorophyll *a* has been found along with the xanthophyll violaxanthin. The reserve food appears to be chrysolaminarin, although this is open to question (Hibberd, 1980c). The cell wall is either quite thin or lacking in zoospores but is obvious in the coccoid forms and is only one piece. All of the present members are coccoid in the vegetative state, and a number are found in the marine environment (*Nannochloris, Monallantus;* Antia *et al.,* 1975). Hibberd (1980c) has reviewed the flagellated forms of the class as well as the taxonomy (1981).

CLASS 6. BACILLARIOPHYCEAE

Of all the classes of the Chrysophyta, the diatoms are the most diverse with over 12,000 described species. In addition, they are the most important group in the

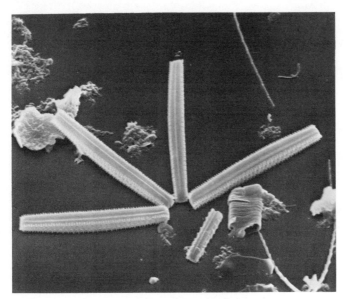

Figure 8-19   The pennate diatom *Thalassionema nitzschioides* forms stellate chains. Part of a chain is seen in this scanning electron micrograph. × 500. (Courtesy of Greta Fryxell, Texas A&M University, College Station.)

marine environment because of their role in primary productivity (see Chapter 20). Werner (1977) has estimated that diatoms supply 20–25% of the world's net primary production. Diatoms are uninucleate and unicellular. Their pigments and reserve food are the same as those of the division. A number of studies (e.g., Borowitzka and Volcani, 1978) have shown that the chloroplast and pyrenoid structures are also typical Chrysophyta forms (Figure 3-5; 8-24). Oil and chrysolaminarin are common food reserves; the oil may become so abundant that it fills the cell with droplets. Although most species are photosynthetic, some diatoms are heterotrophic and even colorless, living in such places as the mucilage of *Fucus*. Two major forms of unicells are distinguished based on cell symmetry, the Pennales with bilateral symmetry (Figure 8-19; shoe-box form) and the Centrales with irregular or radial symmetry (Figure 8-20; hat-box form).

The most characteristic feature of the diatom is the cell wall or *frustule,* which consists of two halves, the epitheca and the hypotheca (Figure 8-21). The frustule can be compared to a box with a cover and a bottom, the top and bottom being the "valves" (e.g., valve view) and the overlapping sides being the region called the "girdle" (e.g., girdle view). The frustule is outside the plasmalemma and consists of a hydrated form of silicon dioxide (opaline or hydrated type) and organic material (pectic acid). Studies on the formation, biochemistry and structure

Figure 8-20 The centric diatom *Actinoptychus senarius* from Capetown, South Africa, in valve view. Note the arrangement of punctae on the valve. × 1,300. (Courtesy of Greta Fryxell, Texas A&M University, College Station.)

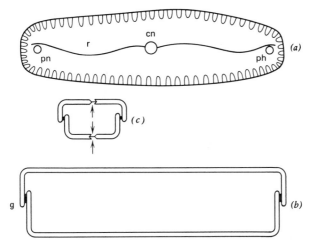

Figure 8-21 Diagram of a pennate diatom frustule. (*a*) valve view with a raphe (*r*) and central and polar nodules (*cn*, *pn*). (*b*) Girdle view showing the connecting band at the overlapping halves (*g*). (*c*) Cross section of the frustule shows the outer and inner fissure of the raphe (arrows).

of the diatom frustule are numerous (see Chiappino and Volcani, 1977). Frustule formation is the result of valve initiation within a silica deposition vesicle during cell division. The frustule may be highly perforated (puncta, areolae) or ornamented (costae, nodules, raphes, spines). Some benthic diatoms produce a stalk that apparently consists of chitin and serves to attach the diatom to algae or some other substrata.

The gliding motility of the vegetative cells is found only in members of the Pennales, which are bilaterally symmetrical and possess a raphe. The raphe extends from the two polar nodules at either end of the valve to a central nodule. The nodules have pores transversing them (Figure 8-20). Apparently the cytoplasm does not extend beyond the frustule, although the raphe is continuous as a slit from the outer to the inner side of the frustule. There are a number of theories explaining the mechanism of diatom movement. The flow of cytoplasm in the raphe, like the tread of a tank, could result in friction, pushing the cell forward. Indeed, cytoplasmic streaming in the raphe is always in the direction opposite to the diatom motion, supporting this concept. In addition, the observed expulsion of mucilage at the nodules would furnish more substrate for the cytoplasm to move against. A mucilage trail is well known for motile diatoms. In the centric or radially symmetrical diatoms, a uniflagellated (pleuronematic) sperm has been described for a number of species. The flagellum of this sperm is unique in that only nine outer paired microtubules occur; the inner two are lacking.

Asexual reproduction is by vegetative cell division. Diatoms possess a highly organized and irregular spindle (Pickett-Heaps and Tippit, 1978) and the nuclear envelope is ruptured during mitosis. After mitosis two new hypothecae are formed within the old valves, and the daughter cells separate. In most diatoms studied, the old hypotheca becomes the new epitheca for one of the two daughter cells, with a concurrent diminution in size. When a cell becomes quite small, enlargement back to the original size can occur by the extrusion of the protoplast, swelling, and subsequent formation of a new frustule.

More commonly, the smaller-sized frustules become sexually active, producing gametes after meiosis. All of the life histories studied indicate that the diatoms are diplontic. Upon fusion of the gametes, the zygote protoplast usually swells and forms a unique and distinctive frustule that will not look like the frustule of a typical vegetative cell. These are resting cells, or zygotes, and are called *auxospores*. Auxospores may also be formed without fusion, acting as an asexual resting stage. The protoplast of the auxospore will ultimately form the typical vegetative diatom frustule characteristic of that species. Members of the Pennales will produce up to four amoeboid gametes per cell (isogamy, anisogamy), whereas members of the Centrales will produce one-two nonmotile eggs and up to two uniflagellated sperm (oogamy).

Siliceous frustules are well preserved in the fossil record and large diatom deposits date back to the Cretaceous period. Diatomaceous earth is mined in a number of areas of the world (especially in California, Figure 2-18) and is used in industrial filtration, polishing, insulation, and as a material for absorbing nitroglycerin to make dynamite (Alfred Nobel, founder of the Nobel prizes discovered the last-named use).

Ecologically, diatoms are everywhere, as plankton, epiphytes, or as benthic communities on rocks, sand, or mud surfaces. Intertidal, benthic diatoms play an important role in the stabilization of soft substrata and as a food source for benthic fauna (Hendey, 1964). Some intertidal species show a cyclic rhythm in the vertical movement to and away from the surface of estuarine mud (Hopkins, 1966). The importance of diatoms as primary producers in aquatic food chains has long been recognized (Chapter 20), and vertical migrations of phytoplankton have been described (Hendey, 1964).

A good reference on marine diatoms is the text edited by Werner (1977) and another by Hendey (1964). The latter author did not recognize the two morphological orders presented in this text. The retention of these two orders here, although intermediate forms are well established, is to facilitate the presentation of major features of the centric and bilateral species.

Centrales

Members placed in this order are circular (*Coscinodiscus*, Figure 8-22*a*) or irregular (*Isthmia*, Figure 8-22*b*) in shape. The ornamentation is always radial in symmetry. In many species, the cells can be joined together by mucilage pads to produce filamentous-type chains (*Bacteriastrum*, Figure 8-23). This order is best represented in the marine environment.

All species studied reproduce sexually through eggs and sperm (oogamy). However, only a few life histories are known. Egg formation involves meiotic division of the diploid nucleus. Degeneration of some of the haploid nuclei then

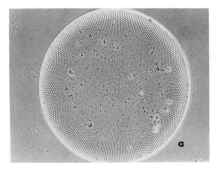
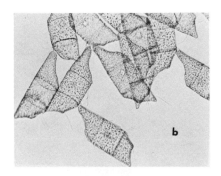

Figure 8-22 Two examples of centric diatoms. (*a*) *Coscinodiscus centralis* is a large centric diatom with a diameter of 150 μm (valve view). (*b*) *Isthmia enervis* is an irregular-shaped diatom with a width of 212 μm (girdle view), note overlapping values. (Courtesy of Lana Tester, Florida Department National Research Marine Laboratory, St. Petersburg.)

occurs so that only one-two haploid eggs result. Cells producing the male gametes first divide mitotically, forming 2–64 small diploid spermatogonial cells. Meiosis occurs in these cells after the spermatogonial protoplasts have moved out of the frustule. Four uniflagellated, haploid sperm result from each meiotic division. Fertilization is achieved when the sperm swim to eggs that have been exposed in the old frustule. Exposure is either through the separation of the two parts of the mother frustule or by the production of special fertilization pores in the frustule wall. After fusion, the zygote swells and forms the thick-walled auxospore. The auxospore germinates following formation of the typical species frustule within it.

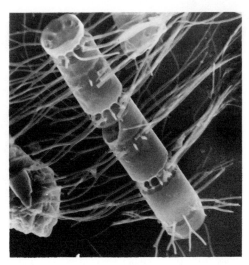

Figure 8-23 *Bacteriastrum furcatum*, a colonial centric diatom. Three entire cells are shown in this chain. Note the setae that aid in floatation. × 700. (Courtesy of Greta Fryxell, Texas A&M University, College Station.)

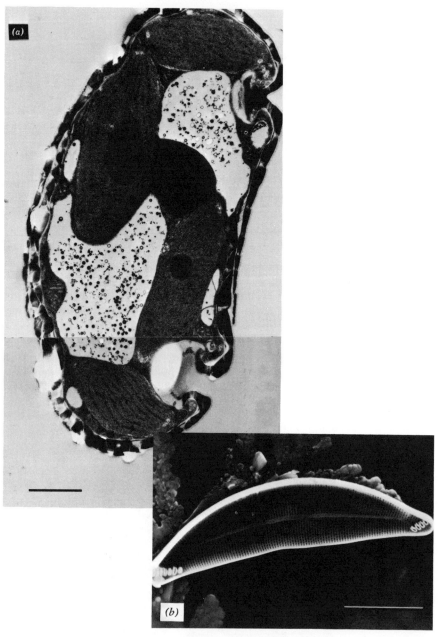

Figure 8-24 *Pseudohimantidium pacificum,* a diatom that is epizoic on copepods. (*a*) A paratransapical section through the entire cell. Four chromatophores (two with pyrenoids), mitochondria, and vacuolar osmiophilic bodies are evident. The unit equals 1 μm. (*b*) The whole valve with striations of the axial area and labiate process at the tips of the diatom where it attaches. The unit equals 2 μm. (Courtesy of R. Gibson, Harbor Branch Laboratory, Ft. Pierce, Florida.)

Pennales

The diatoms referred to in this order have bilateral symmetry (Figure 8-19). Many members have raphes (Figure 8-19), which are longitudinal slits on one or both valves. Raphe-bearing pennate diatoms can glide, apparently through secretion of mucilaginous material from the polar nodules. Most of the members have two large, strap-shaped plastids. Members are equally distributed between freshwater and marine habitats. A number of species are stalked, living on plants (epiphytic) or animals (epizoic). The stalk consists of chitin, which is unique for the plant kingdom. One such stalked species is *Pseudohimantidium,* which is epizoic on copepods. Figure 8-24 shows a thin section of this diatom with the external grove from which the stalk develops. The diatoms are diploid, so that meiosis occurs in the final stages of gamete formation.

In all but one genus, *Rhabdonema,* sexual reproduction, where known, is by nonflagellated isogametes that are somewhat amoeboid. In *Rhabdonema,* however, there are eggs and amoeboid male gametes (oogamy). In all genera of this order the cells that become sexually induced are usually quite small, resulting from continuous diminution of the frustule from continued mitotic divisions. The cells come together in pairs, usually within a mucilaginous mass. The actual production of the amoeboid gametes, whether they are isogametes or anisogametes, varies from one species to another, and sexual reproduction can be grouped into three types. In all cases two naked gametes fuse, either both migrating to each other in the mucilage or one migrating to the other one, which remains near or in the mother frustule. After enlargement to the typical species size, the resulting zygote will produce an auxospore, or resting cell. In a few cases auxospore production can occur asexually through the enlargement of a single, diploid cell.

REFERENCES

Antia, N. J., T. Bisalputra, J. Y. Cheng, and J. P. Kalley. 1975. Pigment and cytological evidence for reclassification of *Nannochloris oculata* and *Monallantus salina* in the Eustigmatophyceae. *J. Phycol.* **11**: 339–343.

Bold, H. C. and M. J. Wynne. 1978. *Introduction to the Algae.* Prentice-Hall, Englewood Cliffs, N.J.

Borowitzka, M. A. and B. E. Volcani. 1978. The polymorphic diatom *Phaeodactylum tricornutum:* Ultrastructure of its morphotypes. *J. Phycol.* **14**: 10–21.

Cattolico, R. A., J. C. Botthroyd, and S. P. Gibbs. 1976. Synchronous growth and plastic replication in the naturally wall-less *Olisthodiscus luteus. Plant Physiol.* **57**: 497–503.

Chiappino, M. L. and B. E. Volcani. 1977. Studies on the biochemistry and fine structure of silica cell formation in diatoms. VII. Sequential cell wall development in the pennate *Navicula pelliculosa. Protoplasma* **93**: 205–221.

Gayral, P. and J. Fresnel. 1976. Nouvelles observations sur deux Coccolithophoracées marines: *Cricosphaera roscoffensis* (P. Dangeard) *comb. nov. et Hymenomonas globosa* (F. Magne) comb. nov. *Phycologia* **15**: 339–355.

Gibbs, S. P. 1962. Nuclear envelope-chloroplast relationships in algae. *J. Cell Biol.* **14**: 443–444.

Gibson, R. A. 1979. Observations of stalk production of *Pseudohimantidium pacificum* Hust and Krasske (Bacillariophyceae: Protoraphidaceae). *Nova Hedwigia* **31**: 899–915.

Hendey, N. I. 1964. *An Introductory Account of the Smaller Algae of British Coastal Waters. Part V: Bacillariophyceae (diatoms)*. Ministry of Agriculture Fisheries and Food. HMS Stationary Office, London.

Heywood, P. 1980. Chloromonads. In Cox, E. R. (ed.). *Phytoflagellates,* Elsevier/North Holland, New York, pp. 351–380.

Herth, W. and P. Zugenmaier. 1979. The lorica of *Dinobryon. J. Ultrast. Res.* **69**: 262–272.

Hibberd, D. J. 1976. The ultrastructure and taxonomy of the Chrysophyceae and Prymnesiophyceae (Haptophyceae): a survey with some new observations on the ultrastructure of the Chrysophyceae. *Bot. J. Linn. Soc.* **72**: 55–80.

Hibberd, D. J. 1980a. Xanthophytes. In Cox, E. R. (ed.). *Phytoflagellates,* Elsevier/North Holland, New York, pp. 243–272.

Hibberd, D. J. 1980b. Prymnesiophytes (Haptophytes). In Cox, E. R. (ed.). *Phytoflagellates,* Elsevier/North Holland, New York, pp. 273–318.

Hibberd, D. J. 1980c. Eustigmatophytes. In Cox, E. R. (ed.). *Phytoflagellates,* Elsevier/North Holland, New York. pp. 319–334.

Hibberd, H. J. 1981. Notes on the taxonomy and nomenclature of the algal class Eustigmatophyceae and Tribophyceae (synom Xanthophyceae) *Bot. J. Linn Soc.* **82**: 93–119.

Hibberd, D. J. and G. F. Leedale. 1970. Eustigmatophyceae—a new algal class with unique organization of the motile cell. *Nature* **225**: 758–760.

Hibberd, D. J. and G. F. Leedale. 1972. Observations on the cytology and ultrastructure of the new algal class, Eustigmatophyceae. *Ann. Bot.* **36**: 49–71.

Hopkins, J. T. 1966. The role of water in the behavior of an estuarine mudflat diatom. *J. Mar. Biol. Assoc. U.K.* **46**: 617–626.

Loeblich, A. R., III and K. E. Fine. 1977. Marine Chloromonads: More widely distributed in neritic environments than previously thought. *Proc. Biol. Soc. Wash.* **90**: 388–399.

Manton, I. and K. Oates. 1975. Fine-structural observations on *Papposphaera* Tangen from the southern hemisphere and on *Pappomonas gen. nov.* from South Africa and Greenland. *Br. Phycol. J.* **10**: 93–109.

Ostroff, C. A. and S. D. Van Valkenburg. 1978. The fine structure of *Pseudopedinella pyriforme* Carter (Chrysophyceae). *Br. Phycol. J.* **13**: 35–49.

Ott, D. W. and R. M. Brown. 1978. Developmental cytology of the genus *Vaucheria* IV. Spermatogenesis. *Br. Phycol. J.* **13**: 69–85.

Paasche, E. 1968. Biology and physiology of coccolithophorids. *Ann. Rev. Microbiol.* **22**: 71–86.

Pickett-Heaps, J. D. and D. H. Tippit. 1978. The diatom spindle in perspective. *Cell* **14**: 455–467.

Pienaar, R. N. 1969. The fine structure of *Hymenomonas* (Cricosphaera) *carterae.* II. Observations on scale and coccolith production. *J. Phycol* **5**: 321–331.

Pienaar, R. N. 1980. Chrysophytes. In Cox, E. R. (ed.). *Phytoflagellates,* Elsevier/North Holland, New York, pp. 213–242.

Pienaar, R. N. and R. E. Norris. 1979. The ultrastructure of the flagellate *Chrysochromulina spinifera* (Fournier) *comb. nov.* (Prymnesiophyceae) with special reference to scale production. *Phycologia* **18**: 99–108.

Sikes, C., R. D. Roer and K. M. Wilbur. 1980. Photosynthesis and coccolith formation: Inorganic carbon sources and net inorganic reaction of deposition. *Limnol. Oceanogr.* **25**: 248–261.

Slankis, T. and S. P. Gibbs. 1972. The fine structure of mitosis and cell division in the Chrysophycean alga *Ochromonas danica. J. Phycol.* **8**: 243–256.

Van Valkenburg, S. D. 1971a. Observations on the fine structure of *Dictyocha fibula* Ehrenberg. I. The skeleton. *J. Phycol.* **7**: 113–118.

Van Valkenburg, S. D. 1971b. Observations on the fine structure of *Dictyocha fibula* Ehrenburg. II. The protoplast. *J. Phycol.* **7**: 118–132.

Van Valkenburg, S. D. 1980. Silicoflagellates. In Cox, E. R. (ed.). *Phytoflagellates,* Elsevier/North Holland, New York, pp. 335–350.

Van Valkenburg, S. D. and R. E. Norris. 1970. The growth and morphology of the silicoflagellated *Dictyocha fibula* Ehrenberg in culture. *J. Phycol.* **6**: 48–54.

Werner, D. (ed.). 1977. *The biology of diatoms.* Botanical Monograph Vol. 13. University of California Press, Berkeley.

West, J. A. 1969. Observations on four rare marine microalgae from Hawaii. *Phycologia* **8**: 187–192.

Wilce, R. T. and D. R. Markey. 1974. *Rhamnochrysis aestuarinae,* a new monotypic genus of benthic marine Chrysophytes. *J. Phycol.* **10**: 82–88.

CHAPTER 9

Cryptophyta, Euglenophyta, and Pyrrhophyta

Division Cryptophyta

This division contains about 24 genera, with many species occurring in marine habitats. Butcher (1967) reviewed the marine species and included elegant colored drawings of many. Gantt (1980) reviewed the division and presented the following general characteristics: (1) cells asymmetric with a dorsiventral flattening (2–20 μm long by 3–16 μm wide); (2) two flagella of equal or almost-equal

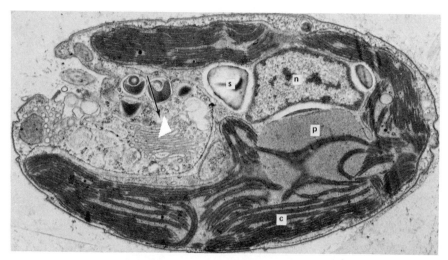

Figure 9-1 Electron micrograph of a species of *Chroomonas,* showing the large lobed chloroplast (c) in which thylakoids are loosely paired, a large pyrenoid (p) occurs, and the phycobiliproteins are found in the interthylakoidal spaces (Gantt *et al.,* 1971). The starch grains (s) are free in the cytoplasm and adjacent to the nucleus (m) in the posterior end. At the anterior end, an oblique section through a flagellum is evident and seen in the grove, and the Golgi complex (arrow) is visible. × 17,500. (Reproduced from *The Journal of Cell Biology,* 1971, volume 48, pages 280–290, by copyright permission of The Rockefeller University Press.)

length that emerge from a shallow reservoir; (3) a periplast with thin plates arranged in regular rows; (4) ejectosomes (trichocysts) along the cell periphery and smaller ones in the reservoir; (5) an envelope that encloses both the nucleus and the chloroplast; (6) a single chloroplast with one to many pyrenoids, the thylakoids occurring singly or in bands; (7) chlorophylls *a* and *c,* and phycobiliproteins as the photosynthetic pigments; (8) a nucleomorph, distinct from the nucleus, that is found between the chloroplast and the chloroplast-nuclear envelope (Gillott and Gibbs, 1980); and (9) mitosis in which there is partial breakdown of the nuclear envelope.

CYTOLOGY

Pigments

Chlorophylls *a* and *c* are present as well as α- and β-carotene. A group of specialized xanthophylls like alloxanthin, and the phycobiliproteins (phycoerythrin and phycocyanin) make this division unique. Furthermore, the phycobiliproteins are not organized into phycobilisomes as found in the red and blue-green algae but occur in the interstitial matrix between the two membranes of the thylakoids (Gantt *et al.,* 1971).

Cytological Structure

The generalized cell structure for the division is shown in Figure 9-1 for *Chroomonas* as seen in thin section with an electron microscope (Gantt *et al.,* 1971). Usually there are one or two chloroplasts, which have loosely paired thylakoids. Pyrenoids are found in the plastid; starch may surround them or may be free in the cytoplasm. A single nucleus, usually with densely staining chromatin, is evident in the posterior portion of the motile cell.

Mitosis (karyokinesis) and cell division (cytokinesis) is distinctive in the few genera studied (Oakley and Dodge, 1976; Oakley and Bisalputra, 1977). In *Chroomonas* and *Cryptomonas,* the nuclear membrane disappears early in mitosis, and a dense, continuous plate of chromatin material forms without distinct chromosomes being evident. At anaphase two plates of chromatin material migrate to the poles. The spindle apparatus is rectangular. Microtubules arise from a broad area at each pole. The process of cell division is accomplished by a cytokinetic furrow that constricts the cell. The differences in mitosis between dinoflagellates supports the separation of these two divisions.

All of the cryptomonads (except *Hillea*) contain ejectosomes. These are coiled cylinders of membraneous material similar to trichocysts (Wehrmeyer, 1970; Hausmann, 1973). The ejectosomes line the gullet, with smaller ones occurring under the outer cell covering (periplast). Ejectosomes are released if the cell is osmotically shocked or begins to dry out. In general, cryptomonad ejectosomes are similar to dinoflagellate trichocysts.

Motility

The motile cells have a pair of apically inserted flagella. The flagella are unique, compared to other groups of algae, because both have mastigonemes (stichonematic) (see Table 3-3).

Cell Wall Structure

The cell is coated with a periplast that can be organized into a series of hexagonal plates. Cellulose has been reported for some members, but its presence is questionable (Gantt, 1980).

Food Storage

True starch, as found in the Chlorophyta, is produced in the chloroplast and may form starch platelets around the pyrenoid. Oil is also stored in the cytoplasm as droplets.

Morphology and Reproduction

The members are mostly motile unicells, although some coccoid forms occur. The cells are dorsiventrally flattened and usually oval to pyriform (Butcher, 1967).

Although mitotic figures have been observed the life histories of members of this division are essentially unknown. One case of sexual reproduction is reported for *Cryptomonas* in which meiosis is zygotic and the organism has a haplontic life history (Gantt, 1980).

Ecology and Taxonomy

Although members of the division are widely distributed, little is known about the division. The treatment by Gantt (1980) is the most detailed account of the marine forms. She points out that members occur in salt marsh pools as well as open coastal and pelagic waters.

The motile and nonmotile (coccoid) species are differentiated as distinct orders. *Cryptomonas* is a good example of the former group. This common flagellate is an oval-shaped, biflagellated (stigonematic flagella) alga (Figure 9-2). The flagella are anteriorly inserted in the gullet. Butcher (1967) illustrates the same species in a variety of colors ranging from green and blue-green to red-green. The cells have a contractile vacuole and ejectosomes line the gullet. Two lobed plastids and numerous starch grains are characteristic of the cell.

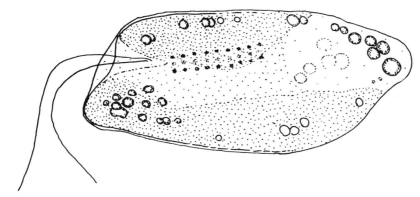

Figure 9-2 *Cryptomonas maculata* is a motile cell about 4–6 μm in diameter and 17 μm in length. The gullet extends 1/2 to 2/3 of the cell and is lined with three-four rows of trichocysts. Two equal-length flagella, both stichonematic, extend from the gullet. Color ranges from deep to pale red.

Division Euglenophyta

Most phycological treatments present the euglenoids as a distinct division adjacent to the Chlorophyta or as a class of that division because of their grass-green coloration and pigments like those of the green algae. A general review of the division has been given by Leedale (1967) and Walne (1980). Elegant drawings of marine forms are presented by Butcher (1961). In the present text, the euglenoids are separated from the green algae and considered protists with "captured" chloroplasts. As Gibbs (1978) has argued, the chloroplasts of euglenoids may have evolved from a symbiotic green alga captured in a protist. The evidence for such a view can be stated as follows:

1   There is a third chloroplast membrane, not derived from the endoplasmic reticulum, which may be the remaining plasmalemma of the symbiotic green alga.

2   The reserve food of the euglenoids is paramylum and is not like the starch of green algae.

3   There are a number of protist or animalistic organelles common to the members of Euglenophyta, including a gullet, a free eyespot or photoreceptor, trichocysts, and a pellicle rather than a cell wall.

4   The process of cell and nuclear division is quite unique when compared to green algae.

5   There are a large number of colorless forms of euglenoids and chloroplast loss can be induced.

CYTOLOGICAL FEATURES

Pigments

Euglenoids have the same chlorophylls as the Chlorophyta (chlorophylls *a* and *b*,) as well as similar accessory pigments (β-carotene and xanthophylls). A specialized xanthophyll, astaxanthin, is so abundant that it can give the cell a blood-red coloration.

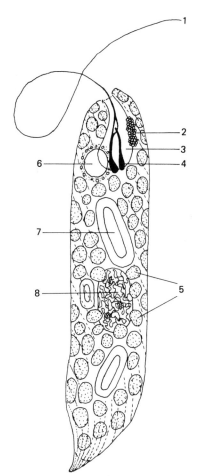

Figure 9-3   The green phytoflagellate *Euglena* can be regarded as a living organism that combines plant and animal characteristics in one cell. The diagrammatic view shown here (*E. spirogyra;* Leedale. 1966) shows some of the animal/plant features: (*1*) a projectory locomotory flagellum, (*2*) an eyespot, (*3*) the reservoir, (*4*) a short, nonemergent flagellum, (*5*) numerous discoid chloroplasts, (*6*) a contractile vacuole, (*7*) many paramylum grains, and (*8*) a single nucleus.

Cytological Structure

The general structure of a euglenoid cell is given in Figure 9-3. The gullet, or reservoir, is a characteristic feature. The cells are uninucleate; one to several chloroplasts are present; they may contain pyrenoids. A photoreceptor, or eyespot, occurs at the anterior end of the cell and is *not* associated with a chloroplast as in other algae. The photoreceptor appears to be associated with a flagellar swelling and may either act as a shading device on the swelling, a light receiver, or a sensor. Trichocysts can be present in the gullet.

Mitosis and cytokinesis are unique in this division. In *Euglena* (Gillott and Triemer, 1978), the nuclear envelope remains intact and the spindle forms within the nuclear envelope (closed mitosis). Chromosomes, which remain condensed even during interphase, do not migrate to the poles at anaphase. Instead, the nucleus elongates, becomes dumbbell-shaped and breaks, carrying a clump of chromosomes to each daughter cell. Kinetochores, organelles characteristic of animal cells, are present during mitosis as well.

Motility

Basically euglenoids are biflagellated, although only one flagellum may emerge from the gullet. At least the emergent flagellum, an acronematic form, has the typical nine-plus-two microtubular structure. A second flagellum can be nonemergent, appressed against the emergent one in the gullet. A swelling on the emergent flagellum is associated with the photoreceptor. Euglenoid movement, a specialized amoeboid type, is characteristic of all members of the division.

Cell Wall Structure

The plasmalemma covers the euglenoid cell and a protein pellicle is composed of helically wound, overlapping strips. The pellicle can contain ferric hydroxide particles, which impart a reddish hue to the cell.

Food Storage

Paramylum, a crystalline β-$(1 \rightarrow 3)$-glucose polymer, is the photosynthate. It occurs around the pyrenoid or as free particles in the cell.

MORPHOLOGY AND REPRODUCTION

Most of the species are unicellular, motile, and photosynthetic, although some may be colorless phagophytes or saprophytes. A few are colonial or coccoid, and these, such as *Colacium,* can form dendroid colonies. When motile, all of the species have an apical opening called a *gullet*. At least in some of the colorless forms, the gullet will serve as the phagotrophic structure, and thus the term *reservoir* is probably not the best choice.

Asexual reproduction appears to be the only method of reproduction, although sexual stages were reported in earlier studies. There is some suggestion of genetic recombination based on studies of cultured material (Leedale, 1967; Walne, 1980). Cell division is longitudional, beginning in the gullet. First the basal bodies replicate, then the periplast develops a longitudinal furrow. As described above, mitosis is unique and basically amitotic, as a type of nuclear fragmentation seems to occur.

ECOLOGY AND TAXONOMY

Although euglenoids are more common in freshwater environments, there are a number in estuarine and intertidal zones. One or more vitamins are required (auxotrophic) even for the photosynthetic forms. All species are apparently able to absorb extracellular organic material. Evidence of ingestment of bacteria has been observed in the gullets of colorless forms but not in those of photosynthetic forms. A number of species have the ability to lose chloroplasts and associated pigments if held in the dark, thus going from autotrophs to heterotrophs.

The classification is unclear. Two orders are recognized. The Eutreptiales are identified by the presence of two emergent flagella. *Eutreptia* (Figure 9-4) is a typical euglenoid of brackish-water environments and can be found on wet mud. The second order (Euglenales) contains the euglenoids with one emergent flagellum (Figure 9-5). A number of genera are common to brackish water and intertidal zones, especially in areas of nutrient enrichment; these genera include *Euglena*, *Lepocinclis, Phacus* (Figure 9-6) and *Trachelomonas* (Figure 9-7).

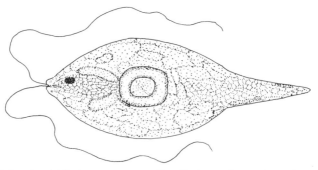

Figure 9-4 *Eutreptia viridis* is common to saline ditches and estuaries. The biflagellated euglenoid (note paramylum body and eyespot) is large (60 × 13 μm) and is easily seen under low magnification in a compound microscope.

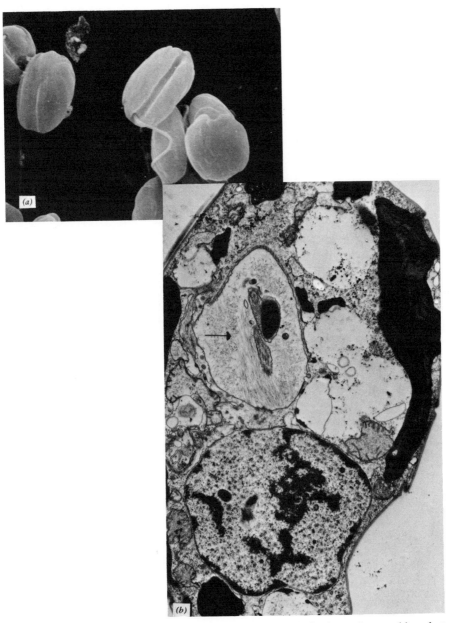

Figure 9-5 *Cryptoglena pigra,* an example of an euglenoid having only one chloroplast but typical euglenoid characteristics (Rosowski and Lee, 1978). (*a*) A scanning electron micrograph showing the uniflagellated (emergent) cell with the unique longitudinal sulcus (× 1,680). (*b*) A transmission electron micrograph showing two lobes of the chloroplast, the large nucleus, a small portion of the reservoir (arrow), with sections through the emergent and nonemergent flagella, and the typical euglenoid periplast surrounding the cell. × 8,000. (Courtesy of James Rosowski, University of Nebraska, Lincoln.)

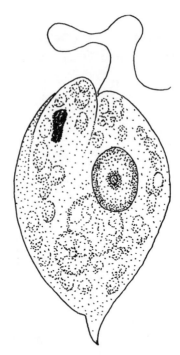

Figure 9-6 *Phacus triqueter* measures 45 × 30 μm, is ovate and quite flattened with a posterior oblique point. The uniflagellated (emergent flagellum) cell is characteristic of brackish, low-salinity water such as that found in ditches.

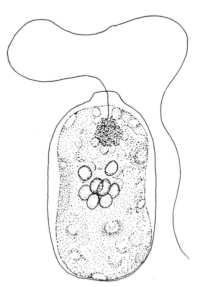

Figure 9-7 *Trachelomonas hispida*, a cell of about 30 × 20 μm, elliptical with rounded ends. Numerous (10–20) chloroplasts and small paramylum bodies are common. Typical of the genus, the cell is encased in a well-developed envelope or capsule (lorica) with a terminal pore through which the flagellum emerges.

Division Pyrrhophyta

This division contains the dinoflagellates and related organisms and forms a major component of marine phytoplankton communities. About 130 genera representing 1200 species are in the division. The dominant group in this division, the class Dinophyceae, is composed of eukaryotic organisms of both plant and animal affinities (Steidinger, 1979; Steidinger and Cox, 1980). Most species are unicellular, free-living motile cells, although some are coccoid, filamentous or saclike parasites. Some species are bioluminescent, some are toxic (red tides), whereas others are intracellular symbionts (zooxanthellae) of invertebrates. Geologically, representatives of this class date back to the Silurian period (about 440 million years ago). There are over 1700 described fossil species, the greatest species diversity occuring in the Cretaceous period.

The most unusual features of dinoflagellates are the distinctive cell shape (Figure 9-8) and the two dissimilar flagella of the class Dinophyceae. Of the two flagella, one is found wrapped around the cell in a girdle grove called a cingulum, whereas the other trails behind (see Figure 3-20). Dinoflagellates are autotrophic, auxotrophic, or heterotrophic. Because of the cytological features of the

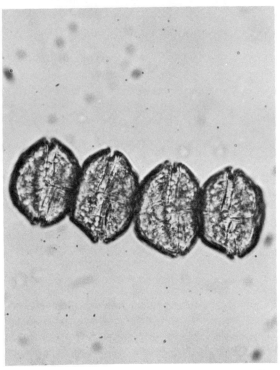

Figure 9-8 A light micrograph of a chain-forming species, *Gonyaulax monilata*. It will reach about 50 μm in diameter (see also Figure 3-18). (Courtesy of Karen Steidinger, Florida Department Natural Resources Marine Laboratory, St. Petersburg.)

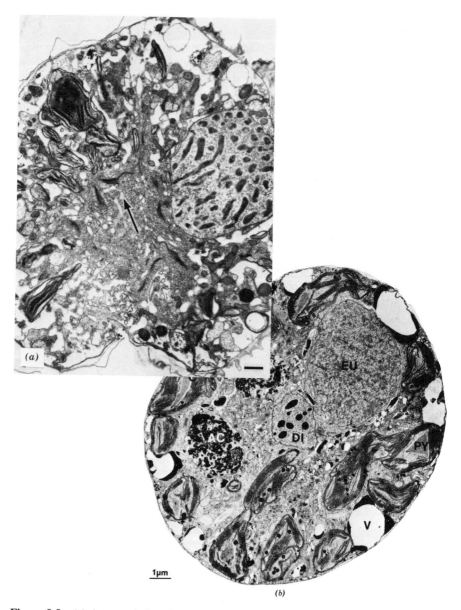

1µm

(b)

Figure 9-9 (a) A transmission electron micrograph of *Gonyaulax polyhedra* showing typical cell structure of a dinoflagellate cell (after Gaudsmith and Dawes, 1972). One lobe of the nucleus is visible and Golgi bodies (arrow) are present in the central region. Portions of discoid chloroplasts are found in the periphery of the cell. (b) A few dinoflagellates, such as *Peridinium balticum* have been shown to have both a eukaryotic nucleus (*EU*) and a dinoflagellate nucleus (*DI*). Chloroplasts (c), an accumulation body (AC), and peripheral vacuoles (V) are also visible. (Tomas and Cox, 1973: courtesy of *Journal of Phycology*.) The unit mark equals 1 µm.

dinoflagellates, Tomas and Cox (1973) and Gibbs (1978) have proposed that the dinoflagellates and the euglenoids are actually protists that have captured photosynthetic symbionts. In the case of dinoflagellates, the "captured chloroplast" would be from a "brown" alga having chlorophylls *a* and *c* and special xanthophylls. Unlike the chrysophytes, fucoxanthin is not a common pigment in dinoflagellates.

CYTOLOGY

Pigments

Chlorophylls *a* and *c, β*-carotene, and the special xanthophylls (peridinin, dinoxanthin, diadinoxanthin) and occasionally fucoxanthin produce the characteristic flame color. Hence the first part of the divisional name is *pyrrho*. Peridinin is the dominant xanthophyll. Because of pigmentation or light scattering by organismal particles, blooms of dinoflagellates in the ocean have been called "red tides."

Cytological Structure

The nuclear cytology and mitosis of dinoflagellates is distinctive and has been the topic of many ultrastructural studies (Spector and Triemer, 1979). When compared to other eukaryotic algae, the Pyrrhophyta have so many unique nuclear features that the term *mesokaryote* was coined by Dodge (1965). The term was meant to suggest a halfway position between the prokaryotic and eukaryotic cell structure, especially in terms of the nuclei (Figure 9-9). The nucleus is large. The chromosomes lack histones or have only low levels of the basic proteins when compared to other eukaryotic chromosomes. The chromosomes remain condensed throughout the life cycle and are easily seen, even with light microscopy (Figures 9-9, 9-10). In mitosis of free-living species, the nuclear membrane and nucleolus remain intact (closed mitosis). Microtubules extend through cytoplasmic channels in the nucleus, attaching to the nuclear membrane adjacent to sites of chromosomal attachment. Colchicine does not affect mitosis. There is nothing even approaching this type of mitosis in other eukaryotic algae. Thus, the term mesokaryote. The term is no longer applied to total cellular structure, since cells have typical eukaryotic organelles, that is, Golgi apparatus, mitochondria, and plastids. Also, DNA studies show eukaryotic sequencing (Loeblich, 1976a). On the other hand, the nucleus of free-living species is not definitely eukaryotic because of mitotic features, low levels of chromosomal proteins and the presence of hydroxymethyluracil, which in one species substitutes for 12% of the thyamine (Loeblich, 1976b). The only other known natural occurrence of this base is in some bacteriophages.

Trichocysts are common and consist of a crystalline protein core surrounded by a membrane. Bouck and Sweeney (1967) found that trichocysts are formed in Golgi vesicles and transported to the outer cytoplasm. The function of trichocysts

Figure 9-10 *Prorocentrum* sp., a large, spherical dinoflagellate that can have a small projection from the apical tip. The colony of cells photographed with the scanning electron microscope shows a distinctive pattern on the cell wall. (Courtesy of Lana Tester, Florida Department Natural Resources Marine Laboratory, St. Petersburg.)

is questionable. They may function in cellular balance, adhesion, osmoregulation, sensory function, or reduction of sinking rates. Mucocysts are also present in some species, and they may contribute to the outer mucopolysaccharide coat.

The chloroplast is discoid or lobed and has a triple membrane with the thylakoids usually arranged in bands of two to three. Only one species, which lacks an algal symbiont, has been shown to have plastids with girdle lamellae as found in the Chrysophyceae. Pyrenoids are common and can be embedded or stalked in the chloroplast. Dodge (1968) has reviewed the structure of dinoflagellate chloroplasts and pyrenoids.

Eyespots are uncommon in dinoflagellates but when present are of four types (Steidinger and Cox, 1980): (1) a mass of lipidlike globules that is independent of the chloroplast and near the sulcus; (2) a single layer of globules that is part of a chloroplast; (3) a membrane-bound eyespot with two layers of globules, separated from the chloroplast; or (4) a specialized ocellus consisting of a lens with a retinoid and associated pigments.

Motility

One of the most important characteristics of this division deals with the flagella (see Figure 3-21), which are either anterior (Desmophyceae) or lateral (Dinophyceae). In the Dinophyceae, one flagellum beats in a transverse plane; it is located in the cingulum and is flattened and pantonematic. The other one is a trailing, long acronematic flagellum. The combination of the whiplash trailing flagellum and the hemihelical beat of the transverse flagellum results in a forward propulsion. The beat of the trailing flagellum results in a corkscrew type of motion, hence the term *dino* which means "whirling." Dinoflagellates are capable of moving 1–2 m/hr and up to 20 m/da. Vertical migrations may occur within the upper 15 m of water.

Cell Wall Structure

The basic covering is a series of membranes, which are called amphiesmas (Loeblich, 1970) or thecae in the class Dinophyceae. The amphiesma is a series of membranes and flattened vesicles or sacs (see Dodge and Crawford, 1970). In the lumens of these sacs, α-cellulose or other polysaccharides may occur, forming plates (armored forms). Naked, or unarmored, forms lacking cellulose also occur. The number and arrangement of plates are used to define genera and species (Figure 9-14). The wall is called a theca and the cell (Figure 9-8) can be divided into an epitheca (above the cingulum) and a hypotheca (below the cingulum). Even species without polysaccharide plates will have a distinctive pattern of flattened vesicles or sacs forming the amphiesma.

Food Storage

The primary reserve food in free-living species is starch, and starch grains are common around the pyrenoids and in the cytoplasm. Oil droplets may also be found in the cytoplasm; the oil reserve is composed of long chained unsaturated fatty acids (C_{14}, C_{16}, C_{18}, C_{22}).

MORPHOLOGY AND REPRODUCTION

Although most vegetative stages of the Pyrrhophyta are motile unicells (Figure 9-8), there are colonial coccoid, and filamentous species as well. Most motile stages are <200 μm; however, several species can be up to 1–2 mm in length.

All the sexual life histories studied thus far (18 out of 1200 free-living species), except *Noctiluca*, are haplontic with meiosis occurring during or after germination of the zygote. The zygote (hypnocyst) can function as a resting stage. Gametes are either isogamous or anisogamous and are usually similar to the parent cell. In *Noctiluca* the gametes are uniflagellated and the zygote develops directly into the new cells. Thus it is assumed that the life history is diplontic. Cyst formation is common in marine species of dinoflagellates; the cyst wall is usually thick and sometimes highly ornamented. Asexual reproduction is by cell division; in unarmored forms the cell pinches in half, and in armored forms, the theca either divides in half or an entirely new one is formed after shedding the old wall (excisement).

ECOLOGY

The collection, preservation, and identification of free-living, especially red tide, dinoflagellates has been reviewed by Steidinger (1979). Selected techniques are presented in Chapter 20. As noted earlier, the cytology of dinoflagellates is unique; they also have a number of interesting ecological features. Dinoflagellates are of major significance to the marine environment for a number of reasons: (1) they are important primary producers; (2) they are the basis of the marine algal blooms termed *red tides;* and (3) they are important symbionts (zooxanthellae) in a variety of invertebrates contain symbiotic algae themselves.

Primary production

Termed "grasses of the sea," dinoflagellates can become the dominant free-living pelagic algae in a phytoplankton community. Steidinger *(personal comment)* has found from 12 years of distribution data that four assemblages of dinoflagellates occur in the Gulf of Mexico, representing estuarine, estuarine/coastal, coastal/open Gulf, and open Gulf populations. Indicator species of confined distribution are found in the first and last assemblages only. She has also found that species diversity increases seaward, whereas total abundance decreases. Estuarine assemblages are characterized by several cosmopolitan species (e.g., *Ceratium hircus, Prorocentrum* sp., *Gymnodinium splendens*) as are the Gulf of Mexico assemblages (e.g., *Ceratium furca, Protoperidinium depressum, Podolampas* spp.). It appears that dinoflagellate populations can be established through cysts that settle out in estuarine or deep-water sites. With the seasonal cycling of the water column through thermal warming or cooling, upwelling, and storm disturbances, these cysts can be stimulated to excyst, and new motile cells will be released (Steidinger, 1975b; Dale, 1977).

Red tides and dinoflagellate toxins

Blooms, or red tides, by dinoflagellates can be toxic or nontoxic depending on the species involved. Red tides are formed by a variety of nontoxic dinoflagellates such as *Noctiluca* in Puget Sound or *Ceratium* in the Gulf of Mexico. Toxic species of dinoflagellates include *Protogonyaulax tamarensis* in New England (Loeblich and Loeblich, 1975), *Protogonyaulax catenella* in Puget Sound (Norris *et al.,* 1973), and *Ptychodiscus brevis (Gymnodinium breve)* in Japan (Iizuka, 1975) and Florida (Steidinger, 1975a).

Ecologically, three basic factors seem to be common to all red tides (Steidinger, 1975a): (1) there is an increase in population size (initiation); (2) there is support for the bloom in the form of suitable salinity, temperature, nutrients, and growth factors; and (3) there is maintenance and movement of the blooms by hydrological and meteorological factors. The residual cyst population in benthic substrata is now being recognized as a source of the red tide organisms (Steidinger, 1975b; Dale, 1977).

The toxin from *Ptychodiscus brevis* (termed GB) appears to depolarize excitable membranes and is distinct from the paralytic shellfish poison (termed PSP) produced by a section (*Catenella*) of the genus *Gonyaulax* (Steidinger *et al.*, 1973; Schmidt and Loeblich, 1979). The toxin from *Ptychodiscus* can be stored in shellfish and if eaten by birds or mammals can cause sickness. No human fatalities from GB toxin have been reported, whereas over 300 human deaths are known to have been due to PSP after eating toxic filter-feeding shellfish (mussels, clams). Unlike GB toxin, PSP (saxitoxinlike toxin) is water soluble.

A dinoflagellate, *Gambierdiscus toxicus* has also been implicated as a causative agent of ciguatera fish poisoning (Bagnis *et al.*, 1979). The highly compressed dinoflagellate cells attach directly to algal thalli or dead coral substrata or lie suspended on the sediment surface and are apparently eaten by grazing fish. The alga produces two toxic substances comparable to fish reference toxins maitotoxin and ciguatoxin.

Symbiotic and trophic features

Two types of symbiosis can be considered here, the symbiosis of dinoflagellates with invertebrates (zooxanthellae) and other algae with dinoflagellates (endophytes). Many dinoflagellates are free-living auxotrophs (require vitamins). In a study of eight photosynthetic dinoflagellates, no examples of heterotrophic growth were found, but the growth of six species was stimulated by glycerol in both dim light and the presence of DCMU, an inhibitor of photosynthesis (Morill and Loeblich, 1979). A wide variety of nonphotosynthetic forms occurs, ranging from saprophytic to parasitic forms. Some parasitic forms are endoparasites, living in the gut of copepods (*Blastodinium*) or annelids (*Haplozoon*). Phagotrophic dinoflagellates can ingest the entire cell of a small eukaryotic cell, or they can attach to cells and suck out the cytoplasm (Pfiester and Popovsky, 1979). Perhaps such organisms can function as herbivores in phytoplankton communities.

The suggestion that dinoflagellates are protists that have captured a photosynthetic alga (Tomas and Cox, 1973; Gibbs, 1978) is supported by the presence of heterotrophic forms and cytological features, as well as their distinctive symbiotic relationships. A number of dinoflagellates have endosymbiotic algae. Typical eukaryotic nuclei have been found in addition to the dinokaryotic nucleus in *Peridinium foliaceum* (Dodge, 1971) and in *Peridinium balticum* (Tomas and Cox, 1973). The two symbionts appear to be chrysophycean algae. Sweeney (1976) also reported a green flagellate symbiont, *Pedimonas,* in the giant dinoflagellate *Noctiluca*. Steidinger and Cox (1980) have proposed a scheme for the evolution of photosynthetic dinoflagellates. Zooxanthellae are species of dinoflagellates that are endozoic in a wide variety of marine invertebrates (jellyfishes, sea anemones, bivalves, turbellarians, protists), especially stony corals. The symbionts seem to be essential for hermatypic coral growth and calcification. Most zooxanthellae are similar to *Gymnodinium* or *Amphidinium* species. However, Loeblich and Sherley (1979) detailed the amphiesma of *G. microadriaticum* and transferred it to the genus *Zooxanthella*. Through the use of labeled photosynthate, it has been shown that glycerol is transferred from the symbiont to the host. Up to 60% of the photosynthate produced by the algal symbiont may be transferred to the host.

Since many free-living species are similar in structure to the zooxanthellae in the free state, they should probably not be taxonomically separated (Taylor, 1973). A good review of symbiosis and zooxanthellae can be found in the text by Trager (1970) and the article by Taylor (1973).

TAXONOMY

There are about 130 genera representing $\sim$ 1200 species in the division. Two classes are recognized here: Desmophyceae and Dinophyceae. For a discussion of other classes, such as those having parasitic forms (Ellobiophyceae and Syndiniophyceae), see Bold and Wynne (1978).

CLASS 1. DESMOPHYCEAE

This class is characterized by motile cells having two apically or subapically inserted flagella, one directed forward and the other beating in a plane at right angles to the first. Of the three orders placed in this class, one contains immobile, palmelloid masses of cells (Order 1, Desmocapsales), another contains colorless cells having a pair of subapically inserted flagella (Order 2, Protaspidales). The third order, Prorocentrales, contains motile cells that are mostly freeliving; they have a theca of two major parts and their flagella are apically inserted. Only one example of the third order will be given.

Prorocentrum

The cell wall, or theca, consists of two halves that are laterally compressed (Figure 9-10). The cells are ovate to pear-shaped. A "tooth" or spinelike projection may occur at the anterior end of the cell. The two anterior flagella consist of one pleuronematic flagellum wrapped around or at right angles to the projecting acronematic one. One to several chloroplasts are usually present. The nucleus is large and has condensed chromosomes.

CLASS 2. DINOPHYCEAE

This class contains the majority of free-living unicellular dinoflagellates. Only 2 of the 12 total orders will be described here. The trophic states range from symbiotic (zooxanthellae) to parasitic, phagotrophic, and auxotrophic species. In turn, morphologies range from unicellular flagellated or amoeboid plants to coccoid or palmelloid colonies, as well as a few filamentous species that reproduce through a dinokontlike zoospore. The main characteristic of the class is the arrangement of the flagella in the motile cell. The pantonematic flagellum is located in the cingulum, whereas the acronematic flagellum trails behind the cell and is located in the sulcus. Both flagella are laterally inserted.

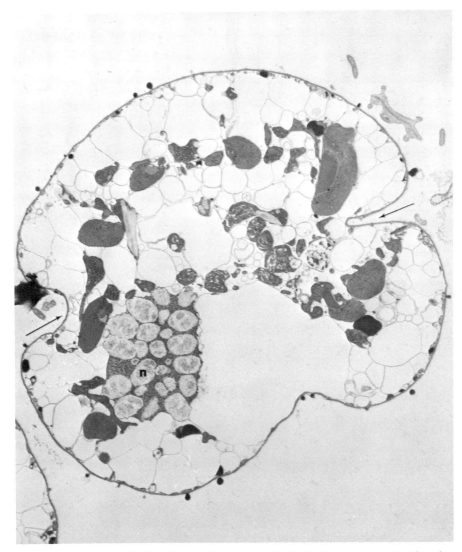

Figure 9-11 A transmission electron micrograph of *Ptychodiscus brevis*, showing the highly vacuolated cell, a lobe of the nucleus with condensed chromosomes (*n*)., and portions of various discoid chloroplasts. Note the cingulum (arrows). (Steidinger *et al.*, 1978). × 4,740. (Couresy of Karen Steidinger, Florida Department Natural Resources Marine Laboratory, St. Petersburg.)

Gymnodiniales

This order contains motile cells (photosynthetic and nonphotosynthetic) that lack polysaccharide thecal plates but have a membrane system termed amphiesma (see Figure 3-31).

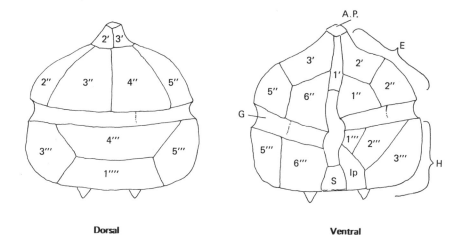

Dorsal   **Ventral**

Figure 9-12 Dorsal and ventral views of *Gonyaulax* sp., demonstrating the plate diagram for the genus. (*E*) Epitheca. (*H*) Hypotheca. (*G*) Cingulum. For plate terminology see text. Figures 9-14 through 9-17 are courtesy of the Florida Department Natural Resources Marine Laboratory, St. Petersburg.

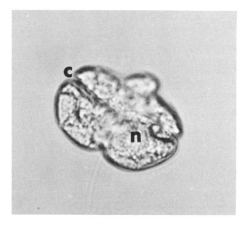

Figure 9-13 A living cell (light micrograph) of the red-tide-forming dinoflagellate *Ptychodiscus brevis (Gymnodinium breve)*, showing the cingulum (c) and nucleus (n). The cell is about 25 μm wide and 13 μm deep. See also Figure 3-21 for a scanning electron micrograph of *Gymnodinium splendens*. (Courtesy of Karen Steidinger, Department Natural Resources Marine Laboratory, St. Petersburg.)

Ptychodiscus (Gymnodinium)

This genus is one of the most intensively studied dinoflagellates because of its role in red tides (Figure 9-11, 9-12). *Ptychodiscus brevis* measures 20–40 μm wide, 10–15 μm deep, and is slightly wider than long (Figure 9-13). The species was previously called *Gymnodinium breve* but has been renamed by Steidinger. The cells are centrally concave and dorsally convex. The cingulum houses the

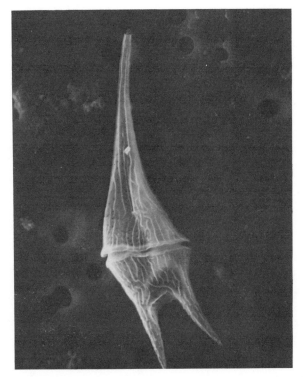

Figure 9-14 A scanning electron micrograph of *Ceratium hircus,* the cell is 130 μm long. Note the elongated epithecal cone.

transverse flagellum and the sulcus houses the longitudinal flagellum. The nucleus is large, 6–9 μm in diameter, and the cell contains 10–20 irregularly shaped chloroplasts (Figure 9-11). Ultrastructural studies (Steidinger *et al.,* 1978) reveal that the organelles are typical of the dinoflagellates, with trichocysts found at the cell periphery. Pyrenoids are common in the chloroplasts (Figure 9-12). Oil droplets are common. The cytoplasm has one unusual feature compared with most other dinoflagellates, that is, it has a highly perforated vacuolar reticulum. Because this is an unarmored form, the thecae lack cellulosic plates. Instead the membrane system consists of an outer membrane, a series of saclike membranes, and an inner membrane termed the plasmalemma. This terminology is in contrast to that of Loeblich (1970). The presence of the flattened sacs in the amphiesma supports the viewpoint that the "naked" forms are similar to the "armored" ones, except for the presence of cellulose plates (see Steidinger *et al.,* 1978).

Peridiniales

In this order, the cells are free-living and motile. The individual membrane sacs of the amphiesma contain polysaccharide plates, and the order is considered to be

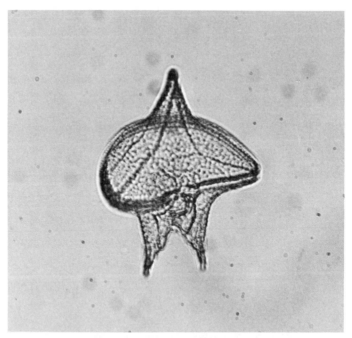

Figure 9-15 A living cell of *Protoperidinium grande* with a cell diameter of about 130 μm. The view is oblique and looking down at the expanded cingular region of the epitheca.

a group of armored dinoflagellates. Both photosynthetic and colorless species are included. Plate number and pattern are critical in the taxonomic placement of species. A Kofoidian code has been established for plate terminology (Figure 9-12). Thus, both the epitheca and the hypotheca can be divided up into three series of plates and associated symbols. The epitheca can have one or more apical ('), precingular ("), and anterior intercalary (*a*) plates. The hypotheca can have one or more posterior intercalary (*p*), postcingular ('''), and antapical (''') plates. In addition, the plates forming the cingulum (*c*) and sulcus (*s*) are important in genus and species identification. An example of a formula for an armored dinoflagellate might be that given for the genus *Gonyaulax* [(3', 0*a*, 6", 6*c*, 6-7*s*, 6''', 1*p*, 1'''')]

Gonyaulax

The plate formula described above is shown in Figure 9-12. The genus is well known because of the large number of species that cause paralytic shellfish poisoning and red tides in Japan, England, New England, Canada, South America, South Africa, and elsewhere. Many of the species are pleomorphic. The first apical plate (1') is narrow, and there is a sharp displacement of the girdle where it joins the sulcus. The cells of *Gonyaulax monilata* are about 57 μm wide and can form catenate chains (Figure 9-8). The ultrastructure of *Gonyaulax*, as well as other armored dinoflagellates, is typical of the class (Gaudsmith and

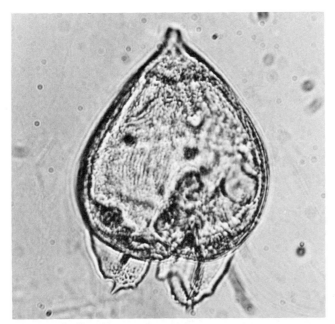

Figure 9-16   A living cell of *Podolampas reticulata* (epithecal view) that is about 70 μm in diameter.

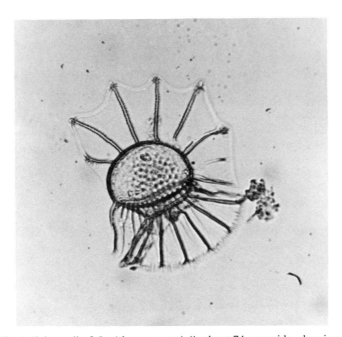

Figure 9-17   A living cell of *Ornithocercus steinii*, about 74 μm wide, showing the highly modified theca that forms a dramatic "keel."

Dawes, 1972). That is, the cytoplasm is finely vacuolated, as it is in most dinoflagellates.

Other types of armored dinoflagellates placed in Peridinales include *Ceratium* (Figure 9-14, also Figure 20-4), *Protoperidinium* (Figure 9-15), *Pyrodinium* (Figure 3-18), and the nonphotosynthetic genus *Podolampas* (Figure 9-16). An example of the order Dinophysales, which is not covered in this text, is given in Figure 9-17 (*Ornithocercus*) and Figure 20-5 (*Dinophysis*).

REFERENCES

Bagnis, R., J-M. Hurtel, S. Chanteau, E. Chungue, A. Inoue, and T. Yasumoto. 1979. Le dinoflagelle *Gambierdiscus toxicus* Adachi et Fukuyo; agent causal probable de la ciguatera. *C. R. Acad. Sc. Paris.* **289**: 671–674.

Bold, H. C. and M. J. Wynne. 1978. *Introduction to the Algae.* Prentice-Hall, Englewood Cliffs, N.J.

Bouck, G. B. and B. M. Sweeney. 1967. The fine structure and ontogeny of trichocysts in marine dinoflagellates. *Protoplasma* **61**: 205–223.

Butcher, R. W. 1961. *An introductory account of the smaller algae of British coastal waters.* Part VIII: *Euglenophyceae = Euglenineae.* Ministry of Agriculture, Fisheries and Food. Fishery Investigations Series IV. HMS Stationary Office, London.

Butcher, R. W. 1967. *An Introductory Account of the Smaller Algae of British Coastal Waters. Part IV: Cryptophyceae.* Ministry of Agriculture Fisheries and Food. HMS Stationary Office, London.

Dale, B. 1977. Cysts of the toxic red-tide dinoflagellate *Gonyaulax excavata* (Braarud) Balech from Oslofjorden, Norway. *Sarsia* **63**: 29–34.

Dodge, J. D. 1965. Chromosome structure in the dinoflagellate and the problem of the mesokaryotic cell. *Excerpta Med. Int. Congr. Ser.* **91**: 339.

Dodge, J. D. 1968. The fine structure of chloroplasts and pyrenoids in some marine dinoflagellates. *J. Cell Sci.* **3**: 41–48.

Dodge, J. D. 1971. A dinoflagellate with both a mesocaryotic and a eukaryotic nucleus. I. Fine structure of the nuclei. *Protoplasma* **73**: 145–157.

Dodge, J. D. and R. M. Crawford. 1970. A survey of thecal fine structure in the Dinophyceae. *Bot. J. Linn. Soc.* **63**: 53–67.

Gantt, E. 1980. Photosynthetic cryptophytes. In Cox, E.R. (ed.). *Phytoflagellates,* Elsevier/North Holland, New York, pp. 381–406.

Gantt, E., M. R. Edwards, and L. Provasoli. 1971. Chloroplast structure of the Cryptophyceae. Evidence for phycobiliproteins within intrathylakoidal spaces. *J. Cell Biol.* **48**: 280–290.

Gaudsmith, J. T. and C. J. Dawes. 1972. The ultrastructure of several dinoflagellates with emphasis on *Gonyaulax polyedra* Stein and *Gonyaulax monilata* Davis. *Phycologia* **11**: 123–132.

Gibbs, S. P. 1978. The chloroplasts of *Euglena* may have evolved from symbiotic green algae. *Can. J. Bot.* **56**: 2883–2889.

Gillott, M. A. and S. P. Gibbs. 1980. The cryptomonad nucleomorph: its ultrastructure and evolutionary significance. *J. Phycol.* **16**: 558–568.

Gillott, M. A. and R. E. Triemer. 1978. The ultrastructure of cell division in *Euglena gracilis. J. Cell Sci.* **31**: 25–35.

Hausmann, K. 1973. Cytologische Studien an Trichocysten. VI. Feinstruktur und Funktionsmodus der Trichocysten des Flagellaten *Oxyrrhis marina* und des Ciliaten *Pleuronema marinum.* Helg. Wiss. Meeresunt. **25**: 39–62.

Iizuka, S. 1975. On occurrence of similar organisms to *Gymnodinium breve* Davis in Omura Bay. *Bull. Plankton Soc. Japan* **21**: 45–48.

Leedale, G. R. 1966. *Euglena:* a new look with the electron microscope. *Adv. Sci.* **May**: 22–37.

Leedale, G. F. 1967. Euglenida/Euglenophyta. *Ann. Rev. Microbiol.* **21**: 31–48.

Loeblich, A. R., III. 1970. The amphiesma or dinoflagellate cell covering. *Proc. N. Amer. Paleontol. Conf. 1969.* Part G: 867–929.

Loeblich, A. R. III. 1976a. Dinoflagellate genetics and DNA characterization. *Stadler Symp.* **8**: 111–128.

Loeblich, A. R. III. 1976b. Dinoflagellate evolution: Speculation and evidence. *J. Protozool.* **23**: 13–28.

Loeblich, L. A. and A. R. Loeblich III. 1975. The organism causing New England red tide: *Gonyaulax excavata*. In *Proc. First Intern. Conf. on Toxic Dinoflagellate Blooms 1974*. Mass. Sci. and Technol. Found., Wakefield, pp. 207–224.

Loeblich, A. R. III, and J. L. Sherley. 1979. Observations on the theca of the motile phase of free-living and symbiotic isolates of *Zooxanthella microadriatica* (Freudenthal) *comb. nov. J. Mar. Biol. Assoc. U.K.* **59**: 195–205.

Morrill, L. C. and A. J. Loeblich III. 1979. An investigation of heterotrophic and photoheterotrophic capabilities in marine Pyrrhophyta. *Phycologia* **18**: 394–404.

Norris, L., K. K. Chew, and A. C. Duxbury. 1973. *Shellfish and the Red Tide. Pacific Search. June, 1973.* Washington Sea Grant Advisory Services, University of Washington, Seattle.

Oakley, B. R. and J. D. Dodge. 1976. The ultrastructure of mitosis in *Chroomonas salina* (Cryptophyceae). *Protoplasma* **88**: 241–354.

Oakley, B. R. and T. Bisalputra. 1977. Mitosis and cell division in *Cryptomonas* (Cryptophyceae). *Can. J. Bot.* **55**: 2789–2800.

Pfiester, L. A. and J. Popovsky. 1979. Parasitic amoeboid dinoflagellates (Dinophyceae). *J. Phycol.* **15**: (suppl.) 27.

Rosowski, J. R. and K. W. Lee. 1978. *Cryptoglena pigra:* A euglenoid with one chloroplast. *J. Phycol.* **14**: 160–166.

Schmidt, R. J. and A. R. Loeblich III. 1979. Distribution of paralytic shellfish poison among Pyrrhophyta. *J. Mar. Biol. Assoc. U.K.* **59**: 479–487.

Spector, D. L. and R. E. Triemer. 1979. Ultrastructure of the dinoflagellate *Peridinium cinctum* F. *ovoplanum*. I. Vegetative cell ultrastructure. *Amer. J. Bot.* **66**: 845–850.

Steidinger, K. A. 1975a. Basic factors influencing red tides. In LoCicero, V. R. (ed.) *Proc. First Intern. Conf. on Toxic Dinoflagellate Blooms 1974*. Mass. Sci. and Technol. Found., Wakefield, Mass.

Steidinger, K. A. 1975b. Implications of dinoflagellate life cycles on initiation of *Gymnydinium breve* red tides. *Environ. Let.* **9**: 129–139.

Steidinger, K. A. 1979. Collection, enumeration, and identification of free-living marine dinoflagellates. In Taylor, F. J. R. and Seliger (eds.). *Toxic dinoflagellate blooms*. Elsevier/North Holland, New York, pp 435–442.

Steidinger, K. A., M. A. Burklew, and R. M. Ingle. 1973. The effects of *Gymnodinium breve* toxin on estuarine animals. In *Marine Pharmacognosy*. Academic, New York, pp. 179–202.

Steidinger, K. and E. R. Cox. 1980. Free-living dinoflagellates. In Cox, E. R. (ed.). *Phytoflagellates*, Elsevier/North Holland, New York, pp. 407–432.

Steidinger, K. A. and E. A. Joyce Jr. 1973. *Florida Red Tides*. State of Florida Dept. Nat. Resources. Educational Series No. 17. Tallahassee, Fla.

Steidinger, K. A., E. W. Truby, and C. J. Dawes. 1978. Ultrastructure of the red tide dinoflagellate *Gymnodinium breve*. I. General description. *J. Phycol.* **14**: 72–79.

Sweeney, B. M. 1976. *Pedinomonas noctilucae* (Prasinophyceae), the flagellate symbiotic in *Noctiluca* (Dinophyceae) in southeast Asia. *J. Phycol.* **12**: 460–464.

Taylor, D. L. 1973. The cellular interactions of algal-invertebrate symbiosis. *Adv. Mar. Biol.* **11**: 1–56.

Tomas, R. N. and E. R. Cox. 1973. Observations on the symbiosis of *Peridinium balticum* and its intracellular alga. I. Ultrastructure. *J. Phycol.* **9**: 304–323.

Tomas, R. N., E. R. Cox, and K. A. Steidinger. 1973. *Peridinium balticum* (Levander) Lemmermann, an unusual dinoflagellate with a mesocaryotic and an eukaryotic nucleus. *J. Phycol.* **9**: 91–98.

Trager, W. 1970. *Symbiosis*. Van Nostrand Reinhold, New York.

Wehrmeyer, W. 1970. Struktur, Entwicklung, und Abbau von Trichocysten in *Cryptomonas* und *Hemiselmis* (Cryptophyceae). *Protoplasma* **70**: 295–315.

Walne, P. L. 1980. Euglenoid flagellates. In Cox, E.R. (ed.). *Phytoflagellates,* Elsevier/North Holland, New York, pp. 5–60.

PART 3

Ecological and Environmental Considerations

CHAPTER 10

Geological Factors

In a study of marine plant communities, it is important to observe various geological features such as topography, characteristics of the substrata, type of shoreline, and sedimentation, if any. All of these geological factors greatly influence the development of a marine plant community.

OCEANS OF THE WORLD

About 72% of the world is covered by oceans. The mean depth is about 3700 m, whereas the average elevation above sea level is only 840 m (Figure 10-1). The size of oceanic waters and the substrata covered by them argue for a basic review of geological factors and their importance to marine plants. Davis (1977) gives an

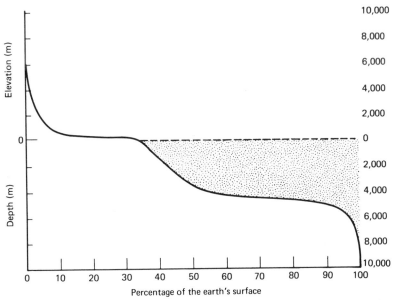

Figure 10-1 A hypsographic curve, shows the percentage area of the earth's surface at a given elevation or depth. The mean elevation of all land is only 840 m above the sea surface (0), whereas the mean depth of the sea is 3729 m below the sea surface.

excellent account of marine geology, and den Hartog (1972) reviews the relationships of substrates and marine plants.

To define a point on the earth's surface, a system of coordinates is needed; latitude, longitude, elevation, and depth are used. Latitude and longitude are angular coordinates; latitude is defined as the distance from the equator and is expressed in lines drawn parallel to the equator (north and south), and longitude is the angular distance from the meridian plane running through the Royal Observatory at Greenwich, England. Longitude is measured from 0° to 180° east and west of the arbitrary starting point at Greenwich Royal Observatory. For example, a lighthouse on the southern tip of Anclote Key, off Tarpon Springs on the west coast of Florida can be more easily pinpointed by the following angular coordinates: 28°10′10″ north latitude and 82°50′42″ west longitude. The degrees, minutes, and seconds are used to indicate distance from the equator (north latitude) and from Greenwich England Royal Observatory (west longitude).

Since the continental land masses extend north and south, there is essentially an antipodal arrangement of land and water masses. A number of oceanic bodies can be described (Figure 10-2). Based on the arrangement of the land masses and the characteristics of the water bodies, five major oceans are usually recognized, the waters surrounding Antarctica (Southern Ocean), the waters surrounding the

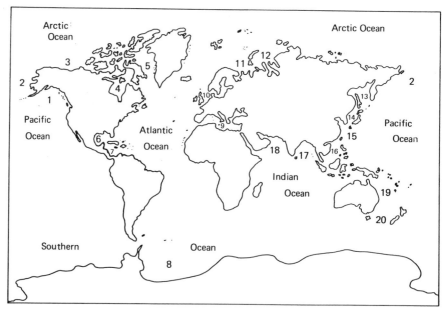

Figure 10-2 A map of the world showing the 5 major oceans. The minor oceans are numbered as follows: (1) Gulf of Alaska, (2) Bering Sea, (3) Beaufort Sea, (4) Hudson Bay, (5) Baffin Bay, (6) Gulf of Mexico, (7) Caribbean Sea, (8) Weddell Sea, (9) Mediterranean Sea, (10) North Sea, (11) Barents Sea, (12) Kara Sea, (13) Okhotsk Sea, (14) Sea of Japan, (15) Yellow Sea, (16) South China Sea, (17) Bay of Bengal, (18) Arabian Sea, (19) Coral Sea, (20) Tasman Sea.

North Pole (Arctic Ocean), and the three central oceans (Pacific, Indian, Atlantic).

The hypsographic curve (Figure 10-1) demonstrates the relief of the earth and shows that the largest portion of the earth lies in depths of from 3500 to 6000 m. The second-largest portion is the low-lying land masses and the shallow waters over the continental shelf. The second region is where the greatest variety of marine plants occurs and where the highest production of organic material can be found (see also Figure 13-5).

GEOLOGICAL HISTORY AND CONTINENTAL DRIFT

Geologists divide the age of the earth into four eras, then each of these into periods and finally the periods into epochs (Table 10-1). The ocean and simple marine life probably existed in Precambrian times (about 3 billion years ago). By the Cambrian or Ordovician periods the ocean was probably similar chemically to the ocean today. However, the geography and climate of the earth has undergone several dramatic changes, including extensive glaciation in the Precambrian, Permian, and Quaternary periods, as well as the most recent separation of the present land masses around 180–190 million years ago. An excellent discussion of this

Table 10-1 Geological Time Scale

Eras (Duration in yr)	When Began (yr)	Epochs	Life Forms	Climates and Major Physical Events
Cenozoic (65 million)	Quaternary (2 million)	Recent Pleistocene	Age of man. Deserts on large scale.	Fluctuating cold to mild. Glacial advances and retreats (Ice Age); uplift of Sierra Nevada.
	Tertiary (63 million)	Pliocene	First known appearance of man-apes. Herbaceous plants abundant.	Cooler. Continued uplift and mountain building. Uplift of Panama.
		Miocene	Whales, apes, grazing animals. Spread of grasslands as forests contract.	Moderate. Extensive glaciation begins again in Southern Hemisphere.
		Oligocene	Large, browsing mammals. Apes appear. Madro-Tertiary geoflora expands.	Rise of Alps and Himalayas. Lands generally low. Volcanoes in Rockies area.
		Paleocene	First-known primitive primates and carnivores.	Mild to cool. Wide, shallow continental seas largely disappeared.

Table 10-1   Continued

Eras (Duration in yr)	When Began (yr)	Epochs	Life Forms	Climates and Major Physical Events
Mesozoic (160 million)	Cretaceous (135 million)		Age of reptiles, extinction of dinosaurs. Angiosperms appear and become abundant.	Lands low and extensive. Last widespread oceans. Elevation of Rockies cut off rain. Africa and South America separate.
	Jurassic (180 million)		Dinosaurs' zenith. Flying reptiles, small mammals. Birds appear. Gymnosperms, especially cycads; ferns.	Mild. Continents low. Large areas in Europe covered by seas.
	Triassic (225 million)		First dinosaurs. Primitive mammals appear. Forests of gymnosperms and ferns.	Continents mountainous. Large areas arid. Separation of northern continents begins the formation of the North Atlantic Ocean.
Paleozoic (400 million)	Permian (270 million)		Reptiles evolve. Origins of conifers, cycads, ginkgos.	Extensive glaciation in Southern Hemisphere. Appalachians formed by end of Paleozoic; most of seas drained from continent.
	Carboniferous Pennsylvanian Mississippian (350 million)		Age of amphibians. First reptiles. Forests, ferns, lycophytes, sphenophytes, gymnosperms. Major groups of fungi in existence.	Warm. Lands low, covered by shallow seas or great coal swamps. Mountain building in eastern U.S., Texas, Colorado. Moist equable climate, temperate to subtropical.
	Devonian (400 million)		Age of fishes. Amphibians appear. Shellfish abundant. Lungfishes. Rise of land plants. Extinction of primitive vascular plants.	Europe mountainous with arid basins. Mountains and volcanoes in eastern U.S. and Canada. Rest of North America low and flat.

Table 10-1   *Continued*

Eras (Duration in yr)	When Began (yr)	Epochs	Life Forms	Climates and Major Physical Events
				Sea covered most of land.
	Silurian (440 million)		Earliest vascular plants. Rise of fishes and reef-building corals. Shell-forming sea animals abundant. Modern groups of algae and fungi.	Mild. Continents generally flat. Mountain building in Europe. Again flooded.
	Ordovician (800 million)		First primitive fishes. Shell-forming sea animals. Invasion of land by plants.	Mild. Shallow seas, continents low; sea covers U.S. Limestone deposits; microscopic plant life thriving.
	Cambrian (625 million)		Age of invertebrates. Trilobites, brachiopods, other animals; marine plants.	Mild. Extensive seas. Seas spill over continents.
Precambrian (4 billion)			First life: blue-green algae and bacteria, eukaryotic cells and multicellularity by close of period.	Dry and cold to warm and moist. Formation of earth's crust. Extensive mountain building. Shallow seas. Glaciation in eastern Canada. Planet cooled. Components of different densities separate under influence of gravity.

Source: Raven, *et. al.* (1976).

history and evidence for the most recent continental drift is found in Chapter 3 of Davis (1977) and in a series of articles published by *Scientific American* (1971). Evidence for earlier periods of continental drift has been presented by Bambach *et al.* (1980).

The concept of continental drift was proposed by von Humboldt in 1810. But only within the last 15 years has combined evidence from magnetism of rocks, the matching of fossil records, and geological deposits of different continents been sufficiently developed to support the theory. One major source of evidence is the deep sea drilling project (DSDP), which has demonstrated the age and type

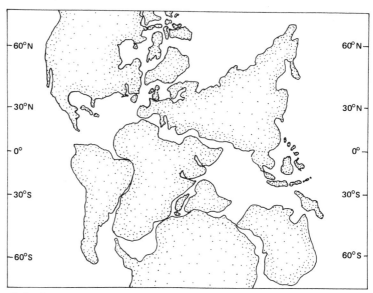

Figure 10-3   Pangea around the mid-Cretaceous, ~110 million years ago. Gondwanaland (Southern Hemisphere) is still intact, whereas the Northern Hemispheric continents are in various stages of separation. The North Atlantic Ocean is just beginning to form. (After Davis, 1977.)

of oceanic sediments. A majority of geologists now believe that during the early Mesozoic period (about 200 million years ago, Table 10-1) there was one super-continent, which was termed Pangea (Figure 10-3) by Edward Seuss in 1915. About 190 million years ago, the northern continents separated to form the North Atlantic Ocean and about 125 million years ago, Africa and South America separated, forming the South Atlantic Ocean. The complete separation at the tropical region occurred about 90 million years ago. India began to move northward about the same time and collided with Asia 45 million years ago. The result of this collision was the initiation of the Himalayan mountains. Australia appears to have separated from Antarctica about 55 million years ago but did not completely separate until 15 million years later. Since most marine algae had already evolved (as early as the Cambrian period) before the separation of Pangea, many of the algal genera are distributed worldwide. Because of continental drift, species were separated, and through mutation and selection new species evolved (speciation), resulting in a number of "paired" or very-similar species that are present in the Atlantic and Pacific Oceans.

The magnetic properties of the rock in the ocean floor indicate that continental drift is still in progress and that it is due to the spreading of the sea floor (Figure 10-4). Australia is continuing on a northward course. The sea floor is moving by convection currents; the movement starts at the midocean ridges, and the substrata flow out in opposite directions. Magnetic patterns, which are oriented parallel the median ridge, indicate the orientation of the molten substrata (magma) at the time of flow; that is, before it cooled and hardened. The field orientation and age of

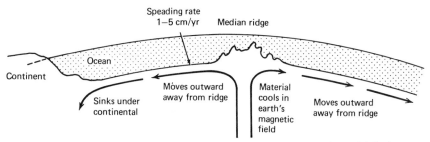

Figure 10-4 Diagram depicts the hypothesis of the spreading sea floor. Molted magma flows out in the area of separating sea floor plates. With the cooling and solidification of the magma, the magnetic minerals in the new substrate are oriented to the earth's magnetic field and so can later be used to determine the age of the sea floor. (After Ross, 1970.)

tiny magnetic particles at the time of cooling can be used to demonstrate the direction of spreading. The older magnetic layers are further from the median ridge and closer to the continents. It is the slow convection currents within the earth's mantle that are gradually bringing volcanic and magnetic rock to the median ridge areas (Figure 10-4).

ZONES OF THE OCEAN

In the first chapter, vertical and horizontal divisions of the ocean were described (Figure 1-4). Ross (1970) identified two main marine geological areas, the continental margin (maritime, shoreline, continental shelf, continental slope and continental rise) and the continental basin (abyssal plain, oceanic ridges, and trenches). The regions are shown in the geological profile in Figure 10-5 (after Ross, 1970). Because benthic plants occur in the more-shallow waters, this text is primarily concerned with the coastal regions and the continental shelf. However, the phytoplankton occur in all of the oceans in the upper waters, and so a brief review of the geological areas of the ocean is useful.

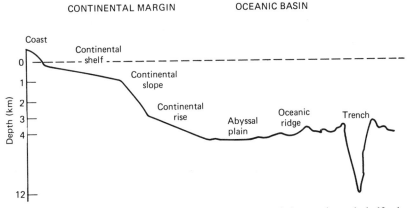

Figure 10-5 A diagrammatic profile of the major zones of the continental shelf, slope, and ocean basin. (Modified from Ross, 1970.)

The Maritime Area or Coast

The meeting of land and water is termed shoreline or coastline. The coast is the area landward from the shoreline, and it can be divided into youthful and mature types of coast. Primary, or youthful, coasts are of nonmarine origin (glacial and volcanic deposits, folding, and faulting) and are usually areas under heavy erosion or active buildup. The secondary, or mature, coasts are those shaped by marine or biological agents. These are typically stabilized structures such as barrier beaches, coral reefs, and marshes.

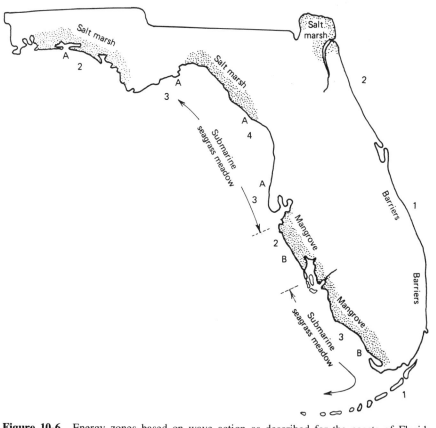

Figure 10-6 Energy zones based on wave action as described for the coasts of Florida (Tanner, 1960). Both major coastal vegetation [indicated by letters: (*A*) salt marsh; (*B*) mangrove swamps] and wave energy levels [indicated by numbers: (*1*) high, ~50 cm; (*2*) moderate, 10–50 cm; (*3*) low, 0–10 cm; (*4*) 0] are indicated. Note that the tidal marsh locations coincide with zero-to-low-energy shorelines and major sand barrier systems with higher-energy shorelines.

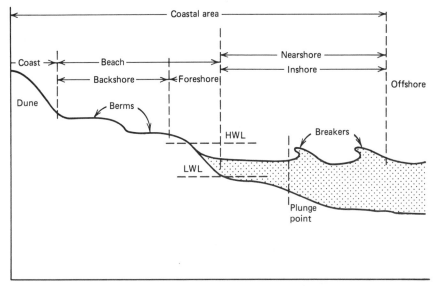

Figure 10-7 A beach profile diagram, demonstrating the major components of a shoreline. Berms are sites of the last major wave action on a beach and are visible at low tide.

Classification of shorelines

One important feature of a shoreline or beach is the energy level, or the *equilibrium beach* concept. The energy level of a beach can be demonstrated by measuring two factors: (1) the average energy at each point on a beach, and (2) the average rate of littoral drift past these points.

For example, Tanner (1960) measured the energy levels on various Florida coasts. Based on average breaker heights, he established four energy levels: high (50 cm or more), moderate (10–50 cm), low (<10 cm), and zero. He further demonstrated that one could relate vegetational and coastal development with energy levels (Figure 10-6). Such energy levels are low when compared to the coasts of New England and western North America. For example, at Santa Barbara, California, the average wave height on a calm day is 50–100 cm. Similarly at the exposed side of the Isle of Shoals off Maine and New Hampshire, wave heights of 30–90 cm are indicative of a calm sea.

Regardless of the site, the effect of waves (Chapter 11) on marine plant communities and coastal geology is enormous. The importance of the shoreline equilibrium is also evident in areas of large littoral drift, where seasonal sand deposition can occur. As Daly and Mathieson (1977) demonstrated, the abundance and distribution of intertidal and subtidal organisms is greatly reduced in areas where sand deposition is a regular factor. The highly abraded rock surfaces along the New Hampshire coast are dominated by opportunistic annual seaweeds *(Enteromorpha)* or psammophytic perennials *(Ahnfeltia, Sphacelaria)*. A lesser species diversity was attributed to the unstable environmental conditions. Several

Table 10-2   Particle Size and Classification, and Average Beach Slope

Sediment Type	Diameter (mm)	Average Slope
Very fine sand	1/16–1/8	1°
Fine sand	1/8–1/4	3°
Medium sand	1/4–1/2	5°
Coarse sand	1/2–1	7°
Very coarse sand	1–2	9°
Granules	2–4	11°
Pebbles	4–64	17°
Cobbles	64–256	24°

adaptive features of psammophytic algae that allow their survival in sand-abraded areas were identified. These include rapid life histories and crustose phases that can withstand sand burial.

Beaches

Beaches are unconsolidated sediments (sand, gravel), usually under the direct influence of waves. Although stable during periods of low wave action, rapid erosion can occur when wave action increases. The slope of a beach is related to the grain size of its sediments, the coarser the grain size, the steeper the beach (Table 10-2). The beach can be viewed as a profile (Figure 10-7), from which the wave energy can be estimated by measuring berm heights and the distance between berms. The berm in a beach profile is the site of sand deposition at the last high tide.

Estuaries

An estuary is a semienclosed coastal body of water having a free connection with the open sea and is diluted by fresh water from land drainage (e.g., submerged river valley). Basically, an estuary is an interface between fresh water and the ocean and can be considered a two-layered transport system. The upper layer is lighter, fresh water with relatively high oxygen levels, and the lower layer is denser salt water with low oxygen levels. Thus the environmental conditions are usually quite diverse. In addition to wide ranges in salinity and temperature, one finds high rates of sedimentation, erosion of soft substrata, and distinct chemical features of the substrata that result in anaerobic and acidic conditions. Ranges in salinity and temperature in an estuary may be quite broad, especially in shallow sites. For example, salt marshes on the northwest coast of Florida have a temperature and salinity range of 12 to 32°C and 2 to 28 ppt respectively, compared with −2 to +22°C and 0 to 30 ppt in a New England salt marsh. Even the water chemistry can be unique because of the levels of calcium and magnesium in the river or stream water. The high calcium content of the spring-fed rivers of the west coast of Florida permits some open coastal species of algae such as *Batophora, Polysiphonia,* and *Gracilaria* to inhabit estuarine waters with reduced salinities. Sometimes these algae are found associated with "fresh water" algae such as *Spirogyra* or *Oedogonium.* In this text the term *estuarine* is defined as having variable and often low salinity whereas *brackish* is considered to mean a body of water that has stable but low salinity (under 5 ppt).

The high levels of organic matter in estuaries are due, in part, to the poor mixing of the overlying fresh water and the salt water, which intrudes from the ocean at high tide. Oxygen levels are usually depressed and salinities higher in the bottom waters of estuaries, unless turbulence or currents increase mixing. If the rate of evaporation is equal to or greater than the rate of influx of fresh water as in the case of a desert estuary, the net result can be that the salt content is increased to a level above that of open coastal waters. Thus not all estuaries have low salinities.

Two major plant communities (tidal marshes) are found in estuaries: (1) salt marshes in temperate latitudes and (2) mangrove swamps in tropical and subtropical latitudes. Both types of tidal marshes are complex, and they support wide varieties of algae and angiosperms. These two major plant communities will be covered in Chapters 18 and 19. Geologically, one can identify three factors that will influence the formation of a marsh: (1) the tidal range, (2) the sedimentation rate and type of substratum, and (3) the changing level of the ocean with relation to the land. As a result of the unique geological features of tidal marshes there are distinctive physiological features of the plants that occupy them.

Marshes tend to develop during times of rising sea levels, such as when salt waters invade river basins. During times of stable sea levels, the filling of marshes occurs. When sea levels are dropping, the rivers and streams may erode away the soft marsh sediments.

Rocky shorelines

The rocky coasts are slow to erode and offer firm substrata for benthic seaweeds. The greatest diversity of algae, both intertidal and subtidal, is found on these shorelines. As will be pointed out in Chapter 15, a number of physical factors play a role in algal community development on rocky coasts, including the chemical and physical features of the actual substratum.

Dawson (1966) listed seven categories of factors that influence algal growth on various substrata (see Table 1-2). After solidarity, both texture and porosity of the substratum are critical factors in attachment (lithophytic algae) and penetration (endolithic algae).

The Continental Shelf

The continental shelves are relatively smooth extensions of the continental land masses, sloping to depths of about 200 to 300 m (100-fathom line). The shelves can be broad, extending 100 km or more, or very narrow, depending upon the geological features of the land masses. Approximately one sixth of the world consists of continental shelves, and all of these are highly productive areas of the oceans. The neritic zone (horizontal classification) is the water covering the shelves, and supports the most planktonic, pelagic, and benthic organisms. In general, the topography of the continental shelf is similar to that of the adjacent land mass.

During the Pleistocene epoch, most continental shelves were alternately exposed and inundated during changes in sea levels. Some of these shelves were glaciated so that today two types of shelves are recognized: glaciated and unglaciated. Usually glaciated shelves consist of coarsely grained sediments. Continental shelves can be extended by the action of large rivers through sediment deposition and the production of submerged deltas. Conversely, many of the deep inshore canyons are due to erosion by rivers during periods of low sea levels.

The Continental Slope and Rise

The average inclination of the continental slope is about 4° (a 7 m drop for every horizontal 100 m). The slope is defined as the area from the edge of the shelf ($\sim$ 200–300 m) to about 3000 m, or the edge of the continental rise. The slope is aphotic, and marine plants are absent. The substratum is usually of continental origin, being sediment such as mud, sand, and some rock outcroppings. The origin of the slope may be due to folding, faulting, reef development, and other factors. About 50% of all slopes are associated with trenches.

The continental slope may end in a gently dropping continental rise, with an inclination of about 1° ($\sim$ 1 m drop in every horizontal 100 m). Many geologists consider the continental rise to be only the lower portion of the continental slope, occurring in depths from about 3000 to 4000 m. In both the slope and the rise, the primary sediment is continental in origin, so these are distinct from the ocean basin.

The Ocean Basin

The ocean basin is usually called the abyssal plain ($\sim$ 4000–6000 m depth) and consists of oceanic sediments. The basin contains trenches (to depths of 12,000 m) as well as oceanic ridges (rising as high as 6000 m) above the abyssal plain. The trenches and oceanic ridges usually occur in the center of ocean basins and are associated with the spreading of the ocean floor and continental drift. An important component of the oceanic sediment found in the ocean basin is skeletal remains of photosynthetic organisms. Extensive deposits of calcareous coccoliths (Prymnesiophyceae) and silicate frustules of diatoms (Bacillariophyceae) (Chapter 8) are common components of the oceanic sediments.

GEOLOGICAL METHODS

In many studies of subtidal or estuarine communities, it would be useful to know the amounts and types of various sizes of sand grains to correlate them with the types of plants present. Two settling tube techniques are useful for these enumerations. In studies of intertidal communities, a determination of elevational levels of various algal communities allows more accurate localization. If a transit is not available, two easy procedures can be employed.

Determination of Elevation

In a study of an intertidal community, line transects are typically used (see Chapter 13) to distinguish algal communities occurring from the spray zone down to the intertidal fringe. Typical problems in such studies include the determination of vertical zones and the location of tidepools. The use of survey tools such as the transit will obviously accomplish this, but such equipment is often not available. Two simple methods can also be employed:

1 **A line level.** If possible, an established reference point above the intertidal zone such as an elevation or bench mark is used. One can also place a permanent mark that can be surveyed at a later date. The permanent mark can be set up by embedding in rock (or a concrete block embedded in the ground) a ⅜ × 2 in. metal rod (a method used on the Isle of Shoals). From the permanent mark, a line is stretched with a line level (standard carpenter's tool). With the line level bubble centered, a meter stick is used to measure the vertical drop from the line level to any point on the transect. This method is repeated from one point to another in a series of steps, with each vertical drop measured and recorded. The result will be a series of steps using a new point each time. This permits the vertical mapping of an intertidal community and determination of the relationship of the height of one zone to another.

2 **The Emery profile method.** Emery (1961) described a simple method to measure beach profiles that requires two wooden rods. These rods are 1½ m tall and are marked every 10 cm. The rods are held vertically, one rod-length apart in a line to be extended across the shore (Figure 10-8). The observer holds the landward rod and sights over the seaward rod to the horizon. The distance from the top of his own rod to the intersection by line of sight is then recorded. Assuming the line of sight is level, the distance on the rod is the difference in elevation between the two points 1½ m apart. If possible, the starting position should be over a bench mark. If the rods tend to sink in a soft substratum, small wooden pads can be attached to the rod bases.

Particle Size Analysis

The importance of determining the particle size and percentage is evident if one is involved in studies of salt marshes and mangrove swamps. Two methods are described: a laboratory procedure and a field procedure are based on Stoke's law of particle mass and velocity in a liquid. The laboratory procedure does not require a set of sieves, whereas the field procedure does.

1 **Emery settling tube.** This procedure can be used in both field and laboratory, but it is most easily conducted in a laboratory. The method consists of intro-

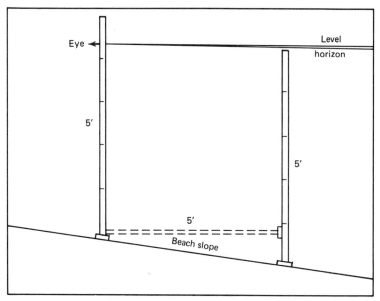

Figure 10-8 The Emery technique for measuring beach slope and determining elevations in an intertidal zone (Emery, 1961). By eying along two transit poles toward the horizon, the drop in elevation can be determined. The two marked poles of known lengths are also used to space one from the other.

ducing a sand sample into a calibrated long (~ 15 cm) glass tube filled with distilled water and measuring the height of the sediment column at predetermined times. A sand grain of large diameter will have a greater settling velocity in water than a sand grain of smaller diameter. Thus it is possible to determine the percentage of a given sand size fraction.

Calibration. Sieve a sand sample to separate the size particles into known fractions (e.g., 1–0.5 mm, 0.5–0.25 mm, 0.25–0.105 mm, etc.). Introduce a 3 g sample of *each* size fraction at the top of the settling tube, and record the time for the entire sample to settle. The settling tube should be a glass cylinder about 1–2 cm diameter and 10–15 cm tall. A stopper can be placed in the bottom. For each fraction, make three determinations of settling time and average them. Construct a curve showing grain size in mm vs. settling times in seconds. Once the settling tube has been calibrated, samples with mixed sizes of grains obtained from the field may easily be analyzed.

Procedure. Place a 3–5 g sample of sand in a glass tube. Fill the tube with water and a drop of detergent; shake several times with thumb over the opening to completely moisten all the grains. Invert the glass tube over the settling tube and, holding the sample tube *just above* the settling tube water level, allow the sample to flow into the settling tube. When the first of the sample enters the water, start a stopwatch.

Read and record the height of the sediment column at the time intervals corresponding to grain sizes determined in the calibration. The percentage by height of the separate size fractions compares very closely with the actual percentage by weight of the size fractions obtained by sieving. Therefore the change in height of the column between readings compared to the total height of the sediment column will yield the percentage of the size fraction involved. Table 10-3 shows a blank sample data sheet and the results of such an analysis.

Table 10-3 A Sample Record Sheet for Sand Analysis (A) and an Example of Such an Analysis (B)

A. *Sample Record Sheet*

Sample From: _____

Date of Analysis: _____

Air Temp: _____

Analyzed by: _____

% Gravel _____   % Sand _____   % Silt _____

Size Grade	Settling Time	Height of Sediment	Δ Height	% Sand Weight	% Total
1–0.5					
0.5–0.25					
0.25–0.125					
0.125–0.062					

B. *Example of a Sediment Analysis*

Sample taken from mouth of Tampa Bay, near Egmont Key, Florida.

% Gravel __25__   % Sand __50__   % Silt/Clay __25__

Size Grade	Settling Time (sec)	Height of Sediment (mm)	Δ Height (mm)	% Sand Weight[a]	% Total
1–0.5	21	12	12	20	10
0.5–0.25	36	27	15	25	12.5
0.25–0.125	105	51	24	40	20
0.125–0.062	201	60	9	15	7.5

[a] % Sand wt. = ht./total ht. × 100 = % ht. = @ % wt.

1–0.5   : 12/60 × 100 = 20
0.5–0.25   : 15/60 × 100 = 25
0.25–0.125 : 24/60 × 100 = 40
0.125–0.062: 9/60 × 100 = 15

% Total:
   % sand weight × % sand in sample = % total sediment
   (e.g., 20% × 50% = 10%)

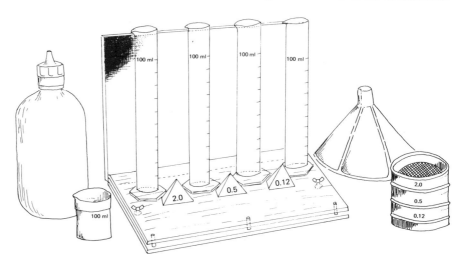

Figure 10-9 A simple portable sediment-sizing apparatus can be used for quick field measurements of coarse sediments. The components are two plexiglass sheets bolted with wing nuts which hold four 100 ml graduated cylinders. A set of four sieves that will separate particles ≥2 mm, 0.5–2 mm, 0.12–0.5 mm, and finer; a funnel; a squirt bottle; and a 100 ml plastic beaker are also shown. The procedure is to take 100 cm³ of a sediment sample using the plastic cup, dump it into the top sieve and wash the sample through the screens using the plastic squirt bottle. The contents of each screen are then washed into the appropriate graduated cylinder using the plastic funnel, and the amounts, are recorded. (Courtesy of R. A. Davis, University of South Florida, Tampa.)

When analyzing a sample by this method, it is well to remember that the tube may have been calibrated on the basis of quartz grains, and the presence of particularly light or heavy minerals will cause results by sieving and by the tube to differ. Large mica flakes will fall more slowly than quartz grains of the same size. Particles of a heavy mineral will fall with the velocity of a larger quartz grain.

2 **Field particle size measurement.** Using four 100 ml graduated cylinders mounted on a plexiglass frame, one can determine relative amounts of four sizes of particles in the field (Figure 10-9). As shown in the figure, the four particle sizes are 2.0 mm or greater, 0.5–2.0 mm, 0.12–0.5 mm and smaller than 0.12 mm. A 100 cm³ sample is poured into the four sieves (7.5 cm diameter) containers and washed through the screens using a plastic squirt bottle. The contents of each screen are washed into the appropriate graduated cylinder using a plastic funnel. The cylinders are mounted to a base by a second sheet of plexiglass with two wing nuts. A black backboard is clamped to the back of the base to facilitate viewing of the cylinders. All components are made of plastic except the bronze wing-nut bolts and the stainless sieves.

The procedure is most effective for coarser sediments since the very fine materials will not settle out in a short period.

REFERENCES

Bambach, R. K., C. R. Scotese, and A. M. Ziegler. 1980. Before Pangea: The geographies of the paleozoic world. *Amer. Sci.* **68**: 26–38.

Davis Jr., R. A. 1977. *Principles of oceanography.* Addison-Wesley, Reading, Mass.

Dawson, E. Y. 1966. *Marine Botany An Introduction.* Holt, Reinhart and Winston, New York.

Daly, M. A. and A. C. Mathieson. 1977. The effects of sand movement on intertidal seaweeds and selected invertebrates at Bound Rock, New Hampshire USA. *Mar. Biol.* **43**: 45–55.

den Hartog, C. 1972. Substrates and multicellular plants. In Kinne, O. (ed.). *Marine Ecology* vol. I, Part 3. Wiley, New York, pp. 1277–1290.

Emery, K. O. 1961. A simple method of measuring beach profiles. *Limnol. Oceanogr.* **6**: 90–93.

Raven, P. H., R. F. Evert, and H. Curtis. 1976. *Biology of Plants,* 2nd ed. Worth, New York.

Ross, D. A. 1970. *Introduction to Oceanography.* Appleton-Century-Crofts, New York.

Scientific American, 1971. *Continents Adrift. Readings from Scientific American.* Freeman, San Francisco.

Tanner, W. F. 1960. Florida coastal classification. *Gulf Coast Assoc. of Geolog. Soc. Trans.* **10**: 259–266.

CHAPTER 11

Physical Factors

A number of physical factors play important roles in the formation and continuation of marine plant communities. Some of the more important ones include light, temperature, waves, tides, and currents. This chapter will review these physical factors and their importance to marine plants.

LIGHT

Visible light (380 to 780 nm; Figure 11-1) is a portion of the electromagnetic spectrum and thus is a form of radiant energy. Light is a prerequisite for life on earth since it furnishes the energy required for photosynthesis, which results in CO_2 fixation by plants. The ability of light to penetrate the ocean defines the photic zone, and the changes in light intensity and quality with depth determine where plants can grow.

Penetration

Sunlight impinging on the ocean surface has a range of wavelengths from 340 nm (ultraviolet light) to about 1100 nm (infrared light), with the maxium intensity at about 500 nm. As light penetrates clear water, both edges of the spectrum are absorbed, but predominantly the longer wavelengths (Figure 11-1). Thus at a depth of 25 m, the range narrows to 400–600 nm, with a maximum intensity at 475 nm (blue light). The changes in light intensity and quality have been used to understand subtidal distributions of marine algae (see Chapter 14). A relationship between light penetration and seagrass standing crops has also been described by Backman and Barilotti (1976). Basically they found a strong correlation between the level of irradiance at depths greater than 0.5 m and the standing crop of eelgrass, *Zostera marina,* in southern California. Leaf density was also a function of light intensity.

Light and Marine Plants

Hellebust (1970) presented a thorough review on light effects on plants. He divided the effects into functional and structural responses. A number of types of func-

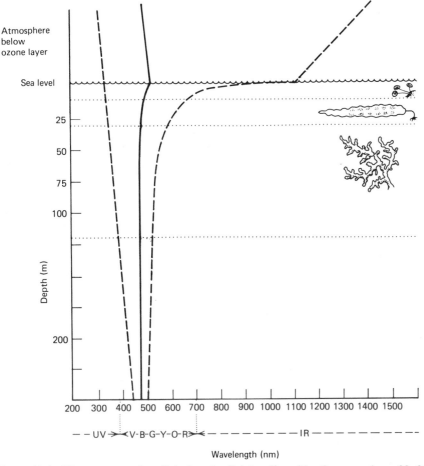

Figure 11-1 The spectrum of sunlight is only slightly affected by the ozone layer 22–25 km above the sea surface. However, when sunlight penetrates the sea, a major loss in infrared light (750–1100 nm) occurs. The solid line shows the wavelength of maximum intensity at any depth; the dashed line shows the wavelength boundaries within which 95% of the solar energy is found at any depth. The algal drawings depict the distribution with depth of green *(Acetabularia),* brown *(Laminaria)*, and red *(Eucheuma)* algae.

tional responses were considered: tolerance, metabolic activity, reproduction and distribution. As might be expected, most algae have a light-intensity-tolerance range, bleaching under high light intensity and ceasing growth under low light intensity. Many metabolic activities are affected by light, including pigment production, photosynthetic rates, movement of chloroplasts (reorientation), bioluminesence, and phototactic responses. Light quality and day length (photoperiod) have been shown to influence reproduction. With regard to day length, the red alga *Porphyra* will only form carpospores under long-day conditions, whereas the *Concocelis* phase of *Porphyra* (filamentous, shell-boring phase) only produces spores with short days. Algal distribution is also affected by light as shown by vertical distribution of phytoplankton and seaweeds (Chapter 15).

Structural responses of algae to light include changes in size, morphological differences, and cytoplasmic changes. Etiolated or elongated algae result when plants are grown under low light, a feature quite common in tank culturing. Plastid displacement will occur in deep-water species. In fact, when the algae are exposed to air, the chloroplasts of the intertidal species *Fucus vesiculosus* align on the interior of the cells, whereas in submerged, plants the plastids occur throughout the cell (Nultsch *et al.,* 1979).

Table 11-1   Light Measurements

Illuminance (brightness)
   1 footcandle (ft–c) = 1 lumen/ft^2 (lu/ft^2) = 10.764 lux (lumens/m^2)
   1 ft–c = 1.076 milliphot

Irradiance (energy units)
   1 watt/m^2 (W/m^2) = 1 Joule/m^2/s (J/m^2/s)
   1 calorie (cal) = 4.19 J
   1 Langley(ly) = 1 cal/cm^2

Photon flux density
   1 erg = 1 quanta/s/m^2/λ^a
   1 Einstein (Ein) = 6.023 × 10^{23} quanta/λ^a
   1 nEin/cm^2/s = 6.023 × 10^{14} quanta/cm^2/s
   1 nEin/cm^2/s = 10uEin/m^2/s
   1 nEin/cm^2/s = $\dfrac{1197}{\lambda(nm)}$ × W/m^2
   1 nEin/cm^2/s = 2 W/m^2 (white light)
   1.66 nEin/cm^2/s = 10^{15} quanta/cm^2/s

Interconversions (for Sylvania cool–white fluorescent light; 400-725 nm)
   1 nEin/cm^2/s = 2.32 ly/day (12:12 L/D photoperiod) = 0.0031 ly/min = 0.1860 ly/h
   1 ly/h = 5.38 nEin/cm^2/s
   1 ly/min = 322.58 nEin/cm^2/s
   1 nEin/cm^2/s = 7047 ft-c
   1 ft-c = 0.0142 nEin/cm^2/s = 4.4 × 10^{-5}ly/m = 2.64 × 10^{-3}ly/h = 3.168 ly/day (12:12 L/D photoperiod)
   1000 lux ≅ 2nE/cm^2/s ≅ 4.2 W/m^2

---

^{a}Lambda (λ) is critical.

Measurement

A number of measuring units and methods are available, and this has resulted in some confusion regarding the interpretation of the effect and quality of light. Table 11-1 presents some of the more commonly used units of light measurements. It is recommended that energy units (ergs, Einsteins, Langleys) be used whenever possible. A number of light-measuring instruments are available such as photometers (irradiance meter, usually with a selenium cell that measures in foot candles), radiometers whose filters allow the selenium cell to be sensitive to specific wavelengths; measures in ergs), and quantum meters (silicon photodiode detectors; measure in ergs or Einsteins). All of these can be obtained and modified for submarine studies.

A very inexpensive method for determing water transparency (Tyler, 1968) and submarine daylight (Weinberg, 1976) is the Secchi disc, which is a 30 cm diameter standard white circle. This disc has been used since 1885 as an international unit of measurement for water clarity. A Secchi disc simply reads the record of depth (D) at which it is just visible. It can be used to determine the extinction coefficient (k) from the following equation:

$$K = \frac{1.7}{D} \tag{1}$$

Calculations demonstrating the usefulness of the Secchi disc are presented by Tyler (1968) and Weinberg (1976). Some typical k factors obtained in oceanic waters are given below:

Date	Depth (m)	K(m)
January	8.5	0.20
March	6.0	0.28
June	9.4	0.18
September	9.4	0.18
December	9.4	0.18

In the above examples, it is apparent that the water transparency is highest during the summer and fall, suggesting low phytoplankton populations. Such measurements are particularly useful for determination of water turbidity due to plankton density.

TEMPERATURE

Of all the physical factors, temperature is the dominant one determining geographic distribution of marine plants (see Chapter 15). For example, mangroves cannot withstand prolonged, freezing temperatures (if any), and algal distribution can be correlated with thermal regimes along the coasts. A new local temperature effect, thermal pollution, is now present because sea water if frequently used as a coolant for power plants and industry.

Measurement

Heat is energy of molecular motion. It can be expressed as calories (1 cal is the amount of heat energy required to raise the temperature of 1 g of water 1°C). Heat is transferred through radiation (radiant energy), convection, or conduction. Temperature is measured in degrees (°C or Celsius scale; °K or Kelvin scale) with the movement of a capillary of mercury (thermometer) or by changes in electrical resistance (thermistor). Absolute zero is where all molecular motion ceases (O°K or −273.15°C).

The surface temperature of the ocean is related to latitude and season. More heat/unit is received at the equator than at the poles, and more is received in the summer than in the winter. Usually three layers can be found in ocean water: (1) a warm, well-mixed surface layer (10–100 m depth); (2) a transition, or thermocline, area where temperature rapidly decreases (~ 500 m depth); and (3) a lower homogeneous cold (2°C) layer where temperature decreases slowly as the bottom is approached (1–3 km depth).

Temperature and Marine Plants

The effects of temperature on marine fungi, bacteria, and blue-green algae were considered by Oppenhiemer (1970), and temperature effects on marine plants were considered by Gessner (1970). Both reviews summarize functional and structural responses of plants.

Tolerance to temperature extremes may be a functional response, especially during periods of extreme cold and heat. Biebl (1970) demonstrated a gradient in temperature tolerance of marine algae; those from cold-water environments were tolerant to lower temperatures, whereas the tropical forms tolerated higher temperatures. As shown in Chapter 14, photosynthesis, respiration, growth (including spore germination), and reproduction are all affected by changes in temperature. For example, *Prophyra* will produce the greatest number of monospores at 10°C.

Structural and chemical responses are also evident. Larger algal forms are usually found in the cooler latitudes; in many cases the protein concentrations are higher in algae from cooler waters.

WAVES

Waves result from the effect of wind on water. Wind is deflected as it blows over the wave profile, causing pressure differences that supply energy to waves. Other types of waves include gravitational (tide), geological (earthquake, landslide), and meteorological (storm) waves. Wave action is a critical factor in determining local seaweed populations. One need only compare algal populations of an exposed and a protected rocky shore to confirm this fact (see Chapter 15).

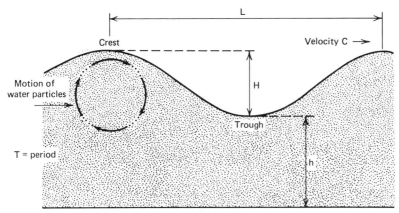

Figure 11-2 Diagram of an ideal wave with its various components, showing the motion of water particles as a wave passes. The length L is the distance between two crests, and height H is the vertical distance from trough to crest of a wave. Velocity of a wave C is the length L divided by the period T or the time it takes a wave crest to pass a point. The depth h of the water influences when a wave will break.

The Ideal Wave

The wave length L is the horizontai distance between two crests measured parallel to wave direction (Figure 11-2). The period T is the time for successive wave crests to pass a fixed point. The wave height is represented by H and depth of water by h. The velocity of the wave C is expressed as $C = L/T$. Thus, a period of 5 sec. and a length of 10 m would result in a velocity of 2 m/sec.

In deep water, the wave forms move forward whereas the water itself moves primarily up and down (circular path) with only a slight forward motion (Figure 11-2). The diameter of the circular motion of water molecules decreases rapidly with depth (h) and is expressed by the formula $h = L/4$. At a depth of 1/4 the wavelength, the size of the circle in which a water molecule circulates is reduced to 1/5 of the surface movement and the water motion is up-and-down rather than circular.

The height and period of wind-generated waves are caused by (1) wind velocity, (2) duration of wind, and (3) distance of open water over which the wind blows (fetch). Wave heights of 25 m have been measured over long exposed regions.

Waves in Shallow Water

All features of the ideal wave except the period T change as the wave moves into shallow water. Wavelength and velocity decrease because of friction on the bottom. When depth h equals half the wave length L, the wave height H rapidly rises. A number of features combine to cause the wave to break. For example, the wave becomes unstable when in internal angle (particle at crest vs particle at substrate) is less than 120°. Another factor is the shift in orbital velocity of water molecules found in the wave crest (high velocity) and at the base of the wave (low velocity). The lower velocity of the water molecules at the wave base is due to friction, or drag, of the water as depth decreases. The wave will break when $h = 4/3\ H$ or when $H = 1/7\ L$.

Measurement

Wave energy can be measured by a dynamometer, which expresses the energy as kg/cm^2. A spring with an attached gauge is firmly anchored to the substratum. A drogue is attached to the spring to record the drag. That is, the drogue moves with the wave and cause the gauge to move (Figure 11-3; Jones and Demetropoulos, 1968).

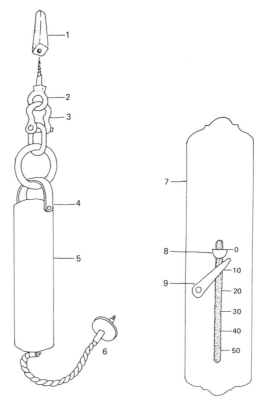

Figure 11-3 A simple spring dynamometer which can be used to measure wave force (after Jones and Demetropoulos, 1968). A hardwood plug (*1*) can be driven into a rock into which an eyescrew is inserted (*2*) and a chain or connecting link (*3*) is attached to the split pin (*4*) holding the dynamometer (*5*). Inside the dynamometer case is a calibrated spring balance (*7*) and a blocker (*8*) preventing the pointer that is attached to the spring (*9*) from returning to zero after a wave has pulled the drogue, or float (*6*).

Waves and Marine Plants

Waves affect plants in a number of ways: (1) by creating drag on plants, resulting in plant removal; (2) by carrying sediment, which erodes or abrades plants; and (3) by impact, causing shearing. The most obvious effect of waves on marine plants can be seen in the distribution of lithophytic algal communities both in relation to species diversity and in zonation. Lewis (1964) has considered the importance of wave action in zonation of rocky shores (see Chapter 15). Southward and Orton (1954) and Kingsbury (1962) have compared algal populations of exposed and protected sites, as have Seapy and Littler (1979). Each of the above papers reports an increase in species diversity in areas of "moderate" to high wave action; however, Southward and Ornton also found less species diversity in areas of "extreme" wave action because of erosion and plant breakage. All four studies found a greatly expanded spray zone in regions of high wave action and a general lifting and broadening of all zones. Charters *et al.* (1973) reported that a number of benthic algal spores do not germinate unless sufficient wave action is present.

The productivity of a shoreline is related to the amount of algal cover. Sheltered sites typically have less diversity but higher biomass and primary productivity, as reported by Seapy and Littler (1979), whereas in exposed areas lower rates of productivity and reduced biomass are common. Highly productive, bladed algae (frondose forms; high surface-to-volume ratios) do not develop well in areas of extreme wave action because of the shearing force; and thus morphological types with lower surface-to-volume ratios (crustose and ropelike thickened algae of lower productivity are more characheristic). For example, Santelices *et al.* (1980) reported that *Lessonia nigrescens* is a plant better adapted to areas with strong wave impact than *Durvillaea antarctica*. Thus the authors found that harvesting of *D. antarctica* is not the basis for its lower cover but rather the condition of high wave impact common to the coasts of central Chile.

TIDES

Tides are the rhythmic rise and fall of sea levels. Tides are waves with periods of about 12 hr and 25 min and wave lengths of about half the earth's circumference (circumference = 46,600 km). Typically there are two highs and two lows each day (Figure 11-4*a*). In smaller ocean basins such as the Gulf of Mexico, a number of different tidal cycles can be found (Figure 11-4*b*).

Causes of Tides

Tides result from the gravitational attraction of the sun and moon on the earth. The strongest of the two is the attraction of the moon, which accounts for the 12 hr and 25 min period and the 14½ da tidal cycle.

The gravitational pull occurs on the side of the earth facing the moon (strongest pull) as well as on the opposite side because of the centrifugal force due to the earth's rotation (Figure 11-5). Physically, the tides are due to attractive and tractive forces. The attractive forces are combined effects of the gravitational and centrifugal pulls of the sun, moon, and earth. The tractive force is the resultant direction of the tidal movement. The result is two bulges of water on opposite sides of the earth that stay aligned with the moon and, to a lesser degree, with the sun. Since it takes the earth 24 hr and 50 min to rotate relative to the moon, there will be two high and two low tides at positions in the central region of the tidal wave. At the fringes of a tidal basin, a variety of mixed semidaily (semidiurnal) and daily (diurnal) tides can occur (Figure 11-5).

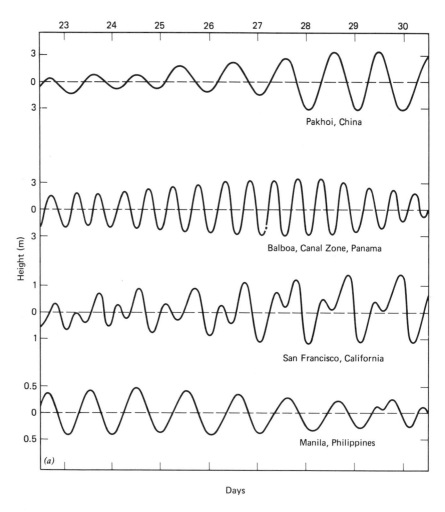

Figure 11-4 Examples of tidal cycles for world sites (a) and sites in the Gulf of Mexico (b) for 8 days. The distinctive features are the actual tidal range, which is quite small (note differences in scale) in the Gulf of Mexico (b) when compared to selected world sites, and

The sun also influences tidal levels on earth, but it only amounts to 46% of the moon's effect. Although much larger than the moon, the sun is farther away and gravitational effects vary inversely with the square of the distance. The principal effect of the sun occurs when both the moon and the sun are aligned parallel to the earth (Figure 11-6). The combined gravitational attraction produces a large *spring tide* every 14½ days. The small-range *neap tides* occur when the sun and moon are at right angles to each other every 14½ days.

The movement of the sun from the Southern to the Northern Hemisphere and back again each year not only is the basis for our seasons but also is the basis for

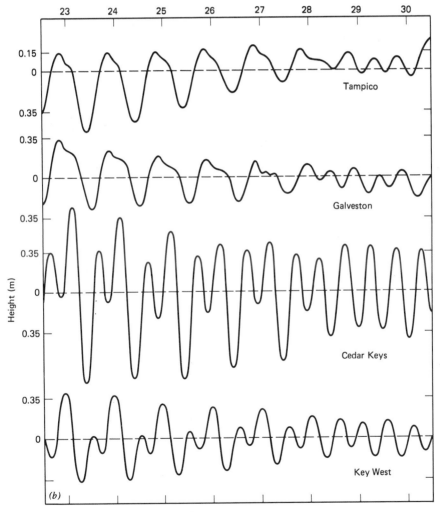

the distinctive tidal cycles. Note that within the same basin (the Gulf of Mexico), the tidal cycles can vary from daily to semidiurnal.

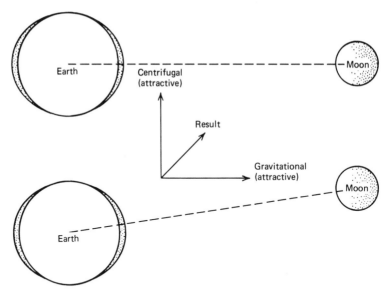

Figure 11-5   The relationship between the earth and the moon with regard to tidal cycles. When the moon is parallel with the earth, the tidal bulges are at the equatorial plane, and when the moon is at a slight angle, the tidal cycles follow. Attractive forces include centrifugal and gravitational pull of the earth and moon respectively with the resulting tidal bulge.

the *equinox tides,* which occur in March and September. Because of the inclination of the earth's axis, the sun and moon are in a higher degree of alignment twice yearly and this combined gravitational pull (attractive forces) results in the largest tides for the year.

Types of Tides

There are a number of features that modify tidal patterns spatially. Thus many coasts do not have two high and two tides each 24 hr and 50 min. (i.e., semidiurnal tides). Basin size and latitude are two features that affect tidal patterns. A coast may have diurnal tides (Pakhoi, China; Tampico, Mexico), equal semidiurnal tides (Balboa, Panama;), unequal semidiurnal tides (San Franciso, California; Cedar Key and Key West, Florida), or mixed tides (Manila, Philippines; Galveston, Texas) as shown in Figures 11-4a and 11-4b. Tides can be determined from a fairly simple ratio:

$$\frac{K_1 + O_1}{M_2 + S_2} \tag{2}$$

Where K_1 is the interaction of the sun and moon, O_1 is the daily pull of the moon, and M_2 and S_2 are the semidiurnal pulls of the moon and the sun, respectively. The type of tide a coast will have can then be predicted from the resulting ratio.

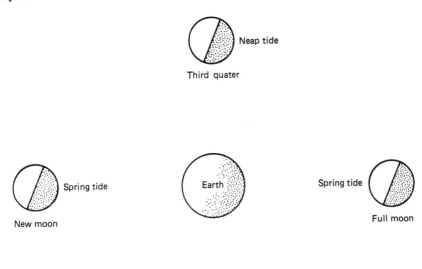

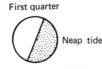

Figure 11-6 The various phases of the moon and the associated spring and neap tides. The strongest force occurs when the moon is new or full for any latitude (spring tides), whereas the weakest force occurs when the moon is in quarters (neap tides).

A semidiurnal (semidaily) tide will have a ratio of 0–0.25; a mixed, predominantly semidiurnal tide will have a ratio of 0.25–0.50; a mixed, mostly diurnal, tide will have a ratio of 1.50–3.00; and a diurnal (daily) tide, a ratio of >3.00. Each ratio is calculated from the gravitational pull created at a specific latitude and longitude by the sun and moon.

There are causes for tides other than the astronomical factors and the centrifugal pull due to the earth's rotation. Meteorological tides, or surges, can result from unusual atmospheric pressures such as sudden cooling due to massive cold fronts. Hurricanes and typhoons are meteorological storms that have strong directional winds that push water ahead of them because of air/wave friction. Geological tides can occur because of submarine earthquakes that result in sudden massive land shifts on the ocean floor. Large-scale slides from coastal mountains or glaciers will also result in massive water displacements and cause local tidal surges. Both meteorological and geological tides can cause extensive damage landward and can extensively damage the intertidal and shallow subtidal communities by sustained exposure and extreme wave action.

A tidal wave is commonly described as a progressive wave, according to the equilibrium theory that has been presented here. However, stationary waves also occur, resulting from the reflection of a tidal wave in smaller basin (as in a bathtub); this is explained as an oscillation of water within a basin. The stationary tidal wave is particularly significant in smaller basins such as the Gulf of Mexico, and basin size must be considered in tide prediction in such areas. The amphidromic wave theory probably best accounts for wave action in large bodies of water such as the North and South Atlantic and Pacific oceans. This theory takes into consideration the deflection of a wave to the right in the Northern Hemisphere and to the left in the Southern Hemisphere (Coriolis force). As the progressive tidal wave moves, it is deflected, and the resulting wave direction is essentially circular in large bodies of water. Thus the tidal wave progresses around a fixed point, an area that will have essentially no tides, for example Tahiti in the Central Pacific (although very small sun tides do occur there).

Tides and Plants

In the chapter on lithophytic communities (Chapter 15) tides will be shown to play a very significant role in intertidal communities. In this text, intertidal zonation is considered to be primarily under the influence of tidal levels and secondarily by other physical factors. Tidal currents, such as those found in confined bodies of water such as Puget Sound, will result in erosion, tidal rapids, and vertical mixing of water currents.

Mathieson *et al*. (1977) described the marine algae found in a New Hampshire tidal rapid (see Figure 1-6). In areas of highest tidal currents (40–80 cm/sec) the shearing force was so pronounced as to remove or reduce the number and size of algal species. Some algae were more sensitive to the tidal currents, *Ascophyllum* being the most sensitive, *Fucus, Chondrus,* and *Gigartina* more tolerant. Several open coast species that were not present in the estuary were common in the tidal rapids, demonstrating the importance of water movement in algal distribution.

CURRENTS

The major oceanic currents are caused by the combined action of wind on the ocean's surface and density differentials between different parts of the sea. Thus horizontal (surface) and vertical currents can occur. An understanding of oceanic currents is useful, because they influence the temperature and nutrient levels, and thus the marine vegetation, of an area. The effects of water movement on plants has been extensively reviewed by Schwenke (1971).

Horizontal Currents

Effect of latitude

Air at the equator and warm latitudes is heated, expands, and rises. Because of expansion and rising, a low-pressure area is produced into which cooler air moves. The rising, heated air is carried north or south where it eventually cools,

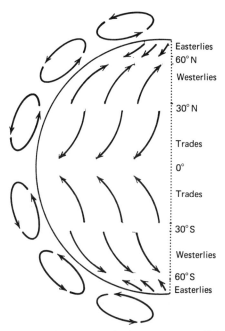

Easterlies
60° N
Westerlies

30° N

Trades

0°

Trades

30° S

Westerlies

60° S
Easterlies

Figure 11-7   The dominant wind patterns (single arrows, parallel to earth's surface) of earth are due to heating and cooling of the earth's atmosphere and the Coriolis force. Thus major wind gyres (circular arrows) are established that in turn produce the major ocean currents.

contracts, and sinks, creating a high-pressure area. There is a deflection of air masses to the right in the Northern Hemisphere and to the left in the Southern Hemisphere because of the Coriolis force.

Instead of a simple rising at the equator and settling at the poles, the winds take curved paths (single arrows, Figure 11-7) and result in a series of gyres (circular arrows, Figure 11-7). The wind system therefore has a pattern of prevailing winds that are the westerlies and trade winds in both the Northern and Southern hemispheres. Areas with no prevailing winds are called doldrums.

Major ocean currents

Just as described for waves, the prevailing winds drive the currents. The easterly trade winds cause the equatorial currents and are common to all oceans. In the Northern Hemisphere, because of land masses, the equatorial currents are deflected north and become the western boundary currents. The western boundary currents are typically strong and are northerly in direction. At about the 40–50° latitudes, the water is deflected eastward by the westerly trade winds and then south after encountering the eastern side of the continents. Figure 11-8 is a map of the world with the major currents. The boldfaced arrows signify the strongest currents, which are equatorial and western boundary currents.

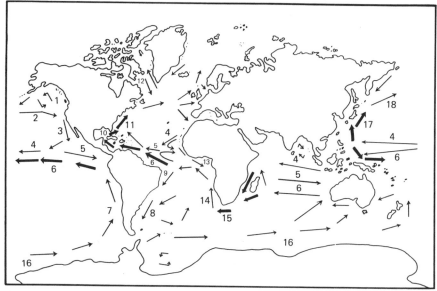

Figure 11-8 The strongest oceanic currents (shown in bold arrows) are the equatorial and western boundary currents. (*1*) Alaska current; (*2*) North Pacific current, (*3*) California current, (*4*) North Equatorial current, (*5*) Equatorial countercurrent, (*6*) South Equatorial current, (*7*) Peru current, (*8*) Falkland current, (*9*) Brazil current, (*10*) Florida current, (*11*) Gulf Stream, (*12*) Labrador current, (*13*) Guinea current, (*14*) Benguela current, (*15*) Agulhas current, (*16*) West Wind Drift, (*17*) Kuro Shio (current), (*18*) Oya Shio (current).

The current patterns in the North Atlantic Ocean will be examined (Figure 11-9) because this region includes the unique Sargasso Sea. The North Equatorial current crosses the Atlantic Ocean in an east-to-west and slightly northern direction, breaking up around the Antilles and terminating around the Yucatán Peninsula. The continuation of these currents results in the Gulf Stream around the southern tip of Florida and the Florida current along the eastern side of the same state; they combine in the western boundary current of the North Atlantic Ocean. The Florida current (and the Gulf Stream) continue in a northeasterly direction until Cape Hatteras, where a more easterly flow is seen, the North Atlantic current. Finally, part of the North Atlantic current is deflected south, and this becomes the Canaries current.

The Sargasso Sea is simply a large eddy, formed by the currents just described. It is about 28.3 million km in area, elliptical, and very small 0.6 m bulge occurs in the center.

Vertical Currents

A number of factors influence the production of rising and falling water currents, including temperature, density, and wind.

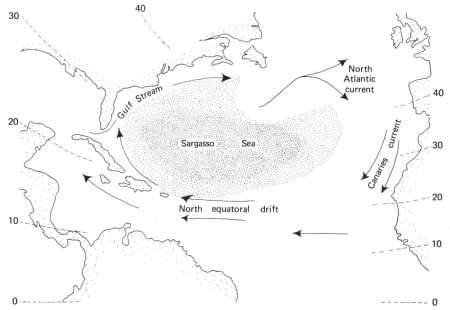

Figure 11-9 The Sargasso Sea, large gyre formed by the currents named in the diagram.

Thermohaline circulation

The most important vertical circulation is a response to density variations caused by salinity and temperature changes in the oceans. These deep-sea circulation patterns are extensive in all oceans and only come to the surface waters in areas of high latitudes where surface water temperatures approach those of deep water.

Coastal upwellings

If a prevailing offshore wind exists, the surface waters can be pushed away, and deeper nearshore waters will rise. The deeper waters are usually nutrient rich and cooler. Areas of coastal upwellings usually result in localized high productivity because of the phytoplankton and benthic algal production. On the Pacific shore of Baja, California, Dawson (1950) found an area of upwelling caused by the prevailing (offshore) westerly winds; this supported a cool-water algal flora with high productivity (phytoplankton and benthic algae). In turn, a large zooplankton and fish population supported a bird population, which resulted in large deposits of guano on the desert coast. Upwelling can have major effects, not only causing increased biomass but also changing local climate.

Divergences

Vertical movement of water can also occur when two currents pass close to one another. At the Atlantic Ocean doldrum (between currents 5 and 6, Figure 11-8) cold, nutrient-rich water is brought up because of the divergence created by the

North and South Equatorial currents. A similar effect occurs in the Gulf Stream when it leaves the Florida current off Cape Hatteras.

Turbulences and convection upwelling

Turbulent water mixing is caused by a variey of disturbances such as strong tidal currents running aginst a wind, high wave activity, or tidal flow over shallow entrances (sills) to bays. Convection water mixing results from differential heating or cooling of the surface waters. Heated surface water expands and allow deep, cool, nutrient-rich water to rise, whereas cooled water becomes denser and sinks. These types of upwellings typically enhance phytoplankton populations or shallow subtidal benthic plant communities.

Wake streams

Upwelling occurs along a major current such as the Florida stream, which pulls adjacent water along with it, causing nutrient-rich water to rise along the edge of the stream. The result is a thin line of nutrient-rich water that supports large planktonic and pelagic populations.

REFERENCES

Backman, T. W. and D. C. Barilotti. 1976. Radiance reduction effects on standing crops of the eelgrass *Zostera marina* in a coastal lagoon. *Mar. Biol.* **34**: 33–40.

Biebl, R. 1970 Vergleichende Untersuchungen zur Temperaturresistenze von Meeresalgen entlang der pazifishchen Kueste Nordamerikas. *Protoplasma* **69**: 71–83.

Charters, A. C., M. Neushul, and D. Coon. 1973 The effect of water motion on algal spore adhesion. *Limnol. Oceanogr.* **18**: 884–896.

Daweson, E. Y. 1950. A note on the vegetation of a new coastal upwelling area of Baja, California. *J. Mar. Res.* **9**: 75–68.

Gessner, F. 1970. Temperature: Plants. In Kinne, O. (ed.). *Marine Ecology* Vol. I, Part 1. Wiley, New York. pp. 363–406.

Hellebust, J. A. 1970. Light: Plants. In Kinne, O. (ed.). *Marine Ecology*, Vol. I, Part 1. Wiley-Interscience, London, pp. 125–158.

Jones, W. E. and A. Demetropoulos. 1968. Exposure to wave action: measurements of an important ecological parameter on rocky shores on Anglesey. *J. Exp. Mar. Biol. Ecol.* **2**: 46–63.

Kingsbury, J. M. 1962. The effect of waves on the composition of a population of attached marine algae. *Bull. Torrey Bot. Club.* **89**: 143–160.

Lewis, J. R. 1964 *The Ecology of Rocky Shores*. English Universities Press, London.

Mathieson, A. C., E. Tveter, M. Daly, and J. Howard. 1977. Marine algal ecology in a New Hampshire tidal rapid. *Bot. Mar.* **20**: 277–290.

Nultsch, W., U. Ruffer, and J. Pfau. 1979. Chromatophorenanordnungen in emersen Thalli von *Fucus vesiculosus* unter verschiedenen Litchtbedingungen. *Helg. Wiss. Meeresunt.* **32**: 228–238.

Oppenheimer, C. H. 1970. Temperature: Bacteria, fungi and blue-green algae. In Kinne, O. (ed.). *Marine Ecology* Vol. I, Part 1. Wiley, London, pp. 347–362.

Santelices, B., J. C. Castilla, J. Canciko, and P. Schmiede. 1980. Comparative ecologia of *Lessonia nigrescens* and *Durvillaeu antarctica* (Phaeophyta) in central Chile. *Mar. Biol.* **59**: 119–132.

Schwenke, H. 1971. Water movement and plants. In Kinne, O. (ed.). *Marine Ecology*. Vol. I. Part 2. Wiley, New York, pp. 1091–1122.

Seapy, R. R. and M. M. Littler. 1979. The distribution, abundance, community structure, and primary productivity of macroorganisms from two central California rocky intertidal habitats. *Pac. Sci.* **32**: 293–314.

Southward, A. J. and H. H. Orton. 1954. The effects of wave-action on the distribution and numbers of the commoner plants and animals living on the Plymouth breakwater. *J. Mar. Bio. Assoc. U.K.* **33**: 1–19.

Tyler, J. E. 1968. The Secchi disc. *Limnol. Oceanogr.* **13**: 1–6.

Weinberg, S. 1976. Submarine daylight and ecology. *Mar. Biol.* **37**: 291–304.

CHAPTER 12

Chemical Factors

The saltiness of seawater has concerned man since early times. If one analyzes seawater it is evident that almost all of the elements are present; more significantly, the major elements are present in a consistent ratio. Seawater influences not only marine life but terrestrial life as well. The intimacy of seawater and marine organisms is apparent in their physiology and biochemistry. Not so obvious is the importance of seawater on terrestrial life. Rainwater for the most part is derived by evaporation from the oceans and the effect of oceanic water along coasts is easily seen by the ecological types of plants (xerophytic) typically present.

This chapter reviews some aspects of chemical oceanography such as the composition and distribution of materials in seawater and some of the processes that influence their composition. Detailed discussions of chemical oceanography are available in Davis (1977) and Martin (1972). Procedures for chemical analyses of seawater can be found in Martin (1970) and Strickland and Parson (1968). The procedures outlined in this chapter are those most commonly used by marine botanists.

PROPERTIES OF SEAWATER

Water Properties

Pure water, although an apparently simple compound (H_2O) has remarkably complex properties. Theoretically, water should freeze at $-150°C$, not $0°C$, and boil at $-100°C$, not $100°C$. Water has a large capacity for absorbing heat, is a strong solvent, and its surface tension is very high when compared to other liquids.

Water molecules are held together by strong (109–111 kcal/mol) covalent bonds. The molecules are polarized, with the two positive hydrogen atoms (H^+) on one side and the single negative (O^-) oxygen atom on the other. In other words, there is an unequal charge distribution. Because of the polarization of the water molecule, they act like small magnets and are associated (bonded) through hydrogen bonding (5.9–10.2 kcal/mol). Because of the hydrogen bonding, pure water requires more heat to raise the temperature than would be required without these bonds. Also, in part because of the bonding, water is more resistant to vaporization (gas formation), boiling at $100°C$ not $-100°C$. This is because more energy is necessary to break the hydeogen bonds between molecules. Hydrogen bonding also means that its molecules tend to bond together to form solid ice at 0°C and not at the theoretical −150°C.

Saltwater Properties

The addition of salt to water causes an alteration in its physical properties. The presence of such ions as sodium and chloride results in a water-molecule-ion association. The ionic bonds between salt and water must be overcome in order for the liquid to freeze or boil. Thus a series of physical changes occurs with increasing salinity as follows:

1 Osmotic pressure increases.
2 Vapor pressure is lowered.
3 Freezing point is lowered.
4 Density increases.
5 Boiling point is elevated.

COMPOSITION OF SEAWATER

The normal salinity of oceanic water ranges from 33 to 37 ppt. The salts entering the oceans come from rivers, estuaries, atmospheric dust, and glaciers. Although much is added, a large amount of salt is lost through inclusion in sediments and biological uptake.

Inorganic Ions

Six major elements (chlorine, sodium, magnesium, sulfur, calcium, and potassium) make up 90%of the total salt in seawater. There is a fairly constant ratio between these six major elements and the three minor elements (bromine, strontium, and boron). Thus if the concentration of one element is known, the others can be calculated. The first nine elements of seawater are usually called the *conservative* ions since their ratios are known (Table 12-1). The remaining ones are called *non conservative* ions as they exist in various proportions throughout the world's oceans. There is some dispute regarding how stable the ratio is between conservative ions, partly because of problems in analysis.

Dissolved Gases

The major gases in sea water are nitrogen, oxygen, and carbon dioxide. Gas solubility depends on three main factors: temperature, the partial pressure of the gas, and the salinity. With regard to carbon dioxide one must also consider chemical reactions.

Table 12-1   The Nine Conservative Ions in Seawater

	Content	
Ion	(g/kg)	(mg at/L)
Major		
Chloride	19.400	548.3
Sodium	10.800	470.2
Magnesium	1.350	53.6
Sulfate	0.885	28.2
Calcium	0.422	10.2
Potassium	0.416	9.9
Minor		
Bromine	0.068	0.8
Strontium	0.008	0.4
Boron	0.004	0.2

Carbon dioxide (CO_2)

This gas is present in considerably higher concentrations in seawater (34–56 ml/l) than in the atmosphere (0.3 ml/l), partially due to the ability of water to absorb more CO_2 than air, volume for volume. More important is the fact that seawater is slightly alkaline (pH 8.2–8.4) owing to the presence of hydroxyl groups and formation of bicarbonates. Thus CO_2 can combine with water to form carbonates and bicarbonates: this is the main source and reservoir of CO_2 in the sea.

The balance of CO_2 in seawater can be seen from the following equations:

$$CO_2 + H_2O \rightleftharpoons H_2CO_3 \tag{1}$$

$$H_2CO_3 \rightleftharpoons HCO_3^- + H^+ \tag{2}$$

$$HCO_3^- \rightleftharpoons CO_3^{2-} + H^+ \tag{3}$$

As can be seen from equation 1, if CO_2 is removed from seawater (through photosynthesis, calcification) the bicarbonates will release CO_2 (a shift in equations 2 and 3). Thus there is a reservoir of carbon dioxide for photosynthetic reactions in seawater. At night, respiration will produce CO_2, and the equation will shift toward the carbonate (right half of equation 3), with bicarbonate (HCO_3) the dominant form.

The above relationship is not as simple as it appears, since both temperature and salinity play a role. Thus availability of the gas CO_2 does not follow Harvey's law of inert gases. The pH will also change with shifts in equations 1, 2, and 3. Essentially pH rises when CO_2 is removed (e.g., during photosynthesis) and falls with increasing concentrations of CO_2 (e.g., during respiration). Thus one convenient method of measuring photosynthetic rates is to monitor changes in pH.

Oxygen (O_2)

As with other gases (Table 12-2), the concentration of oxygen in seawater is variable and is related to temperature, the partial pressure of the gas, salinity, and biological activity. There are two major sources of oxygen in seawater, the atmosphere and plants. Surface waters tend to achieve theoretical levels of oxygen saturation because of air contact. Supersaturation or depletion of oxygen can be the result of photosynthetic and respiratory activities of marine organisms.

Table 12-2 Relationships of Temperature, Salinity and Oxygen Concentration (mg/l)[a]

Temperature (°C)	Salinity		
	0 ppt	18 ppt	36 ppt
0	9.91	8.85	7.78
10	7.64	6.86	6.07
20	6.19	5.58	4.98
30	5.27	4.75	4.24

[a]The values given are concentrations of oxygen (mg/l) for selected combinations of temperature and salinity.

The overall photosynthetic and respiratory pathways involve oxygen as shown in equation 4. Thus rates of photosynthesis and respiration can be measured through oxygen release or uptake:

$$CO_2 + H_2O \underset{\text{energy (chemical)}}{\overset{\text{energy (sunlight)}}{\rightleftharpoons}} (CH_2O) + O_2 \qquad (4)$$

Two major methods are presently employed to measure dissolved oxygen levels. One is a titration method (Winkler) and the other is a conductivity method (oxygen probe).

Titration method or Winkler procedure

Levy *et al.* (1977) have described an automated apparatus for Winkler determinations to avoid human error. Two chemicals are added to the water sample (for details of productivity procedures, see Chapters 14 and 20; see Figure 14-15 for photograph of Winkler setup): manganous sulfate ($MnSO_4$) and alkaline potassium iodide ($KI + KOH$).

These two solutions are heavy and sink rapidly in the biological oxygen demand (BOD) bottle. After addition of the alkaline iodide, the BOD bottle is stoppered without any air bubbles and shaken thoroughly by inversion. The manganous sulfate and the alkali react as follows:

$$MnSO_4 + 2KOH \rightarrow K_2SO_4 + Mn(OH)_2 \qquad (5)$$

The reaction proceeds rapidly to equilibrium, producing the light-colored, pink hydroxide that is oxidized by the dissolved oxygen:

$$2Mn(OH)_2 + O_2 \rightarrow 2MnO(OH)_2 \qquad (6)$$

The brown-colored manganous compound [$MnO(OH)_2$] is allowed to settle in the bottle, shaken a second time and allowed to settle again. The compound is fairly stable and can be stored in a dark cool place for a number of days but is best analyzed as soon as possible. If no organics are present in the seawater, storage can also be done after acidification (equation 7).

Once the precipitate is well settled, concentrated sulfuric acid is added and a manganese sulfate is formed (equation 7). The sulfate compound reacts with the potassium iodide as shown in equation 8. The reactions are

$$MnO(OH)_2 + 2H_2SO_4 \rightarrow 3H_2O + Mn(SO_4)_2 \qquad (7)$$

$$Mn(SO_4)_2 + 2KI \rightarrow MnSO_4 + K_2SO_4 + I_2 \qquad (8)$$

The liberated iodine is chemically equivalent to the amount of dissolved oxygen in the sample and is titrated with a standardized sodium thiosulfate solution, $Na_2S_2O_3$. The basic reaction is an oxidation of thiosulfate to tetrathionate as follows:

$$2Na_2S_2O_3 + I_2 \rightarrow Na_2S_4O_6 + 2NaI \qquad (9)$$

Starch is added to the colorless solution just before all the iodine is bound to aid in determination of the end point. The addition of starch will result in a blue color if iodine is present. The end point occurs when the blue color disappears; this end point can be difficult to determine. Bryan *et al.* (1976) have described a modification of the above procedure using a photometric end point that aids the observer.

Knowing the value of the thiosulfate in terms of oxygen and the volume of the sample titrated, the amount of dissolved oxygen (in ppm) can be calculated.

Oxygen probe

A number of oxygen meters and sensors (probes) are commercially available, including the Beckman (Figure 12-1) and Orbisphere systems. Most meters incorporate an analyzer with an amplifer and controls and a separate sensor. The sensor detects oxygen partial pressure in the sample and causes a signal that is amplified for readout on the meter.

The sensor or probe (Figure 12-2) is placed in the sample (usually a BOD bottle) and a potential of 0.53 volts is applied between the rhodium or gold cathode and silver anode, which are encased in a glass body with a concentrated KCl solution. The cathode is connected electrically to the anode by the film of potassium chloride solution. A gas-permeable Teflon membrane covers the probe, thus separating the cathode-anode assembly from the sample. Oxygen in the sample diffuses through the Teflon membrane covering the probe tip. Diffused oxygen is reduced electrochemically (consumed) at the cathode. The reduction of oxygen causes a current flow proportional to the partial pressure of oxygen in the sample, as expressed in the following reactions:

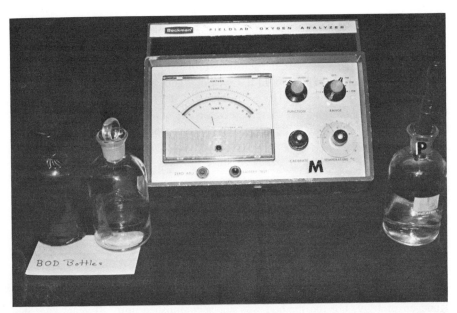

Figure 12-1 Various components of oxygen measurement using an oxygen probe (*P*) and meter (*M*). Darkened (taped) and light (clear) biological oxygen demand (BOD) bottles are also shown. Note that temperature must first be corrected for before determining O_2 levels.

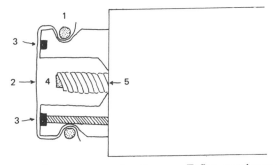

Figure 12-2 Diagram of an oxygen probe. It uses a Teflon membrane (*2*) to contain a saturated KCl solution (*4*). The membrane is held in place by a rubber O-ring (*1*). The probe has a rhodium or gold cathode ring (*3*) and a silver coil for the anode (*5*). Oxygen at a higher partial pressure outside the Teflon membrane diffuses through the membrane by osmosis and accumulates at the cathode, resulting in an increase in current, which is detected by the meter.

$$O_2 + 2H_2O + 4e^- \rightarrow 4OH^- \quad \text{(cathode reaction)} \quad (10a)$$

$$4Ag^+ + 4Cl^- \rightarrow 4AgCl + 4e^- \quad \text{(anode reaction)} \quad (10b)$$

Because the current flow in the sensor is proportional to the partial pressure of the oxygen, no current will flow in the system when oxygen is not present at the sensor tip. There is a linear relationship between the rate of oxygen flow across the Teflon membrane and the current between the anode and cathode.

The readings must be corrected for temperature since membrane permeability varies with temperature (changing at the rate of about 4%/°C). Atmospheric pressure must also be constant or balanced since the process depends on the rate of gas diffusion.

Salinity

The saltiness of the ocean is usually termed salinity, with the concentration usually expressed as grams of dissolved salts per kilogram of seawater or parts per thousand (ppt). Salinity is defined as the weight of solids obtained by drying 1 kg of water under standard conditions. This definition was proposed by Martin Knudsen in 1899, but the method was not accomplished until 70 years later because of the technical difficulties. A discussion of the primary standard and the interesting history of standard seawater is presented by Culkin and Smed (1979). There are now a number of rapid and accurate alternative procedures for salinity determinations.

Measurement of density

Two methods are available to measure seawater density, from which salinity is calculated. One is volumetric, using a pycnometer bottle. The other is the use of a Cartesian diver, or hydrometer. The pycnometer is a volumetric flask of known weight to which exactly 1 l of seawater is added at a standard temperature and pressure. The pycnometer bottle is then weighed and the difference between the weight of 1 l of distilled water and the 1 l of seawater is the salt content. The hydrometer is a float that indicates the specific gravity of a solution (Figure 12-3); subsequently under standard temperature and pressure conditions, the salinity can be calculated.

Measurement of conductivity

The more salt that is dissolved in salt water, the more current will flow across a Wheatstone (electrical) bridge. Thus, a salinometer measures the amount of current (Figure 12-4). The amount of current depends on water temperature; thus conductivity meters (salinometers) must first be adjusted for temperature. The meter is calibrated with standard seawater of known salinity. The conductivity-chlorinity relationship in seawater can be seen from the following equation:

$$\text{Cl (ppt)} = -0.7324 \, R_{15} \quad (11)$$

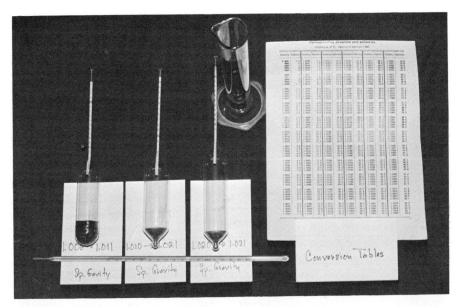

Figure 12-3 Photograph of three hydrometers (Cartesian divers) that are calibrated for different ranges of specific gravities. The graduated cylinder is filled with water, the water temperature recorded, and the specific gravity obtained. Corrected salinity is obtained from the conversion tables.

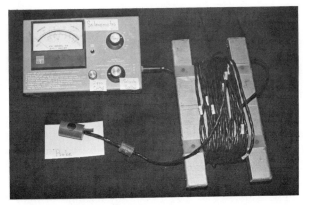

Figure 12-4 A salinometer functions on the principle of conductivity of salt in water and must first be corrected for the water temperature. The probe is a Wheatstone bridge across which a current is measured.

where R_{15} is the ratio of the electrical conductivity of a given sample to that of one where salinity = 35 ppt with both samples held at 1 atm and 15°C. Table 12-3 presents some factors that can be used to calculate salinity from a known conductivity.

Table 12-3   *Relationship of conductivity (millimhos) to salinity (ppt) at different temperatures*

	TEMPERATURE DEGREES C					
SALINITY	0	5	10	15	20	25
20.	17.395	20.105	22.924	25.858	28.900	32.046
21.	18.193	21.024	23.967	27.033	30.209	33.494
22.	18.987	21.938	25.005	28.202	31.512	34.935
23.	19.778	22.849	26.038	29.366	32.808	36.368
24.	20.566	23.755	27.066	30.525	34.097	37.794
25.	21.351	24.659	28.090	31.678	35.381	39.213
26.	22.133	25.558	29.109	32.827	36.659	40.626
27.	22.912	26.454	30.124	33.970	37.931	42.033
28.	23.689	27.347	21.134	35.108	39.198	43.433
29.	24.463	28.236	32.140	36.240	40.458	44.826
30.	25.233	29.122	33.142	37.367	41.713	46.213
31.	26.001	30.004	34.139	38.487	42.961	47.593
32.	26.766	30.881	35.131	39.600	44.202	48.966
33.	27.527	31.754	36.117	40.706	45.435	50.330
34.	28.284	32.622	37.098	41.804	46.661	51.637
35.	29.038	33.484	38.074	42.803	47.878	53.034
36.	29.787	34.341	39.042	43.972	49.086	54.372
37.	30.532	35.192	40.004	45.041	50.284	55.700
38.	31.272	36.036	40.958	46.098	51.470	57.015
39.	32.006	36.872	41.903	47.142	52.645	58.318

Measurement of refraction

The refractive index (n) of a medium is the ratio of the velocity of light in a vacuum to that in the medium. For example, light passing from air to a vacuum has a refractive index of 1.0003. To obtain salinity, one measures the angle of refraction of light passing through a thin film of water (seawater) and then into a glass prism. The refraction of light through distilled water and then through the glass prism can be calibrated to read 1.000; any readings of light through waters of various salinities will be higher.

Refractometers (Figure 12-5) are small, hand-held instruments that can be quite precise and are usually temperature compensated (15–40°C). Refractometers are now available with direct salinity readout scales imprinted on the viewing glass. However, older refractometers or units produced for medical uses (e.g., for urine analysis) can also be used, providing a scale for refractive index is given on the viewing glass. In the latter case all one needs to do is to determine the slope of calibration and then use the slope to convert the refractive index of the unknown sample to salinity. One major advantage of the medical refractometers is that they are one half to one third the cost of oceanographic refractometers.

Figure 12-5 The use of a hand-held refractometer allows rapid measurements of salinity in the field or laboratory through measurement of refraction of light in water.

The slope of calibration is determined as follows. Usually one will insure that the refractometer is zeroed (distilled water will have a refractive index of 1.3330). If not, one can zero according to the manufacturer's instructions (usually adjust the angle of the prism by a small screw that is sealed with wax when done). However, one can determine the salinity without zeroing as shown in the following example where distilled water reading = 1.3330.

The formula is

$$\text{slope of calibration} = \frac{\text{reading of standard salinity} - \text{reading of distilled water}}{\text{Salinity of standard (known)}}$$

An example might be: standard salinity
reading = 1.3394 (refractive index)
standard salinity = 35 ppt
distilled water reading = 1.3330 (refractive index)

$$\text{Slope} = \frac{94 - 30}{35} = \frac{64}{35} = 1.8$$

Note that only the differing digits are necessary and the decimal point is ignored. The slope can now be used to determine the salinity of an unknown sample as follows:

$$\text{salinity} = \frac{\text{reading of unknown sample} - \text{reading of distilled water}}{\text{slope of calibration}}$$

Thus in this example

reading of unknown sample = 1.3385
reading of distilled water = 1.3330
slope of calibration = 1.8

$$\text{Salinity} = \frac{85 - 30}{1.8} = \frac{55}{1.8} = 30.56 \text{ ppt}$$

Again note that only the last two digits are used.

Measurement of chlorinity

The international method for salinity determination is titration for chlorinity. Chlorine, bromine, and iodine can be precipitated as silver salts by titration with silver nitrate ($AgNO_3$). In this method, it is assumed that salinity (S ppt) bears a constant relationship to chlorinity (*Cl* ppt) as follows:

$$\text{salinity} = 0.03 + (1.805 \times \text{chlorinity}) \tag{12}$$

To obtain salinity, the chloride content of a known volume of water at 20°C is determined by titration with silver nitrate, and the results are expressed as grams of chloride per liter at 20°C, a value known as *chlorosity*. *Chlorinity* is then calculated by dividing the chlorosity by the density of the water sample at 20°C to give chlorinity in grams per kilogram of water, or is obtained from the Knudsen tables.

The basic reaction for the titration procedures is as follows:

$$AgNO_3 + NaCl \rightarrow \underset{\text{(white ppt)}}{AgCl} + NaNO_3 \tag{13}$$

Silver chloride is a white precipitate. An end point color change is used to determine when all of the chlorine has been used up. The indicator is potassium chromate and the reaction is as follows:

$$K_2CrO_4 + 2AgNO_3 \rightarrow \underset{\text{(red-orange)}}{Ag_2CrO_4} + 2KNO_3 \tag{14}$$

Silver chromate is a red-orange compound that is easily seen at end point. Titration is terminated when this orange pigment does not disappear with continued (1 min) stirring.

Two procedures will be given in this section for determination of salinity by titration. The first method is the Knudsen technique (Figure 12-6); the second method is for small samples, and is easily carried out in field or laboratory by an investigator or a class.

Knudsen procedure

This is the standard international method for the determination of salinity. Booklets available from the U. S. Department of the Navy and most oceanographic supply houses, summarize the procedure and include the tables for calculation of chlorinity, salinity, and the correction factor. Only a brief outline will be presented here. It should be noted that the preciseness of the procedure far exceeds requirements for salinity determinations in most biological studies.

REAGENTS

1. Distilled water. Must be free from chlorides.

2. Silver nitrate solution. Add 37.11 g of pure silver nitrate in 1 l of distilled water. Store in brown bottle.

3. Potassium chromate solution. Dissolve 5 g of potassium chromate in 100 ml of distilled water.

4. Standard seawater. Supplied by Laboratoire Hydrographique, Charlollenlund Slot, Denmark. Chlorinity is given to 3 decimal places (e.g., 19.381 ppt).

5. Tap grease. Do not use a silicone grease but other commercial stop greases are acceptable.

EQUIPMENT (SHOWN IN FIGURE 12-6)

1. Knudsen pipet (15 ml).

2. Knudsen buret. Note that the gradations do not begin before 19 ml. Therefore there must be sufficient chlorinates (nearly 16 ppt or $S = 28$ ppt). If the salinity is less than 29 ppt a buret graduated continuously from 0 ml must be used.

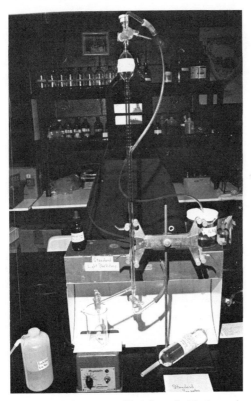

Figure 12-6 The Knudsen titration setup. Titration of chlorine using silver nitrate is the most accurate method for salinity determination. The Knudsen buret is calibrated after the first 19 ml (note bulge on top). A standard white light backup is important. Potassium chromate is used as the indicator. An unopened flask of Copenhagen seawater is also shown.

3. Titration vessel. A 200 ml high-side flask and a magnetic stirrer with a Teflon-covered rod are used. Use a Teflon rod to disperse the precipitate, and recover absorbed chloride ions, and avoid splashing. Titrate against a standard light background.

PROCEDURE

1. Draw out 15 ml of seawater with Knudsen pipet and discard. Repeat the process, obtaining the sample from center of the container and drain into a tall (200 ml) beaker. (Caution, body temperature will influence volume, so do not hold pipet for a long period.) Discharge completely.

2. Add 6 drops of 5% chromate solution.

3. Fill the Knudsen buret with the silver nitrate solution.
Note: this solution has been checked against a normal seawater so that a correction factor is known (see calibration procedure below).

4. Slowly open the buret to a slow drain; avoid splashing. Rinse off droplets on side of beaker with distilled water.

5. As end point approaches, slow down buret to drop-by-drop regulation (end point is reached when red coloration will not disappear for 1 min). Crush up lumps and stir well.

6. Read off the buret at the end point. The calculation of unknown salinity is carried out in the same manner as the calculations for the calibration of the silver nitrate solution (see below). The example given here used the Knudsen tables to obtain k and salinity.

CALCULATIONS

Given: the correction value, $\alpha = -0.13$. The amount of silver nitrate titrated to end point on an unknown sample of seawater was 21.10, and this is called a. The specific correction k for the amount titrated a (e.g., 21.10) varies with that amount; thus one must go to the Knudsen tables and determine k based on a of each titration. If $\alpha = -0.13$ and $a = 21.10$, then $k = -0.19$. The chlorinity and salinity are then calculated.

$$Cl = k + a = 21.10 + (-0.19) = 20.91 \text{ ppt} \qquad (15)$$

$$\text{Salinity} = 0.03 + (1.805 \times Cl)$$

$$\text{Salinity} = 37.77 \text{ ppt}$$

Note that the Knudsen tables were used first to obtain k. These same tables could have then been used to obtain chlorinity (Cl) and salinity (S) without the above calculations.

STANDARDIZATION

Standardization of the silver nitrate solution (obtaining α) is easily done. A sample of seawater with a known chlorinity can be purchased (standard seawater, Copenhagen seawater) in sealed glass vials. This seawater sample is carefully titrated with the silver nitrate solution; at least three replicates are run. The following calculations are then made:

The standard seawater N is used to determine α in the following formula: $\alpha = N - A$, where A is the actual amount of silver nitrate solution titrated. The example below yields a negative α:

$$N = 19.38 \text{ ppt (given on normal seawater vial)} \qquad (16)$$

$A = 19.52$ ppt (obtained by titration or buret reading of normal seawater)

$$\alpha = -0.134$$

Alpha is then used in determining k for each titration. The above procedures are given in detail in Strickland and Parsons (1968) and in the Knudsen booklet.

SMALL SAMPLE PROCEDURE

For most studies in physiological ecology, the precision of a Knudsen titration is not justified. Also, many readings may have to be made, and the investigator may prefer to make these while in the field or in a small field lab. The procedure outlined here was designed by the author for this reason and is very useful in the field or by classes. Three concentrations of silver nitrate solution should be available. The concentration of silver nitrate selected is based on the estimated salinity of the water to be sampled (high, medium, low). All solutions are stored in dark bottles.

$AgNO_3$ Solutions

(g/l)	Molarity	Equivalent to Chloride (mg/ml)
23.95	0.14	5 (high salinity)
19.16	0.11	4 (medium salinity)
9.58	0.05	2 (low salinity)

In this case it is best to use a 5 ml Mohr pipet, graduated to 0.01 ml. The pipet stirring rod, and beaker must all be very clean. To the 1 ml of sample add 10 ml of glass-distilled water and 3 drops of 5% potassium chromate (K_2CrO_4) indicator. Titrate with silver nitrate while swirling or stirring (glass rod) the beaker until the first faint but distinct orange tinge remains. It is necessary to break clots of the white AgCl precipitate with a small stirring rod to ensure complete reaction in the higher range of salinities. Do titrations in duplicate or triplicate and average.

The calculations to determine salinity are as follows:

$$\text{molarity of Cl} = \frac{(\text{volume AgNO}_3) \ (\text{molarity of AgNO}_3)}{\text{sample volume in ml}} \tag{17}$$

Note: 1 mol of $AgNO_3 = 169.87$ g

$$\text{Chlorosity} = (\text{molarity of Cl}) \ (\text{atomic weight of Cl}) \tag{18}$$

Note: atomic weight of Cl $= 35.5$.

$$\text{chlorinity} = \frac{\text{chlorosity}}{\text{density of sample at 20°C}} \tag{19}$$

Note: If density of distilled water $= 1.000$, then
the following densities can be used:

| Density | | | |
@ 4°C	@ 20°C	Salinity	Chlorosity
1.008	1.006	9.9	5.53
1.016	1.014	19.9	11.21
1.025	1.024	30.8	17.50
1.032	1.029	39.9	22.78

$$\text{salinity} = 0.03 + (1.805 \times \text{chlorinity}) \tag{20}$$

NUTRIENTS

The available nutrients in ocean waters are of interest to a marine botanist because of the enormous influence they have on plant communities. The four primary elements necessary for plant growth are oxygen, carbon, nitrogen, and phosphorus, with the ratios 212 0 : 106 C : 15 N : 1 P in seawater (Davis, 1977). Since oxygen (0_2) and carbon (CO_2) have already been considered, this section will be concerned with nitrogen and phosphorus as well as silicon. The latter element is included because of its significance in diatom cell wall (frustule) formation. Many other elements (for example, boron could be included as well, but they are of less importance. Detailed techniques for analysis and determination of concentration of the above elements are available in Martin (1972) and other chemical handbooks.

The low concentration of plant nutrients is a well-known feature of marine waters; this is especially true in tropical regions. The highest concentrations occur in neritic waters and benthic areas of the oceans. In many regions, especially the cold temperate waters, a dramatic seasonal cycle of elements such as nitrogen and phosphorus occurs because of the cyclic growth and death of the phytoplankton and zooplankton. In such regions, nutrient levels rise to the highest peaks in the winter, then drop to almost zero during the spring, and fluctuate during the summer because of growth of plankton communities.

In the measurement of nutrients the standard units include parts per million (ppm), milligram per liter (mg/l); microgram atoms per liter (μg atoms/l), and micromoles (μmoles).

Nitrogen

The element nitrogen (atomic weight 41) occurs in the sea both as compounds and as dissolved gas. Molecular nitrogen (N_2) is the most abundant form of nitrogen (64–95%) in ocean water, but it is not directly usable by plants and animals. A basic constituent of all life, nitrogen is also present in organic material found in oceanic waters at levels ranging from 0.1 to 10.0 μg atoms/l. There are two theories regarding the origin of nitrogen in our atmosphere. One postulates that the primitive atmosphere was composed of ammonia, and this broke down to nitrogen that remained in the atmosphere. The hydrogen combined with oxygen to form the earth's water (hydrosphere). The second theory suggests that nitrogen has come from volcanic activity (8% of volcanic gas is nitrogen).

The compounds commonly found in marine environments include nitrate (NO_3: lower oceanic levels have 0.1–43 μg atoms/l, higher neritic and benthic levels have 1–600 μg atoms/l), nitrite (NO_2: lower oceanic levels here 0.01–3.5 μg atoms/l, higher neritic and benthic levels here 0.1–50 μg atoms/l), and ammonia (NH_3: lower oceanic levels have 0.35–3.5 μg atoms/l, higher neritic and benthic levels here 5–50 μg atoms/l). The nitrogen cycle will be examined here in order to better understand the utilization and source of organic nitrogen in the marine environment.

The nitrogen cycle

As pointed out in the Appendix on marine bacteria (Appendix B), there is now evidence that the nitrogen cycle exists in the marine environment. The nitrogen cycle can be divided into five parts: (1) nitrogen fixation, (2) amino acid synthesis, (3) ammonification, (4) nitrification, and (5) denitrification. A generalized diagram of the cycle is presented in Figure 12-7.

In oceanic waters the levels of fixed nitrogen are low, being produced by lightning and nitrogen-fixing organisms. In neritic waters the levels of fixed nitrogen are higher and are derived from runoff, streams and rivers, benthic sediments, and nitrogen fixers. It is now apparent that benthic bacteria such as *Desulfovibrio* as well as blue-green algae are active in nitrogen fixation in the marine environment. Nitrogen fixing organisms are known to occur in the sediments of most marine plant communities including sea grass beds, salt marshes, mangrove swamps, phytoplankton communities, and lithophytic sites. The importance of nitrogen fixation in coral reef communities has also been shown, and blue-green algae seem to play a significant role in all communities.

Amino acid synthesis occurs in plants, as well as animals dependent on the marine plants for amino acids. Ammonification is the result of decomposition of organic matter (feces, dead organisms) by bacteria and fungi, and there are many examples of this process in the sediments of marine communities. The Appendi-

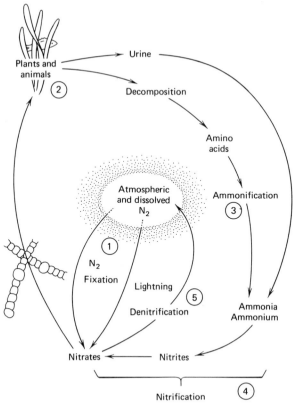

Figure 12-7 Five major components of the nitrogen cycle. Nitrogen fixation (*1*) can occur through bacterial and blue-green algal activity or lightning. Nitrates, and to a lesser
extent nitrites and ammonia, can be incorporated by plants (*2*) that in turn are eaten by
animals. Through decomposition and excretion, various compounds containing fixed nitrogen are converted to ammonia (*3*) and the ammonia or ammonium ion nitrified (*4*) to
nitrates. Denitrification (*5*) can also occur, releasing nitrogen back into the atmosphere.
Whereas nitrogen fixation and incorporation of nitrates are energy requiring, the remaining
steps (*3–5*) are energy releasing. See also Appendix B for a discussion of the role of
bacteria in the nitrogen cycle.

ces on marine fungi (A) and marine bacteria (B) give a number of generic examples of organisms involved in decomposition.

Nitrification is the one area in the nitrogen cycle that is not well documented
in the marine environment. Recent evidence indicates that ammonia (NH_3) can be
directly absorbed by some phytoplankton, however, nitrate (NO_3) is apparently
preferred. Chemosynthetic bacteria such as *Nitrosomonas* (NH_3 to NO_2 + energy)
and *Nitrobacter* (NO_2 to NO_3 + energy) are now known to occur in shallow
marine waters. Deeper-water sediments, lacking adequate levels of oxygen, do
not appear to have such organisms, and thus ammonia may be directly absorbed
or lost.

Denitrification is also energy yielding and a number of facultative as well as obligate anaerobic bacteria are known from deep-water sediments that produce free nitrogen (N_2) or nitrous oxide (N_2O).

Nitrogen assimilation

Nitrogen metabolism in macroscopic marine algae is a relatively new area of research, and only a few species have been examined. For example, nitrogen uptake in the green algae has been studied in *Acetabularia* (Adamich *et al.*, 1975), *Codium* (Hanisak, 1979), the brown alga *Laminaria* (Harlin and Cragie, 1978), and the red algae *Gelidium* (Bird, 1976) and *Gracilaria* (Bird *et al.*, 1980; DeBoer *et al.*, 1978). A number of studies on the uptake and selection of nitrogen sources have been carried out on diatoms (Lui and Roels, 1972; Malone *et al.*, 1975; Glover *et al.*, 1975). Most of the studies seem to indicate that the algae prefer nitrate, but can use nitrite or ammonia.

Growth rates of the macroscopic red algae *Gracilaria* and *Neoagardhiella* can be shown to be related to the types and concentrations of nitrogen enrichment (DeBoer *et al.*, 1978) with ammonia producing the highest growth rates. Similar findings were published for the green alga *Ulva* (Mohsen *et al.*, 1974). However the green alga *Acetabularia* did not show normal growth or cap production when ammonia was used (Adamich *et al.*, 1975). Studies on the utilization of nitrogen compounds in low concentrations suggest that all the algae studied are efficient in scavenging nitrogen (e.g., *Codium*, Hanisak and Harlin, 1978). Bird (1976) found that pools of both NH_4^+ and NO_3^- were simultaneously available for algal assimilation in the agar-producing red alga *Gelidium*. Similar findings by Hanisak and Harlin (1978) were reported for *Codium* and all three nitrogen forms. Bird (1976) also found that blue light will enhance nitrate uptake and amino acid production in *Gracilaria,* suggesting that light quality in the subtidal regions may play a critical role in nitrogen metabolism.

It is now becoming evident that macroscopic and probably microscopic algae can store nitrogen reserves during periods of slow growth, for example in the winter, and then utilize them during the period of rapid spring growth. Such patterns have been shown for *Codium* (Hanisak, 1979) and *Laminaria* (Harlin and Craigie, 1978).

Determination of nitrogen

Chemical procedures for nitrogen determination are described in other texts (Strickland and Parsons, 1968) and only the general procedures are given here. Commercial kits for these tests are available from water-quality supply houses.

Nitrate

Nitrate is most readily tested by cadmium reduction of nitrate to nitrite. Then nitrite is determined by diazotizing with a sulfanilamide and coupling with a dye such as *N*-(1-napthyl)ethylenediamine to form a highly colored azo dye. Because this test is actually designed for nitrite, a second test specific for nitrite must first be carried out (omitting cadmium reduction) and those results subtracted from the nitrate test (e.g. $NO_3^- + NO_2^- = X$; $NO_3^- = X - NO_2^-$).

Nitrite

Direct determination of nitrite is carried out by omitting the cadmium reduction in the previously described procedure so that nitrate will not be reduced. The remainder of the test is as described for nitrate

Ammonia

Although distillation can be used, a color-change procedure is available. The ammonia of seawater is oxidized to nitrite (Nessler's reagent) at room temperature as shown in equation 21.

$$HgI_4^{2-} + NH_4 + OH \rightarrow IHgOHg + NO_2^- + H_2O + I^- \qquad (21)$$

The excess oxidant is destroyed by the addition of a sodium arsenite solution. Then the nitrite is determined by the addition of sulfanilamide and diazotizing as described for nitrate. A probe that can be used with an expanded-scale pH meter is also available for NH_3 determination.

Phosphorus

The element phosphorus is present in seawater as $H_2PO_4^-$ and HPO_4^{2-}, with levels ranging from 0 to 0.003 mg atoms/l at the surface and 0.09 mg atoms/l in deeper waters. The simple phosphate ion (PO_4^{3-}) apparently is rare. Martin (1970) and others have demonstrated the existence of a phosphorus cycle (Figure 12-8 and Appendix B). Distinct seasonal levels, especially in the higher latitudes, of phosphorus are well established (see Chapter 20). The cycle can be divided into three areas: addition (*a*), removal (*b*), and uptake or concentration by organisms (*c*).

About 14 million metric tons of phosphorus are carried to the sea through erosion and land runoff (Figure 12-8). The ocean contains a reserve of about 12 $\times 10^{13}$ metric tons of dissolved phosphorus from dissolution of particulate inorganic phosphate, by precipitation of insoluble phosphate and by plant uptake. In turn, plants are consumed by animals. Soluble organic phosphate will be released from organisms after death or excretion. The death of both plants and higher organisms will result in particulate phosphorus deposition.

Dissolved organic phosphates may be released from particulate complexes through decomposition (see also Appendix B) or deposited in the ocean sediments. Most of the phosphorus is particulate phosphorus that will settle out in oceanic sediments, and some will be returned through turbulance to active ocean pools of dissolved inorganic phosphates. Some dissolved organic phosphorus will be converted to inorganic forms by bacteria. A small amount ($\sim$ 10,000 metric tons) of phosphorus (mainly bird guano) will be returned to the land.

Plants typically absorb and incorporate orthophosphate, and thus this is the form that is measured. The standard procedure is a reaction of orthophosphate

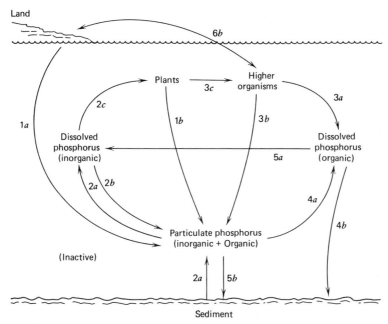

Figure 12-8 The prosphorus cycle can be divided into three major portions: addition (a), removal (b) and uptake or concentration by organism (c). The original source of phosphorus is the land (1a) in the form of organic and inorganic phosphorus; this forms the basic pool in seawater. Dissolved phosphorus in seawater is therefore either organic (4a) or inorganic (2a). With the conversion of organic to inorganic phosphorus (5a), uptake by plants (2c) can occur, and the phosphorus can then in turn be incorporated into higher organisms (3c). Excretion by and decomposition of plants (1b) and higher organisms (3b, 3a) can release organic phosphorus that can be precipitated (4b, 5b). Particulate phosphorus can also be released from the sediment (2a). Because of feeding by terrestrial animals on marine organisms and death of terrestrial animals and plants in the sea, some phosphorus is lost or gained directly from land (6b) (Modified from Martin, 1970).

with an acidified molybdate solution to form a phosphomolybdate heteropoly acid. The phosphomolybdate is then reduced to an intensely colored phosphomolybdenum blue. The procedure incorporates a strong acid and ammonium molybdate, which are added to seawater. If orthophosphate is present, a faint yellow color will occur. After addition of a reducing agent such as stannous chloride or an amino acid, an intense blue color occurs. The color can be compared against a standard color chart or measured photometrically. Ammonium molybdate is usually used in solution, whereas stannous chloride is used as a powder. Strickland and Parsons (1968) recommend the use of a solution of potassium antimonyltartrate in place of the stannous chloride to avoid interference from any arsenic that may be present. Arsenic interferes since it will substitute for phosphate and reduce molybdate, producing a false reading.

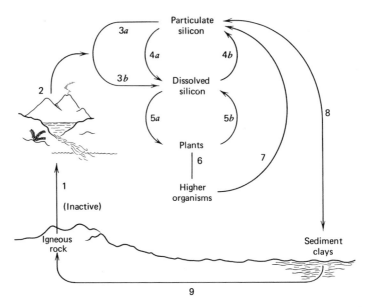

Figure 12-9   The silicon cycle in the ocean parallels the phosphorus cycle but is simpler. Silicon is released from terrestrial and marine sediments (*8, 2*). Marine sediments are converted to various terrestrial substrates (*9*) from which erosion will carry silicon back to the oceans (*1, 2*). Both particulate (*3a*) and dissolved silicon (*3b*) are present in seawater and can be converted back and forth (*4a, 4b*). Plants will utilize dissolved silicon (*5a*) and release it (*5b*) and in turn animals can feed on plants (*6*) and release particulate silicon (*7*) (Modified from Martin, 1970).

Silicon

This element is critical to cell wall formation in diatoms, which form the single most important group of phytoplankton in the oceans. Silicon can be found in concentrations of about 0 to 0.05 mg/l in clear pelagic waters. Since the element is common to clay particles, most neritic waters will have higher levels, especially in the undissolved form of SiO_2 or dissolved as $Si(OH)_4$. The silicon cycle can be divided into the active and inactive pools (refer to Figure 12-9 for numbers). Weathering of igneous rock (*1*, Figure 12-9) causes silicon clays and dissolved silicon to be carried to the sea by rivers. Particulate silicon or dissolved silicon occur in the ocean and these are exchangable. Dissolved silicon is taken up by plants, especially diatoms and released after death. Since zooplankton and higher organisms feed on diatoms and in turn die, dissolved silicon is again released. Particulate silicon can settle out in the oceanic sediments forming clays that can be incorporated into igneous rock.

The method for determining levels of soluble forms of silicon uses the reaction of orthosilicic acid (H_4SIO_4) with molybdate to form the 1:12 silicomolybdic acid molecule. The absorbance of either the intensely yellow product or, after reduction with p-methylaminophenolsulfate(Metol), the intensely blue product (molybdenum blue) can be measured with a spectrophotometer.

REFERENCES

Adamich, M., A. Gibor, and B. M. Sweeney. 1975. Effects of low nitrogen levels and various nitrogen sources on growth and whorl development in *Acetabularia* (Chlorophyta). *J. Phycol.* **11**: 364–367.

Bird, K. T. 1976. Simultaneous assimilation of ammonium and nitrate by *Gelidium nudifrons* (Gelidiales: Rhodophyta). *J. Phycol.* **12**: 238–241.

Bird, K. T., C. J. Dawes, and J. T. Romeo. 1980. Patterns of non-photosynthetic carbon fixation in dark held respiring thalli of *Gracilaria verrucosa. Zeitschrift Pflanzenphysiol.* **98**: 359–364.

Bryan, J. R., J. P. Rilery, and P. J. LeB. Williams. 1976. A Winkler procedure for making precise measurements of oxygen concentration for productivity and related studies. *J. Exp. Mar. Biol. Ecol.* **21**: 191–197.

Culkin, F. and J. Smed. 1979. The history of standard sea water. *Ocean. Acta* **2**: 355–364.

Davis, R. A., Jr. 1977. *Principles of Oceanography,* 2nd ed. Addison-Wesley, Reading, Mass.

DeBoer, J. A., H. J. Guigli, T. L. Israle, and C. F. D'Elia. 1978. Nutritional studies on two red algae: Growth rate as a function of nitrogen source and concentration. *J. Phycol.* **14**: 261–266.

Glover, H., J. Beardall, and I. Morris. 1975. Effects of environmental factors on photosynthesis patterns in *Phaeodactylum tricornutum* (Bacillariophyceae). I. Effect of nitrogen deficiency and light intensity. *J. Phycol.* **11**: 424–429.

Hanisak, M. D. 1979. Nitrogen limitation of *Codium fragile* sp. *tomentosoides* as determined by tissue analysis. *Mar. Biol.* **50**: 333–337.

Hanisak, M. D. and M. M. Harlin. 1978. Uptake of inorganic nitrogen by *Codium fragile* subsp. *tomentosoides* (Chlorophyta). *J. Phycol.* **14**: 450–454.

Harlin, M. M. and J. S. Craigie. 1978. Nitrate uptake by *Laminaria longicurris* (Phaeophyceae). *J. Phycol.* **14**: 464–467.

Levy, E. M., C. C. Cunningham, C. D. W. Conrad, and J. D. Moffatt. 1977. A titration apparatus for the determination of dissolved oxygen in seawater. *J. Fish. Res. Bd. Can.* **34**: 2218–2220.

Lui, N. S. T. and O. A. Roels. 1972. Nitrogen metabolism of aquatic organisms. II. The assimilation of nitrate, nitrite, and ammonia by *Biddulphia aurita. J. Phycol.* **8**: 259–264.

Malone, T. C., C. Garside, K. C. Haines, and O. A. Roels. 1975. Nitrate uptake and growth of *Chaetoceros* sp. in large outdoor continuous cultures. *Limnol. Oceanogr.* **20**: 9–19.

Martin, D. F. 1970. *Marine Chemistry,* Vol. 2. Marcel Dekker, New York.

Martin, D. F. 1972. *Marine Chemistry,* Vol. 1, 2nd ed. Marcel Dekker, New York.

Mohsen, A. F., A. F. Khaleata, M. A. Hashem, and A. Metwalli. 1974. Effect of different nitrogen sources on growth reproduction, amino acid, fat and sugar contents in *Ulva fasciata* Delil. *Bot. Mar.* **17**: 218–222.

Strickland, J. D. H. and T. R. Parsons. 1968. *A practical Handbook of Sea Water Analysis.* Bull. 167. Fisheries Res. Board Can., Ottawa.

CHAPTER 13

General Ecology: Concepts and Methods

This book incorporates an ecological approach to the study of marine plants. Thus, some basic concepts regarding the interaction of organisms and their environment (e.g., ecology) are presented here. This chapter reviews general ecological concepts (levels of organization, dynamics) and methods (measurement of community structure, taxonomic procedures), whereas the following chapter concentrates on physiological and chemical considerations. The three preceding chapters reviewed the abiotic factors in preparation for consideration of the biological factors in ecological studies. All five chapters of Part 3 provide a foundation for understanding the marine plant communities covered in Chapters 15–20. The reader is referred to A.R.O. Chapman's (1979) text on levels of organization in seaweeds for further ecological information.

GENERAL ECOLOGY

In order to understand an ecological system, we must first look at the various levels of organization and identify the critical features of each. The levels are presented starting with the largest unit, the ecosystem, then progressing through the smaller units, communities, niches, and populations. Following this, selected dynamic concepts will be considered, including productivity, food webs, biological interactions, and man's influence on marine plant communities.

Ecosystem

The largest unit of biological organization, the ecosystem, includes interrelationships of environmental factors and communities of organisms. Tansley introduced this term in 1930 when he became impressed with the interplay between organisms and the physical factors of an area. It is the most difficult level to study because of the complexity of the interactions. For example, an ecosystem might be an estuary including the tidal marsh (salt marsh or mangrove swamp), the rivers and adjacent land, and the nearby coastal waters (Figure 13-1). Included in

Figure 13-1   The ecosystem found at Tampa Bay, Florida (27°01′ north latitude and 82°30′ west longitude) is shown for November 13, 1973, by a NOAA satellite. The estuarine system includes not only Tampa Bay proper but also the rivers entering it and the offshore coastal sea grass communities.

such a system would be a number of distinctive communities: the salt marsh or mangrove swamp, the riverine community, and the seagrass community or benthic community of the coastal waters. At this level, one is concerned not only with what lives in the region and the environmental parameters but also with the flow of energy (fixed carbon, food web) and the various mineral cycles (nitrogen, phosphorus, see Chapter 12).

Within an ecosystem one will find autotrophs (self-supporting, primary producers) such as photosynthetic organisms and chemosynthetic bacteria. These organisms produce organic compounds from inorganic ones. The plants, the major primary producers, are the sources of energy for the food webs and thus are the basis of the food chains. Yet even these primary producers are involved in a complex food chain, since they rely on bacteria for nitrogenous compounds and perhaps some vitamins.

The remaining organisms of an ecosystem utilize living or decomposing organisms including plants and thus are called heterotrophs. Grazers (primary consumers) such as phagotrophic dinoflagellates and zooplankton in a plankton community or benthic invertebrates (sea urchins) and fish feed on plants, and are called herbivores. The animals that feed on the grazers or each other are the third major tier in the food pyramid and are called carnivores (secondary consumers). Detritus feeders (sea cucumbers, starfish) feed on the decomposing plant materials or dead animals. Saprophytic fungi and bacteria act as decomposers, breaking down dead organic material.

In addition to the typical autotrophic and heterotrophic organisms, organisms with more specialized symbiotic relationships are found in an ecosystem. Parasitic fungi (Appendix A), bacteria (Appendix B) and algae (see Chapter 7) attach and grow on seaweeds. Commensal relationships, in which one organism benefits but does not damage its host, are common in the marine environment. Examples of commensalism are known between fungi and algae (composite "species," lichens, Appendix A). Mutualistic relationships, in which both organisms benefit from a symbiotic association, are well known among marine plants. They are best exemplified by the dinoflagellates (zooxanthellae) found in stony corals and in other invertebrates (Chapters 8, 16; but see also the section on biotic interactions in this chapter). Thus an ecosystem includes organisms from all trophic levels, resulting in a highly complex food web and a number of nutrient cycles.

Communities

A community is an aggregation or group of organisms that occurs in a particular environment, having constant interactions with each other and the environment. Chapters 15–20 deal with six distinct marine plant communities and present dominant plant types and their adaptations, both abiotic and biotic characteristics. Community structure can be assessed in a number of ways (Chapman, 1979), including classification of dominant species, numerical relationships, or floristic methods. All of these methods are useful, and at least some aspects of each method are presented in the final section of this chapter.

The classification of community structure by its dominant species or zones is one of the more popular methods. Dayton (1975) divided the subtidal kelp com-

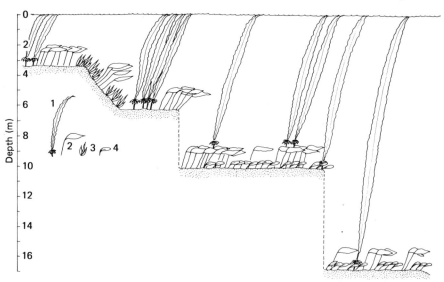

Figure 13-2 A profile of the kelp canopies formed at three different depths at Amchitka Island, Alaska (Dayton, 1975). *Alaria fistulosa* (*1*) forms the major canopy and extends to the surface. Various species of *Laminaria* (*2*) form the next canopy layer, whereas *Laminaria longipes* and *Agarum cribrosum* (*3*, *4*) form the third canopy layer. A fourth layer, not diagrammed, consists of a turf formed by red and green algae. Dayton found in experimental studies that removal of the various layers, or canopies, influenced the flora beneath them as well as the herbivore populations.

munity at Amchitka Island (Figure 13-2), Alaska, into four vertical zones. The large kelp *Alaria fistulosa* (up to 22 m tall) formed the uppermost canopy below which a second layer of *Laminaria* spp. developed. A third layer, dominated by the kelp *Agarum cribosum*, covered a fourth layer of turf vegetation of small red and green algae. The division of intertidal areas into zones or bands, such as the universal zonation schemes of Stephenson and Stephenson and Lewis (see Chapter 15), also relies on dominant species. As shown in Figure 15-7, three universal zones (littoral, eulittoral, sublittoral) and their dominant organisms have been proposed for the intertidal classification. A detailed discussion of these zones is given in Chapter 15.

The second method, numerical analysis, avoids the circular argument of zone recognition by dominant species, at least for the intertidal environment (Russell, 1972). In his study of intertidal zonation and community structure in Great Britain, Russell showed species relationships and community identity through cluster analysis. By using total species diversity, he concluded that the littoral fringe of Lewis was the only distinctive vegetational unit (see Chapter 14).

Floristic classification, or phytosociology, attempts to parallel typical taxonomic hierarchy (e.g., species to genera, etc.) in a classification of plant associations and other vegetational units. The method has been used in Europe and Japan to classify the seaweed floras for entire coastlines in parts of the Mediterranean and most of Europe. Phytosociology identifies communities by the dominant types of vegetation (species and morphology). The most popular school of phytosociology has been developed from the Braun-Blanquet approach (Zurich-Montpeller school) and is presented in the methods section of this chapter. The approach uses two basic numerical measurements, the cover-abundance and sociability classifications. The numerical rankings obtained from these items can then be used to determine dominant associations and zonation of seaweed communities, as shown from den Hartog's (1959) study of the Netherlands coasts (Table 13-1).

Table 13-1   Marine Plant Associations Found on the Coasts of the Netherlands

Zone	Association	Type
Littoral fringe	*Verrucaria maura*	Lichen
	Calothrix scopulorum	Blue-green alga
	Pelvetia sp.	Brown alga
	Fucus spiralis	Brown alga
Eulittoral	*Fucus vesiculosus*	Brown alga
	Ascophyllum nodosum	Brown alga
	Fucus serratus	Brown alga
	Polysiphonia sp.	Red alga
	Chaetomorpha spp.	Green alga
Sublittoral	*Laminaria* spp.	Brown alga

Source: After den Hartog, (1959).

Niche

The specific habitat of a species within a community is termed a niche. The habitat is not just the physical location of the plant, but all of the environmental and physiological aspects of the organism's interaction with the environment. Grinnell (1924) referred to the "ecological or environmental niche" as the ultimate distributional unit of a species or subspecies. Vandermeer (1972) reviewed this concept and described three types of niches: fundamental, partial, and realized. A fundamental niche is the species' potential habitat if no competition existed—that is, its ideal one. A partial niche is the habitat a species occupies if competition is not fully realized. The realized niche is the actual site and extent of occupation of a given species. The realized niche results from species and environmental interactions.

Population

A population is a group of organisms belonging to a single species and occupying a particular area. Thus, in a mangrove swamp, a niche might be the pneumatophores of the black mangrove *Avicennia germinans*, with a population of the red alga *Bostrychia binderi* occupying a distinct level on the aerial roots. Some of the most thorough ecological studies of marine plants have dealt with populations and their relations to the seaweed community. Dayton's study on the palmlike brown alga *Postelsia palmaeformis* and its competitive role with mussels in highly exposed intertidal rocks of the state of Washington is one example (Figure 15-2).

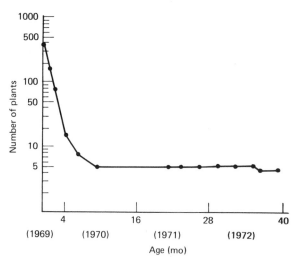

Figure 13-3 A survivorship curve demonstrating the loss of young sporophytes of the giant kelp *Macrocystis pyrifera* through time (Rosenthal *et al.*, 1974). In a study lasting 5.7 years in California it was apparent that almost all sporophyte losses occurred within the 4 mo. following spore germination. Well over 95% of all sporophytes never reached 1 yr in age, those that did usually survived (note flattening of curve) throughout the study.

Another is the dominance of the red algal population of *Gastroclonium coulteri* and its relationship with adjoining populations of *Gigartina* and the surf grass *Phyllospadix* by Lynn Hodgson in the intertidal zone at Monterey, California. Both examples are detailed in Chapter 15 and demonstrate the need to simultaneously understand physiological and life history data.

Chapman (1979) presents three major types of population parameters that should be measured: size or density, age distribution, and the spatial distribution of individuals. The size or density of a population is affected by such factors as spore germination, death, and growth. In a study of the giant kelp *Macrocystis pyrifera*, Rosenthal et al. (1974) found that more than 60% of the newly established sporelings died within 1 mo; after 9 mo only five plants of the original 387 *Macrocystis* sporophytes remained (Figure 13-3). Genetic studies of populations are uncommon in seaweed research. Cheney and Babbel (1978) compared the genetic similarities of several populations of the red alga *Eucheuma* spp. in Florida by electrophoretic analyses of isozymes (alleloenzymes). Several different groups were established from this data as well as from more traditional morphological and taxonomic studies (Cheney, 1975). Another genetic method being used is controlled crossing of cultured strains of the same species collected from different geographies. Müller has done this with *Ectocarpus* (see Chapter 6) and West with *Gigartina* (see Chapter 7).

ECOLOGICAL DYNAMICS

A number of active areas of research in marine botany can be considered under this topic, including evolution and succession of communities, food webs, plant-animal interactions, and productivity. Utilizing various marine communities, specific examples of these ecological processes will be described in Chapters 15–20.

Ecological Evolution and Succession

Terrestrial ecologists suggest that there is an evolutionary pattern in the development of an ecosystem and its communities. Clements suggested in the early twentieth century that *succession* is a worldwide process and that any ecosystem will show a series of developmental stages toward stable, *climax* communities. Furthermore, if a climax community is destroyed, succession will begin again. Although there is some argument in the literature, succession toward a mature or climax community appears to occur in the marine environment. The initial phase in a succession of stages is called the *pioneer* community, and it is characterized by having few species (low diversity) and high numbers of individuals (high abundance). Most algal pioneer species are opportunistic, plants that can take advantage of numerous environments, have simple life histories, and are usually annuals rather than perennials.

A good example of a marine pioneer community was published by Maxwell Doty (1967), who had a unique opportunity to study algal recolonization after a lava flow. He began his study on the island of Hawaii, starting with sterile, newly formed rock. Within a few years, a series of successional stages of algal populations had cycled through. A climax type of algal community was evident within 10 yr. When Doty compared the 10-yr-old community with those found on 100-yr-old lava flows, he found some distinctions, suggesting that mature intertidal communities would still show minor population changes.

Disturbances of a regular nature can also aid in the establishment of a climax community. Sousa (1979) found that it took about 3 yr for a mature algal community dominated by *Gigartina canaliculata* to form on cleared boulders of a southern California shoreline. The early successional alga *Ulva lactuca* could prevent recruitment of the perennial red algae by outcompeting for settling space. However selective grazing on *Ulva* by a crab *(Pachygrapsus crassipes)* accelerated succession to a community of long-lived red algae.

In general, later stages of succession will have increasingly greater diversity of species, with a corresponding decline in abundance of individuals per species. Also, perennial forms and those developing from residual holdfasts (perennating) will be more common in subsequent longer-lived stages of succession.

Interaction of the plants in a community with their physical environment becomes increasingly complex as the more advanced successional stages occur. Climax or mature communities demonstrate a balance between the physical and biological parameters. Since a climax community is self-sustaining, this stage has a well-organized energy pathway driven by the sun. A climax community usually has the greatest diversity of species and the highest degree of interaction between biological and physical components. Littler's studies on Hawaiian crustose corallines of fringing reefs are a good example of the complexity of climax communities (Chapter 16).

A number of marine communities can be viewed as climax communities, including coral reefs, rocky shores, salt marshes, mangrove swamps, and seagrass beds, although some botanists will disagree. All of these communities show successional stages toward their ultimate development and have a high degree of interaction between physical factors and their biological components. All of these communities can have a highly diverse flora that occupies a variety of niches.

Odum (1969) presented a summary of features to distinguish successional (developmental) and climax (mature) communities. At least three ecological facets of species diversity* were considered:

1 Diversity of species is reduced whereas abundance of a species is increased in less-than-optimum environmental conditions (e.g., pioneer and early stages of succession).

2 Diversity of species in a community increases with increasingly diverse conditions. Usually one can find more niches in a stable climax community than in a successional stage.

*The term species diversity is taken here to include the two components of Odum (1969): (1) variety of species as expressed in species-number ratios; (2) equitability or evenness in apportionment of individuals among the species. Both components should be high in a mature, or climax, community.

3 Diversity of species and the complexity of a community is highest when the balance between environmental and biological factors is stable.

Most successional studies on lithophytic algae utilize new surfaces placed in the water such as bricks, cement plates, or other materials (Wilson, 1925; Tsuda and Kami, 1973), or denuded and cleaned existing rock (Northcroft, 1948). In all cases, algal regrowth over a relatively short period of a few years showed a series of successions from microscopic growths to more-complex communities of macroscopic and microscopic algae.

Food Webs

An ecosystem is made up of primary producers (plants), primary consumers (herbivores), secondary consumers (carnivores), decomposers (fungi and bacteria), and parasites. These organisms can be arranged into a *food chain,* and within this sequence one can find distinct energy or trophic levels. The entire complex structure can be displayed as a food web. In earlier studies this was usually diagrammed as a pyramid with the largest biomass or available-energy levels, the primary producers, placed at the base, and the smallest biomass or available-energy, the secondary consumers, at the pymmid apex. However, such displays do not consider overlap, or feedback, as well as possible sources of input or loss of energy.

The more typical method of displaying a food web today is to utilize the terminology and symbols of Odum and his associates. An example of energy transfer is taken from a study of a sea grass bed (*Posidonia*) in the Mediterranean (Figure 13-4). Ott and Maurer (1977) showed that in a 4 m deep *Posidonia*

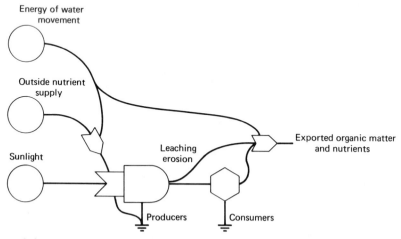

Figure 13-4   The paths of energy flow between a primary producing *Posidonia* sea grass community and the consumers is shown from the study of Ott and Maurer (1977). Energy sources are symbolized as circles, input and output as arrows, primary producers as rectangles with a rounded edge, and consumers as hexagons. Note that the net loss through exported organic matter and nutrients is smaller than amount consumed (sea urchins).

oceanica bed, a 1 m² area had a biomass of 1.15 kg dry weight and a net production of 1.52 g C/m² da (grams of carbon per square meter per day) during the summer months. About 50% of this production was consumed in the system and about 30% was exported out of the system on a continuous basis. The major herbivore, a sea urchin called *Paracentrotus lividus*, consumed about 10% of the production, which was primarily detritus.

Energy drains from the *Posidonia* community occurred through leaching out by the water currents, loss of photosynthetic activity because of the extensive epiphytism on the leaves, and a constant loss of fragments through mechanical action of water movement (Figure 13-4). In this study, no secondary consumers were identified; however, the sea urchins must have had predators. About 20% of the energy was not accounted for in the system, and it was thought to be lost through leaching. This conclusion was based on the findings of another study, by Miller and Mann (1973), who found a large loss in photosynthetic energy through the leaching of photosynthates from the seaweeds of a rocky coast of Canada. The photosynthates were located in the seawater. Other food webs are presented for salt marsh (Chapter 18) and mangrove (Chapter 19) communities.

Biotic Interaction

Both plant-plant (competition) and plant-animal (grazing) interactions are now well known in the marine environment and are the subjects of some of the most exciting research. A number of examples of competition and grazing are presented in Chapter 15 for intertidal and subtidal lithophytic seaweed communities. Climax community structure and composition is largely the result of such biotic interaction. Two other examples of biotic interaction are presented here, namely symbiosis and the utilization of natural plant products.

Symbiosis

There are a number of types of symbiosis (mutualism, commensalism, parasitism), and these have been considered in Chapters 7 (red algal parasites), 9 (endosymbiotic algae in dinoflagellates, symbiotic origin of euglenoids and dinoflagellates, zooxanthellae) and will also be included in Chapters 16 (hermatypic corals and zooxanthellae) and in Appendices A (composite fungialgal relationships, algal diseases) and B (endosymbiotic bacteria, diseases). Another aspect of symbiosis is the incorporation of chloroplasts by nudibranchs.

A number of species found in the order Ascoglossa are known to retain chloroplasts in their bodies and to obtain photosynthates from the plastids (Greene, 1970; Trench, 1975). Most of these ascoglossans feed on species of the coenocytic green alga *Caulerpa*, and have a modified radula that permits the animal to rupture the algal cell wall and suck out the cytoplasm and its plastids. The chloroplasts are shunted into a digestive direrticulate of the gut, which forms a netlike pattern in the pseudopodia. In this way, the chloroplasts are exposed to light. The ascoglossan *Tridachia crispata* will retain chloroplasts for long periods (50% chlorophyll retention over 58 days), whereas other species of ascoglossans show a much shorter half-life for chlorophyll retention—*Oxynoe antillarum*, 15 days; *Elysia cauze*, 11 days; *E. tuca*, 5 days (Green, 1970).

Incorporation of plant products

The production of special natural products by algae and the incorporation or use of these products by the grazers is a new area of study in biotic interactions. Collins (1978) presents a review of algal toxins. The concept of chemical defense systems against grazers has been well documented for terrestrial plants but until recently little was known about algal adaptations. Since marine algae have been in existence for a much longer period than terrestrial plants, one might assume that there has been a chemical "evolutionary arms race" between coevolving grazers and algae.

For example, a number of species of the coenocytic green alga *Caulerpa* produce the hydroxyamide compound caulerpicin, which is a mixture of sphingosine derivatives. The proposed structure of compound, which was first discovered by Doty and Santos (1966), has been revised by Mahendran *et al.* (1979). It has a molecular formula of $C_{42}H_{83}NO_3$. The proposed structure is as follows:

$$\text{Methane } (CH_2)_nCH = CHCHOHCHCH_2OH \qquad (1)$$
$$\underset{\displaystyle NHCO(CH_2)_n\text{methane}}{|}$$

in which $n = 12, 14, 20,$ or 22.

It is interesting that caulerpicin is a sphingosine derivative, because this is a common long-chain base of all animal sphingolipids and until recently was considered to be solely of animal origin. A similar product has been isolated from a red alga, *Amansia glomerata* (Cardellina and Moore, 1978).

As pointed out in the previous section on symbiosis, a number of nudibranchs feed on such coenocytic green algae as *Caulerpa*. Some of these ascoglossans include *Elysia cauze*, *Oxynoe antillarum*, and *Lobiger souverbiei*. Vest (1981) found that these three nudibranchs only feed on species of *Caulerpa* that contain caulerpicin (*E. cauze: Caulerpa racemosa, C. ashmeadii, C. sertularioides; O. antillarum* and *L. souverbiei: C. racemosa, C. paspaloides*) and that confirms the original ideas of Doty and Santos (1970) for Phillipine species. Caulerpicin may serve as a feeding attractant or as a distasteful repellent if passed on the egg masses of the ascoglossan.

Another nudibranch, *Aplysia*, is known to accumulate a number of red algal compounds, the use of which is unknown. The bile pigment of the ink (aplysioviolin) that *Aplysia* expels is derived from phycoerythrin monomethyl esters obtained from its red algal diet (Chapman and Fox, 1969). Irie *et al.* (1969) have shown that various sesquiterpenoid aromatic bromocompounds found in *Aplysia* also occur in one of the commonly eaten red algae, *Laurencia* sp. A number of studies have suggested that some of the bromocompounds are distasteful to predators of *Aplysia* (Kinnel *et al.*, 1977; Dieter *et al.*, 1979) and therefore are accumulated by the animal. Ambrose *et al.* (1979) has reviewed this literature and disagrees, saying the distastefulness is due to acid skin glands and not the sesquiterpenoids. However, Dieter *et al.*, (1979) found a brominated sesquiterpenoid that was also a potent feeding deterrent. The compound was found in the red alga *Chondria cnicophylla* upon which *Aplysia brasilana* feeds.

Antifouling and antimicrobial compounds of algae

There is a large literature dealing with natural products of algae (for review see Faulkner, 1977), especially with relation to antimicrobial effects (for review see Henriquez *et al.*, 1979). Lipid extracts, especially from brown algae (Dictyotales) are particularly effective antimicrobial and antiviral agents (Caccamese *et. al.*, 1980). In addition, tannins, which are known from brown algae, may be important as antiherbivore compounds similar to those of terrestrial plants. In this regard, Sieburth (1964) has found that tannins are strong antibacterial and antiepiphyte compounds in brown algae.

Organic Production

Both the amount of organic material that is produced (standing crop) and the rate of carbon fixation (productivity) should be considered in an energy study. The standing crop, or biomass, is usually expressed as grams of organic matter or dry weight per square meter (m^2). Biomass is an effective method to determine the importance of organisms in a community. Such data can be converted to kcal/m^2 or to joules (1 J = 4854 kcal; 1 kcal is the amount of energy required to raise one l of water 1°C at 1 atm).

Productivity and production are rate-measuring units for describing the amount of carbon fixed in photosynthesis. The procedures employed may measure oxygen production (Winkler titration using BOD bottles, oxygen probe using BOD bottles, manometric method measuring oxygen release or uptake), carbon dioxide uptake or release (pH measurement), or carbon fixation (labeled carbon incorporation). Because plant respiration utilizes a portion of the chemical energy produced in photosynthesis, and plants are consumed by animals or die; one must consider both gross and net primary production. Gross primary production represents the total amount of fixed carbon, before respiration; whereas net primary production is the amount remaining after grazing, death, and respiration by the plant.

How productive are marine plants and their communities? The biomass and productivity of six major marine plant communities are described in Chapters 15–20. In Table 13-2, a wide variety of plant communities are compared from various productivity studies on marine plants. Figure 13-5 shows productivity for the world's oceans. As one might guess, there is wide variation between communities and studies, some rates being higher than terrestrial communities whereas others are lower.

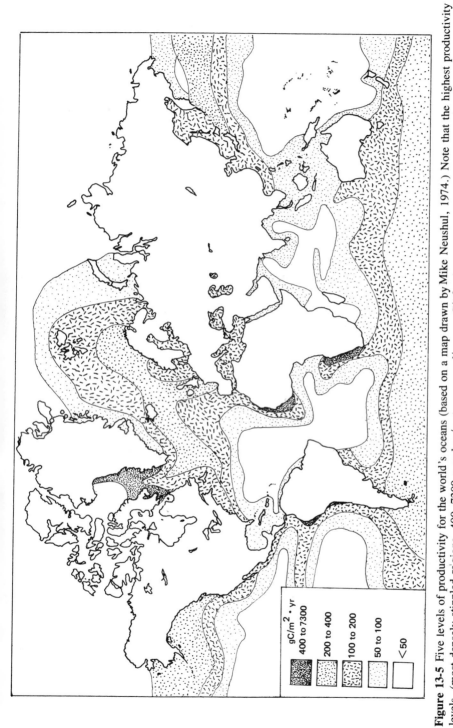

Figure 13-5 Five levels of productivity for the world's oceans (based on a map drawn by Mike Neushul, 1974.) Note that the highest productivity levels (most densely stippled regions, 400–7300 g carbon/meters squared/year: g $C/m^2 \cdot yr$) occur along the coasts on the eastern sides of the oceans where upwelling is found. In contrast, the lowest productivity (clear regions, < 50 g $C/m^1 \cdot yr$) of all oceans, where there is little water mixing.

Table 13-2   A Comparison of Primary Productivity of Selected Plant Communities

Community	Region	Net Rate $(g/C/m^2 \cdot day)$	Reference
Phytoplankton			
(Coastal)	Denmark	0.01–0.85	Bakus (1969)
(Bay)	Puerto Rico	0.52	Bakus (1969)
(Bay)	Long Island	0.50	Steidinger (1973)
(Bay)	Tampa Bay	0.3 –0.5	Steidinger (1973)
(Oceanic)	Sargasso Sea	0.2	Steidinger (1973)
(Oceanic)	Gulf of Mexico	0.1	Steidinger (1973)
(Oceanic)	Caribbean	0.1–0.2	Steidinger (1973)
Reef atolls	Eniwetok	1.6–7.2	Littler and Murray (1974)
Fringing reef (crustose coralline algae only)	Hawaii	0.5–2.6	Littler and Murray (1974)
Macroalgae: intertidal			
(18 species)	Southern California	0.4–3.1	Littler and Murray (1974)
(*Ascophyllum Fucus*)	Nova Scotia	2.1–3.9	Bakus (1969)
Macroalgae: subtidal			
(*Macrocystis*)	California	12–13	Bakus (1969)
(*Macrocystis*)	California	17	Towle and Pearse (1973)
(9 species)	California	1.0–9.0	Littler and Murray (1974)
(Seaweed bed)	Nova Scotia	4.8	Mann (1973)
(Seaweed bed)	Canary Islands	1.5–3.0	Johnston (1969)
Seagrasses			
(*Thalassia*)	Florida and Texas	5.7–17.8	Bakus (1969)
(*Thalassia*)	Puerto Rico	5.6	Bakus (1969)
(*Thalassia* and *Cymodocea*)	Laccadives	5.8	Qasim and Bhattadhiri (1971)
(*Posidonia*)	Malta	2–5	Drew (1971)
(*Zostera*)	Denmark	2–7	Peterson (1914)
(*Zostera*)	Alaska	3.3–3.8	McRoy (1970)
Salt marsh	Georgia	4.2	Bakus (1969)
Mangrove swamp	West Indies	3.8	Bakus (1969)

If one combines the major communities into marine habitats, the primary productivity within the biosphere can be summarized (Table 13-3). It is apparent that intertidal and subtidal algal beds are highly productive even when compared to such terrestrial communities as tropical rain forests and terrestrial swamps. It is also apparent that the open oceans have very low productivity when compared on a square meter basis; they are often described as "wet deserts" (Figure 13-5).

Table 13-3   Net Primary Productivity of Major Regions of the World

	Area Worldwide 10^6 km^2	Net Primary Productivity	
		Mean (g C/m^2 · yr.)	Total (10^9 tons C/yr)
Marine habitats			
Open ocean	332.0	125	41.5
Upwelling zones	0.4	500	0.2
Continental shelf	26.6	360	9.6
Intertidal algal beds and subtidal reefs	0.6	2000	1.2
Estuaries	1.4	1800	2.5
Total marine	361.0	152	55.0
Selected terrestrial habitats			
Tropical rain forest	17.0	2000	34.0
Temperate deciduous forest	7.0	1200	8.4
Tundra and alpine	8.0	140	1.1
Swamp and marsh	2.0	2500	5.0
Extreme desert	24.0	3	0.07
Total terrestrial	149.0	721	107.4

Finally, the energy loss between each trophic level is high. Up to 90% of the chemical energy contained within a given biomass will be lost rather than transferred to the level above. Furthermore, in seaweeds and phytoplankton, as much as 70% of the chemical energy produced can be leaked or actively excreted. It is truly impressive that even with the trapping of only 1–2% of the sunlight energy, a 90% loss of chemical energy between each trophic level, and the high loss of photosynthates through leakage from marine plants, there is still sufficient food to support a complex set of food webs in the marine environment that support a number of consumer levels.

THE MARINE ENVIRONMENT AND MAN

As our technology has grown, so has the use and misuse of the marine environment by man. People are becoming aware of the delicate balance in various marine ecosystems and the effect terrestrial activities have on marine communities. Mangrove swamps and salt marshes, for example, are no longer considered wastelands ready for development, but highly productive communities that play essential roles in the offshore production of sea life.

Because of the increasing dependence on the marine environment, it is essential that marine resources be conserved and protected. Overharvesting of these natural resources is now evident. During the last four centuries man has traveled farther and farther to obtain sufficient seafood as the local supplies become depleted or extinguished. One example of such worldwide ranging for seafood can be found in the giant, floating fish factories or processing ships. Only recently have attempts been successful on an international scale to limit the use of such processing ships. The misuse of the marine environment in the past is now evident, and efforts are increasing to rectify such damage, caused by dredge filling, organic and thermal pollution, and careless introduction of new organisms. Maritime and coastal laws are being examined and rewritten to preserve the critical sea life of our continental shelves. This section will review some of the various interactions of man in the marine environment. The reader is referred to Andrews (1976) for a review of most of these factors.

Dredge Filling

Major changes have occurred in the coastlines of industrial countries because of engineering projects (dredging, filling, jetty production, harbor improvements, navigational channel formation). The filling and channelization of tidal marshes have resulted in their deaths because of loss of needed sediments and nutrients, as well as the erosion of the soft sediments. For example, in a small bay off the mouth of Tampa Bay, Florida, 20% of the sediments (i.e., 1400 ha) were removed by dredge-and-fill operations by 1950 and another 20% by 1970. The mangrove fringe was destroyed so that canals, homes, and marinas could be built. The estimate of fish loss was 73 metric tons at a value of $1.4 million annually. Fisheries ecologist, S. H. Thompson (1961) recognized this problem and presented three important steps that should be taken for Boca Ciega Bay and other west coast bays of Florida:

1 Declare a moratorium on the sale of submerged lands.
2 Produce inventories of all submerged-land ownership and use.
3 Establish subtidal governing boards to control the use and sale of subtidal lands.

Such steps were unique and unheard of in 1961; however, all three are in common practice today. Similar problems are known around the world. Kaneohe Bay, Hawaii, is another example where dredge filling has resulted in extensive sedimentation and eutrophication.

Organic Pollution: Sewage

The addition of domestic and industrial pollutants may stimulate or inhibit growth (Murray and Littler, 1978). For example Littler and Murray (1975) found that the algal populations near a sewage outfall in southern California were less diverse and had a highly reduced community complexity. The dominant outfall biota were smaller growth forms of macroalgae, typically pioneer-type species, representing early successional stages. Such pioneer species have simpler and shorter life histories than nearby macroscopic algal communities unaffected by sewage outfall.

In a number of studies dealing with domestic pollution, key macroscopic algae were identified as *phytometers* or indicator species of polluted conditions (Bellamy *et al.*, 1968; Burrows and Pybus, 1971; Skulberg, 1970). The kelp forests of the British Isles were used as phytometers; Bellamy *et. al.* (1968) presented a summary of factors affecting the ecosystem, including organic pollution (Figure 13-6). They found a highly complex set of relationships between pollutants added and the productivity within kelp forest ecosystems. The main factors were grazing pressure and the effects of nutrients and toxins on the grazers. They found that four factors were affected by the organic pollution: nutrient levels, suspended material (ash, sewage colloids, oil), sediment, and toxic substances. The nutrients caused growth stimulation, especially for opportunistic species, and helped maintain large populations of browsing organisms that in turn, prevented the replacement of kelp forest plants. Sediments acted as light filters and coated the plants, preventing high rates of photosynthesis as well as masking sites for algal settlement. Toxic substances were incorporated into the food chain, concentrated by plants, and passed on to the grazers and the secondary consumers.

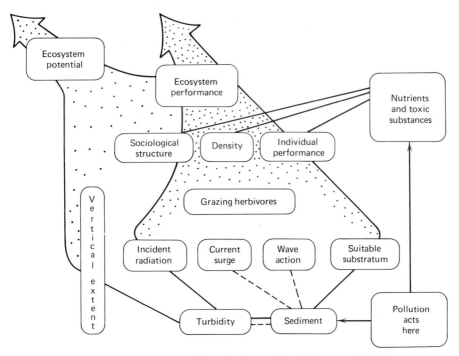

Figure 13-6 An energy web summarizes the factors that influence a kelp forest ecosystem. The diagram is from a study by Bellamy *et al.* (1968) of a *Laminaria* forest in England and shows the effect of pollution on the ecosystem. Note the different directions the ecosystem will take with regard to its potential and its actual performance if pollutants are introduced.

Toxic substances abound in industrial wastes. Effluents from pulp kraft mills include chlorinated phenols, quinones, sulphides, mercaptans, and fatty acid residues. Acids or tailings from mining deposits can be quite toxic. For example, in Tampa Bay there is a 2 ha, hard, white deposit of calcium fluoride as a result of seepage from a nearby phosphate plant. The unique fluoride deposit has killed off the benthic plants throughout the immediate region.

Organic Pollution: Oil and Dispersants

Oil spills and the effects of dispersants used to break down the oil are now a fact of life in the marine environment. In a number of studies on the effects of oil on marine life, it has been found that the dispersants are more toxic than the oil itself. In addition, dispersal of the oil usually results in a broader contamination rather than a localized contamination. The breakup of the supertanker Torrey Canyon off the coast of England demonstrated how destructive oil dispersants can be to coastal vegetation.

Damage is greater to intertidal floras than to subtidal ones because the intertidal region can become coated with the oil or dispersants. In a study of the extensive oil leakage off the coast of Santa Barbara, California, Foster et al., (1971) found a massive intertidal kill, but relatively little damage to the offshore kelp beds. They found that the intertidal flora had recovered within two years to prespill levels of standing crop and species diversity. The direct effect of oil spills is one of increasing toxicity along a chemical gradient from paraffins to naphthalene, olefins, and aromatic compounds. The primary effect is an increase in cell permeability, since the cell membrane has a high level of phospholipids in its makeup.

Organic Pollution: Pesticides and Chemicals

Many man-made chemicals are now at detectable levels in the sea, including aromatic hydrocarbons, polychlorinated biphenyls (PCBs) and substituted quinones. Little is known regarding the effect of such chemicals on the algae, but swelling or shrinkage of the plants (hypoplasia, hyperplasia) and abnormal growth forms have been demonstrated in the laboratory. Cell division and metabolism in phytoplankton is inhibited by PCBs even when they are present at only a few parts per million. Boney (1974) demonstrated that at very low concentrations, polycyclic aromatic hydrocarbons stimulate abnormal cell proliferation in red algae. The effect of DDT is well known on land, especially with regard to bird hatching. In the ocean, the effect on phytoplankton is variable; however, even low levels seem to cause toxicity to most species.

Heavy Metals and Radioisotopes

Heavy metals are normal constituents of the oceans and may be required by plants in trace amounts. Radioisotopes also occur naturally. Problems arise when excess amounts of either occur in estuaries or open oceans as marine plants will concentrate them. Some algae are accumulators of radioisotopes and heavy-metal pollutants. However, these same algae are also a part of the food chain, passing the concentrations along to the herbivores.

The direct effects of heavy metals and radioisotopes on algae are poorly understood. The various studies indicate that excessive accumulation (3–5 ppm) of lead, zinc, mercury, copper, or others is sufficient to cause changes in cell appearance, growth, and metabolism of phytoplankton. With regard to filamentous algae, Edwards (1972) found that relatively small concentrations of copper (0.01 mg/l) depressed the growth of the red alga *Callithamnion*. At present, most findings are based on laboratory studies, which may not be directly related to field conditions. Radioisotopes are accumulated in algae as well, but there seems to be little direct effect on the plant.

Thermal Pollution

The use of marine waters as coolants in power plants is becoming common along most coasts (Vadas, 1979). Since temperature is known to be a critical factor in algal distribution (see Chapter 15), it is not surprising that even a 2–3°C change in immediate water temperature can result in widespread damage to the marine plants. Vadas (1979) summarized several effects of thermal damage on marine plants. The main effects of heated-water discharges are as follows: (1) stratified flow conditions occur, and the warmer water is isolated in the upper levels, not mixing with the cooler, lower waters; (2) water turbidity usually rises because of the stratification and mixing of sediments along the cooling channels; (3) The capacity of the heated water to hold oxygen drops and this can result in semianerobic conditions; and (4) shift in the ratio of photosynthesis to respiration occurs because of the increased metabolism of the algae.

Temperature injury causes discoloration, brittleness, necrosis, and distorted growth of algae, as shown in the giant kelp *Macrocystis*. The most common effect is a disjunct population of lower diversity. Usually the species more tolerant to higher temperatures are pioneer forms, and they increase in abundance. Species diversity can even drop to a near monoculture status. Blue-green algae typically are the most abundant group of algae that will collect in a thermopolluted intertidal region, even at higher, cooler latitudes. In one tidal marsh, the biomass of the dominant salt marsh plants decreased by 48% after one year and another 40% after the second year because of thermal loading (Vadas *et al.,* 1976). Blue-green algae were common on the substrate of that marsh as well. Recovery of the salt marsh plants can be rapid (within 3 yr) after removal of the thermal pollutant (Keser *et al.,* 1978; Vadas *et al.,* 1978).

Most of the biological-impact studies carried out on both temperate and subtropical environments have demonstrated that the water used for cooling must be brought back to ambient coastal water temperatures. Otherwise productivity of the local marine plant communities will be reduced and the long-range productivity will drop. Immediate damage is usually found in the resident fish and invertebrate populations.

Biological Pollution

Two examples of introduced "nuisance" species of seaweeds are described. There are many examples of terrestrial animal and plant introductions, including the "killer bees" of South America and the water hyacinth of Florida. In the marine environment two introduced seaweeds have caused problems, namely *Codium fragile* ssp. *tomentosoides* in New England and *Sargassum muticum* along the west coast of North America and the west coast of Europe.

Codium fragile ssp. *tomentosoides* (Figure 13-7) is a coenocytic green alga that apparently was introduced into the northeastern coast of North America from either Holland or the state of Washington during the transplantation of oysters in 1957 (Fralick and Mathieson, 1973). Within 10 yr of the first report, the subspecies was established from New Jersey to Maine and had become quite abundant in southern New England. The branched, dark-green felty plants attach as germlings or fragments to the oyster shells and cover the oysters, essentially smothering them. Two separate studies were initiated on the growth of *Codium fragile* ssp. *tomentosoides*. Fralick and Mathieson (1973) found that the plant reproduced primarily by fragmentation. Based on respiratory and photosynthetic rates, they found the plant was well adapted to the prevailing salinity, light, and water temperatures. Hanisak (1979) found that the growth pattern of *C. fragile* ssp. *tomentosoides* was controlled seasonally by an interaction of temperature and irradiance.

Figure 13-7 *Codium fragile* ssp. *tomentosoides* has become a nuisance alga in much of Long Island Sound and southern New England. The introduced coenocytic green alga will attach to rocks or oyster beds in the intertidal zone and can attain masses of more than $12 kg/m^2$ wet weight. The white tips are due to exposure to freezing temperatures. Fralick and Mathieson (1973) and Hanisak (1979) found that a form of fragmentation (swelling and constriction of the tips) in the winter resulted in massive vegetative reproduction.

The second example is *Sargassum muticum,* a brown seaweed (Fucales) that has a highly differentiated plant body (blades, stipe, floats, holdfast and apical meristem, Figure 13-8). The species is a native of Japan and apparently was first introduced to the west coast of North America on oyster shells. Scagel (1956) suggested that the introduction was repetitous possibly between 1902 and 1940, when the Japanese oyster *Crassostrea gigas* were imported. Like *Codium, S. muticum* is an opportunistic species and a rapidly growing plant. The species is now established from Puget Sound to southern California, forming large beds off Santa Catalina Island, with individual plants reaching 2–10 m in height.

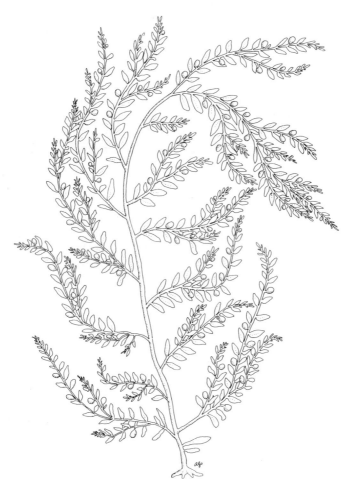

Figure 13-8   The brown seaweed *Sargassum muticum* can attain lengths of over 3 m and will dominate shallow subtidal regions of southern California and boat harbors in England. The species has a wide tolerance to water temperatures and salinities and is able to reproduce rapidly. The species is a native of Japan and was apparently introduced through oyster transplants to Puget Sound.

The success of *S. muticum* in a new area appears to be primarily due to its ability to spread to new areas, colonize them and establish new populations. Norton (1980) has found that drifting fragments are responsible for much of its invasiveness, but perhaps equally important are the dense local settlements of germlings around established colonists. These dense germling developments essentially crowd out all other species. Certainly the plant tolerates a wide range of temperatures as demonstrated by its spread into the British Isles, where it has become a nuisance weed in the shallower boat harbors of southern England.

METHODS OF GENERAL ECOLOGY

Fieldwork in marine botany is usually the most enjoyable part of a study, and it is important that the investigator be prepared in advance in order to make the most of a field effort. The methods presented here include ecological techniques basic to most field studies and do not require extensive equipment. The importance of statistical analysis and the monitoring of various physical factors should not be ignored in any serious field investigation, but such procedures are beyond the scope of this chapter. A number of procedures and a description of equipment used to monitor marine plant communities and the environmental factors are outlined in Chapters 10–12.

Algal Identification and Preservation

Both dissecting and compound microscopes are needed, since cell and filament measurements (width and length) are required for most smaller algae. The instruments should be equipped with calibrated ocular micrometers. Freehand sections can be made with a sharp single-edged razor blade (e.g., to determine if a medulla exists). Standard microscopic slides, coverslips, and lens paper should be available. The following solutions should be available in dropper bottles: distilled water, seawater, 10% HCl for decalcification of sections or calcified epiphytes, Gram's iodide (I_2KI: 2 g KI + 2 g I_2 in 25 ml distilled water) for starch determination.

Except for immediate study, all collections of algae should be preserved in a 5% formalin (40% formaldehyde) and seawater. The preserved material should be kept in a cool, dark place to prevent bleaching out. It is recommended that as much identification as possible be made before making pressed herbarium specimens, since identification of internal structure is easier before pressing. All algae that will be mounted for herbarium specimens should be preserved; otherwise mildew or other contamination can occur, even in the herbarium cases.

Two types of algal collections should be made in a study of a marine plant community: herbarium (pressed on paper) and wet-stack (stored in liquid in bottles) specimens. Both are extremely useful for later study. Herbarium specimens can be made by floating the alga out in normal tap water (after 24 hours of preservation in seawater and 5% formalin) in a shallow pan such as cafeteria-type

Figure 13-9 Herbarium specimens. Most macroalgae make excellent herbarium specimens when they are floated out in a large tray (right of photograph) the herbarium sheet is brought up from underneath. The plant press consists of outer wooden boards, cardboard sheets to allow air movement, absorbant blotters, and newspaper folded over the herbarium sheet with the wet alga, which is covered with a sheet of muslin. This organization is repeated for each algal specimen.

trays (Figure 13-9). The herbarium paper used should have a high rag content (100% is best) and should be of standard herbarium size 11½ in. × 12½ in.). The need for a high rag content is obvious, because the paper is totally immersed in a tray under the floating alga and then is removed gently from the pan with the alga spread over the paper. For smaller specimens, the paper can be cut in half. Some types of good-quality drawing paper can also be used; this paper may be more easily available as well as much less expensive.

Each herbarium sheet should be numbered (in pencil) before mounting the alga. The number is recorded in a record book where all the data regarding the collection is kept. Such data should include date, location, zone (of the intertidal region) or depth, physical features of the site, algal name including the authors, and any remarks regarding the plant or its characteristics. From this recorded data, the herbarium label can be made up. Figure 13-10 is a photograph of a typical herbarium specimen. The label is glued onto the lower right-hand corner of the herbarium paper after the alga has been pressed and is thoroughly dry.

After the alga and herbarium paper have carefully been removed from the tray, the alga is pressed. The procedure is similar to the construction of a sandwich. First a sheet of cardboard (with fluting to allow air flow through it) is placed on the press board (Figure 13-9), then a blotter, then an open one-to-two sheet-thick piece of newspaper, and then the herbarium sheet with the wet alga. The alga is covered with either wax paper, a sheet of thin muslin, or cheesecloth to prevent sticking. The newspaper is folded down, a blotter is added and another corrugated cardboard for final cover. The process can now be repeated so that the plant press may contain 25–40 sheets of herbarium paper. The top of the press is added and the straps are tightened sufficiently to press the alga. The press is then

Figure 13-10 An example of a herbarium specimen with a herbarium label. Note that the label is in the lower right-hand corner.

placed on edge on a plant dryer or over a few light bulbs placed in a wooden box. It is important that the corrugated cardboard fillers have air tubes running at right angles to the press length so that the warm air can flow up through the press. Drying in a plant dryer usually takes about two to three days or about four to five days over a few light bulbs. Blotters should be changed daily if possible. Since wet algal specimens will shrink and curl, it is important to have the specimens dry before removing them from the press. Specimens that do not stick well to the herbarium paper (e.g., *Sargassum,* coralline alga) can be glued down with a commercial, clear-drying glue such as Elmer's glue. Small specimens of a coralline alga can be stored in envelopes mounted on the herbarium sheets. Bulky specimens such as crustose corallines can be stored in small boxes with the herbarium labels glued to them.

Semipermanent slides of a preserved alga can be made as well. The alga is stained with 1% aniline blue or acid fuchsin on the slide. After about a minute the slide is rinsed gently with 5% formalin in distilled water. The excess water is removed carefully and a drop of two of a mixture of 50% (aqueous) Karo syrup containing 5% formalin is added. The slide is allowed to dry overnight. The next day a drop of 80% Karo syrup is added and the slide is covered with a coverslip. The slide can be placed in a slot cut into a cardboard strip for protection and then mounted in an envelope on the herbarium sheet.

Wet-stack, or liquid-preserved, specimens are also important since any microscopic study will require these. If a wet-stack specimen is made, it should be noted on the herbarium label and given the same collection number. Small specimens are easily preserved and available if kept in wet-stack collections. The alga can be stored in a buffered 5% formalin solution in the dark. The solution should be changed once a year.

Sectioning of most algae can be done by hand using a clean, sharp single-edge razor blade. Hold the alga on the slide with the index finger and mince the extended portion of the alga, using the index finger as a guide. This type of sectioning can be carried out under a dissecting microscope (10–25×). If available, a freezing microtome can be used; however, for most studies it is not necessary.

Decalcification of calcareous algae is accomplished by adding a few drops of 10% HCl to the algal section or fragment placed on a slide. After bubbling has ceased, the slide is rinsed in distilled water and covered with a coverslip. Usually the decalcified material can then be gently mashed by tapping with a pencil, eraser down. Larger calcified forms can be decalcified by using Perenjis solution: 4 parts 10% nitric acid, 3 parts 70% ethanol, 3 parts 0.5% chromic acid. The fragments are allowed to soak in the solution for 4 to 6 hrs.

Field Collection

The general field gear needed includes a snorkle, mask, collecting shoes (e.g., tennis shoes) if warm water or boots if cold water. Also, gloves, scraper or crowbar, knife, pail, plastic bags with water-resistant labels, and a record book should be available. It is always recommended to obtain as much physical data as possible; several items of specialized equipment are described in Chapters 10–12 (radiometer, quantum light meter, Secchi disc, refractometer, hydrometer, oxygen meter). Minimum equipment for intertidal studies should include a line transit marked in meters, a thermometer, and a line level for use in elevation determinations. A water sample should be taken and kept on ice for pH and salinity determinations (Chapter 12) and, if possible, nutrient analysis (Chapter 12) either in the field or in the laboratory. A sample can be taken for sediment analysis (Chapter 10) or chemical determination of the components. Topography, wave action (energy level of the beach), and a general list of maritime vegetation can be described in the record book (see also Chapter 10).

Community Measurements

In addition to a general species list based on the collections, one may wish to determine zonation (transects: line, band, quadrat) or dominance and statistical features of a uniform area (quadrat: square or point method).

Transect

If a study is to be made in an area of changing vegetation that would be found in an intertidal zone or gradation between salt marsh and freshwater regions, a line transect is usually the best choice (Figure 13-11). Such a procedure can be used in intertidal or subtidal regions (Croley and Dawes, 1969).

To carry out this procedure, secure well-marked stakes in a straight line perpendicular to the coast, and run a metered rope between the stakes. For the intertidal region, start at a level above the highest tide mark. Depending, on how rapid the intertidal area slopes, make a general collection every centimeter, meter or 10 m unit. These collections will serve as plant identifications for each zone and should be placed in identified bags (marked according to meter level). The elevation of each collection in the intertidal zone can be determined by the line level or Emery techniques (Chapter 10).

Determine the percentage of species present by counting all species found along the line. For each unit length, a zone, add up all the species present and divide by the total number of species for the transect.

Measure the length a given species occupies along the rope for each measuring unit. This amount will vary as different intertidal regions are reached. By dividing the transect length a species covers by the total transect length of a zone, percentage cover of that unit can be obtained.

The entire transect can be summed up by a general species list that gives (1) species names; (2) frequency (number of subunits within which a species occurs divided by the total number of subunits), (3) abundance (number of plants occurring along the entire transect line divided by the number of occupied subunits (4) general zonation of each species with relation to sealevel; and (5) physiogamy or form of the species.

Profiles can be made based on the transects for visual displays of plant zonation (Figure 1-9). Representative herbarium specimens should also be made for each zone.

Quadrat

This method is best used in regions of uniform physical factors (no zonation). A good example would be a uniform sublittoral plant community (algal, seagrass). The quadrat procedure permits a much more accurate use of such statistical measurements as percentage cover, frequency, and abundance than does the line transect procedure. More complex methods are in statistic texts.

A quadrat can be any size or shape providing the area covered is known. Usually it is a square meter that is subdivided into 10 cm units by string (Figure

Figure 13-11 Use of line transects and square-meter quadrats is recommended for field studies of marine plant communities. The students are collecting algae from selected subunits of the quadrat. Line transects are best used in regions where zonation may exist, whereas quadrats are most useful within a uniform (lacking zonation) community. The lines can be marked in centimeters or meters, whereas the quadrats can be square meters subdivided by strings. The quadrat used here was made from electrical conduit, and the corners were made out of outlet boxes.

13-10). To use it, if possible locate the central region of the plant community, set up a random block design for quadrat location, or simply throw the quadrat in various directions around a 360° radius.

At least 5 quadrats or fractions thereof should be measured. Count all the plants, by species found in the quadrat or fraction thereof, and determine percentage cover of each species. If a subfraction of a quadrat is used, then multiply the numbers obtained to equal the whole quadrat. The following statistics can be determined for each species:

$$\text{frequency} = \frac{\text{number of occupied quadrats}}{\text{number of quadrats}}$$

$$\text{density} = \frac{\text{total number of individual plants}}{\text{total number of quadrats}}$$

$$\text{abundance} = \frac{\text{total number of individual plants}}{\text{number of occupied quadrats}}$$

Survey

This technique is based on the phytosociological techniques of Braun-Blanquet and promoted by the Zurich-Montpeller group. Van den Hoek and den Hartog modified this technique for marine algal studies (den Hartog, 1959).

The size of the survey region should be quite large, for example, a few kilometers. Most studies employing the procedure are concerned with expanses of coastlines. Sample plots (1 m², 10 m²) are selected based on visual observation.

Based on numerous observations, tables are prepared and associations of sea-weeds deduced. The survey method is designed for rapid coverage of a large area and not only requires rapid recognition of the species, but also requires use of the transect method (to determine if any zonation is present) and quadrat method (to determine statistical data).

Lower abundance is determined for *each* species and is usually expressed with regard to the area covered. The symbols used in the tables are as follows:

r   Rare, usually only one specimen per plot

+   Sparse, a few specimens only in the plot

1   A good number of individuals, numerous but covering less than half of the plot

2   Many individuals, covering 25–50% of sample plot

3   Covering 50–75% of sample plot

4   Covering 75–100% of sample plot

Species sociability can also be determined (how the plants grow).

1   Single plants

2   In small clusters or tufts

3   In small patches and distinct groups

4   In a carpet or mats, dense but not in completely closed vegetation

5   In a homogeneously, continuous dense carpet, little else able to penetrate

REFERENCES

Ambrose, H. W., R. P. Givens, R. Chen, and K. P. Ambrose. 1979. Distastefulness as a defense mechanism in *Aplysia brasiliana* (Mollusca: Gastropoda). *Mar. Behav. Physiol.* **6**: 57–64.

Andrews, J. H. 1976. The pathology of marine algae. *Biol. Rev.* **51**: 211–253.

Bakus, G. J. 1969. Energetics and feeding in shallow marine waters. *Intern. Rev. Gen. and Expt. Zool.* **4**: 273–369.

Bellamy, D. J., D. M. John, and A. Whittick. 1968. The "kelp forest ecosystem" as a "phytome-ter" in the study of pollution of the inshore environment. In *Underwater Assoc. Rep. 1968*, London. pp. 79–82.

Boney, A. M. 1974. Aromatic hydrocarbons and the growth of marine algae. *Mar. Pollu. Bull.* **5**: 185–186.

Burrows, E. M. and C. Pybus. 1971. *Laminaria saccharina* and marine pollution in Northeast England. *Mar. Pollu. Bull.* **2**: 53–56.

Caccamese, S., R. Azzolina, G. Furnari, M. Cormaci, and S. Grasso. 1980. Antimicrobial and antiviral activities of extracts from Mediterranean algae. *Bot. Mar.* **23**: 285–288.

Cardellina, J. H. and R. E. Moore. 1978. Sphingosine derivatives from red algae of the Ceramiales. *Phytochemistry* **17**: 554.

Chapman, A. R. O. 1979. *Biology of Seaweeds. Levels of Organization*. University Park Press, Baltimore.

Chapman, D. J. and D. L. Fox. 1969. Bile pigment metabolism in the sea hare *Aplysia*. *J. Exp. Mar. Biol. Ecol.* **4**: 71–78.

Cheney, D. P. 1975. *A Biosystematic Investigation of the Red Algal Genus Eucheuma (Solieriaceae) in Florida*. Ph.D. dissertation. University of South Florida, Tampa.

Cheney, D. P. and G. R. Babbel. 1978. Biosystematic studies of the red algal genus *Eucheuma*. I. Electrophoretic variation among Florida populations. *Mar. Biol.* **47**: 251–264.

Collins, M. 1978. Algal toxins. *Microbiol. Rev.* **42**: 725–746.

Croley, F. C. and C. J. Dawes. 1969. Ecology of the algae of a Florida Key. Part I. A preliminary list of the marine algae including zonation and seasonal data. *Bull. Mar. Sci.* **20**: 165–185.

Dayton, P. K. 1975. Experimental studies of algal canopy interactions in a sea-otter dominated kelp community at Amchitka Island, Alaska. *U. S. Dept. Commer. Fish. Bull.* **73**: 230–237.

den Hartog, C. 1959. The epilithic algal communities occurring along the coast of the Netherlands. *Wehtia* **1**: 1–230.

Dieter, R. K., R. Kinnel, J. Meinwald, and T. Eisner. 1979. Brasodol and isobrasodol: two bromosesquiterpenes from the sea hare *(Aplysia brasiliana)*. *Tet. Lett.* **19**: 1645–1648.

Doty, M. S. 1967. Pioneer intertidal population and the related general vertical distribution of marine algae in Hawaii. *Blumea* **25**: 95–105.

Doty, M. and T. Santos. 1966. Caulerpicin, a toxic constituent of *Caulerpa*. *Nature* **211**: 990.

Doty, M. and T. Santos. 1970. Transfer of toxic algal substances in marine food chains. *Pac. Sci.* **24**: 351–355.

Drew, E. A. 1971. Botany, In Woods, J. D. and J. N. Lythgoe (eds.). *Underwater Science*, Oxford University Press, London, pp. 75–90.

Edwards, P. 1972. Cultured red alga to measure pollution. *Mar. Pollution Bull.* **3**: 184–188.

Edwards, P. and C. van Baalen. 1970. An apparatus for the culture of benthic marine algae under varying regimes of temperature and light-intensity. *Bot. Mar.* **13**: 42–43.

Falkner, D. J. 1977. Interesting aspects of marine natural products chemistry. *Tet. Lett.* **33**: 1421–1443.

Foster, M., M. Neushul, and R. Zingmark. 1971. The Santa Barbara oil spill. 2. Initial effects on intertidal and kelp bed organisms. *Environ. Pollution* **2**: 115–134.

Fralick, R. A. and A. C. Mathieson. 1973. Ecological studies of *Codium fragile* in New England, U.S.A. *Mar. Biol.* **19**: 127–132.

Greene, R. W. 1970. Symbiosis in sacoglossan opisthobranchs: Functional capacity of symbiotic chloroplasts. *Mar. Biol.* **7**: 138–142.

Grinnell, J. 1924. Geography and evolution. *Evol.* **5**: 225–229.

Hanisak, M. D. 1979. Growth patterns of *Codium fragile* spp. *tomentosoides* in response to temperature, irradiance, salinity, and nitrogen source. *Mar. Biol.* **50**: 319–332.

Henriquez, P., A. Candia, R. Norambuena, M. Silva, and R. Zemelman. 1979. Antibiotic properties of marine algae. II. Screening of Chilean marine algae for antimicrobial activity. *Bot. Mar.* **22**: 451–453.

Irie, T., M. Suzuki, and Y. Hayakawa. 1969. Isolation of aplysin, debromoaplysin and aplysinol from *Laurencia okamurai*. *Bull. Chem. Soc. Japan* **42**: 843–844.

Johnston, C. S. 1969. The ecological distribution and primary production of macrophytic marine algae in the Eastern Canaries. *Int. Rev. Ges. Hydrobiol.* **54**: 473–490.

Keser, M., B. R. Larson, R. L. Vadas, and W. McCarthy. 1978. Growth and ecology of *Spartina alterniflora* in Maine after a reduction in thermal stress. In Thorp, J. H. and J. W. Gibbons (eds.). *Energy and Environmental Stress in Aquatic Systems*, DOE Symposium Series (Conf-771114), Nat. Tech. Infor. Serv., Springfield, Va., pp. 420–433.

Kinnel, R, A. J. Duggan, and T. Meinwald. 1977. Panacene, an aromatic bromoallene from a sea hare *(Aplysia brasiliana) Tet. Lett.* **44**: 3913–3916.

Littler, M. M. and S. N. Murray. 1974. The primary productivity of marine macrophytes from a rocky intertidal community. *Mar. Biol.* **27**: 131–135.

Littler, M. M. and S. N. Murray. 1975. Impact of sewage on the distribution, abundance and community structure of rocky intertidal macro-organisms. *Mar. Biol.* **30**: 277–291.

Littler, M. M. and S. N. Murray. 1978. Influence of domestic wastes on energetic pathways in rocky intertidal communities. *J. Appl. Ecol.* **15**: 583–595.

Mahendran, M., S. Somasundaran, and R. H. Thomson. 1979. A revised structure for caulerpicin from *Caulerpa racemosa. Phytochemistry* **18**: 1885–1886.

Mann, K. H. 1973. Seaweeds: Their productivity and strategy for growth. *Science* **182**: 975–981.

McRoy, C. P. 1970. Standing stocks and related features of eelgrass populations in Alaska. *J. Fish. Res. Bd. Can.* **27**: 1811–1821.

Miller, R. J. and K. H. Mann. 1973. Ecological energetics of the seaweed zone in a marine bay on the Atlantic coast of Canada. III. Energy transformations by sea urchins. *Mar. Biol.* **18**: 99–114.

Murray, S. N. and M. M. Littler. 1978. Patterns of algal succession in a perturbated marine intertidal community. *J. Phycol.* **14**: 506–512.

Neushul, M. 1974. *Botany.* Hamilton, Santa Barbara, Calif.

Northcraft, R. D. 1948. Marine algal colonization on the Monterey Peninsula, California. *Amer. J. Bot.* **35**: 396–404.

Norton, T. A. 1980. The varied dispersal mechanisms of an invasive seaweed, *Sargassum muticum.* In *Intern. Phycol. Soc. Meet. Glascow. 1980 abstr.* 54 pp.

Odum, E. P. 1969. The strategy of ecosystems development. *Science* **164**: 262–270.

Ott, J. and L. Maurer. 1977. Strategies of energy transfer from marine macrophytes to consumer levels: The *Posidonia oceanica* example. In Keegan, B. F., P. O. Ceidigh, and P. J. S. Boaden (eds.). *Biology of Bethnic Organisms,* Permagen, New York, pp. 493–502.

Peterson, C. G. J. 1914. Om Baendeltangens *(Zostera marina)* Aarsproduktion i de danske Farvande. In Jungerson, F. E. and J. E. B. Warming (eds.). *Mindeskrft i anledning of Hundrerdaaret for Papetus Steenstrups Fodsel.* B. Lunos Publ., Copenhagen.

Qasim, S. Z. and P. M. A. Bhattadhiri. 1971. Primary production of a seagrass bed on Kavaratti Atoll (Laccadives). *Hydrobiologia* **38**: 29–38.

Rosenthal, R. J., W. D. Clarke and P. K. Dayton. 1974. Ecology and natural history of a stand of giant kelp, *Macrocystis pyrifera,* off Del Mar, California. U. S. Dept. Commer. Fish. Bull. **72**: 670–684.

Russell, G. 1972. Phytosociological studies on a two-zone shore. I. Basic pattern. *J. Ecol.* **60**: 539–545.

Scagel, R. F. 1956. Introduction of a Japanese alga, *Sargassum muticum* into the Northeast Pacific. *Fish Res. Pap. Wash. Dept. Fish.* **1**: 49–58.

Sieburth, J. McN. 1964. Antibacterial substances produced by marine algae. *Develop. Ind. Microbiol.* **5**: 124–134.

Skulberg, O. M. 1970. The importance of algal cultures for the assessment of eutrophication of the Oslofjord. *Helgo. Wiss. Meeresunt.* **20**: 111–125.

Sousa, W. P. 1979. Experimental investigations of disturbance and ecological succession in a rocky intertidal algal community. *Ecol. Monog.* **49**: 227–254.

Steidinger, K. A. 1973. Phytoplankton ecology: A conceptual review based on eastern Gulf of Mexico research. *CRC Crit. Rev. Microbiol.* **3**: 49–68.

Thompson, S. H. 1961. What is happening to our estuaries? In *Trans. 26th North Amer. Wild and Nat. Res. Conf.*, Wildlife Mang. Inst., Washington, D.C. **26**: 318–322.

Towle, D. W. and J. S. Pearse. 1973. Production of the giant kelp *Macrocystis* estimated by in situ incorporation of ¹⁴C in polyethylene bags. *Limnol. and Oceanogr.* **18**: 155–159.

Trench, R. K. 1975. Of 'leaves that crawl': Functional chloroplasts in animal cells. In *Symbiosis*. Cambridge University Press, London, pp. 229–265.

Tsuda, R. T. and H. T. Kami. 1973. Algal succession on artifical reefs in a marine lagoon environment in Guam. *J. Phycol.* **9**: 260–264.

Vadas, R. L. 1979. Abiotic disease in seaweeds: thermal effluents as causal agents. *Experientia* **35**: 435–437.

Vadas, R. L., M. Keser, P. C. Rusanowski, and B. R. Larson. 1976. The effects of thermal loading on the growth and ecology of a northern population of *Spartina alterniflora*. In Esch, G. W. and R. W. MacFarlane (eds.). *Thermal Ecology*, ERDA Symposium Series (Conf-750425) Augusta, Ga., pp. 54–63.

Vadas, R. L., M. Keser, and B. Larson. 1978. Effects of reduced temperatures on previously stressed populations of an intertidal alga. In Thorp, J. H. and J. W. Gibbons (eds.). *Energy and Environmental Stress in Aquatic Systems*. DOE Symposium Series. (Conf-77114). Nat. Tech. Infor. Serv., Springfield, Va., pp. 434–451.

Vandermeer, J. H. 1972. Niche theory. *Ann. Rev. Ecol System.* **3**: 107–132.

Vest, S. 1981. Distribution of cauleupin and caulerpicin, in Floridean Caulerpas, and their relation to the feeding selectivity of two ascoglossons. Master's thesis, University of South Florida, Tampa.

Wilson, O. T. 1925. Some experimental observations of marine algal successions. *Ecology* **6**: 303–311.

Yarish, C., K. W. Lee, and P. Edwards. 1979. An improved apparatus for the culture of algae under varying regimes of temperature and light intensity. *Bot. Mar.* **22**: 395–397.

CHAPTER 14

Physiological Ecology: Concepts and Methods

Physiological ecology is the study of the physiological mechanisms that allow organisms to respond to their environments. Once something is known about the operations of these mechanisms, they can be considered in relation to the environment, and predictions can be made about the success of a given phenotype in any environment.

Lawrence (1981) has argued that physiological ecology can bridge the "gap" between molecular and whole-organism biology. To do this, he suggests that studies must be carried out at various levels of organization: molecular, physiological, and organismal. His main point is that if suborganismal functions are truly adaptive, they must be demonstrated to be so. Thus the studies needed the most in physiological ecology are time-course experiments that follow sub-organismal and entire-organismal activities. Such studies of marine plants are lacking.

Some of the more common physiological studies in plants have looked at photosynthesis and respiration, especially with regard to physical factors, morphological adaptations, changes in plant chemistry, and comparative enzymatic studies. The most commonly studied mechanisms are the respiration and photosynthesis of plants, since these processes are basic and can be easily monitored. The linking of carbon gain through photosynthesis with growth and effects of physical factors can illustrate how the strategic needs of plants can be met (Mooney, 1976). Through studies of the effects of limiting photosynthesis in a population, one should be able to determine the environmental constraints on energy capture and allocation. The results of such studies are very important in our attempts to cultivate marine plants.

A number of examples of bridging the gap between cultivation and the physiological studies of marine plants are now available. One such example is a study of the effect of temperature on carrageenan production and the effect of nitrogen on the growth response of the red algae *Chondrus crispus*. Simpson and Shacklock (1979) found that if grown at lower temperatures or with excessive levels of fixed nitrogen, the red alga would show a major drop in the accumulation of carrageenan. Thus they recommend that at least for the final phases of a

Chondrus mariculture, the plant be "starved" for nitrogen and growth under higher temperatures.

This chapter will introduce a number of physiological mechanisms and will describe how botanists have used them to better understand plants' responses to a set of environmental conditions. The chapter is divided into four sections, the first three dealing with concepts (physiological, morphological, and chemical responses of marine plants) and the fourth presenting methods. The goal of the chapter is to make the reader aware of the importance of physiological and chemical tools in studies of marine plant communities, populations, or species, and how these tools will aid in ecological studies. Several texts are recommended for further reading: Kinne (1970, 1971, 1976), Stewart (1974), and Hellebust and Craigie (1978).

PHYSIOLOGICAL RESPONSES TO ENVIRONMENTAL FACTORS

Two areas will be considered in this section: the effects of light quality and quantity, and the photosynthetic and respiratory responses of plants to variations in physical factors. In addition, the process of photorespiration will be introduced.

Light and Algae

The adaptative mechanisms of algae to light have been debated since Englemann's (1884) theory of complimentary chromatic adaptation. Brody and Brody (1962) showed that the concentrations of chlorophyll and phycoerythrin may be varied by irradiating the unicellular red alga *Porhyridium cruentum* with either green or blue light. In the blue-green alga *Anacystis nidulans* Jones and Myers (1965) also found that qualitative variations in pigments could be induced by these wavelengths of light. However, none of the above studies were quantitatively correlated.

Oltmanns (1892) refuted Englemann's ideas and proposed that the vertical distribution of algae was controlled by light intensity. Responses to light quality observed in *Porphyridium cruentum* (Brody and Emerson, 1959; Brody and Brody, 1962) support Oltmanns's ideas. The plant will show a decrease in pigment content under high monochromatic or white light intensities ($>$ 100 ergs/cm^2 · sec).

The importance of quality and intensity of submarine illumination is supported by the information available from light studies in the ocean. Submarine illumination is considerably different from that in the atmosphere, as demonstrated in Chapter 11 (e.g., Figure 11-1). Most red light is absorbed by a depth of 25 m in clear oceanic water. The orange, yellow, and green wavelengths are absorbed rapidly in succession. At a depth of about 200 m in clear water only the midwavelengths of blue occur, centering at about 475 nm. Not only is there a loss in various wavelengths with depth, but light intensity also decreases. Thus over the years algal distribution with depth has been explained either as (1) complementary chromatic adaptation due to spectral changes (Englemann, 1884), or (2) a direct effect of light intensity (Oltmanns, 1905).

Complementary chromatic adaption

This concept is often used to explain the vertical distribution of red algae in the deepest waters as a result of the abundance of the water-soluble phycobilin phycoerythrin. The pigment will absorb blue light (495 nm) as shown by Haxo and Blinks (1950) and is very abundant in deep-water red algae. For example, *Eucheuma,* a large, fleshy tropical red alga, shows an increasing concentration of phycoerythrin in deep-water sites during the summer off the west coast of Florida. The loss of red wavelengths and increasing dominance of blue light in shallow waters (15 m) is attributed to an increase in plankton. The change in wavelengths may cause an increase in phycoerythrin (Moon and Dawes, 1976). Both studies (Haxo and Blinks, 1950; Moon and Dawes, 1976) suggest that red algae can chromatically adjust to the deep benthic environment by changes in pigment concentration. As noted earlier, this has also been shown in the laboratory by Brody and Emerson (1959) using the unicellular red alga *Prophyridium.*

Chromatic adaptation has also been invoked to explain the existence of a number of green algae that occur at depths of 100 m or more in clear tropical waters where the light quality is predominantly greenish blue. Yokohoma *et al.* (1977) found the deep-water species of several green algae such as *Ulva, Cladophora, Valonia, and Codium* contained a specialized xanthophyll, siphonoxanthin, which absorbs around 540 nm. Thus these species were chromatically adapted to the blue-green light regime, and the occurrence of siphonoxanthin among these taxonomically distinct green algae demonstrates its ecological significance rather than any taxonomic relevance (Yokohoma *et al.,* 1977). Another pigment characteristic of coenocytic green algae, siphonein, was also found to absorb in the green wavelengths and was considered to be ecologically important for the deep freshwater species *Dichotomosiphon tuberosus* (Kageyama and Yokohama, 1978).

Light intensity

The other characteristic of light penetration into deeper water is the decrease in intensity (Zvalinsky *et al.,* 1978). The view first outlined by Oltmanns proposed that vertical zonation was affected by the amount of energy available. In this regard, Larkum *et al.* (1967) found that the distribution of marine algae on vertical submarine cliffs in the clear water off Malta had the following zonation. The upper 15 m of subtidal zone was dominated by brown algae, green algae dominated from 15 to 75 m, whereas red algae predominated below 75 m and into a cave. Within the depth range (30–150 m) the major change in light was found to be intensity, not spectral quality. Also, in a laboratory study on a filamentous red alga *Griffithsia pacifica,* Waaland *et al.* (1974) demonstrated that the change in ratio of phycoerythrin to chlorophyll was controlled by light intensity. Zvalinsky *et al.* (1978) compared the absorption spectra and secondary derivatives of chlorophyll *a* obtained from green *(Ulva fenestrata)*, red *(Grateloupia turuturu, Porphyra ochotensis)*, and brown *(Punctaria punctata, Sargassum pallidum)* algae from sunny and dim sites in shallow waters and at depths of 25 m. The algae obtained from low light intensities had mainly long-wave forms of chlorophyll *a* and a predominance of the longwave photochemical system. They cited light intensity as the causative factor. Similarily, Dring (1981) has argued that changes in pigment composition in marine algae, observed with increasing depth, are largely adaptations to low irradiance and not the spectral composition of underwater light.

It now appears, as is often the case when there are opposing theories, that both theories are partially correct: light quality and intensity are both important for benthic algae. Tore Levring (1966) suspended red, green, and brown algae at different depths and measured their photosynthetic rates. The plants had been collected from different depths; each species was studied at all depths. He summarized the relationship between submarine light and vertical distribution as follows: "Algal types are not only adapted to different irradiance intensities, but also to different spectral composition of the radiant energy." In another study on changes in pigment concentrations in seaweeds, Ramus *et al.* (1976) came to the same conclusion, namely, "Seaweeds modify their photon-gathering photosynthetic antennae (e.g., pigment systems) to ambient light fields in the water column by both intensity adaptation and complimentary chromatic adaptation." Dring's (1981) more recent study suggests that chromatic adaptation may be in response to light intensity rather than light quality and that this "illustrates the mean tricks that Nature sometimes plays on us, such as dressing up one form of adaptation to look like another".

Responses to Single Environment Factors

Any number of features of the marine plant's environment can influence photosynthetic and respiratory rates. Four of the most obvious physical factors are considered here, namely light intensity, temperature, salinity, and desiccation. By determining the optimal photosynthetic and lowest respiratory rates of marine plants under varying physical conditions one can begin to characterize the "optimal" environment for different plants.

Light

Algae show differing photosynthetic responses to light intensity on a population basis (Dawes *et al.*, 1976), seasonal (Mathieson and Norall, 1975), and morphological basis (King and Schramm, 1976a; Arnold and Murray 1980). Raven and Smith (1977) believe that certain algae are adapted to low light (shade algae) while others require high irradiance (sun algae). In a later paper Raven *et al.* (1979) have decided that giant-celled algae such as *Chaetomorpha darwinii* are shade plants and small-celled algae such as *Chaetomorpha linum* are sun plants and "canopy-dominants." Just how light adaptation occurs in plants is open to question. A likely physiological mechanism would be in the control of production of photosynthetic pigments or in the case of cell size, the rate of cytoplasmic streaming. Since photolysis of pigments occurs under increasing light intensitities, a rapid rate of pigment synthesis may result in higher rates of photosynthesis under higher levels of light.

Although not directly related to light, photorespiration may also play a significant role in the control of photosynthetic rates. Photorespiration in plants results in the breakdown of glycolic acid produced in the Calvin cycle of photosynthesis, utilizing oxygen in the process. The breakdown results in a loss of fixed carbon. An example of the significance of photorespiration can be seen in a graph (Figure 14-1) showing the rate of oxygen released by the red alga *Hypnea musciformis* with and without a substrate inhibitor. The compound, potassium glycidate, is an analogue of glycolic acid, the first compound of the photorespiration pathway, and acts as a competitive inhibitor of glycolic acid oxidase. The graph shows an increased photosynthetic output of oxygen with the inhibitor, suggesting that a

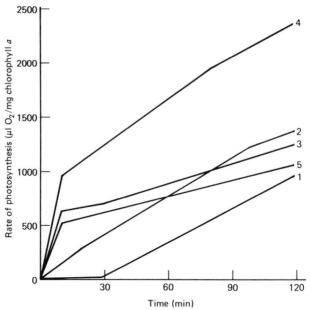

Figure 14-1 Photorespiration can be demonstrated in the red alga *Hypnea musciformis*. The five curves represent photosynthetic responses of the red alga after exposure to various concentrations of an inhibitor of photorespiration, potassium glycidate. The graph shows response (μl oxygen released/mg chlorophyll *a*) over time. (*1*) No inhibitor (control); (*2*) 2.85 mM (millimolar) potassium glycidate; (*3*) 14.28 mM; (*4*) 28.56 mM; (*5*) 57.12 mM. Note that most concentrations of the inhibitor show some increase in photosynthetic rates as measured by oxygen release due to reduction of photorespiration, but the optimum level of inhibitor was 28.56 mM (*4*). The highest concentration of the inhibitor, 57.12 mM (*5*), apparently damaged the plants since photosynthetic response dropped below the rates at lower concentrations.

photorespiratory pathway exists in *Hypnea*. The evidence also suggests that the direct result of photorespiration in *Hypnea* is a reduction of its productivity by as much as 50%. As light intensity is increased, photorespiratory rates can also increase, and this results in an algal loss of fixed carbon.

Populations of the same species collected at distinctive sites can show quite distinctive adaptations to light intensity. The fleshy red alga *Hypnea musciformis* is common in a variety of shallow-water sites along the west coast of Florida (Dawes *et al.*, 1976) on both exposed rocky communities (Point of Rocks) and protected, low-light mangrove communities (Cockroach Bay) within 54 km of each other.

Figure 14-2 shows that the low-light populations from the mangrove swamp have a much higher photosynthetic response to light increase than plants from the exposed coastal community growing under high submarine illuminations. When levels of chlorophyll *a* were compared, the opposite was true; the exposed community had twice the level of chlorophyll *a* that the mangrove community plants had. Both populations of *Hypnea* showed a leveling off of photosynthetic re-

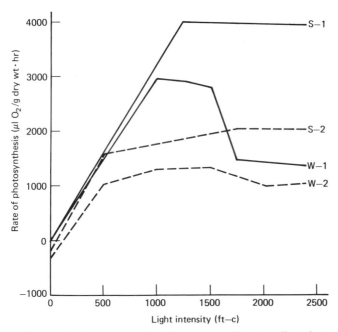

Figure 14-2 Photosynthetic responses for *Hypnea musciformis* collected at a mangrove swamp (*1*, solid line) and an open coastal site (*2*, dashed line) in the summer (*S*) and winter (*W*), with relation to light intensity (foot candles). The plants collected from the mangrove swamp (*1*) show higher responses over their coastal counterparts for each season. Both populations showed saturation at higher light levels. There is little shift of the saturation point over the seasons, but the point remains distinct for the two populations (Dawes *et al.*, 1976).

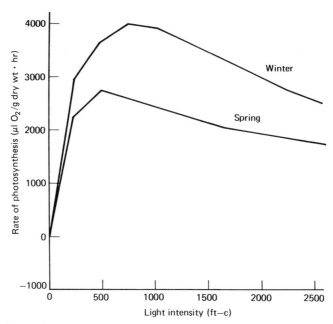

Figure 14-3 A winter and spring photosynthetic response to light levels by the subtidal alga *Phycodrys rubens* from New England. There is a shift in highest photosynthetic re- sponses from the winter (750–1000 ft-c) to the spring (500 ft-c). This shift can be correlat- ed with increased plankton development during the spring and resulting lower water clarity (Mathieson and Norall, 1975).

sponses at relatively low light intensities. The studies indicated that the plants collected at the mangrove swamp had the highest rates of photosynthesis, al- though they had lower levels of chlorophyll *a* and that both populations appeared to be adapted to relatively low light levels in spite of their subtropical location.

In a seasonal study of deep-water algae in New England, Mathieson and Norall (1975) showed a distinct seasonal shift in photosynthetic rates from winter to spring (Figure 14-3). The spring plants of the red alga *Phycodrys* showed a reduction (inhibition) of photosynthesis at a lower light intensity (above 200 ft-c) than winter plants (above 750 ft-c). Seasonal effects may be attributed to the health of the plants, and *Hypnea musciformis* showed lower responses in winter than in summer (Figure 14-2), as shown by Dawes *et al.* (1976).

The plant segment sampled and/or its morphology will also influence response to light intensity. King and Schramm (1976a) showed that the kelp *Laminaria* and the rockweed, *Fucus,* had their lowest rates of photosynthesis at the base of the plant and their highest rates in the blades. In turn, Littler and Murray (1974) and Arnold and Murray (1980) found that the highest rates of photosynthesis occurred in sheetlike plants (e.g., *Ulva*) and the lowest rates in crustose or complex forms. They also found that the sheetlike plants were usually opportunistic annual spe- cies, whereas the crustose or more-fleshy plants were usually perennials.

Temperature

Photosynthetic responses to temperature are often quite useful in suggesting seasonal growth patterns (Chock and Mathieson, 1978) or distinctions between populations (Dawes *et al.*, 1978). In addition, the resistance of algae to extremes in temperature has been studied (Schwartz and Almodovar, 1971; Biebl, 1970), as have growth rates under varying temperatures (Prince and Kingsbury, 1973; Hanisak, 1979; Yarish *et al.*, 1979a). The physiological mechanisms responsive to changes in temperature are usually considered to be enzymatically controlled. Thus, some marine plants contain enzymes that cannot function or will be damaged at extreme warm or cold temperatures. Therefore these plants are restricted to specific tmperature ranges. Seasonal changes in respiratory or photosynthetic responses within the optimal temperature range may be due to changes in the types of enzymes (alleloenzymes, isozymes) produced in a plant during the year. Since most enzymes occur in a number of forms (isozymes) and these can differ among populations or species, it is hypothesized that the isozymes may respond in different ways to various temperatures or other physical factors (see Chapter 13). One example of isozyme-temperature relationships has been shown for the

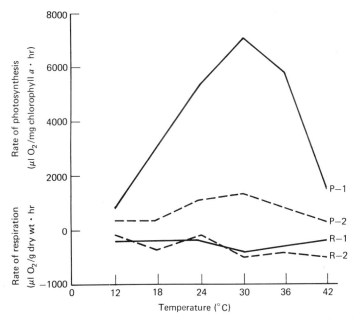

Figure 14-4 Photosynthetic (*P*) and respiratory (*R*) rates for the red alga *Gracilaria verrucosa* responding to various temperatures. Plants from a salt marsh (*2*, dashed line) and a mangrove swamp (*1*, solid line) are compared. Although both populations of *Gracilaria* showed maximum photosynthesis at 30°C, it is apparent that the mangrove plants had much higher rates, even when expressed in terms of chlorophyll *a*. Respiratory rates increased (dropped lower on graph) with temperature with little difference between the two populations (Dawes *et al.*, 1978).

thermal sensitive isozymes of malate dehydrogenase in cattail (*Typa,* Liu *et al.,* 1978). This is an area in which significant research is yet to be carried out in algae.

A comparison of different populations of the same seaweed from two tidal marshes on the west coast of Florida showed distinct photosynthetic responses to changes in temperature (Dawes *et al.,* 1978). Even when the photosynthetic rates were expressed as microliters of oxygen per ml of chlorophyll *a,* it was evident that the algae collected in mangrove swamps had much higher response levels (Dawes *et al.,* 1978). Similar optimal temperatures were found for populations of the same species *(Gracilaria verrucosa)* when photosynthetic rates were compared (Figure 14-4). It was concluded that the two populations of these species were at least acclimatized (if not adapted) to distinct environmental regimes.

On a seasonal basis, one of the New England salt marsh ecads, *scorpioides* of the brown algal species *Ascophyllum nodosum,* showed distinct temperature responses in midsummer and midwinter (Chock and Mathieson, 1978). When the marsh ecad was compared to the typical species, no difference could be found; both forms showing the same seasonal responses to temperature (Figure 14-5).

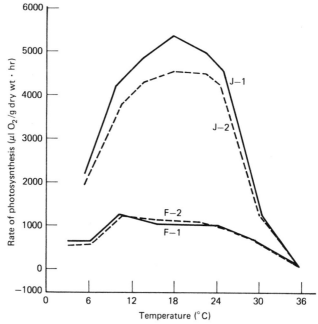

Figure 14-5   A comparison of photosynthetic rates, showing the temperature responses of the brown alga *Ascophyllum nodosum* (*1*, solid line) and its salt marsh ecad, *scorpioides* (*2*, dashed line) in midsummer (*J* = July) and midwinter (*F* = February) in the Great Bay Estuary of New Hampshire. There is an identical photosynthetic response of the species and its ecad for each season, but a quite distinct response for the two seasons. It appears that the winter material was damaged and could not respond as well as the rapidly growing summer plants (Chock and Mathieson, 1978).

Some of the data gathered by monitoring photosynthetic rates at various temperatures must be questioned since higher responses may only be short term (Dawes, 1979). For example, when plants of the red alga *Eucheuma denticulatum* were monitored at 30 and 35°C, the highest photosynthetic rates were obtained. Yet when the same plant samples were returned to 25°C, the photosynthetic responses were lower than those of plants held at 25°C (Figure 14-6). Such results suggest enzymatic damage at the higher temperatures, which was not initially evident during the first few hours.

Algal resistance to extremes of high temperature is also shown by studies of reef algae (Schwartz and Almodovar, 1971). Temperature tolerance of algae held in test pans in the sun or shade depended on the plants' original location. The reef-flat algae were the most resistant to high temperatures and the seaward algae the least. In a similar fashion, Biebl (1970) used algal death (changes in color, turbidity, plasmolysis, coagulation of cellular content) to measure temperature re-

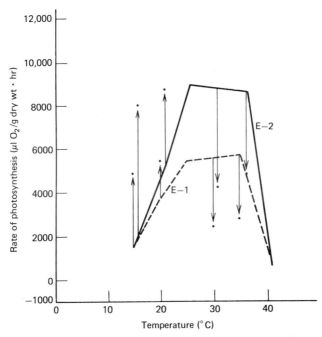

Figure 14-6 Photosynthetic temperature responses for two species of the tropical red alga *Eucheuma denticulatum* (*E-2*, solid line) and *E. spinosum* (*E-1*, dashed line). Although grown at the same pilot farm in Hawaii by Maxwell Doty, the two species still showed different responses to temperatures. Furthermore, major drops in photosynthetic rates occurred when plants held at 30 and 35°C were returned to 25°C, (arrows directed downward) and major increases in rates occurred when plants held at 15°C and 20°C were returned to 25°C (arrows directed upward). No change in photosynthetic rate occurred for plants held continuously at 25°C. This suggests that the effect of higher temperatures, although raising photosynthetic rates, also caused damage to the plant's enzyme system (Dawes, 1979).

sistance. He found that low and high temperature hardiness decreased with plants from the intertidal to subtidal regions, and that heat hardiness increased from north to south. Plants were subjected to identical periods of submersion in seawaters of various temperatures, subsequently they were examined for signs of death. Such information has been used to explain geographical distributions of marine algae as described in Chapter 15.

Probably the most effective, but relatively slow, method of determining thermal tolerances and optima is to measure plant growth and reproduction under different temperatures. In a culture study of the red alga *Chondrus crispus* Prince and Kingsbury (1973) found a sharp break in sporling growth between 21 and 24°C. Higher temperatures caused morphological abnormalities and all growth ceased at 26°C. Spores died at 35–40°C, but it was evident the critical "ecological" temperature was between 21 and 24°C. Simpson and Shacklock (1979) reported that the amount of carrageenan produced increased with higher temperatures, whereas the amount of nitrogen accumulated increased with lower temperatures during growth of *Chondrus crispus*. Hanisak (1979) further found that the green alga *Codium fragile* (Figure 13-7), was tolerant to a broad range of temperatures. He also noted that both irradiance and temperature interacted to control the seasonal growth pattern of *Codium fragile* ssp. *tometosoides*. Hanisak suggested that the plant has become a nuisance in Long Island Sound because of its tolerance to a broad range of light and temperature. Growth of the articulated coralline *Corallina* was found to be optimal at 12°C (2.9 mm/6 wk) and death would occur only at much higher temperatures (25°C; Colthart and Johansen, 1973), although growth had ceased by 20°C. This last study argues against determining distribution of a species by temperatures at which death occurs.

Salinity

Salinity can be an important factor in many cases of local distribution of marine algae. For example, Munda (1978) described a salinity-dependent distribution of seaweeds in Icelandic fjords, starting from the mouth and progressing into fresh water. Structural and physiological tolerance mechanisms that allow for changes in salinity (s occur in an estuary) are variable. Wiencke and Lauchli (1980) found an increase in number and size of cell vacuoles in the red alga *Porphyra* in both hyper- and hypotonic seawater. Amino acid content has been shown to be important in controlling osmotic balances in salt marsh angiosperms (de la Cruz and Poe, 1975), in the red alga *Porphyridium* (Gilles and Pequeux, 1977) and in Black Sea Algae (Perlyuk *et al.*, 1973). Osmoregulation can apparently occur in a number of ways, utilizing amino acids or various carbohydrates. The amino acids most commonly associated with osmoregulation seem to be the ones associated with glycolysis and Krebs cycle enzymes. When an alga is subjected to lower salinities, free amino acids are removed from the cell sap, resulting in lower cell osmolarity. More amino acids are released as the surrounding salinity rises.

Use of various sugars in osmoregulation occurs in unicellular algae such as the diatom *Cylindrotheca* (Paul, 1979) and the flagellated green alga *Dunaliella*

(Wegmann, 1971, 1979). The former utilizes mannose while the latter produces glycerol in response to higher salinities. A number of macrophytic red algae *(Porphyra, Iridaea)* may use an isofluoridoside (αD-galactopyranosyl-glycerol) as an osmoregulator (Kaus, 1968). In addition, various other organic solutes can contribute to osmotic balance, including sorbitol and proline in *Klebsormidium marinum*, sucrose and glutamic acid in *Ulothrix fimbriata* and mannitol in two eustigmatophytes (Brown and Hellebust, 1980). Thus organic solutes were found to be functional in both vacuolated and nonvacuolated algae.

In a study of 17 species of red, green, and brown algae, Kirst and Bisson (1979) found that maintenance of turgor pressure over a wide range of osmotic concentrations (470–1860 mOsm/kg) was achieved by changing concentrations of Na^+, K^+, and Cl^-. They divided the 17 species into three groups: those having high Na^+ concentrations, those with high K^+ concentrations, and those in which both cations occurred in equal amounts. They proposed that a Cl^- pump was involved in turgor regulation; this was further demonstrated in another detailed study of the red alga *Griffithsia monilis* (Bisson and Kirst, 1979). They found that the concentrations of low-molecular-weight organic compounds, mostly photosynthates, also changed with osmolarity. Kirst (1979) in his study of osmoregulation in the green phytoflagellate *Platymonas subcordiformis* concluded

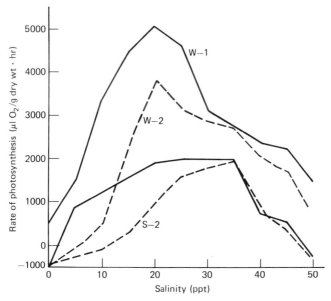

Figure 14-7 The effects of salinity on photosynthetic rates is *Hypnea musciformis* collected from an estuarine mangrove swamp (*1*, solid line) and a stable high-salinity coastal site (*2*, dashed line) in the summer (*S*) and winter (*W*). The estuarine plants were more active than the open-coastal plants on a dry weight basis. Both populations showed broad tolerances to salinity, but the open-coastal plants did have narrower tolerances (Dawes *et al.*, 1976).

that transient increases in ion levels following osmotic shocks were only used to "bridge the concentration gap." This temporary ion balance covered the time period during which sufficient mannitol was synthesized to allow osmotic adaptation.

Photosynthetic rates are influenced by salinity, as shown by comparisons of populations from different salinity regimes and within a population over a series of seasons (Dawes *et al.*, 1976). The red alga *Hypnea musciformis* showed a seasonal response to salinity with a lower, broader (5–35 ppt) photosynthetic response in the summer and a sharper, stronger peak (20 ppt) in the winter (Figure 14-7). A comparison of the red alga *Gracilaria verrucosa* from a salt marsh (salinities 0–15 ppt) and a mangrove swamp (salinities 15–30 ppt) showed distinct differences in salinity tolerances (Dawes *et al.*, 1978). *Gracilaria* from the salt marsh showed low photosynthetic responses at all salinities tested (highest rate at 30 ppt) (Figure 14-8), even tolerating 3 da in distilled water. The same species from the mangrove swamp had a higher response from 10 to 40 ppt but showed respiration and died if held in distilled water. Thus the two populations of the same species showed distinctive adaptations or acclimation to salinity regimes.

Whether an alga is acclimatized or is genetically adapted to a salinity range can only be determined by growing the plant through at least one generation

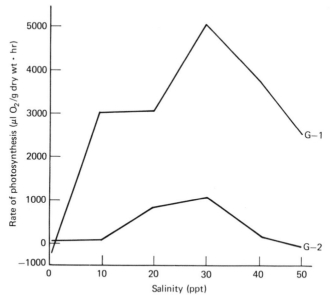

Figure 14-8 Photosynthetic responses of the red alga *Gracilaria verrucosa* to various salinities. The responses are distinct for plants collected in a mangrove swamp (*G-1*) and a salt marsh (*G-2*). The populations from the mangrove swamp showed significantly higher rates of photosynthesis and peaked at 30 ppt. Although both populations showed very broad tolerances to salinities, the salt marsh population actually carried on photosynthesis after being held 3 da in distilled water (Dawes *et al.*, 1978).

under controlled conditions (Yarish *et al.*, 1979a; Lawrence, 1981; Bolton, 1979). Russell and Bolton (1975) grew three isolates of *Ectocarpus siliculosus* for 4 years in 34 ppt salinity and then exposed these isolates to differing salinities. In spite of the 4 yr acclimatization, all three isolates showed a response pattern to reduced salinities that could be correlated with the salinity regimes of their original habitats. Bolton (1979) grew the brown alga *Pilayella littoralis* in high salinities, and demonstrated a genetic basis for salinity tolerances of the isolates based upon the original habitats. Yarish *et al.* 1979a) found that in spite of long-term culturing, the resulting tetraspore germlings of *Bostrychia radicans* and *Caloglossa leprieurii* still grew best in salinities similar to those of the original habitat. However, photosynthetic responses demonstrated the euryhaline nature of the New Jersey estuarine red algae (Yarish *et al.*, 1979b). Similar findings have been reported for brown algae from the west coast of Sweden (Nygren, 1975) and for populations of *Fucus* from Scotland (Khfaji and Norton, 1979).

Desiccation

In the past, intertidal algae were thought to "tolerate" exposure periods, and it was believed that this period was one of "stress." Several studies, however, indicate that many intertidal species are more photosynthetically active under such conditions or may require exposure for normal growth. Schonbeck and Norton (1979) found that *Fucus serratus* grew fastest when submerged for 11 out of every 12 hr and that nutrient depletion was a major problem during long periods of exposure. Maximum gross and net photosynthesis was found to be highest during periods of exposure for *Fucus distichus* (Quadir *et al.*, 1979), *Bostrychia binderi* (Dawes *et al.*, 1978), and species of *Iridaea, Porphyra* and *Endocladia* (Johnson *et al.*, 1974). Desiccation resistance seems to be determined primarily by the ratio of evaporating surface area to volume of the organs and not by mucilage content or wall thickness (Dromgoole, 1980). Thus it seems that tidal exposure itself cannot be considered a detrimental stress condition and that rates of carbon fixation should be determined both in the air and in the water if primary production of intertidal species is desired.

The Synergistic Effect of Physical Factors

Seaweeds are not subjected to a single factorial change but to an entire set of environmental parameters. For example, during the changing tides, in an estuary, subtidal and intertidal species may be sequentially exposed to the higher-salinity, cooler, and clearer waters of the incoming tide and then to lower-salinity, warmer, and more-turbid waters of the outgoing estuarine tide. Thus the combination of physical factors such as light, temperature, and salinity can have a synergistic effect.

In a study of photosynthetic and respiratory responses to 48 combinations of light (three levels), temperature (four levels) and salinity (four levels) regimes, distinctive synergistic responses were noted for the red alga *Hypnea musciformis*

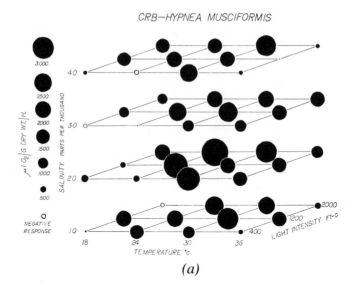

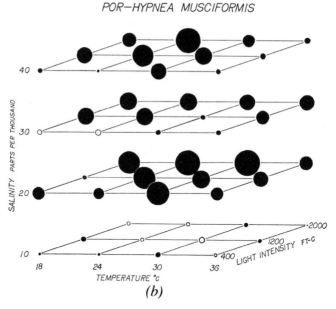

Figure 14-9 The mean net photosynthetic rates for the red alga *Hypnea musciformis*. (*a*) Rates from a mangrove swamp (CRB). and (*b*) rates from an open coastal site (POR). The plants, collected in the summer, were exposed to 48 combinations of light, temperature, and salinity. The size of the spheres indicates level of photosynthetic response (*a*). The estuarine plants (*a*) exhibited broad tolerances to various light, temperature, and salinity combinations with the highest photosynthetic responses occurring at the lower salinities, intermediate temperatures (24–30°C), and high light levels (1200–2000 ft-c). On the other hand, the coastal population (*b*) showed optimal responses within a narrow range of physical factors (2000 ft-c, 24–20°C, 20 ppt) (Dawes *et al.*, 1976).

(Dawes *et al.*, 1976). Two populations were compared, one from a mangrove swamp (Cockroach Bay, or CRB,) and the other from an open coastal site (Point of Rocks, or POR,) from the west coast of Florida (Figure 14-9). When the results of these two figures are compared it is evident that open coastal population (POR-*Hypnea*) cannot tolerate low salinities (10 ppt), and shows low photosynthetic rates under reduced light intensities (400 and 1200 ft-c). The highest responses occurred with 20 ppt, and the highest light intensities (Figure 14-9*b*). In contrast, the plants collected from the mangrove swamp (CRB-*Hypnea*) showed high photosynthetic rates at all salinities under low-to-medium light intensities (Figure 14-9*a*). The populations did not show distinctive temperature responses; this may be because the two sites were near each other (54 km) and had similar ambient water temperatures.

A more effective, but slower, method to measure the synergistic effects of light, temperature, and salinity is through the use of a crossed gradient (Figure 14-10 *a*; Edwards and Van Baalen, 1970; Waaland, 1973; Yarish *et al.*, 1979c). In a study to determine the optimal growth conditions for the carrageenophyte *Gigartina exasperata* Waaland (1973) was able to determine the optimum combination of light (16×10^3 ergs/cm^2) and temperature (18–20°C), as shown in Figure 14-10*b*.

The terms commonly used to describe an organism's ability to live in an environment are adaptation and acclimation. The former considers genetic limits (tolerances) while the latter describes plasticity (ability to adjust). Future physiological studies should consider enzymatic (isozymic) features of differing populations to determine their importance in acclimative ability and adaptation tolerances.

MORPHOLOGY AND PHYSIOLOGICAL RESPONSES

Littler and Murray (1974), Littler (1980), and Arnold and Murray (1980) showed that encrusting and complex morphologies were the lowest producing types (e.g., gC/m^2), whereas sheetlike algae were the highest producers in a rocky intertidal community. King and Schramm (1976b) also observed the highest productivity in intertidal sheetlike or filamentous algae that were annuals and a lower rate of productivity in coarser, perennial, subtidal algae. A model has now been proposed describing early and late successional macroalgae in which simpler forms are contrasted with more complex morphologies (Littler and Littler, 1980). The model argues that survival strategies change with regard to adaptations (genetic) of the macroalga to grazing, reproduction, morphology, and life histories.

CHEMICAL LEVELS AND ALGAL STRATEGIES

Chemical analyses have been carried out on marine algae, especially those having economic importance. Some studies have concentrated on single constituents (phenolic production, carrageenan, alginate) whereas others have given general analyses of major constituents (carbohydrate, lipid, protein, ash). Apart from eco-

nomic reasons, an evaluation of the seasonal variation of the chemical components of an alga should yield insight into the physiological processes occurring in
a plant and identify distinctions in the growth, lag, or reproductive phases. Thus,
chemical studies can aid the investigator in determining the growth and reproductive strategy of a plant.

Photosynthetic Pigments

The basic structures and distribution of the major photosynthetic pigments are
summarized in Chapter 3. Methods for extraction and determination of levels can
be found in the final section of this chapter. There are a number of chapters
dealing with algal pigments in the text on algal physiology and biochemistry edited by Stewart (1974).

Chlorophylls

These are the basic pigments in light absorption and photochemistry in plants and
photosynthetic bacteria. Four major algal chlorophylls occur: chlorophylls *a, b, c,*

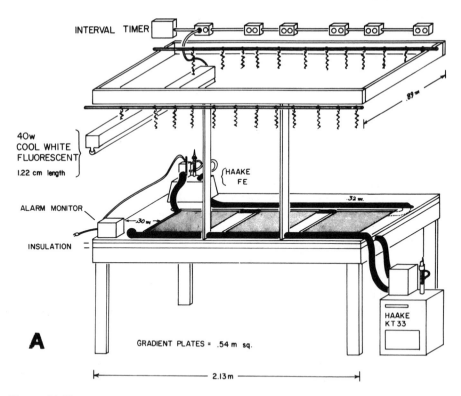

Figure 14-10 (*A*) A diagram of a crossed gradient growth table. The individual light
banks can come on for various periods of time and there is a temperature gradient ranging
from a warm side (left) to a cold side (right) controlled by heating and cooling water
systems. (*B*) An example of spore germination of the red alga *Gigartina exasperata* on a

and *d*. Table 14-1 summarizes some of the physical features of these four chlorophylls and Figure 11*a* illustrates the absorption spectrum of chlorophyll *a* obtained from *Eucheuma isiforme* (Moon and Dawes, 1976).

Chlorophyll levels can be used to calculate the biomass of a phytoplankton community (Chapter 20) or to express rates of photosynthesis in distinct species or populations (μl O$_2$/mg Chl *a* $\cdot$ hr). An expression of photosynthetic activity relative to amounts of chlorophyll *a* is recommended when various species of algae are compared. Morphologies of algae differ so that a large, fleshy alga will have many cells lacking chloroplasts, and a good percentage of the thallus not involved in photosynthesis. Comparison of photosynthetic rates between fleshy and filamentous forms of algae based on grams dry weight would, therefore, yield erroneous results. If photosynthesis is expressed in terms of chlorophyll *a*, a more common basis of comparison is available. Extraction is usually with 80–90% acetone or 90% methanol in the cold.

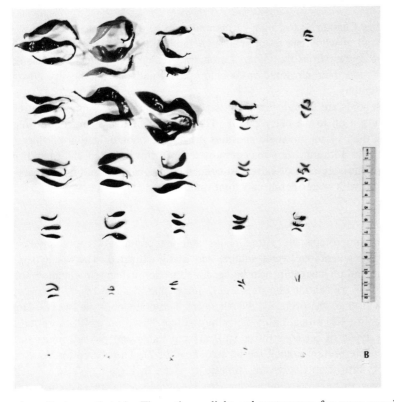

crossed gradient growth table. The optimum light and temperature for spore germination and growth are evident based on plant size. (Figure 14-10*A* courtesy of Charles Yarish, University of Connecticut, Stamford; Figure 14-10*B* courtesy of Robert Waaland, University of Washington, Seattle.)

Table 14-1   Solubility Functions and Absorption Bands of Chlorophylls

Pigment	Distribution	Function	Solubility	Absorption band (nm) Soret region	Absorption band (nm) Redband region
Chlorophyll *a*	All algae	Photosynthetic	Alcohols, diethylbenzene, acetone	430	660, 665
Chlorophyll *b*	Green algae	Light harvesting	Same as Chlorophyll *a*	435	645
Chlorophyll *c*	Brown algae	Accessory to photosystem II	Ether, acetone, methanol, ethyl acetone	c_1 444 c_2 452	583 634 586 635
Chlorophyll *d*	Red algae	Minor accessory	Ether, acetone, alcohol, benzene	400, 456	696

Carotenoids

As noted in Chapter 3, the algal carotenoids can be divided into xanthophylls and carotenes all of which are tetraterpenes. A large number of carotenes and xanthophylls are known from the algae. Carotenoids can be extracted with acetone or methanol, and then separated from chlorophylls and one another by absorption chromatography.

Carotenoids are thought to serve as accessory pigments, trapping light energy and passing it on to the chlorophylls. They may also have a protective function, accepting the oxygen molecule and thus preventing photooxidation of chlorophyll molecules. In addition, they may remove energy from excited chlorophyll molecules. The measurement of cartenoid concentrations can be used to compare algal distribution with water depth and light quality.

Biliproteins

These chromoproteins are tetrapyrroles and are commonly called phycobilins. Biliprotein pigments are water soluble and easily extracted. The two types, phycoerythrin and phycocyanin, can be separated by ion exhange chromatography on a column of tricalcium phosphate gel. Increasing concentrations of phosphate buffers will elute the proteins, the phycoerythrin pigments usually preceding the phycocyanins. Classification of the biliproteins is based on their absorption spectra; three types of phycoerythrins (reflects red light) and phycocyanins (reflects blue light) are present among the algae (Table 14-2). The absorption spectrum of *r*-phycoerythrin from *Eucheuma isiforme* is given in Figure 14-11*b*.

Changes in biliprotein and chlorophyll content can be used to follow algal acclimatization to changes in submarine light levels. Thus in a seasonal study on a tropical red alga, *Eucheuma isiforme,* the concentrations and ratios of pigments were used to show a chromatic acclimatization to the benthic environment (Moon

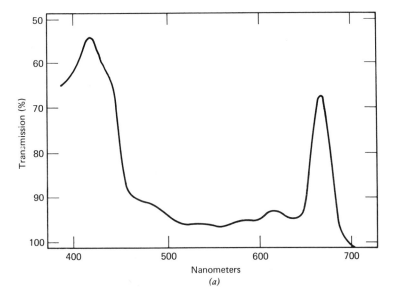

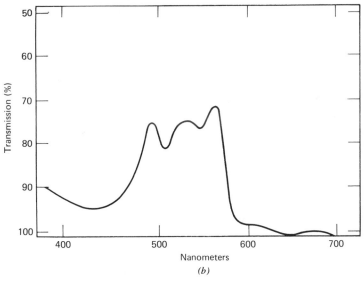

Figure 14-11 Chlorophyll absorption spectra. (a) spectrum of chlorophyll a and (b) r-phycoerythrin, both extracted from the red alga *Eucheuma isiforme*. Phycoerythrin content of the water extract was determined by using the absorption coefficient of 80.2 g/1 and the optical density of 565 nm to calculate mg of phycoerythrin per g dry wt. The pellet from the above extraction was resuspended in 2 ml of 80% spectroanalyzed acetone, centrifuged at 45,000 × gravity for 3 min at 4°C and the supernatant measured at 663 and 430 nm. The absorption coefficient of 84.0 g/1 was used to calculate mg of chlorophyll a per g dry weight (Moon and Dawes, 1976).

Table 14-2   Comparison of the Biliproteins: Phycoerythrin and Phycocyanin

	Group of Algae	Maximum absorption (nm)
Phycoerythrin		
r	Florideae, Rhodophyta	550
b	Bangioideae, Rhodophyta	575
c	Cyanophyta	550
Phycocyanin		
r	Rhodophyta	553
c	Cyanophyta	600
Allocyanin		650

and Dawes, 1976). The ratio of phycoerythrin to chlorophyll *a* changed from 20:1 in the spring to 1:1 in the summer. The change in ratio correlated with a decrease in water clarity and the loss of red wavelengths in the shallow waters due to an increase in plankton abundance in the summer.

Chemical Constituents of Algae

There are few determinations of major chemical constituents of marine algae; most have been carried out on economically important seaweeds. Knowledge of seasonal changes in lipid, carbohydrate, protein, and ash levels can be of importance in determining the plant's physiological state. In a study of species of the red alga *Eucheuma* in Florida, Dawes *et al.* (1974) demonstrated that the ratio of protein to carbohydrate could be used to determine if the plants were in a rapid growth phase (1:10), a slow growth phase with high photosynthetic activity (1:16), or a reproductive phase with no growth (1:25). It was also found that deep-water plants had higher levels of protein (6-8% based on dry weight) than shallow-water plants (2-5% based on dry weight). In a study of the red alga *Pterocladia,* a seasonal cycle in agar accumulation was found (Abdel-Fattah *et al.,* 1973). Protein was high throughout the year, reaching 35% in the spring, whereas the lipid content was low (0.6–1.3%).

Phycocolloid Extracts

Because of the economic importance of phycocolloids (Chapter 2), there is more information about them than about other cellular constituents. In a seasonal study of the giant kelps *Macrocystis* and *Nereocystis* from Puget Sound, Wort (1955) found a general rise in phycocolloids in the summer, whereas dry weight and ash were maximal in the fall. He also found a reciprocal relationship between lipid and algin content.

Seasonal variations have been reported in agar yields and gel strengths in the red alga *Gracilaria* (Oza, 1978; Hoyle, 1978). Both studies indicated that algal harvesting was optimal in later summer. Hoyle (1978) found that periods of high growth and high protein levels corresponded with the periods of lowest agar yields.

There are distinct types of carrageenan (κ, λ, μ, ν, and ι), which differ in their levels of sulfation and in the ratios of galactose to 3,6-anhydrogalactose, resulting in distinct physical properties (gel strength, viscosity, binding properties). Carrageenan-yielding red algae belong primarily to the order Gigartinales. Some genera in this order contain both κ-and λ carrageenan (*Iridaea, Chondrus, Gigartina;* see Figure 2–8); κ-carrageenan is restricted to the haploid gamethophyte (cystocarpic and spermatogonial plants) whereas λ-carrageenan is only found in the diploid tetrasporophyte (Dawes *et al.*, 1977a, 1977b). Other genera, regardless of the reproductive phase, only contain κ-(*Hypnea musciformis, Eucheuma cottonii*) or ι-carrageenan (*Eucheuma isiforme, E. uncinatum, E. denticulatum, Agardhiella tenera;* Dawes, 1979).

Seasonal studies have demonstrated that although yields may change, the quality and type of ι-carrageenan were constant in species found in the waters off Florida (Dawes *et al.*, 1977a). In Florida, the optimal time for harvesting *Eucheuma isiforme* was from mid to late summer. There was an inverse relationship between carrageenan yield and plant growth in *Eucheuma*, the highest yields occurring during periods of slow growth. Such studies on phycocolloid ecology are important both from a harvesting viewpoint and a taxonomic basis.

PHYSIOLOGICAL AND CHEMICAL METHODS

One of the goals of this text is to equip the student with simple yet useful methods that can be applied in the field and laboratory. Thus some general methods of describing seaweed growth and reproduction in chemical terms are presented here. The reader is also referred to the text on physiological and biochemical methods edited by Hellebust and Craigie (1978). In the present section, selected procedures found useful by the author are presented.

Dry Weight Determination

Percentage dry weight is one of the simplest, yet most-useful methods for determining the biomass or dominance of a population. Plants are weighed after blotting (wet weight) and then dried. Drying may be carried out in an oven at 60°C or, if chemical analyses are to be carried out, for 2–3 da in an evacuated desiccator jar that contains concentrated sulfuric acid. Various studies have used methods that included rinsing the fresh plants in distilled water to remove the surface salt. However, when data such as ash levels are compared, it is apparent that each rinsing method causes some variation in ash levels. Therefore it is recommended that no rinsing be done except to remove epiphytes and debris, and this should be done in ambient seawater. All dried material should be stored in capped vials in a desiccator, and analysis, if any, should be carried out within a few days to avoid possible breakdown.

Pigment Extraction and Determination

Levels of chlorophylls and phycobilins in macroscopic algae are useful in expressing photosynthetic rates at various depths, seasons, and sites. Procedures for the extraction and estimation of chlorophylls from phytoplankton employ slightly different (volumetric) calculations, and these are presented in Chapter 20.

Percentage dry weight is used to calculate the amounts of pigments in the tissue. Take at least five pieces of tissue that are similar to those from which the pigment will be extracted. Obtain the wet and dry weights (normal drying procedure). Divide the dry weight by the wet weight to obtain the percentage dry weight; this will be used in the final calculation for phycoerythrin and chlorophyll.

Phycoerytherin

Determination of the amounts of this pigment in red or blue-green algae is particularly useful if one is concerned with pigment production and ratios under differing light regimes. Five to 10 pieces of tissue, usually about 0.5 to 1 g wet weight, similar to the material used for determination of percentage dry weight are disrupted in 2 ml of 0.1 M phosphate buffer (pH 6.5) in the cold (ice bath) using a ground glass homogenizer. The homogenate is centrifuged at 45,000 $\times$ for 20 min at 4°C. The supernatant is decanted (pellet saved, see below) into a centrifuge tube and the optical density (D) measured on a spectrophotometer. A visible scan (700 to 380 nm) of this extract should yield three peaks, 565, 540, and 496 nm, characteristic of phycoerythrin.

Phycoerhythrin content can be determined by measuring the optical density at 565 nm and using the absorption coefficient (α) of O'Carra (1965) for *Ceramium* (80.2). Other absorption coefficients for specific red algae are also available (O'Carra, 1965). The formula for determining the amount of phycoerythrin present in the tissue is shown in equation 1:

$$12.4^* \text{ mg/l} \times OD_{565} \times \frac{\text{volume (ml)}}{\text{fresh weight (g)} \times \text{precentage dry weight} \times 1000 \text{ (mg/l)}}$$

$$= \frac{\text{phycoerythrin (mg)}}{\text{dry weight tissue (g)}} \tag{1}$$

Chlorophyll a

Five to ten pieces of tissue, usually about 0.5–1 g wet weight similar to the material used for determination of percentage dry weight, or the pellet used for phycoerythrin analysis are disrupted with 2 ml of 80% spectroanalyzed acetone in the cold (ice bath) using a ground-glass homogenizer. All steps are carried out in the dark and at a low temperature. After careful transfer to a centrifuge tube, the suspension is centrifuged at 45,000 $\times$ g for 30 min at 4°C. The acetone superna-

---

*The number 12.4 is the inverse of the absorption coefficient for phycoerythrin.

tant is scanned between 700 and 380 nm to yield a sharp peak near 663 nm; the pellet is discarded.

Chlorophyll content in the tissue is determined by measuring the optical density at 663 nm (or nearby) and is calculated by using the equation

$$11.9 \, \text{mg/l} \times OD_{663} \times \frac{\text{volume (ml)}}{\text{Fresh weight (g)} \times \text{percentage dry weight} \times 1000 \, (\text{ml/l})}$$

$$= \frac{\text{chlorophyll } a \, (\text{mg})}{\text{dry weight (g)}} \tag{2}$$

The number 11.9 is the inverse of the absorption coefficient for chlorophyll a (84) published by Smith and Benitez (1955). Arnon (1949) used a slightly lower absorption coefficient (82.04) for higher plants. Meeks (1974) presents absorption coefficients for all the chlorophylls. Note that by using the inverse of the absorption coefficient, one can multiply it directly by the OD obtained.

Measurement of Oxygen Production

Photosynthetic and respiratory rates of marine algae can be extremely useful in determining tolerances to environmental parameters and aiding in the understanding of a species niche in a community. Thus three methods are presented in this section, manometric (pressure changes in a closed system), oxygen probe (osmotic movement of oxygen across a membrane), and the Winkler test (titration against iodine that is equivalent to the amount of dissolved oxygen). The chemistry of the latter method was explained in the chapter on chemical factors (Chapter 12). Use of light and dark bottles is the final procedure presented in this section.

Manometric measurement of oxygen production

This measurement requires either a Warburg apparatus or a respirometer (Figure 14–12) and thus is the most expensive procedure. A five-day experiment is presented here with a series of steps to be used in preparing algae and testing photosynthetic and respiratory rates under various light, temperature, and salinity conditions using a respirometer. The detailed steps for calibration of a manometric system as well as the valving procedures and cleaning of equipment should be obtained from the manufacturer's instruction sheets.

DAY 1. PREPARATION OF THE PLANTS

1. Sort and wash the plants in ambient seawater; determine salinity and temperature of ambient seawater at the site of collection.

2. Prepare the algae for phycoerythrin (if desired) and chlorophyll a determinations. It is recommended that all rates of oxygen evolution in photosynthesis be expressed as microliters of oxygen per milligram of chlorophyll a per hour (μl O_2/mg chl $a \cdot$ hr).

3. Prepare marine plants for protein analysis. Rates of respiration can be expressed as microliters of oxygen consumed per milligram of protein per hour (μl O_2/mg protein $\cdot$ hr) or on the basis of dry weight.

Figure 14-12 One tool used in measuring photosynthetic or respiratory rates is the respirometer, a manometric device. The reference (large) flask is used to balance the gases in the 20 smaller reaction flasks (one shown). Each reaction flask has an individual manometer with a digital readout (*d*). All flasks are held inside the water bath where temperature and lights (underneath) are controlled. Other gases, such as oxygen, can be vented into the system.

4. Set up the salinity run by placing about 0.5 g (wet weight) of cleaned algae in each of the prepared salinity flasks (125 ml Erlenmeyer flasks with 100 ml filtered seawater of the salinities to be studied). Place these flasks on a shaker in a growth chamber and allow to gently shake. Water should be changed at least once per day.

5. Prepare a carbon dioxide source (1%) for the runs to be carried out on day 2. The CO_2 releasing solution is that of the Pardee method (Umbriet *et al.*, 1972) and consists of the following chemicals:

Diethanolamine	12	ml
Thiourea	30	mg
Potassium bicarbonate		
($KHCO_3$)	6	g
Hydrocloric acid (6 *N*)	4.4	ml
Distilled water	13.6	ml

Add the ingredients in the order listed; otherwise a precipitate will form. Use a 50 ml Erlenmeyer flask, stir with a magnetic stirrer, and allow to stand stoppered overnight.

DAY 2. TEMPERATURE SERIES

If necessary, prior to doing a temperature series a biomass and longevity run can be carried out (Figure 14–13). This will aid the investigator in selecting the correct amount of plant material for each flask and avoid the effects of shading as well as determining how sensitive the plants are to long periods in a reaction flask. A graded series of plant material by wet weight is placed in a series of reaction flasks (e.g., 0.1, 0.5, 1.0, 1.5 g wet weight), and the plants are allowed to photosynthesize at a standard temperature and light intensity for up to 4 hr with readings every 15 min. The resulting graph (Figure 14–13) can then be used to determine biomass and length of experiment.

Usually five to seven temperatures are used, bracketing the ambient temperature. For example, a series might be in 5°C steps ranging from 5–30°C.

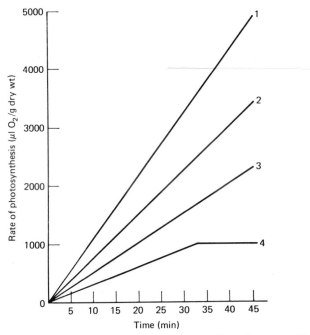

Figure 14-13   A standard procedure in manometric studies is to first run a biomass and longevity study on the plant material. The graph depicts photosynthetic responses of the red alga *Hypnea musciformis* with releation to time and amount of plant material held in each reaction flask of the respirometer. Each reading is an average of five replicate flasks: (*1*) o.1 g fresh weight of material; (*2*) 0.5 g; (*3*) 1.0 g; (*4*) 1.5 g. A distinct drop in photosynthetic responses occurred with increasing amounts of plant material; the first three sets of flasks with plants showed a linear photosynthetic response over time. The drop in response of (*2*), (*3*), and (*4*) is due to shading, and the lack of linearity in (*4*) is because of removal of available carbon dioxide in the flask.

1   Fill the respirometer water bath with distilled or deionized water to the appropriate level. Use of tap water will usually result in some coating of the glass bottom and thus reduce light quality.

2   Open all valves, turn on instrument according to operating procedures, and set desired temperature. Note: in a temperature series it is usually most rapid to start at the lower range by adding ice to speed cooling and then add hot water to increase the temperature to the next higher setting.

3   Arrange the flasks so that in a 20-flask experiment, 18 flasks will contain plants and 2 will be blanks (controls). Close unoccupied channels and adjust the gas volume of the reference flask.

4   To each reaction flask add 10 ml of filtered seawater to the main flask and 0.7 ml CO_2 source solution to the sidearm. The CO_2 source solution will maintain a concentration of 1% CO_2 during photosynthesis, removing CO_2 above 1% during respiration.

5   Weigh out about 1–2 g (wet weight) of healthy filamentous algae or growing tips of fleshy forms of a standard size ($\sim$ 2 cm lengths) and place in the seawater in the flasks.

6   Insure that all manometric units have open valves, that water temperature is correct, and digital readout dials are adjusted to a center value (piston is in median position).

7   Attach reaction flasks to valving system, lower flasks into circulating water bath, and allow to equilibrate for about 30 min. Lights should be on.

8   After equilibration, close the valves. The photosynthetic run should last about 30 min to 1 hr. Actual length of time is dependent on linearity of photosynthetic rate. This can be determined by the biomass run (Figure 14-13).

9   Readings should be taken every 10 min. After final reading for the photosynthesis run, open the valves, turn off the lights on respirometer and room and allow to equilibrate for about 30 min. Readjust manometric dials to center point. Then initiate respiration run for another 30 min to 1 hr. If desired, foil-covered flasks can be used as dark respiration flasks during the light run; however, this procedure is less satisfactory because of the possibility of light leaks and difficulty of equilibrating the temperature in covered flasks.

10   At the termination of the respiration run, open the valves, remove the flasks and adjust the instrument to the new (usually higher) temperature. A more rapid rise in temperature can be accomplished by adding hot water and siphoning off the excess from the water bath. Remove specimens from the reaction flasks and dry them (60°C for 12 hr) in preweighed drying pans; reweigh next day. The dry weight can be used to calculate photosynthetic rates per gram dry weight or per milligram chlorophyll *a*. New specimens should be used for each temperature run.

11 At the end of the last temperature run, clean up the equipment, remove as much grease as possible from the flask opening and stopper. Flasks can be cleaned by soaking in hot soapy water (if petroleum jelly is used) or heating in a concentrated KOH solution (if vacuum grease is used). Vacuum grease should be used for all temperatures above 30°C. All flasks are then washed in hot tap and distilled water, rinsed in a dilute HCl solution, rerinsed and air dried. A new CO_2 source should be prepared for day 3 runs.

DAY 3. LIGHT SERIES

1 The light series can be carried out by modifying the respirometer with a rheostat to control the floodlight intensity or by simple removal of some of the floodlights to produce a lower intensity. Since dimming the floodlights tends to increase red light, it is best to remove some of the bulbs for lower intensities or to add broad spectrum fluorescence bulbs overhead (Figure 14–12).

2 The operating procedure is the same that is used for the temperature series except that a standard, optimum temperature is used (determined from previous day's run). Respiration rates should also be followed after each light series.

3 New specimens should be used for every light intensity. After completion of a series, clean up and prepare a new CO_2 source for day 4. Record dry weights from the temperature series of day 2.

DAY 4. SALINITY SERIES

1 This series uses plants placed in various salinities on day 1; with optimum temperature and light conditions determined on days 2 and 3. Determine photosynthetic rates, followed by respiration rates.

2 After completion of the series, clean up and record dry weights from light series of day 3.

DAY 5. CLEAN UP

1 Drain the respirometer and remove all excess grease on stoppers.

2 Record dry weights from salinity series of day 4 and complete all calculations.

Winkler procedure for oxygen production

The actual chemistry of the titration procedure is described in Chapter 12; only the measurement technhques are presented here. Commerical Winkler-type field and lab kits are available as well. Oxygen levels are expressed as milligram atoms of oxygen per liter but this can be converted to ppm.

CHEMICALS

Manganous sulfate solution:

	$MnSO_4 \cdot 4H_2O$	480 g/l distilled water (or)
	$MnSO_4 \cdot 2H_2O$	400 g/l distilled water (or)
	$MnSO_4 \cdot H_2O$	365 g/l distilled water

Alkaline iodide solution:

	NaOH	500 g/500 ml distilled water
	KI	300 g/450 ml distilled water

Concentrated sulfuric acid: H_2SO_4 (36 N).

Starch indicator solution: Suspend 2 g of soluble starch in 300–400 ml of distilled water. Titrate with 20% NaOH and stir vigorously until the solution clears or is slightly opalescent. Allow to stand for 1–2 hr. Add concentrated HCl until solution is just acid to litmus paper. Then add 2 ml glacial acetic acid, dilute to 1 l with distilled water, and store in rdfrigerator.

Standard 0.01 N thiosulfate solution: Dissolve exactly 2.9 g $Na_2S_2O_3 \cdot 5H_2O$ in 1 l of water. Store in a dark bottle.

Potassium iodate standard solution: (0.01 N): Dry KIO_3 at 100° C for 1 hr, then weigh out exactly 0.3567 g. Dissolve in 200–300 ml distilled water, warm slightly, and bring to the 1000 ml mark (in volumetric flask) with distilled water.

PROCEDURES

Studies should be run in glass-stoppered BOD Bottles (300 ml) that have been acid cleaned and thoroughly rinsed and dried (see procedure 4 below). With this size BOD bottle, use 1 ml each of the three reagents in the order manganous sulfate, alkaline iodide, and sulfuric acid. The steps are as follows:

1   Add 1 ml of manganous sulfate *followed* by 1 ml of alkaline iodide, *well below* the surface. Replace stopper, displacing water at the top.

2   *Shake well* for about 1 min by inverting bottle several times to distribute precipitate (if allowed to settle too soon, dissolved O_2 at top of bottle may not be absorbed). Allow floc to settle to half the volume. Sample can be stored for 1–3 da in a dark, cool place at this stage.

3   Acidify with 1 ml of concentrated sulfuric acid by letting acid run down the neck of bottle. Stopper bottle without displacing floc.

4   Shake well. Since liberated iodine diffuses slowly, see that it is distributed evenly before withdrawing a sample for titration. The bottles may be stored in the dark and cold at this stage for a maximum of 24 hr. Precipitate will dissolve rapidly. The titration can be carried out within an hour after the precipitate dissolves. Any macroscopic algae should be removed immediately following the dissolving of this precipitate.

5 Measure out 50 ml for titration (100 ml if O_2 concentrations is < 0.1 mg atoms oxygen per liter).

6 Titrate with 0.01 N thiosulfate until the iodine fades to a pale straw color. Add 5 ml of starch solution and titrate rapidly to the first disappearance of the blue color. Disregard reappearance of color as this is probably due to traces of iron salts or nitrites.

CALCULATIONS

If a 50 ml sample is used from a 300 ml BOD bottle, the following calculation can be used to determine mg atoms of oxygen per liter:

$$\text{mg}-\text{at } O_2/\text{l} = 0.1006 \times F \times V$$

where F is the standardization factor and V is the amount titrated. If other than a 50 ml sample is used then an X aliquot is taken from a Y ml bottle and the calculation used is:

$$\text{mg}-\text{at } O_2/\text{l} = \frac{Y}{Y-2} \times \frac{5.00}{X} \times F \times V$$

STANDARDIZATION OF THE THIOSULFATE (FACTOR F)

Fill a 300 ml BOD bottle with distilled water, and in reverse order, add 1.0 ml concentrated H_2SO_4 and 1.0 ml alkaline iodide solution, and *mix* thoroughly. No oxygen should be fixed. Add 1.0 ml manganese sulfate solution and mix. Take a 50 ml aliquot—add 5 ml. 0.01 N iodate solution (careful measurement). Allow iodine liberation to procede at room temperature for 2–3 min. Then titrate with 0.01 N *thiosulfate*. If V is the titration in millileters then

$$F = \frac{5.00}{V} \text{ for } 0.01 \, N \text{ thiosulfate}$$

CONVERSION

1 ml O_2 (@ NTP)/l $= 11.20 \times$ mg$-$at $O_2/$l

2 mg $O_2/$l $= 16.00 \times$ mg$-$at $O_2/$l

3 ml $O_2/$l $= 1000 \times \mu$l $O_2/$l

4 mg $O_2/$l $=$ ppm

Oxygen probe

The basic workings of an oxygen meter and probe are presented in Chapter 12 (see also Figures 12-1 and 12-2). The procedures outlined here are for an oxygen-measuring system that requires calibration. Some of the newer models do not.

Calibration of the oxygen meter such as a Beckman Field Lab usually includes the following three steps regardless of the instrument manufacture:

1   Adjust zero position of the needle when the instrument is off.
2   Depress battery-test button and insure battery is operational.
3   Calibrate instrument by one of the following means: air calibration, saturated water, or standardized water using Winkler procedure.

Only the last method is given here, since it is the most accurate. See the manufacturer's procedures for the other ones.

METHOD

Prepare two identical water samples, titrate one sample using this Winkler procedure and then place sensor in the second sample and set temperature control to temperature of water. Always calibrate instrument before use, if possibly by the Winkler procedure, if not then by air saturated water procedure.

Light/dark bottle procedure (Winkler and oxygen probe)

Acid-cleaned 300 ml clear-glass BOD bottles as well as opaque-glass BOD bottles are required. A minimum of three clear and two dark bottles will be needed; however, it is recommended that three bottles (dark and light) contain algae in addition to the two blank light and one blank dark bottles. The dark bottles as well as the stoppers, should be painted black and the bottles should be covered completely with aluminum foil. As with the manometric procedures, the BOD bottles can be exposed to a variety of temperatures (water bath), light intensities (by shading with neutral-density filters) and salinities. One important advantage of BOD bottles is that *in situ* studies can be carried out by placing the bottles in thin wire cages and suspending them at the site. The water should be clear, all epiphytes removed, and the plants cleaned of debris.

PROCEDURE

Fill three light and two dark bottles (or any number of replicates of these) to overflowing (Figure 14-14). Place sufficient macroscopic algae in one clear and one dark bottle to insure response but not crowding (shading effect). Pair a dark and a light bottle without algae (blanks) with the algal-containing dark and light bottles in the holding rack and leave for 30 min−3 hr (time depending on rate of photosynthesis). This rack can be placed *in situ,* suspended at various depths, or run in a growth chamber. Series of these light/dark bottle pairs can be used with different salinities, depths (light intensity), or in temperature-controlled systems.

Test the remaining light bottle for level of oxygen (Winkler titration or oxygen meter) at the beginning of the experiment (background oxygen level). At the end of each run test the remaining dark and light bottles (blank and algal contain-

14-14 A Winkler set-up including light and dark BOD bottles (300 ml) and the titration apparatus. The dissolved oxygen in a BOD bottle is fixed with the addition of $MnSO_4$ and KI in NaOH. Addition of concentrated H_2SO_4 will release free iodine. A 0.01 N solution of sodium thIosulfate is then used to titrate the iodine, and a starch indicator is added to aid in determination of the end point. Note the standard light backup and magnetic mixer.

ing bottles), using either an oxygen meter or the Winkler procedure. If the Winkler procedure is used, remove the plant material immediately after acidification and clearing of the solution. Dry the plant in preweighed pans, and use this dry weight to express photosynthetic and respiratory rates relative to the dry weight.

CALCULATIONS FOR BENTHIC ALGAE

Based on the levels of oxygen found in each bottle, the following determinations can be made. The numbers given are examples of dissolved oxygen (ppm).

1 Gross photosynthesis   *Example:*

   Light bottle + algae: 15 ppm (after photosynthesis)
   Dark bottle + algae: 2 ppm (after respiration)
                               13 ppm (net total photosynthesis)

2 Net photosynthesis

   Light bottle + algae: 15 ppm (after photosynthesis)
   Blank bottle   7 ppm (background oxygen)
                                8 ppm (algae and background)

3. Net respiration

   Blank bottle   7 ppm (background oxygen)
   Dark bottle + algae: 2 ppm (after respiration)
                                5 ppm (net total respiration)

CALCULATIONS FOR BACKGROUND EFFECTS

The above determinations include background photosynthesis and respiration by plankton and bacteria. The calculations below separate out the production by the macroscopic algae from these background rates. The numbers given are examples.

4 Background photosynthesis due to plankton

   Light bottle (blank)   9 ppm (after photosynthesis)
   Blank bottle   7 ppm (background oxygen)
                               2 ppm (background photosynthesis)

5 Background respiration due to plankton & bacteria

   Blank bottle   7 ppm (background oxygen)
   Dark bottle (blank)   6 ppm (after respiration)
                               1 ppm (background respiration)

6 Net photosynthesis of algae alone

                               8 ppm (net total photosynthesis)
                               2 ppm (equation 4)
                               6 ppm (background respiration)

7 Net respiration of algae alone

                               5 ppm (equation 3)
                               1 ppm (equation 5)
                               4 ppm (net algal respiration)

Standard chemical analyses

The procedures presented are standard, simple and rapid, allowing the biologist to sample a large number of individuals in a population. The accuracy of the tests are more than adequate for determining ranges and distinctions between populations or species but they are not absolute, thus a number of samples should be analyzed and standard deviations be determined. It is stressed that the procedures employed will detect trends (seasonal, sexual distinctions) as well as differences between populations.

Protein analysis

Two standard procedures are presented here with references to two more recent modifications of these procedures. Both the Lowry and Layne procedures (Folin-phenol and biuret) are excellent.

Lowry procedure (Folin-phenol reagent)

The procedure was first published in 1951 by Lowry et al. There are two distinct steps which result in a final color with protein in this reaction: (1) reaction with copper in alkali; and (2) reduction of the the phosphomolybdic reagent by the copper-treated protein. The first reaction of copper in alkali and protein results in a dissociation of up to 1 mole of chromogenic protein-bound copper in alkali and protein results in a dissociation of up to 1 mole of chromogenic protein-bound copper for every 7 or 8 amino acid residues. The reduction of the Folin reagent occurs at a pH of about 10 and must be in a highly buffered solution. This procedure measures protein, not individual amino acids. However, the test is most sensitive to proteins containing tryptophan and tyrosine.

PROCEDURE

1 Weight out 10 mg homogenized dried algal material and place in test tube.

2 All 5 ml of 1 N NaOH, cap with marble, allow to stand at room temperature for 24 hr).

3 Pipet 0.5 ml of the extract into a test tube.

4 Add 5 ml alkaline copper tartrate reagent (prepare fresh each day).

2% $CuSO_4 \cdot 5H_2O$	2 ml
4% Sodium tartrate	2 ml
3% Na_2CO_3 in 0.1 N NaOH	96 ml

Note: Mix together only in the listed order or a precipitate will result.

5 Add 0.5 ml 1 N Folin-Ciocalteu phenol reagent (ready-made or see reference) diluted 1:1 with water. Mix and let stand for 20 min.

6 Place in spectrophotometer and read at 660 nm.

7 Calculations are expressed as percentage protein, as determined from optical density using a standard curve.

$$\% \text{ Protein} = \frac{(\text{mg protein}) \ (10^*)}{\text{mg of tissue}} \times 100$$

STANDARD CURVE

1 Weigh out 25 mg bovine serum albumin (reagent grade).

2 Wash into a 50 ml volumetric flask and bring to 50 ml. with 1 N NaOH.

3 Resulting foam can be suppressed by using a weak airstream into mouth of flask to dry out bubbles.

4 Make a dilution series:

Standard Protein Solution (ml)	1 N NaOH (ml)	Resulting Protein(mg)
0.5	0.0	0.25
0.4	0.1	0.20
0.3	0.2	0.15
0.2	0.3	0.10
0.1	0.4	0.05
0.0	0.5	0.00

5 To each tube of the above dilution series (use duplicates as well) add 5 ml alkaline copper tartrate reagent (see above). Mix and allow to stand for 10 min.

6 Add 0.5 ml 1 N Folin-Ciocalteu phenol reagent (see above) that has been diluted 1:1 with water. Mix and let stand for 20 min.

7 Place in a spectrophotometer and read at 660 nm.

8 Plot the optical density against known amount of protein for standard curve.

Hartree (1972) has published a modification that gives a linear photometric response and does away with the two disadvantages of the Lowry procedure (Lowry, *et al.*, 1951): (1) varying color yields with different proteins and (2) the nonlinear relationship between color yield and protein concentration. Use of a more-concentrated alkaline copper tartrate reagent at temperatures above ambient will increase the color yield (removing the effect of side chains that effectively shield the polypeptides).

Layne's biuret method

A different colorimetric method is based on a purple complex that results between substances containing two or more peptide bonds and copper salts in an alkaline solution (Layne, 1957). Since no substances that give the biuret reaction, other

---

*Dilution of original sample.

than protein are normally present in biological materials, this is an excellent test. Although color development with various (specific) proteins is not identical, deviations are encountered less frequently than with the Folin-phenol reaction. However, more material is required for the assay. The main advantage to the assay is the reaction site, peptide bonds, thus making the analysis most effective on all proteins.

PROCEDURE

To obtain biuret reagent; dissolve 1.5 g of cupric sulfate ($CuSO_4 \cdot 5H_2O$) and 6.0 g of sodium potassium tartrate ($NaKC_4H_4O_6 \cdot 4H_2O$) in 500 ml of water. Add with constant stirring 300 ml of 10% NaOH (from carbonate-free solution). Dilute to 1 l with water, and store in plastic-lined bottle. Any signs of black or reddish precipitete indicates contamination.

1 Weigh out 100–200 mg sample and add 5 ml of 1 N NaOH. Let stand for 24–72 hours at room temperature. Stir the solution occasionally by tapping the base of the test tube.

2 Take a 1 ml aliquot of the extract and add 4.0 ml biuret reagent. Mix on Vortex shaker or by tapping bottom of test tube rapidly and allow to stand for 30 min at room temperature.

3 Determine optical density at 540 nm. Use a blank (control tube having 1 ml 1 N NaOH and 4 ml biuret reagent).

4 Determine percent protein by using formula below and a standard curve as described in Lowry procedure but with a standard solution of 100 mg/50 ml.

$$\frac{(mg\ protein)\ (5)}{mg\ tissue} \times 100$$

5 To obtain the standard curve, dissolve 25 mg of bovine serum albumin in 1 N NaOH and make a dilution series of 0.0, 0.2, 0.4, 0.6, 0.8, and 1.0 ml of the standard; add to 4 ml of the biuret reagent. Make up the difference with 1 N NaOH.

Dorsey et al. (1977) have published a protein assay requiring heating the sample with biuret for 100 min at 100°C before addition of the Folin reagent. Different proteins give almost identical optical densities on a nitrogen basis with this procedure. Thus this procedure combines the biuret and Folin reagents to give a protein assay with the specificity of the biuret method and the sensitivity of the Lowry method.

Carbohydrate analysis

The phenol-sulfuric acid method of Dubois et al. (1956) is a simple, colorimetric procedure for dissolved carbohydrates. Simple sugars, oligosaccharides, polysaccharides, and derivatives having free or potentially free reducing groups give an orange-yellow color when treated with phenol and concentrated sulfuric acid. The reaction is sensitive and stable as well as inexpensive. Insoluble carbohydrate (cellulose) can be determined by subtraction after determining protein, lipid, ash, and soluble carbohydrate.

PROCEDURE

1   Weigh out 5 mg homogenized dried algal material and place into a 10 ml centrifuge tube.

2   Fill with 10 ml 5% TCA (trichloroacetic acid) in water (mark level on tube in order to return to original volume after heating).

3   Heat in a hot water bath for 3 hr (80–90°C). Shake tubes gently at least twice to insure adequate extraction. This step should be standardized for all runs.

4   Remove tubes from bath, cool, return to volume (mark) with distilled water. Centrifuge for 10 min at the highest speed on a standard tabletop centrifuge.

5   Pipette 0.2 ml aliquot into a test tube. Place a finger over pipet and insert the tube through the surface before taking sample to avoid scum.

6   Add 1 ml 5% phenol and mix by tapping test tube.

7   Rapidly add 5 ml concentrated H_2SO_4 and mix by tapping. This is an exothermic reaction.

8   Allow the tube to cool (30 min) and then read at 490 nm (blue bulb, no filter).

9   Calculations:

$$\% \text{ Carbohydrate} = \frac{\text{mg carbohydrate}}{\text{mg tissue used}} \times 100$$

Note: mg carbohydrate = (mg carbohydrate in 0.2 ml) (50).* The milligrams carbohydrate in 0.2 ml is obtained from the optical density and standard curve.

STANDARD GLYCOGEN CURVE

1   Weigh out 25 mg glycogen (reagent grade).

2   Wash into 50 ml volumetric flask with 5% TCA.

3   Make a dilution series:

*Dilution of original sample.

Standard Solution (ml)	5% TCA (ml)	resulting glycogen (mg)
0.20	0.00	0.100
0.15	0.05	0.075
0.10	0.10	0.050
0.05	0.15	0.025
0.00	0.20	0.000

4 To each tube (0.2 ml) add 1 ml of 5% phenol.

5 Add 5 ml H_2SO_4, wait 30 min, and read at 490 nm.

6 Plot a standard curve showing milligrams carbohydrate vs. optical density.

Lipid analysis

This gravimetric procedure for total lipid determination in dried algal material is modified from Freeman *et al.* (1957) and Sperry and Brand (1955). The procedure requires no standard curve, since total lipid weight is determined from the dried sample. The method consists of homogenizing the tissue in a 2:1 chloroform-methanol mixture and washing the extract with water. The resulting mixture separates into two phases. The lower phase is total lipid extract. The washing removes essentially all the nonlipid contaminants from the extract.

PROCEDURE

1 Weigh out ~100 mg of homogenized dried algal material and put into a 30 ml screw-cap vial that has been marked at 25 ml level. A large amount of algal material is necessary because of low lipid levels.

2 Fill the vial to the 25 ml mark with a 2:1 (volume: volume) chloroform-methanol mixture.

3 Loosely cap the vial and heat at 60°C for 15 min.

4 Cool and return to volume at mark with chloroform-methanol mixture.

5 Tighten the cap, shake, and filter the solution through a Whatman number 541 filter paper.

6 Collect to the mark in a 30 ml screw-cap vial that has been marked at 20 ml level (two thirds of original sample).

7 Add 4 ml of distilled water to filtered aliquot, cap, and shake for 5 min.

8 Centrifuge the aliquot to separate phases (~ 1500 rpm), then decant the upper aqueous phase.

9 Evaporate the solvent (chloroform-methanol) at 60°C under stream of air. Note: Always filter the air from a line through a cotton-filled vacuum flask to remove any oil from the compressor.

10 Transfer the lipid extract by minimal chloroform-methanol rinses into preweighed shell-glass vials.

11 Evaporate the solvent at 60°C under stream of air (see note in step 9).

12 Cool the glass vial and weigh. Percentage lipid is obtained by multiplying by 3/2 and dividing by the original dry weight, and multiply by 100.

Ash determination

The inorganic fraction can be easily determined by weight difference before and after heating in a muffle furnace at 500°C for 4 hr. Longer periods or higher temperatures should be avoided in ashing so that carbonate is not removed from the sample (Paine, 1971). Insure that the sample to be ashed is dry, as should be samples for all analyses. A minimum of five samples should be run to determine ash and the standard deviation.

Calorie determination

Calories can be calculated using the conversion factors of Brody (1964): protein (g) $\times$ 5.65; starch and soluble carbohydrate (g) $\times$ 4.1, fat (g) $\times$ 9.45. Joules (international) are obtained by multiplying kcal by 4854.

REFERENCES

Abdel-Fattah, A. F., N. H. Abed, and M. Edress. 1973. Seasonal variation in the chemical composition of the agarophyte *Pterocladia capillacea*. *Aust. J. Mar Freshwat. Res.* **24**: 177–181.

Arnold, K. E. and S. N. Murray, 1980. Relationships between irradiance and photosynthesis for marine benthic green algae (Chlorophyta) of differing morphologies. *J. Exp. Mar. Biol. Ecol,* **43**: 183–192.

Arnon, D. I. 1949. Copper enzymes in isolated chloroplasts. Polyphenoloxidase in *Beta vulgaris*. *Plant Physiol.* **24**: 1–15.

Biebl, R. 1970. Verleichende Untersuchungen zur Temperaturresistenze von Meeresalgen entlang der pazifschen Küste Nordamerikas. *Protoplasma* **69**: 61–83.

Bisson, M. A. and G. O. Kirst. 1979. Osmotic adaptation in the marine alga *Griffithsia monilis* (Rhodophyceae): The role of ions and organic compounds. *Aust. J. Plant Physiol.* **6**: 523–538.

Bolton, J. J. 1979. Estuarine adaptation in populations of *Pilayella littoralis* (L.) Kjellm. (Phaeophyta, Ectocarpales). *Estuar. Coast. Mar. Sci.* **9**: 273–280.

Brody, M. and S. S. Brody. 1962. Induced changes in the photosynthetic efficiency of *Porphyridium cruentum*. *Arch. Biochem. Biophys.* **96**: 354–359.

Brody, M. and R. Emerson. 1959. The effect of wave length and intensity of light on the proportion of pigments in *Porphyridium cruentum*. *Amer. J. Bot.* **46**: 433–440.

Brody, S. 1964. *Bioenergetics and growth*. Hafner, New York.

Brown, L. M. and J. A. Hellebust. 1980. The contribution of organic solutes to osmotic balance in some green and eustigmatophyte algae. *J. Phycol.* **16**: 265–270.

Chock, J. S. and A. C. Mathieson. 1978. Physiological ecology of *Ascophyllum nodosum* (L.) Le Jolis and its detached acad *scorpioides* (Hornemann) Hauck (Fucales, Phaeophyta). *Bot. Mar.* **22**: 21–26.

Colthart, B. J. and H. W. Johansen. 1973. Growth rates of *Carollina officinalis* (Rhodophyta) at different temperatures. *Mar. Biol.* **18**: 46–49.

Dawes, C. J. 1979. Physiological and biochemical comparisons of *Eucheuma* spp. (Florideophyceae) yielding *iota*-carrageenan. In Jenson, A. and J. R. Stein (eds.) *Proc. Ninth Intern. Seaweed Symp.* Science Press, Princeton, N.J. pp. 188–207.

Dawes, C. J., J. M. Lawrence, D. P. Cheney, and A. C. Mathieson. 1974. Ecological studies of Floridian *Eucheuma* (Rhodophyta, Gigartinales). III. Seasonal variation of carregeenan, total carbohydrate, protein, and lipid. *Bull. Mar. Sci.* **24**: 286–299.

Dawes, C. J., R. Moon, and J. LaClaire. 1976. Photosynthetic responses of the red alga *Hypnea musciformis* (Wulfen) Lamouroux (Gigartinales). *Bull. Mar. Sci.* **26**: 467–473.

Dawes, C. J., N. F. Stanley, and D. J. Stancioff. 1977a. Seasonal and reproductive aspects of plant chemistry, and *iota*-carrageenan from Floridian *Eucheuma* (Rhodophyta, Gigartinales). *Bot. Mar.* **20**: 137–147.

Dawes, C. J., N. F. Stanley, and R. E. Moon. 1977b. Physiological and biochemical studies on the *iota*-carrageenan producing red alga *Eucheuma unicnatum* Setchell and Gardner from the Gulf of California. *Bot. Mar.* **20**: 437–442.

Dawes, C. J., R. E. Moon, and M. A. Davis. 1978. The photosynthetic and respiratory rates and tolerances of benthic algae from a mangrove and salt marsh estuary: A comparative study. *Estuar. Coast. Mar. Sci.* **6**: 175–185.

de la Cruz, A. A. and W. E. Poe. 1975. Amino acid content of marsh plants. *Estuar. Coast. Mar. Sci.* **3**: 243–246.

Dorsey, T. E., P. W. McDonald, and O. A. Roels. 1977. A heated biuret-folin protein assay which gives equal absorbance with different proteins. *Anal. Biochem.* **78**: 156–164.

Dring, M. J. 1981. Chromatic adaptation of photosynthesis in benthic marine algae: An examination of its ecological significance using a theoretical model *Limnol Oceanogr* **26**: 271–284.

Dromgoole, F. I. 1980. Desiccation resistance of intertidal and subtidal algae. *Bot. Mar.* **23**: 149–159.

Dubois, M., Gilles, K. A., Hamilton, J. K., Rebers, P. A. and F. Smith. 1956. Colorimetric method for determination of sugars and related substances. *Anal. Chem.* **28**: 350–356.

Edwards, P. and C. Van Baalen. 1970. An apparatus for the culture of benthic marine algae under varying regimes of temperature and light intensity. *Bot. Mar.* **13**: 42–43.

Englemann, T. W. 1884. Untersuchungen über die Quantitativen Beziehungen Zwischen Absorption des Lichtes und Assimilation en Pflanzenzellen. *Bot. Ztg.* **42**: 81–93.

Freeman, N. K., F. T. Lindgren, Y. C. Ng, and A. V. Nichols. 1957. Infrared spectra of some lipoproteins and related lipids. *J. Biol. Chem.* **203**: 293–304.

Gilles, R. and A. Pequeux. 1977. Effect of salinity on the free amino acids pool of the red alga *Porphyridium purpureum* (= *P. cruentum*). *Comp. Biochem. Physiol.* **57A**: 183–185.

Hanisak, M. D. 1979. Growth patterns of *Codium fragile* ssp. *tomentosoides* in response to temperature, irradiance, salinity, and nitrogen source. *Mar. Biol.* **50**: 319–332.

Hartree, E. F. 1972. Determination of protein: a modification of the Lowry method that gives a linear photometric response. *Anal. Biochem.* **48**: 422–427.

Haxo, F. T. and L. R. Blinks. 1950. Photosynthetic action spectra of marine algae. *J. Gen. Physiol.* **33**: 389–422.

Hellebust, J. A. and J. S. Craigie (eds.). 1978. *Handbook of Phycological Methods. Physiological and Biochemical Methods.* Cambridge University Press, New York.

Hoyle, M. D. 1978. Agar studies on two *Gracilaria* species (*G. bursapastoris* (Gmelin) Silva and *G. coronopifolia* J. Ag.) from Hawaii. II. Seasonal aspects. *Bot. Mar.* **21**: 347–352.

Johnson, W. S., Q. Gigon, S. L. Gulmon, and H. A. Mooney. 1974. Comparative photosynthetic capacities of intertidal algae under exposed and submerged conditions. *Ecology* **55**: 450–453.

Jones, L. W. and J. Myers. 1965. Pigment variations in *Anacystis nidulans* induced by light of selected wavelengths. *J. Phycol.* **1**: 7–14.

Kageyama, A. and Y. Yokohama. 1978. The function of siphonein in a siphonous green alga *Dichotomosiphon tuberosus. Jap. J. Phycol.* **26**: 151–155.

Kaus, H. 1968. Alpha-Galaktosyl-glyzeride und Osmoregulation in Rotalgen. *Z. Pflanzen Physiol.* **58**: 428–443.

Khfaji, A. K. and T. A. Norton. 1979. The effects of salinity on the distribution of *Fucus ceranoides. Estuar. Coast. Mar. Sci.* **8**: 433–439.

King, R. J. and W. Schramm. 1976a. Determination of photosynthetic rates for the marine algae *Fucus vesiculosus* and *Laminaria digitata. Mar. Biol.* **37**: 209–213.

King, R. J. and W. Schramm. 1976b. Photosynthetic rates of benthic marine algae in relation to light intensity and seasonal variations. *Mar. Biol.* **37**: 215–222.

Kinne, O. (ed.). 1970. *Marine Ecology,* Vol. 1. Part 1.; 1971, Vol. 1, Part 2; 1971, Vol. 1, Part 3; 1976, Vol. 3, Part 1., Wiley, New York.

Kirst, G. O. 1979. Osmotische Adaptation der marinen Planktonalgae *Platymonas subcordiformis* (Hazen). *Ber. Deutsch. Bot. Ges.* **92**: 31–42.

Kirst, G. O. and M. A. Bisson. 1979. Regulation of turgor pressure in marine algae: Ions and low-molecular-weight organic compounds. *Aust. J. Plant Physiol.* **6**: 539–556.

Larkum, A. W. D., E. A. Drew, and R. N. Crossett. 1967. The vertical distribution of attached marine algae in Malta. *J. Ecol.* **55**: 361–371.

Layne, E. 1957. Spectrographic and turbidimetric methods for measuring proteins In Kolowich, S. P. and N. O. Kaplan (eds.). *Methods in Enzymology.* Vol. 3. Academic, New York, pp. 447–454.

Lawrence, J. M. 1981. Multi-level measurement of the time course of acclimation. *Proc. USSR-USA Workshop Physiol. Ecol. 1979.* Inst. Mar. Biol. For East Sci. Ctr., Vladivostock.

Levring, T. 1966. Submarine light and algal shore zonation. In Bainbridge, R., C. G. Evans, and O. Rackham (eds.). *Light as an Ecological Factor,* Symp. of British Ecol. Soc., Wiley, New York, pp. 305–318.

Littler, M. M. and D. S. Littler. 1980. The evolution of thallus form and survival strategies in benthic marine macroalgae: Field and laboratory tests of a functional form model. *Amer. Nat.* **116**: 25–44.

Littler, M. M. 1980. Morphological form and photosynthetic performances of marine macroalgae: Tests of a functional/form hypothesis. *Bot. Mar.* **22**: 161–165.

Littler, M. M. and S. N. Murray. 1974. The primary productivity of marine macrophytes from a rocky intertidal community. *Mar. Biol.* **27**: 131–135.

Liu, E. H., R. R. Sharitz, and M. H. Smith. 1978. Thermal sensitivities of malate dehydrogenase isozymes in *Typha. Amer. J. Bot.* **65**: 214–220.

Lowry, O. H., N. J. Rosebrough, A. L. Farr, and R. J. Randall. 1951. Protein measurement with the folin phenol reagent. *J. Biol. Chem.* **193**: 265–275.

Mathieson, A. C. and T. L. Norall. 1975. Physiological studies of subtidal red algae. *J. Exp. Mar. Biol. Ecol.* **20**: 237–247.

Meeks, J. C. 1974. Chlorophylls. In Stewart, W. D. P. (ed.), *Algal Physiology and Biochemistry.* Botanical Monographs 10. University of California Press, Berkely, pp. 161–175.

Moon, R. E. and C. J. Dawes. 1976. Pigment changes and photosynthetic rates under selected wavelengths in the growing tips of *Eucheuma isiforme* (C. Agardh) J. Agardh var. *denudatum* Cheney during vegetative growth. *Br. Phycol. J.* **11**: 165–174.

Mooney, H. A. 1976. Some contributions of physiological ecology to plant population biology. *System. Bot.* **1**: 269–283.

Munda, I. M. 1978. Salinity dependent distribution of benthic algae in estuarine areas of Icelandic fjords. *Bot. Mar.* **21**: 451–468.

Nygren, S. 1975. Influence of salinity on the growth and distribution of some Phaeophyceae on the Swedish west coast. *Bot. Mar.* **18**: 143–147.

O'Carra, P. 1965. Purification and N-terminal analysis of algal biliproteins. *Biochem. J.* **94**: 171–174.

Oltmanns, F. 1892. Ueber die Kultur-und Lebensbedingungen der Meeresalgen. *Jb. Wiss. Bot.* **23**: 349–440.

Oza, R. M. 1978. Studies on Indian *Gracilaria* IV: Seasonal variation in agar and gel strength of *Gracilaria corticata* J. Ag. occurring on the coast of Veraval. *Bot. Mar.* **21**: 165–167.

Paine, R. J. 1971. The measurement and application of the calorie to ecological problems. *Ann. Rev. Ecol. System.* **2**: 145–164.

Paul, J. S. 1979. Osmoregulation in the marine diatom *Cylindrotheca fusiformis*. *J. Phycol.* **15**: 280–284.

Perlyuk, M. F., V. S. Zlobin, and T. A. Orlova. 1973. Variations in the amino acids distribution in algae inhabiting a region of variable salinity. *Soviet Pl. Physiol.* **23** (Part 1): 40–45.

Prince, J. S. and J. M. Kingsbury. 1973. The ecology of *Chondrus crispus* at Plymouth, Massachusetts. III. Effect of elevated temperature on growth and survival. *Biol. Bull.* **145**: 580–588.

Quadir, A., P. J. Harrison, and R. E. DeWreede. 1979. The effects of emergence and submergence on the photosynthesis and respiration of marine macrophytes. *Phycologia* **18**: 83–88.

Ramus, J., S. I. Beale, D. Mauzerall, and K. L. Howard. 1976. Changes in photosynthetic pigment concentration in seaweeds as a function of water depth. *Mar. Biol.* **37**: 223–229.

Raven, J. A. and F. A. Smith. 1977. "Sun" and "Shade" species of green algae: Relation to cell size and environment. *Photosynthetica* **11**: 48–55.

Raven, J. A., Smith, F. A. and S. M. Glidewell. 1979. Photosynthetic capacities and biological strategies of giant-celled and small-celled macroalgae. *New Phytol.* **83**: 229–309.

Russell, G. and J. J. Bolton. 1975. Euryhaline ecotypes of *Ectocarpus siliculosus* (Dillw.) Lyngb. *Estuar. Coast. Mar. Sci.* **3**: 91–94.

Schonbeck, M. and T. A. Norton. 1979. The effects of brief periodic submergence on intertidal fucoid algae. *Estuar. Coast. Mar. Sci.* **8**: 205–211.

Schwartz, S. L. and L. R. Almodovar. 1971. Heat tolerance of reef algae at La Parguera, Puerto Rico. *Nova Hedwigia* **21**: 231–240.

Simpson, F. J. and P. F. Shacklock. 1979. The cultivation of *Chondrus crispus*. Effect of temperature on growth and carrageenan production. *Bot. Mar.* **22**: 295–298.

Smith, J. H. C. and A. Benitez. 1955. Chlorophylls: analysis in plant material. *In* Paech, K. and M. V. Tracey (eds.), *Modern Methods of Plant Analysis*. Springer-Verlag, Berlin, pp. 142–196.

Sperry, W. N. and F. C. Brand. 1955. The determination of total lipids in the blood serum. *J. Biol. Chem.* **213**: 69–76.

Stewart, W. D. P. (ed.). 1974. *Algal Physiology and Biochemistry*. Botanical Monographs 10. University of California Press., Berkeley.

Umbriet, W. W., R. H. Burris, and J. F. Stauffer. 1972. *Manometric Techniques for the Study of Tissue Metabolism*. Barges, Minneapolis, Minn.

Waaland, J. R. 1973. Experimental studies on the marine algae *Iridaea* and *Gigartina*. *J. Exp. Mar. Biol. Ecol.* **11**: 71–80.

Waaland, J. R., S. D. Waaland, and G. Bates. 1974. Chloroplast structure and pigment composition in the red alga *Griffithsia pacifica:* Regulation by light intensity. *J. Phycol.* **10**: 193–199.

Wegmann, K. 1971. Osmotic regulation of photosynthetic glycerol production in *Dunaliella*. *Biochem. Biophys. Acta* **234**: 317–323.

Wegmann, K. 1979. Biochemische Anpassung von *Dunaliella* on wechselnde Salinität und Temperature. *Ber. Deutsch. Bot. Ges.* **92**: 43–54.

Wienchke, C. and A. Lauchli 1980. Growth, cell volume and fine structure of *Porphyra umbilicalis* in relation to osmotic tolerance. *Planta* **150**: 303–311.

Wort, D. J. 1955. The seasonal variation in chemical composition of *Macrocystis interfrigolia* and *Nereocystis luetkenana* in British Columbia coastal waters. *Can. J. Bot.* **33**: 323–341.

Yarish, C., Edwards, P. and S. Casey. 1979a. Acclimation responses to salinity of three estuarine red algae from New Jersey. *Mar. Biol.* **51**: 289–294.

Yarish, C., P. Edwards, S. Casey. 1979b. A culture study of salinity responses in ecotypes of two estuarine red algae. *J. Phycol.* **15**: 341–346.

Yarish, C., K. W. Lee, and P. Edwards. 1979c. An imporved apparatus for the culture of algae under varying regimes of temperature and light intensity. *Bot. Mar.* **22**: 395–397.

Yokohama, Y., A., Kageyama, T. Ikawa, and S. Shimura. 1977. A carotenoid characteristic of chlorophycean seaweeds living in deep coastal waters. *Bot. Marina* **20**: 433–436.

Zvalinsky, V. I., N. A. Ivanov, and S. I. Chernova. 1978. Native state of the chlorophyll *a* in marine algae in dependance on light conditions of their habitat. In *Ecological Aspects of Seaweed Photosynthesis*, Acad. Sci. USSR Far East. Sci. Cen. Inst. Mar. Biol. Trans. No 11 pp. 88–101.

PART 4

Marine Plant Communities

CHAPTER 15

Lithophytic Communities

Rocky shores support some of the most productive and extensive algal populations in the marine environment. Thus many lithophytic communities are well
documented, especially those in the more accessible intertidal areas. The edge of
the sea, a comparatively narrow zone, is one of the best habitats for plant growth,
especially in temperate latitudes (Mann, 1973). Here annual net production may
reach 1000–2000 g C/m² subtidally and 500–1000g C/m² intertidally (Mann and
Chapman, 1975). For example, primary productivity was found to range from 0.4
to 3.1 g C/m²·da in 18 algal populations around the island of San Clemente off
the southern California coast (Littler and Murray, 1974).

A conspicuous zonation of marine plants and animals is evident in all intertid-

Figure 15-1 An example of intertidal zonation of seaweeds, particularly fucoid brown
algae, from southern California. *Porphyra* sp. (*P*) dominates the upper intertidal zone,
whereas *Laminaria* sp. (*K*) can be found in the lower intertidal zone. (Courtesy of Kimon
Bird, Harbor Branch Foundation, Fort Pierce, Florida.)

al regions, with a variety of environmental factors cited as causal agents (Figure 15-1). In the past, much discussion and effort has centered on comparison of lithophytic algal communities from diverse coasts. There has been considerable debate about the role of marine organisms in defining these zones. In the present chapter, several environmental factors that influence the distribution and abundance of lithophytic algal communities are examined. In addition, community features of exposed and protected sites are compared. Several important factors influencing marine plant distribution on rocky shores were outlined in Chapter 1 (Table 1-2), whereas aspects of succession and community development were considered in Chapter 13.

ENVIRONMENTAL PROPERTIES

The primary factors influencing intertidal and subtidal zonation are variable. By determining the most critical factors at a given site, one can better understand the associated plant communities, whether they are intertidal, subtidal, coastal, estuarine, temperate, or tropical. The first section of this chapter will consider some of these major factors.

Factors Affecting Intertidal Zonation

In a discussion of exposed and protected intertidal communities, Lewis (1964), states that "plants and animals are probably distributed in accordance with their differing abilities to endure exposure to areal conditions." For intertidal communities, the major factor must be tides; all other factors can be considered *modifying* (exposure, substrata, climatic and biotic factors). Because water temperature is also considered to be important in determining geographical distribution, it will be dealt with in a separate section.

Tides

Tides were discussed in Chapter 11. Here we will consider such factors as rhythms, range and size, and frequence of submersion and emersion. All of these combine to produce the "tidal environment" as proposed by Lewis (1964). When the tide is high, the temperature is essentially uniform and factors such as water loss, gas exchange, gas availability, and nutrient depletion appear to be minimal problems for an alga. However, when the tide is low, such factors as air and substratum temperature, light intensity, and desiccation play major roles in the selection of tolerant intertidal species. A gradient exists, with these factors becoming increasingly more critical in the upper tidal zones because of longer exposures.

Tidal amplitudes show temporal variation at any one site, being the greatest during the spring tides (i.e., full- or new-moon periods) and smallest during neaptides. If the spring tides (see Chapter 11) coincide with hot, dry, or windy weather, severe damage to the intertidal organisms will occur. For example, it is common to find bleached intertidal algae such as *Laurencia papillosa* (Rhodophyta), *Dictyosphaeria cavernosa* (Chlorophyta) and *Padina santae-crucis* (Phaeophyta) in May and June in the Florida Keys when unusually low tides occur at midday. Bleaching and death in Pacific coast populations of *Corallina officinalis* var. *chilensis, C. vancouveriensis* and *Laurencia pacifica* (Rhodophyta) are common during the fall when the lowest tides (LLW-lowest low water) occur during the midday or early afternoon.

The tidal frequency or cycle is also critical for intertidal algal survival. Semidaily tides mean that two highs and two lows occur each 24½ hour period, whereas daily tides mean that only one high and one low tide occur each day. Thus the more frequently submerged intertidal zone with a semidaily tidal pattern can support a more diverse algal population. On the other hand, a daily tidal cycle can result in the intertidal zone's being exposed for a longer period of time, which can result in bleaching and desiccation of the organisms. Information on amplitudes and frequencies of tides is available from local tidal data such as those in publications of the U.S. Coast and Geodetic Survey.

Exposure

The exposure of a shoreline results from a combination of factors: wave action (activity and height) and shore topography (steepness and ruggedness). As discussed in Chapter 11, waves are caused by wind, and their size (height) is determined primarily by fetch (uninterrupted distance of travel). A jutting headland will encounter larger waves than the shoreline behind it. Also, if the shore is angled to the prevailing wave pattern, its exposure will be less than if it is parallel to the waves. The flatter and more extensive the shore (gentle slope) and the more irregular the shore profile, the more gentle will be the effect of waves. Steeper topographies result in greater wave action. Beach or shore profiles (described in Chapter 10), as well as the morphology and composition of the dominant seaweeds, can be used to assess the extent of wave action. In areas subjected to high wave action, there is a tendency for the dominant seaweeds to be coarse and structurally complex, as are *Eisenia* and *Postelsia* (Phaeophyta) on the west coast of the United States and *Alaria* (Phaeophyta) and *Rhodymenia-Palmaria,* Rhodophyta) on the east coast. In areas of high wave energy, there is a general elevation of zones (see Chapter 11). Shoreline energy levels, similar to those described by Tanner (1960) for Florida (see Chapter 10), can also be used to determine the importance of wave action. The spray zone of an intertidal community is under the direct influence of wave action, and it can be an excellent zone in which to ascertain the level of wave activity.

Seapy and Littler (1979) carried out a comparative study of a wave-exposed sea stack and a protected boulder beach on the California coast. As reported in other studies, there was an upward shift (i.e., expansion) of comparable vertical zones and an increase in diversity of algal species on the wave-exposed sea stack. One effect of a high-energy shore is the rapid removal of competitive organisms, producing space for other species and often causing higher species diversity.

Substratum

The composition and texture of a rocky substratum will greatly influence its plant communities. For example, a white, porous shell hash, which is common in the tropics, reflects light and has a high water-holding capacity. Thus blue-green algae such as *Entophysalis* and green algae such as *Valonia* and *Dictyosphaeria* can penetrate the substratum and grow in semiprotected niches. In tropical intertidal regions, the algae are predominantly attached under the partially dissolved and fragmented limestone substrata. In cooler, temperate areas where insolation is lower, the color and shape of the substrata are less important, as macroscopic seaweeds like *Fucus* and *Ascophyllum* even grow on smooth, dark-colored granitic rocks. In general, the more mosaic the surface topography, the more diverse are the number of intertidal algae (Seapy and Littler, 1979). Tide pools and irregular boulders offer a variety of vertical and horizontal algal-attachment surfaces that are differentially exposed to the sun and sea.

Climatic factors

Climatic factors such as the amount and intensity of sunlight, amount of rainfall, and ranges of air and water temperatures affect the geographical distribution of algae. The same factors may also contribute to their local distribution, but these effects are more subtle. Usually more subtidal algal species (higher diversity) occur in tropical and subtropical waters than in temperate waters. Conversely the highest numbers of intertidal algal species are found in temperate climates. One need only compare the number of species identified in New England (Taylor 1962) and the Caribbean (Taylor 1960) to see how true this statement is. The numbers of subtropical and tropical species are in part due to the higher light intensity and longer photoperiod year-round, as well as higher water temperatures. The enhanced diversity and biomass of intertidal species in temperate latitudes is probably due to reduced desiccation that is partially caused by cooler air temperatures and lower insolation.

Such critical climatic factors as temperature, light, and desiccation are interrelated and directly affect algal distribution in the intertidal zone (Carefoot, 1977). Several investigators have evaluated the effects of desiccation on the recovery and photosynthetic tolerances of intertidal seaweeds. For example, Kanwisher (1957) demonstrated that *Fucus* could lose 91% of its water during a normal intertidal exposure in a Massachusetts beach. Brinkhuis *et al.* (1976) demonstrated that *Ascophyllum* from Long Island Sound showed enhanced photosynthesis with a water loss up to 25%; further water loss caused a dramatic drop in photosynthetic rates. Dawes *et al.* (1978) found that several intertidal seaweeds characteristic of salt marshes and mangrove swamps *(Caloglossa, Bostrychia, Catenella)* exhibited enhanced photosynthetic and reduced respiratory rates when monitored in the air compared with the rates when submerged. Accordingly, they as well as Johnson *et al.* (1975) concluded that desiccation is not always a detrimental condition for intertidal algae. In addition, Dawes *et al.* suggested that productivity calculations for intertidal species should include information for aerial conditions.

High temperatures will increase desiccation rates as well as damage enzyme systems. Likewise, low to freezing temperatures, which are common in cool temperate intertidal zones, can result in internal damage to seaweeds. As noted by Carefoot (1977), "There is no such thing as consistency in an intertidal habitat."

The levels of light intensity on the shore can be quite high on a clear, sunny day and most intertidal seaweeds show some adaptation to prevent light damage. Typically the upper intertidal algae have masking compounds (fucosan or polyphenol granules in brown algae; high levels of phycobilins in blue-green algae) that may protect the chlorophyll molecules from photolysis (breakdown by light). Chloroplast movement is also possible (see Chapter 11). The importance of such adaptations is evident when an unusually low tide occurs at midday under clear skies. At this point some species, especially lower intertidal forms, bleach and die. Other algae have avoided this problem by having perennating basal portions from which upright fronds can regrow (e.g., *Corallina* spp. in southern California; *Chondrus crispus* in New England).

Biological factors

Grazing and competition for space are important biological factors influencing lithophytic algal communities. Examples of both grazing and competition will be considered here and in the next section dealing with subtidal algal communities. With regard to the intertidal zone, an elegant example of interspecific competition has been shown for the red algae *Gigartina papillata* and *Gastroclonium coulteri* and the surfgrass *Phyllospadix torreyi* (Hodgson, 1980) on the Pacific coast in California. *Gastroclonium* forms a dense turf at about the 0.3 m level above mean low tide. If the typical higher intertidal red algae *Gigartina papillata* and *Rhodoglossum affines* are removed, *Gastroclonium* does not expand into the newly opened space. However, *Gigartina* grows slower and replaces *Gastroclonium* if the latter is cleared away. Below the *Gastrocolonium* band, the surfgrass *Phyllospadix torreyi* usually occurs. *Gastroclonium* colonizes the surfgrass area if the grass is removed, but the surfgrass cannot occupy a *Gastroclonium* site. Hodgson found, through desiccation experiments, that *Gastrocolonium* could not tolerate more than a 50% water loss, and thus a physical factor limited its upper boundary. The lower boundary of *Gastroclonium* was a result of a biological factor, competition with the surfgrass.

Another example of interspecific competition has been demonstrated between fucoid algae *Pelvetia canaliculata* and *Fucus spiralis* in Scottish bays (Schonbeck and Norton, 1980). *Pelvetia* is normally confined to the highest intertidal zone, whereas *Fucus* occupies the zone immediately below. Although a very slow grower in the high zone, *Pelvetia* grew rapidly when transplanted into the *Fucus* zone. It was determined that *Fucus* sporlings outcompeted *Pelvetia* sporlings in the *Fucus* zone.

Grazing is another biological factor that can increase or decrease species diversity within the intertidal zone. Lawrence (1975) summarized over 50 examples in which sea urchins denuded the bottom, resulting in barren grounds. Moderate grazing usually allows the establishement of new species and the continuance of competitively "inferior" algal species as well as a continuation of mature forms of algae, thus providing maximal diversity (Lubchenco, 1978). Slocum (1980) suggests that crustose phases in a heteromorphic life history may be grazier-dependent and allow a species to occupy a niche during periods of grazing activity (e.g., *Gigartina papillata/Peterocelis* crust). On the other hand, too little or too extensive grazing can reduce species diversity. For example, Rafaelli (1979) found that the density of grazers (molluscs) was inversely related to the species diversity and abundance of algae in the lower intertidal zone, based on selected removal of molluscs.

Another example is the interrelationship between the periwinkle, *Littornia,* the red alga *Chondrus crispus,* and the green alga *Enteromorpha.* Lubchenco and Menge (1978) found that *Chondrus crispus* and *Enteromorpha* sp. will compete for space in many lower and mid intertidal tide pools of New England shores. *Enteromorpha* is an opportunistic, "weedy" species that can overgrow and shade out the slower-growing perennial alga *Chondrus.* However, the periwinkle, *Littorina,* selectively grazes on *Enteromorpha,* passing over the holdfasts and erect blades of *Chondrus.* Thus if the periwinkle is present, *Chondrus* will dominate the tide pools. So effective is this grazing that Cheney (personel communication) has used periwinkles to remove *Enteromorpha* in contaminated *Chondrus* cultures. Sze (1980) has found that in high tide pools, where *Chondrus* does not occur, *Enteromorpha* dominated pools if *Littorina* was absent. The extension of *Littorina* into the higher tide pools in turn was controlled by wave action, the periwinkle not being present in areas of low wave action.

Lubchenco (1980) found that the fucoid zone would extend into the lower intertidal zone if *Chondrus* was removed and herbivores such as the periwinkle snail were prevented from grazing the fucoid germlings. She also noted that the lower limit of *Chondrus* can be controlled by sea urchin grazing whereas the upper limit was controlled by tolerance to desiccation.

Vance (1979) found that by excluding the sea urchin *Centrostephanus coronatus* from subtidal rocky surfaces in California, the algal population rapidly converged to a flora similar to that of an ungrazed (control) site. As described by Lubchenco's model (1978), taxonomic diversity increased in the exclusion sites because of the replacement of a small number of taxa that could withstand urchin grazing by a larger number of intolerant taxa. Vance's conclusion was that overgrazing by the urchin *Centrostephanus* decreased diveristy but permitted the establishment of competitively inferior species, which were resistant to grazing.

Another example of biological competition in the intertidal zone can be drawn from Dayton's (1973) study of the sea palm, *Postelsia palmaeformis* (Phaeophyta). The relatively small (5–30 cm) erect plant (Figure 15-2a) competes for space on exposed rocky shores along the North Pacific coasts of the United States. A typical site will have uplifted tidal zones because of the intense wave action. In such areas, *Postelsia* competes with barnacles *(Chthamalus dalli, Balanus glandula,* and *B. cariosus),* mussels *(Mytilus californianus)* and various other algae. The barnacles, and especially the mussels, outcompete most algae and form nearly continuous colonies. Dayton (1973) suggested that the spores from *Postelsia* can settle and germinate on other algae, mussels, and barnacles, producing the gametophytes that bear eggs and sperm. The resulting sporophytes then develop and grow on the fronds of the algae, barnacles, and mussels. As the sporophytes mature, they offer a much-higher resistance to the waves, so that the young *Postelsia* plants and their attached hosts are ripped off the rocks by the surf, leaving bare rocks. Young remaining sporophytes are then able to develop. These plants previously had been shaded out by the previous mussels and barnacles (Figure. 15-2*b*).

Mussels dominate shores with a gradual slopes; apparently the only way *Postelsia* can replace them is by this competitive interference. Removal of mussel colonies by mechanical abrasion of drift logs is another way *Postelsia* communities can establish on such shores. Paine (1979) found that *Postelsia* communities occurred primarily in overhanging or steep rocky faces or where the shearing force of waves was common enough to cause mussel removal. He concluded that although mussels are a superior competitor on exposed shores, *Postelsia* is able to colonize the same shores using a natural disturbance such as wave shearing or a mechanical disruption such as log abrasion.

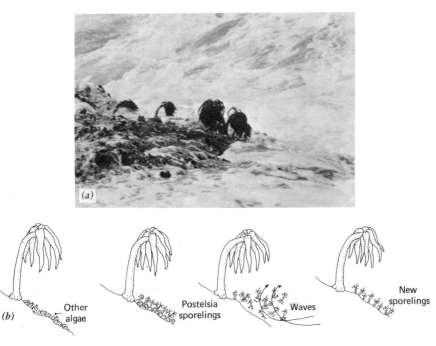

Figure 15-2 (*a*) The sea palm *Postelsia palmaeformis*. It can become dominant in the lower intertidal zone of high-wave energy areas of the northwest coasts. (*b*) Dayton (1975) and Paine (1979) have found that Postelsia cannot directly outcompete mussels, but by having sporelings attach to other algae and mussels and then be cleared away, new attachment sites are produced for the remaining germlings.

The large brown kelp *Hedophyllum sessile* will competitively displace a number of fugitive algal species and furnish a protective habitat for other obligate understory algae on high-wave-energy intertidal rocks of San Juan Island, Washington (Dayton, 1975). However, in the most-exposed sites Dayton found that *Hedophyllum* lost its dominance because of two other kelps, *Laminaria setchellii* and *Lessoniopsis littoralis*. When the urchin *Strongylocentrotus purpuratus* overgrazes areas dominated by *Hedophyllum, Laminaria,* or *Lessoniopsis,* a succession of algal communities is initiated. Dayton found that rates of succession and ultimate recovery of *Hedophyllum* canopies could take up 3 or more years and even then only 25% canopy recover occurred if desiccation was severe. If wave action was high, and therefore desiccation low; up to 66% of the *Hedophyllum* canopy formed within a year. In areas of very high wave energy, *Hedophyllum* quickly lost its dominance because of the other two kelps.

Factors Affecting Subtidal Distribution

Marine plant zonation occurs subtidally as well, but here community zonation is usually more subtle. One example of subtidal zonation, which was described by Dr. M. Neushul for western Washington (see also Hodgson and Waaland, 1979), was presented in Chapter 1. A number of physical factors are important in subtidal distribution; many of them probably act synergistically. These factors include light (intensity and quality), salinity, wave action, and water movement. As suggested by Haxo and Blinks (1950), the quality of light may be critical in determining the distribution of subtidal algae. Ramus *et al.* (1976) found that light quality was less important than quantity, and the concept of chromatic adaptation in algae (see Chapter 14) was not useful in explaining algal distribution. The conflicting data and interpretations regarding light quality and algal distribution are reviewed in Chapter 14.

Water movement is important with regard to nutrient availability and the removal of toxic substances. Neushul (1972) has reviewed the physical factors involving the formation of boundary layers around algae as well as algal morphologies as adaptations to various types of water movements. He pointed out that seaweeds can occupy one or more regions of water motion Figure 15-3*b*). In a kelp forest the upper current zone has a water velocity of about 1 m/sec, whereas closer to the bottom there is a slower oscillating water movement due to the surge from waves. In addition, there is a boundary layer about 2 cm thick along the substratum. Spore dispersal with relation to water movement has also been considered by Charters *et al.* (1973); they emphasized that water motion is a dominant factor in the life histories of the most benthic algae. Santelices (1977) found a direct correlation between species composition and the horizontal movement of water across a Hawaiian reef. Mathieson *et al.* (1977) demonstrated that species composition and abundance were directly related to water movement in a New Hampshire tidal rapids. The latter study showed that "open coastal species" of marine algae populated regions of a tidal stream, well inland, because of increased water movement.

A

B

Figure 15-3 There is competition for light in the forests of the giant kelp, *Macrocystis pyrifera*. (*A*) Reed (1979) demonstrated a three tiered kelp community off Stillwater Cove, California in part based on light levels. There is also competition for nutrients that can be limiting because of water movement. (*B*) Neushul (1972) demonstrated that the *Macrocystis* plants are exposed to three different types of water movement: (*a*) the primary surface currents, which may have a 1 cm/sec flow rate; (*b*) the surge due to wave action; and (*c*) the boundary layer where there is an almost complete lack of water movement along the substratum. (Photograph (*A*) courtesy of Dan Reed, Moss Landing, California.)

The importance of long-term studies of subtidal seaweeds can be shown in a five-year analysis of seasonal collections of 12 coastal sites in northern New England (Mathieson, 1979). Wide variations in species diversity were noted at nearby locations with similar hydrographic conditions; this was attributed to substratum availability, sand scouring, and wave action. From this detailed study, Mathieson made a number of generalizations: (1) there were few species restricted to the deepest subtidal zone; (2) the majority of the deeper-water algae were limited by substratum or light; (3) most of the deepest subtidal species were crustose or fleshy forms that favor low-light conditions; (4) the ratio of perennial to annual species changed in favor of the former in the greater depths and this was considered to be due to the stable hydrographic parameters; a similar shift in the ratio between red and green algae occurred with the reds increasing with depth; and (5) the tetrasporic phases of several red algae were the dominant phases in the deeper waters.

Since many physical factors have been discussed in the section on intertidal algae and also in chapters 10–12, this section will emphasize biotic factors within the subtidal zone. The concept of a species *niche* was presented in Chapter 13; it is an important one with regard to species competition. A simple definition of competition might be the ability of one species to control a factor that another also needs. Examples of such factors might include sufficient space for attachment and growth, light for adequate photosynthesis, or sufficient nutrients. Competition occurs when there is a limited supply of any factor.

Examples of competition can be found in kelp beds (Figure 15-3*a;* 15-3*b*) and seagrass communities (Figure 15-4*b*). Since kelp beds are one of the most conspicuous components in temperate, shallow subtidal habitats, a number of studies of kelp beds have been published. Dayton (1975) observed that the canopy effect of Alaskan kelps *Laminaria* spp. and *Alaria fistulosa* was important for sea urchin refuge (see Figure 13-2). Furthermore, the smaller kelps *Laminaria longpipes* and *Agarum cribrosum* could outcompete *Alaria*.

In the Macrocystis beds of California, there is competition for light, with the understory algae adapted to lower light levels. If the dominant giant kelp, *Macrocystis pyrifera,* is removed, then the understory algal community changes in species and diversity. Reed (1979) described a giant kelp forest at 10–20 m depth at Stillwater Cove in California. He found a two-layered kelp canopy, with a surface layer of *Macrocystis pyrifera* (Figure 15-3*a*) and a dense subsurface layer of *Pterygophora californica*. An understory of articulated and encrusting coralline algae was also present. After removal of the kelps from a 100 m² area, Reed followed floristic changes using permanent quadrats. A rapid recruitment of new kelp plants occurred with only a moderate recruitment of other brown algae including *Desmarestia* spp. and *Nereocystis luetkeana*. No recruitment occurred in regions where the canopy was not disturbed. Accordingly, Reed suggests that competition for light is the major factor determining algal recruitment and growth in the kelp forest understory. Dr. L. Hart (University of Southern California) has found similar light relationships between *Macrocystsis pyrifera Pelagophycus porra* (Figure 15-4*a*). The giant blades of *P. porra* appear to be adapted to the low light levels beneath the *Macrocystsis* canopy.

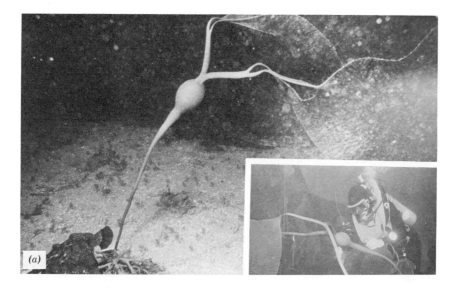

Figure 15-4   (a) Competition for light can be seen in the giant bladed kelp *Pelagophycus porra* that grows under the *Macrocystsis* canopy in southern California. Dr. L. Hart (insert) has followed the development of this annual. (b) Psammophytic algae such as *Udotea* spp. and *Halimeda* spp. compete for space and light with seagrasses along the west coast of Florida. (Figure 15-4a courtesy of Lizabeth Hart, University Southern California, Los Angeles; Figure 15-14b courtesy of Kimon Bird, Harbor Branch Foundation, Ft. Pierce, Fl.)

In seagrass beds, there is competition for both space and light between the seagrasses and macroscopic algae. In the extensive beds of seagrasses found along the west coast of Florida (Figure 15-4*b*), if the grasses are removed, then psammophytic seaweeds (sand-loving species) become established and form nearly pure stands.

A conspicuous example of grazing in the subtidal zone is the effect of sea urchins on the giant kelp, *Macrocystis pyrifera*. In southern California, there has been a progressive reduction in the giant kelp forests from the early 1900s to the 1960s (Wilson *et al.*, 1977). For example, kelp beds in the Palos Verdes area near Los Angeles covered 627 hectares (1 ha is 10,000 m² or 2.47 acres) in 1911; by 1968 essentially no plants remained. The reasons for this dramatic decrease are many, including "natural flutuations." One of the most important causes was domestic pollution, which caused a dramatic increase in the numbers of two sea urchin species, *Strongylocentrotus franciscanus* and *S. purpuratus*. The increased organic content of local waters enhanced larval development and resulted in larger-than-normal populations of urchins in the kelp area. A combination of natural population fluctuations of the kelps, water pollution, and unusually high urchin grazing decimated the kelp beds. In addition, populations of the urchin's natural predator, the sea otter, were decimated by man. Abalone *(Haliotis* sp.), a prized marine food, also disappeared during this time, since it depended on kelps for food and protection. A combination of private and state organizations and many scientists, especially Dr. Wheeler North, began a *Macrocystis* restoration project in 1956. The sewage outfall near Palos Verdes was carefully monitored, kelp outplanting methods were developed, and the sea urchin populations were destroyed or reduced by crushing or the selective application of quicklime (CaO). Competitive algae, such as *Pterygophora californica* (the elkhorn kelp), were cut down by divers to allow more space for the *Macrocystsis spores*. By 1977 the Palos Verdes *Macrocystis* bed covered 13 ha, and abalone had returned. An even more dramatic return of giant-kelp beds has been engineered at Point Loma, San Diego (Wilson *et al.*, 1977).

The effectiveness of sea urchin grazing was demonstrated in a study by Pearse and Hines (1979) at Monterey, California. They observed mortality of sea urchins *(Stronglylocentrotus sp.)* that fed on kelps in the subtidal zone. Four species of brown algae, especially the giant kelp, showed a rapid expansion on the seaward side of the existing kelp forest after urchin populations were reduced. Pearse and Hines believe that the sea urchins were preventing the brown algae from spreading seaward through extensive grazing on the outer edge of the algal beds.

It appears that brown algae are the preferred food of sea urchins throughout the world. *Strongylocentrotus franciscanus* feeds extensively on brown algae in the subtidal and intertidal regions of Friday Harbor Island in Puget Sound (Paine and Vadas, 1969). When subtidal algal populations were caged to prevent grazing, brown algae quickly dominated during a three-year study. *Hedophyllum sessile* became the dominant alga in the intertidal zone (Dayton, 1975; see previous section on intertidal zonation), whereas *Laminaria* dominated in the subtidal region (-7.3 to -8.3 m). Paine and Vadas suggested that occasional urchin browsing could make a major contribution to the variety of algae co-existing within limited areas on the rocky shores. Similar findings regarding the dominance of subtidal crustose corallines and the brown algae *Agarum cribrosum* and *Desmarestia* spp. were suggested in New Foundland because of extensive grazing of the sea urchin *Strongylocentrotus drobachiensis* (Mathieson et al., 1969). The sea urchin apparently consumed all the subtidal algae in many of the smaller bays with the exception of the corallines and the two brown algae.

Factors Affecting Geographic Distribution.

The most frequently cited factor determining the geographic distribution of shallow-water biota is water temperature. Other factors such as salinity, substratum, and wave action are primarily important in determining local distribution and abundance of communities. The geographic limits of distribution are established where temperatures are too extreme for survival or where they exceed critical temperatures for reproduction, or the completion of life histories. Gessner (1970) has proposed four critical temperatures for any alga:

1 Minimum temperature for survival.
2 Minimum temperature for repopulation (growth and sporulation).
3 Maximum temperature for repopulation (growth and sporulation).
4 Maximum temperature for survival.

However, Valentine (1966) in his study of Pacific coast molluscan ranges, found that generalized seasonal isotherms can control major regional or provincial boundaries but do not correlate well with distributions within a region. Similarly, in a recent phytogeographic study of California, Murray et al. (1980) found that the algal flora increased southward and could be correlated with latitude. Rather than specific isotherms, they proposed two distributional barriers for many algae, one near Monterey Bay and the other near Point Conception. The two sites were considered natural break points in the Californian algal flora as 115 species did not extend north or south of Monterey Bay, whereas another 21 species had northern or southern end points near Point Conception. They cited water temperature and the dominant currents as the major factors in producing these break points.

Setchell (1920) believed that isotherms (average annual temperature) alone could explain most algal distribution. He proposed five phytogeographical zones based upon 5°C isotherms: upper boreal (2–10°C), lower boreal (10–15°C), temperate (15–20°C), subtropical (20–25°C), and tropical (25–30°C). He argued that most algae only reproduce within a 5°C range and that the use of isotherms (lines on a map showing mean-temperature ranges) would demonstrate this pattern. However, temperatures can vary much more than the isotherms for any given site (Valentine, 1966). A single extreme low or high temperature can have a greater effect than a series of more-moderate temperatures.

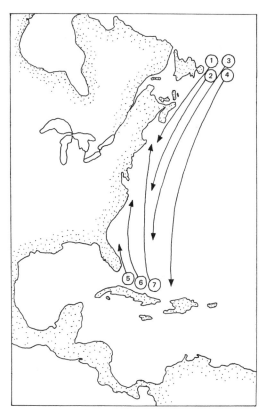

Figure 15-5 Humm (1969) has proposed seven distributional groups of inshore marine algae for the Atlantic coast of North America. He divides the algae into two floras, one centered in tropical waters, the other in cold waters north of Cape Cod. Each of these floras has species that extend north or south from the distributional centers, Capes Cod, Hatteras, or Kennedy, and one (7) is cosmopoliton.

Similarly Michanek (1979) outlined five phytogeographical zones throughout the world, based upon isotherms and known algal floras: artic, subartic, cold temperate, warm temperate, and tropical, both in the Northern and Southern hemispheres. However, the scheme does not appear to consider the significance of endemic species and is primarily concerned with the shallow-water benthic floras.

Pielou studied latitudinal spans (geographic ranges) of seaweeds along the eastern (1977) and western (1978) seaboards of the United States. She concluded that the latitudinal spans of congeneric species overlap to a significant degree. In addition, she suggested that congeneric species overlap much more than nonrelated species, and closely related congeneric species pairs are much more likely to be found near one another than to be widely separated. She believes that geographical ranges of seaweeds are not affected by interspecific competition.

Humm (1969) also noted the difficulty of relying on one major factor, water temperature, to explain species distribution. He suggested that the actual compo-

sitions of inshore floras should be considered rathern than isotherms. Accordingly, he proposed seven inshore algal associations along the eastern (Atlantic)
seaboard of the North American continent (Figure 15-5). Six of these were suggested as originating from either cold-or-warm-water regions. Humm made five
basic assumptions: (1) the algal zones are independent of latitudes; (2) there is
much overlapping of zones; (3) indicator species are present for each group; (4)
one should expect some irregularity in the distribution of the various groups since
hydrographic conditions may be irregular; and (5) endemic species are limited to
the northern most or southern most populations. The three inshore populations of
northern origin as well as the three populations of southern origin are limited by
natural barriers (e.g., Capes Cod, Hatteras, and Kennedy). The seventh population consists of cosmopolitan species such as *Ulva lactuca,* that extend along the
entire range. Even so, arguments can be made for "physiological species of
Ulva" (Rhyne, 1973). It is interesting that Van den Hoek (1979) found a similar
distribution pattern for 43 species of *Cladophora* in the nothern Atlantic Ocean. It
appears that many green algae are cosmopolitan species; this is also evident in the
study by Murray *et al.* (1980) regarding algal distribution in California. Another
factor to be considered in species distribution is the effect of stable temperatures
found in deeper waters. In general, deep-water algal species do not appear to be
as limited geographically as shallow-water algae, probably because of the greater
extent and stability of their environment. Thus much of the deep-water (to 40 m)
flora on the North Carolina shelf is tropical in nature and the species are an extension of the Caribbean flora (Schneider, 1976.)

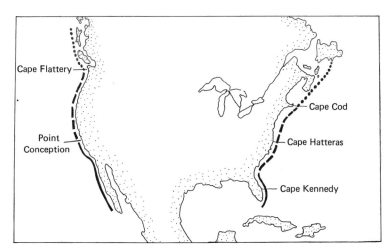

Figure 15-6 Stephenson and Stephenson (1972) have proposed three main geographic
groups of intertidal algae on the west and east coasts of North America: southern assemblages (solid lines), central assemblages (dashed lines), and northern assemblages (dotted
lines). Major break points for the two coasts with regard to floristic changes were proposed
to be specific capes or points where oceanic currents diverge. In turn, these major assemblages can be subdivided into more specific groups.

Stephenson and Stephenson (1972) have identified comparable intertidal algal distributions along the eastern and western coasts of North America (Figure, 15-6), with regard to temperature. The flora of each coast was divided into four intertidal groups (northern, central, southern, and cosmopolitan), with the central group differentiated into a northern and southern component. Again, they found major break points or barriers (Capes Cod and Kennedy on the east coast; Puget Sound and Point Conception on the west coast).

Classification of the Intertidal Zone

There are two major texts describing the biological classification of intertidal zones (Stephenson and Stephenson, 1972, and Lewis, 1964). In both, the authors have considered several physical and biological factors that influence zonation patterns. The generalized schemes proposed in each text are presented in Figure 15-7. Although they have somewhat different terminologies the two schemes both emphasize the importance of wave exposure and substratum in determining zonation patterns. An important difference between the two authors' approaches to the intertidal zones lies in their interpretations of the upper limits of the intertidal zone, that is, whether or not a supralittoral zone exists. Both schemes utilize the same dominant organisms in establishing zones (periwinkle, barnacles, kelps).

In this text, specific organisms have not been employed as markers (see Chapter 1) as many of these are restricted to temperate or tropical waters. Instead, a simple, universal zonation scheme is used: spray (supralittoral fringe), intertidal (midlittoral), intertidal fringe (infralittoral fringe), and subtidal (sublittoral). Although the influence of tides is considered of paramount importance, there is no doubt that the effects of air-water interfaces are highly significant (exposure,

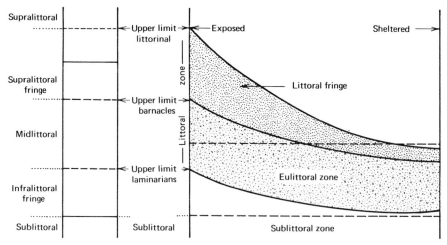

Figure 15-7 Comparison of the intertidal schemes of Stephenson and Stephenson (1972), left, and Lewis (1964), right. Both use dominant cold-water species to identify the zones. In Lewis's scheme, when the effects of exposure and sheltering are reduced, as in sheltered regions, a compression of the zones occurs (far right).

etc.). As noted by Stephenson and Stephenson and Lewis, variations in moisture, light penetration, substratum, and sedimentation must also be considered.

EXAMPLES OF ZONATION

Four major lithophytic communities will be considered here: open coastal, sheltered coastal, tidal pools, and subtidal regions.

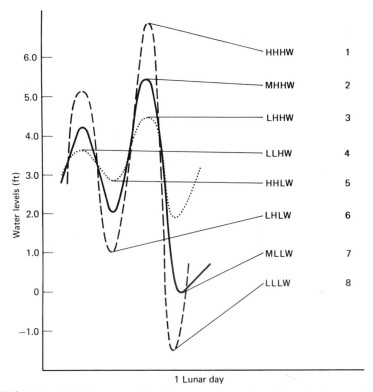

Figure 15-8 Doty (1946) proposed that the most important factor in the zonation of marine algae of an intertidal region is tides. In studies of marine algae found intertidally on the west coast of North America, he identified eight possible tidal levels (expresed as Ft in graph) starting with highest-high high water (HHHW: 6.5 ft (2.5 m), never submerged) to lowest-low low water (LLLW: −1.2 ft (−0.6 m), never exposed). Above zone 1 is a spray zone and below zone 8 is the subtidal region. For each zone of the Oregon coast, Doty was able to determine length of exposure for a single tidal change and the dominant alga as follows: (*1*) Spray zone (above 2.5 m, never submerged); (*2*) Marine lichens (1.5–2.5 m exposed 120 hr or infrequent submersion); (*3–4*) *Endocladia* (1.2–1.5 m exposed 18–24 hr); (*4–5*) *Pelvetia* (1–1.2 m exposed 6–9 hr); (*5–6*) *Iridea* 0.3–1 m submerged 12 hr); (*6–7*) *Egregia* (0.0–0.3 m, submerged most of the time); (*7–8*) *Prionitis* (−0.3 to 0.0 m, submerged almost always); (*8*) *Cystoseira* (always submerged). The solid line is mean tidal levels, the dotted line is the minimum and the dashed line the maximum tidal levels over a 14½ da tidal cycle.

Open Coastal

In order to demonstrate the difference between a predominantly tidal classification and one relying primarily on algal populations, two classic studies are presented: the study of the lithophytic communities along the Pacific coast by Doty (1946) and that of the Florida Keys by Stephenson and Stephenson (1950).

Tidal zones

Figure 15-8 is taken from Doty's (1946) study of the marine algal distribution of an Oregon coast. Doty postulated that specific levels of the tidal range can be correlated with the vertical limits (breaks) of many prerennial algae. The elevational breaks could be related to other environmental factors, but in his opinion tidal levels were of primary importance. Both elevation and duration of exposure are important (Figure 15-8). He described the resulting exposure as the "critical tide factor." In a later paper, the critical tide factor hypothesis was tested experimentally (Doty and Archer, 1950). They concluded that elevation and time of exposure were the most important features controlling vertical limitations of the intertidal biota. They proposed that no one secondary factor (light intensity, salinity, air temperature, desiccation) limited an organism throughout its natural range.

Community Zones

Stephenson and Stephenson (1972) here summarized their numerous studies of intertidal sites throughout the world. One of their earlier studies (1950) in the Florida Keys is illustrated in Figure 15-9. The boundaries between the principle zones were defined by organisms rather than tidal levels. A combination of factors, such as substratum, desiccation, and especially exposure, were noted as controlling factors for the intertidal populations. Accordingly, they described several associations of plants and animals they believe have evolved the same requirements and tolerances to exposure.

Sheltered Coasts

The present discussion is limited to open coastal areas with stable and/or higher salinities (for estuaries, see Chapters 18 and 19). Although no site totally lacks wave action, sheltered conditions with reduced wave activity are common, especially in rugged coastlines. In such areas, algal zones are usually narrow and reflect the tidal levels directly. The reduced wave action of a "semiexposed site" will usually allow a more luxuriant and diverse algal community (Lewis, 1964) with some algal species characteristic of high-energy shores absent. Similarily, Seapy and Littler (1979) found a higher diversity on a wave-exposed sea stack than on a protected boulder beach. However, the productivity at the boulder-beach site was higher than that at the exposed site by about one third because of the more-massive and less-efficient species. It should also be noted that a marked decrease in both species diversity and abundance can occur in sheltered coves due

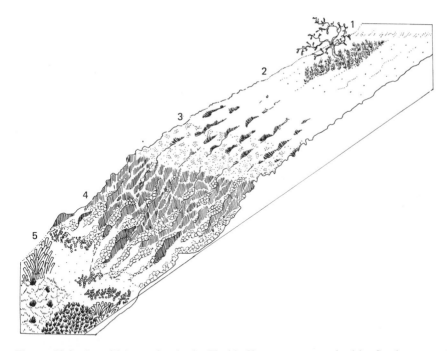

Figure 15-9 Intertidal zonation in the Florida Keys was summarized by Stephenson and Stephenson (1972) in this figure. Zone 1 is gray zone with various xerophytic angiosperms such as *Batis* and the black mangrove, *Avicennia*. Zone 2 is a black zone with *Littorina* spp. (periwinkle) as well as some blue-green algae. Zone 3 is a yellow zone containing *Chthamalus* sp. (barnacle), various blue-green algae, and some filamentous green algae. Zone 4 is the lower reef platform; the upper edge is the low-water level at spring tides and contains *Valonia*, various small corals, *Halimeda*, and the sea urchin *Echinometra*. Zone 5 is the subtidal region, where corals will become quite abundant along with sea urchins, the sea grass *Thalassia* and various coenocytic green algae.

to environmental conditions (Lewis, 1964). In protected areas with very low wave action, sedimentation as well as variable salinities can be found because of rainfall, runoff and poor water movement.

Lewis (1964) and Stephenson and Stephenson (1972) have noted that the upper and middle levels of sheltered shores show transitions from the open coastal to protected waters. There is a compression of the upper communities and a more-luxuriant development of specialized epiphytic communities (see Seapy and Littler, 1979). The most important environmental conditions would seem to be the availability of a stable substratum, the rate of sedimentation, the absence of abundant water movement, and reduced nutrient availability.

Figure 15-10 gives a comparison of an open coast site (Peggy's Cove) and a protected site (Mason's Cove) in Nova Scotia, Canada, based upon the studies by Stephenson and Stephenson (1972). A compression of zones and a reduced number of species were recorded at the protected site.

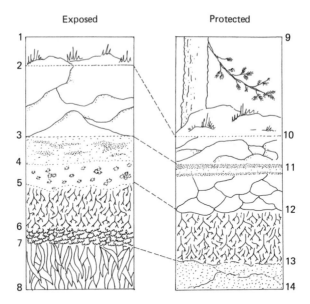

Figure 15-10   A comparison of an exposed (Peggy's Cove) and sheltered (Mason's Cove) intertidal region in Nova Scotia, Canada. The dashed lines connecting the two diagrams demonstrate the compression of the major subunits in a protected site. The numbers refer to major organisms found in each zone: At Peggy's Cove: *(1–2)* land lichens, grasses and bushes, *(2–3)* nearly bare rock, *(3–4)* rock with some *Codium* and blue-green algae, *(4–5)* barnacle zone, *(5–6)* fucoid zone, *(6–7)* *Chondrus* zone, *(7–8)* kelp zone. At Mason's Cove *(9–10)* forest with conifers, bushes, land lichens, *(10–11)* nearly bare rock, *(11–12)* rock with blue-green algal banding, *(12–13)* fucoid zone, *(13–14)* mud, gravel, and stones. (From Stephenson and Stephenson, 1972.)

Tide Pools

Even to the casual observer, it is evident that tide pools vary greatly with regard to their species diversity and abundance of algae (Figure 15-11). Low interidal pools of similar dimensions as upper ones usually have more plants and animals. Many factors are important in determining the presence and abundance of biota, including climatic factors such as rainfall (salinity dilution), sunlight, and air temperature, and such biotic features as photosynthetic and respiratory rates. (Femino and Mathieson, 1980). A rise in water temperature, which is associated with increased respiration rates, can cause a drop in pH due to an increase in dissolved CO_2. Ambler and Chapman (1950) found that the two most critical factors for tide pools in the upper intertidal zone were salinity and temperature. They also noted a distinct rise in oxygen and pH during the day due to photosynthesis.

According to Klugh (1924), the single most critical factor for tide pool organisms is temperature. His studies showed that tide pools with large temperature ranges had low species diversity and abundance. He compared several tide pools

Figure 15-11 A tide pool in southern Australia will contain well over 100 species of macroscopic algae if water exchange is high.

in New Brunswick, Canada, with similar size but at six levels of duration of exposure, temperature, pH, and abundance of algae. As shown in Table 15-1, which is taken from Klugh's study, the highest tide pools had the greatest increase in water temperature (14.7°C maximum), whereas pH showed no major change.

Table 15-1   Comparison of Tide Pool Factors[a]

Height Above Mean Low Water (m)	Monthly Exposure (hr)	Temperature of Water at End of Exposure (°C)	pH of Water at End of Exposure	Species Diversity
1.0	13	14.0	8.1	High
1.6	88	16.5	8.0	High
2.1	175	17.6	8.4	Moderate
3.6	285	20.3	8.1	Moderate
4.2	420	22.7	7.9	Low
5.1	560	24.3	8.1	Low

Source: After Klugh (1924).
[a]Seawater temperature was 14.7°C.

Subtidal Communities

With the advent of scuba, and now submersibles, marine botanists have been able to initiate careful studies of subtidal benthic algal communities (Neushul, 1965). Numerous marine botanists employ these techniques. The subtidal floras in many locales are both very rich and complex. For example, in a study of the subtidal flora of Puget Sound, Washington, Neushul (1967) described three major communities (Figure 15-12). A shallow-water *Laminaria* association was most common

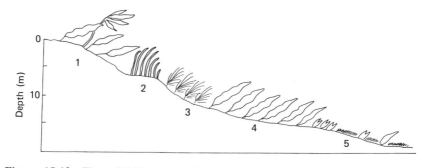

Figure 15-12 The subtidal community at San Juan Island (Friday Harbor) in Puget Sound, Washington, studied by Dr. M. Neushul (1967). The transect shown here was in an area with a steep drop so that zonation was more rapid. The physiographic profile includes various species of *Laminaria* and *Nereocystis* (*1*), some shallow *Zostera* beds (*2*), an area of more-fleshy red algae such as *Agardhiella* (*3*), a major bed of *Agarum* (*4*), and finally smaller, more-turfy red algae such as *Rhodoptilum* and *Callophyllis* (*5*).

in subtidal areas with gradual slopes, and it was comparable to the lowest intertidal fringe community described by Muenscher (1951; See. Chapter 1 and Figure 1-9). A region dominated by *Agarum* (Laminariales) was present at middepths (5–15 m), and a red algal association (*Rhodoptilum, Callophyllis*) with abundant hydroids at the lower depths (17–20 m).

In a study of subtidal marine algae off southern Cape Cod, Massachusetts, Sears and Wilce (1975) classified algae according to their thallus persistence. Four groups were described: seasonal annuals, aseasonal annuals, pseudoperennials, and perennials. A total of 142 subtidal species were described and grouped into 10 algal associations based on two types of substrata. Pronounced seasonal changes in species diversity were found from 5 to 22 m. At greater depths, the ratio of red to brown algae increased, and there were four times as many reds as browns below 22 m.

Dawes and van Breedveld (1969) demonstrated that a large (157 species) variety of macroscopic algae occurred in deep waters (20–60 m) off the west coast of Florida. The flora was tropical in nature and was seasonally stable, which coincided with stable light, salinity, and temperature regimes. Seasonal fluctuations were especially common among the brown algae (dominant in spring). Use of deep submersibles such as the Johnson Sea Link (Figure 1-10) has allowed phycologists to extend direct observations to even-deeper subtidal flora.

REFERENCES

Ambler, M. P. and V. J. Chapman. 1950. A quantitative study of some factors affecting tide pools. *Roy Soc. N. Zeal Trans*. **78**: 394–409.

Brinkhuis, B. H., N. R. Tempel, and R. F. Jones. 1976. Photosynthesis and respiration of exposed salt-marsh fucoids. *Mar. Biol*. **34**: 349–359.

Carefoot, T. 1977. *Pacific Seashores. A Guide to Intertidal Ecology*. University of Washington Press, Seattle.

Charters, A.C., M. Neushul, and D. Coon. 1973. The effect of water motion on algal spore adhesion. *Limnol. Oceanog.* **18**: 884–896.

Dawes, C. J., R. E. Moon, and M. A. Davis. 1978. The photosynthetic and respiratory rates and tolerances of benthic algae from a mangrove and salt marsh estuary: a comparative study. *Estuar. Coast. Mar. Sci.* **6**: 175–185.

Dawes, C. J., and J. F. van Breedveld. 1969. *Benthic Marine Algae. Memoirs of the Hourglass Cruises*. Marine Res. Laboratory, State of Florida, Dept. of Natural Resources. St. Petersburg.

Dayton, P. K. 1973. Dispersion, dispersal, and persistence of the annual intertidal alga, *Postelsia palmaeformis* Ruprecht. *Ecology* **54**: 433–438.

Dayton, P. K. 1975. Experimental evaluation of ecological dominance in a rocky intertidal algal community. *Ecol. Monog.* **45**: 137–159.

Doty, M. S. 1946, Critical tide factors that are correlated with the vertical distribution of marine algae and other organisms along the Pacific Coast. *Ecology*, **27**: 315–328.

Doty, M. S. and J. G. Archer. 1950. An experimental test of the tide factor hypothesis. *Amer. J. Bot.* **37**: 458–464.

Femino, R. J. and A. C. Mathieson. 1980. Investigations of New England marine algae IV. The ecology and seasonal succesion of tide pool algae at Bald Head Cliff, York, Maine, USA. Bot. Mar. **23**: 319–332.

Gessner, F. 1970. Temperature and Plants. In Kinne, O. (ed.). *Marine Ecology*, Vol 1, Part 2. Wiley, New York, pp. 362–406.

Haxo, F. T. and L. R. Blinks. 1950. Photosynthetic action spectra of marine algae. *J. Gen. Physiol.* **33**: 389–422.

Hodgson, L. M. 1980. Control of the intertidal distribution of *Gastroclonium coulteri* (Harvey) Kylin, in Monterey Bay, California. *Mar. Biol.* **57**: 121–126.

Hodgson, L. M. and J. R. Waalond. 1979. Seasonal variation in the subtidal marcroalgae of Fox Island, Puget Sound, Washington. *Syesis* **12**: 107–112.

Humm, H. J. 1969. Distribution of marine algae along the Atlantic Coast of North America. *Phycologia* **7**: 43–53.

Johnson, W. S., A. Gigon, S. L. Gulmon, and H. A. Mooney. 1975. Comparative photosynthetic capacities of intertidal algae under exposed and submerged conditions. *Ecology* **55**: 450–453.

Kanwisher, J. 1957. Freezing and drying in intertidal algae. *Biol. Bull.* **113**: 275–285.

Klugh, A. B. 1924. Factors controlling the biota of tide-pools. *Ecology* **5**: 180–196.

Lawrence, J. M. 1975. On the relationships between marine plants and sea urchins. *Oceanogr. Mar. Biol. Ann. Rev.* **13**: 213–286.

Lewis, J. R. 1964. *The Ecology of Rocky Shores*. English Universities Press, London.

Littler, M. M. and S. N. Murray. 1974. The primary productivity of marine macrophytes from a rocky intertidal community. *Mar. Biol.* **27**: 131–135.

Lubchenco, J. and B. A. Menge. 1978. Community development and persistence in a low rocky intertidal zone. *Ecol. Monog.* **59**: 67–94.

Lubchenco, J. 1978. Plant species diversity in a marine intertidal community: importance of herbivore food preference and algal competition abilities. *Amer. Nat.* **112**: 23–39.

Lubchenco, J. 1980. Algal zonation in the New England rocky intertidal community: an experimental analysis. *Ecology* **61**: 333–344.

Mann, K. H. 1973. Seaweeds: Their productivity and strategy for growth. *Science* **182**: 975–981.

Mann, K. H. and A. R. O. Chapman. 1975. Primary production of marine macrophytes. In Cooper, J. P. (ed.). *Photosynthesis and productivity in different environments*, IBP vol. 3, Cambridge University Press, New York, pp. 207–248.

Mathieson, A. C. 1979. Vertical distribution and longevity of subtidal seaweeds in northern New England U.S.A. *Bot. Mar.* **30**: 511–520.

Mathieson, A. C., E. Tveter, M. Daly, and J. Howard. 1977. Marine algal ecology in a New Hampshire tidal rapid. *Bot. Mar.* **20**: 277–290.

Mathieson, A. C., C. J. Dawes, and H. J. Humm. 1969. Marine algae from the coast of Newfoundland. *Rhodora.* **71**: 110–160.

Michanek, G. 1979. Phytogeographic provinces and seaweed distribution. *Bot. Mar.* **22**: 375–391.

Muenscher, W. L. C. 1951. A study of the algal associations of San Juan Island. *Puget Sound Mar. Sta. Publ.* **1**: 59–84.

Murray, S. N., M. M. Littler, and I. A. Abbott 1980. Biogeography of the California marine algae with emphasis on the Southern California Islands. In D. M. Powes (ed.), *The California Islands: Proceedings of a Multidiscilpinary Symposium.* Santa Barbara Museum of Natural History, Santa Barbara, Calif., pp. 325–339.

Neushul, M. 1965. SCUBA diving studies of the vertical distribution of benthic marine plants. *Proc. 5th Mar. Biol. Symp. Goteburg, Botanica Gotoburgensia Pt. III* pp. 161–176.

Neushul, M. 1967. Studies of subtidal marine vegetation in western Washington. *Ecol.,* **48**: 83–94.

Neushul, M. 1972. Functional interpretation of benthic marine algal morphologies. In I. A. Abbot and M. Kurogi (eds.). *Contributions to the Systematics of Benthic Marine Algae of the North Pacific. Jap. Soc. Phycol.,* Kobe, pp. 47–73.

Paine, R. T. 1979. Disaster, catastrophe, and local persistence of the sea palm, *Postelsia palmaeformis. Science* **205**: 685–686.

Paine, R. T. and R. L. Vadas. 1969. The effects of grazing by sea urchins, *Stronglocentrotus* spp., on benthic algal populations. *Limnol. Oceanogr.* **14**: 710–719.

Pearse, J. S. and A. H. Hines. 1979. Expansion of a central California kelp forest following the mass mortality of sea urchins. *Mar. Biol.* **51**: 83–91.

Pielou, E. G. 1977. The latitudinal spans of seaweed species and their patterns of overlap. *J. Biogeog.* **4**: 299–311.

Pielou, E. C. 1978. Latitudinal overlap of seaweed species: evidence for quasisympatric speciation. *J. Biogeog.* **5**: 227–238.

Raffaelli, D. 1979. The grazer-algae interaction in the intertidal zone on New Zealand rocky shores. *J. Exp. Mar. Biol. Ecol.* **38**: 81–100.

Ramus, J., S. I. Beale, and D. Mauzerall. 1976. Correlation of changes in pigment content with photosynthetic capacity of seaweeds as a function of water depth. *Mar. Biol.* **37**: 231–238.

Reed, D. 1979. Canopy structure in a *Macrocystis* bed. 60th Annual Meeting of West. Soc. Nat. Abst. San Francisco.

Rhyne, C. F. 1973. *Field and Experimental Studies on the Systematics and Ecology of Ulva curvata and Ulva rotundata.* University of North Carolina Sea Grant Publ.. UNC-SG-73-09. Chapel Hill.

Santelices, B. 1977. Water movement and seasonal algal growth in Hawaii. *Mar. Biol.* **43**: 225–235.

Seapy, R. R. and M. M. Littler. 1979. The distribution, abundance, community structure, and primary productivity of macroorganisms from two central California rocky intertidal habitats. *Pac. Sci.* **32**: 293–314.

Sears, J. R. and R. T. Wilce. 1975. Sublittoral, benthic marine algae of southern Cape Cod and adjacent islands: seasonal periodicity, associations, diversity, and floristic composition. *Ecol. Monog.* **45**: 337–365.

Schneider, C. W. 1976. Spatial and temporal distributions of benthic marine algae on the continental shelf of the Carolinas. *Bull. Mar. Sci.* **26**: 133–151.

Schonbeck, M. W. and T. A. Norton. 1980. Factors controlling the lower limits of fucoid algae on the shore. *J. Exp. Mar. Biol. Ecol.* **43**: 131–150.

Setchell, W. A. 1920. The temperature interval in the geographical distribution of marine algae. *Science* **52**: 187–190.

Slocum, C. J. 1980. Differential susceptibility to grazers in two phases of an intertidal alga: advantages of heteromophic generations. *J. Exp. Mar. Biol. Ecol.* **46**: 99–110.

Stephenson, T. A. and A. Stephenson. 1950. Life between tide-marks in North America, I. The Florida Keys. *J. Ecol.* **38**: 354–402.

Stephenson, T. A. and A. Stephenson. 1972. *Life Between Tidemarks of Rocky Shores.* Freeman, San Francisco.

Sze, P. 1980. Aspects of the ecology of macrophytic algae in high rockpools at the Isles of Shoals (USA). *Bot. Mar.* **23**: 313–318.

Taylor, W. R. 1960. *Marine Algae of the Eastern Tropical and Subtropical Coasts of the Americas.* University of Michigan Press, Ann Arbor.

Taylor, W. R. 1962. *Marine Algae of the Northeastern Coast of North America.* University of Michigan Press, Ann Arbor.

Valentine, J. W. 1966. Numerical analysis of marine molluscan ranges on the extratropical northeastern Pacific shelf. *Limnol. Oceanogr.* **11**:198–211.

Vance, R. R. 1979. Effects of grazing by the sea urchin, *Centrostephanus coronatus*, on prey community composition. Ecology. **60**: 537–546.

Van den Hoek, C. 1979. The phytogeography of *Cladophora* (Chlorophyceae) in the northern Atlantic Ocean, in comparison to that of other benthic algal species. *Helgo. Wiss. Meeresun.* **32**: 374–393.

Wilson, K. E., P. L. Haaker, and D. A. Hanan. 1977. Kelp restoration in southern California. In R. W. Krauss (ed.). *The Marine Plant Biomass of the Pacific Northwest Coast*, Oregon State University Press., Corvallis, pp. 183–202.

CHAPTER 16

Coral (Biotic) Reefs

Coral reefs have attracted much attention ever since Charles Darwin presented a hypothesis regarding their evolution and formation. Biologists have noted the important role that organisms, especially corals (Coelenterata), play in coral reef development. Even so, a proposal has been made that the name be changed to biotic reefs because of the critical role of coralline algae (Womersley and Bailey, 1969). Van den Hoek *et al.* (1975) disagrees with the proposed name change and argues that the original term, "coral reef," should be retained because of the dominance of the coral animals. At any rate, the formation of a coral reef is based on a balanced association of organisms, plant and animal, alive and dead. Although there is a wide diversity of plants and animals making up a coral reef, the paradox is that the surrounding waters of an open ocean reef are usually very low in nutrients and phytoplankton. One of the critical questions today is how coral reefs can show such high productivity in oceanic waters that are essentially deserts with regard to sustaining nutrients.

PHYSICAL AND GEOLOGICAL CONSIDERATIONS

A coral reef is a biological community rising from the sea floor that consists of a solid limestone ($CaCO_3$) structure strong enough to withstand the force of waves. The dominant organisms are coral animals, that have a calcified exoskeleton, and algae, many of which also are calcified. In earlier studies, coral animals were assigned the major role in reef formation; however, more recent botanical studies indicate that algae may actually dominate reef formation, at least in the shallow waters.

Reef Characteristics

The four major factors that influence reef development are temperature, light, salinity, and immersion (Davis, 1977). Coral reefs are found within the 21–22°C isotherms of the world, and they are confined to the tropical oceans (Figure 16-1). Coral growth is most rapid in depths of less than 35 m and in waters of stable, oceanic salinities. The upper edge of the reef is limited by exposure to air. The

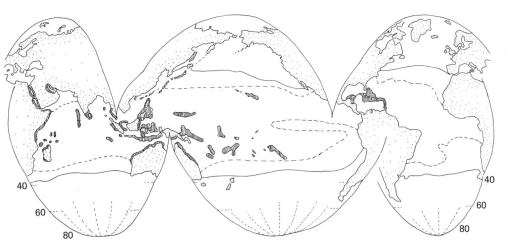

Figure 16-1   The world distribution of coral reefs (dark hatched regions). The reef areas are exaggerated because of the scale. Isotherms show the extended 70°F (summer, solid) and restricted (winter, dashed) 70°F (21°C) ranges for the northern and sourthern hemispheres.

limestone base of the coral reef is derived from the fossilized skeletons of the corals (invertebrates) and calcified algae.

Calcareous Algae

Calcification and calcareous algae are essential in the formation of coral reefs since it is from the calcareous (aragonite) skeleton of coral animals and the calcareous (aragonite or calcite) deposits on or in the cell walls of many algae that the limestone reef is formed. The most important group of algae in terms of their contribution to reef formation, is the red, coralline algae that deposit a form of calcium carbonate called calcite (Table 16-1). The crustose coralline algae form the outer algal ridges and cement together the various coral skeletons. A variety of calcified green algae found in the Caulerpales and Dasycladales deposit another form of calcium carbonate called aragonite. Much of the sand and lagoonal sediment in coral atolls is from these green algae. Littler (1976) reviewed the roles of calcareous algae, and Wray (1977) has described the geological and paleological importance of these plants.

   Calcareous algae, like all calcified organisms, are most common in the tropics, where oceanic water is supersaturated with calcium carbonate. The distribution of calcareous algae, especially the corallines, seems to be related to such physical factors as light, temperature, and wave action as well as competition and grazing (Littler, 1976). Such factors are important in controlling the development and structure of fringing and atoll reef systems (Littler, 1976; Adey, 1978). Some of the advantages of calcification of an alga suggested by Littler include mechani-

Table 16-1   Calcareous Algae

Taxonomic Group (Chapter)	Generic Example	Type of Calcium Carbonate	Site of Deposition
Chrysophyta (8)			
Prymnesiophyceae	*Hymenomonas*	Calcite and aragonite	Coccoliths that are formed in cell
Cyanophyta (4)			
Nostocales	*Oscillatoria*	Calcite and aragonite	In gelatinous sheath and on surface of cell
Chlorophyta (Chlorophyceae) (5)			
Caulerpales	*Halimeda*	Aragonite	Surface of cell
Dasycladales	*Acetabularia*	Aragonite	Surface of cell
Chlorophyta (Charophyceae) (5)			
Charales	*Chara*	Calcite	Surface of cell
Rhodophyta (7)			
Nemalionales	*Galaxura*	Aragonite	Surface of cell
Gigartinales	*Titanophora*	Aragonite	Surface of cell
Cryptonemiales	*Corallina*	Calcite	In cell wall and on surface
Phaeophyta (6)			
Dictyotales	*Padina*	Aragonite	Surface of cell

Source: Modified from Littler (1976).

cal support of the alga, resistance to abrasion, wave shock and grazing, and protection against fouling epiphytes and high or intense light.

Types of Reefs

There are three reef types, the fringe, barrier, and atoll (Figure 16-2) reefs. The fringing reef grows out from a land mass but is still attached to it. A good example is the fringing reef of the Florida Keys. A barrier reef is separated from the land mass by a lagoon. The two largest examples of barrier reefs are the Great Barrier Reef off Australia and the barrier reef off Belize, Central America (Figure 16-1). An atoll is a circular-to-oval-shaped reef that surrounds a central lagoon; it is not associated with any obvious land mass. The small islands that may form on the atoll, usually occur on the windward (seaward) side.

The most impressive feature of atolls is size. In the Pacific, the lagoons of Eniwetok and Kwajalein atolls are 16–60 km in diameter and have a number of small fringing islands. Eniwetok Atoll consists of about 40 islands; almost all are on the windward side. In the Caribbean, the lagoon of Glover's Reef is about 12 km long and 9 km wide with four small islands on the southeastern side (windward side) as shown in Figure 16-3. One can move from the windward edge of a

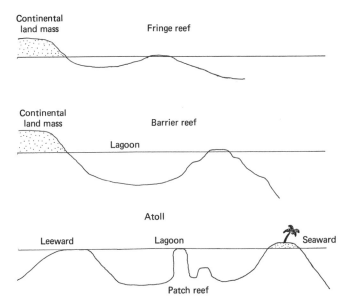

Figure 16-2 The three basic types of coral reefs. The first two are shown developing on a continental shelf, whereas the third usually is formed from either a subsiding volcano or a fault line.

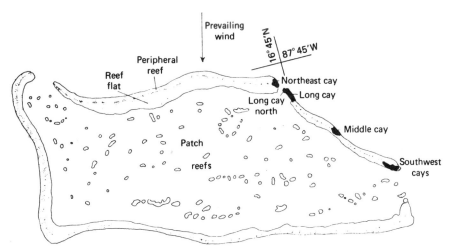

Figure 16-3 A diagrammatic map of the atoll Glover's Reef off Belize, Central America, in the western Caribbean. The reef measures 12 × 9 km with a lagoon about 10–15 m deep. All the islands occur on the eastern, or windward, side of the atoll. Note the numerous patch reefs in the lagoon.

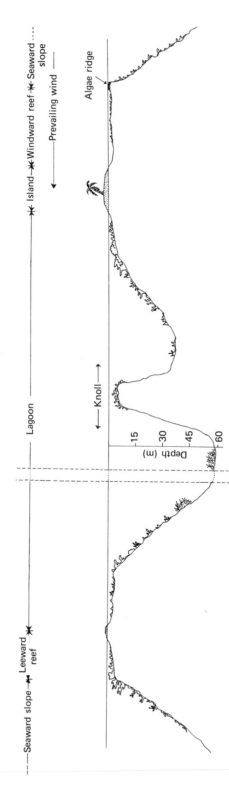

Figure 16-4 A cross section of a Pacific atoll, showing a much-deeper lagoon than those found in the Caribbean. A well-defined algal ridge is especially prominent on the prevailing-wind side and may be absent on the leeward reef. The lagoon, much shortened in this diagram, may reach 30–50 km in diameter.

reef island, where wave action is almost always high and surf is a pronounced feature, over the island and into the quiet lagoon. In the lagoons of all atolls, one finds smaller individual "patch" reefs or knolls. A cross section of a Pacific atoll in Figure 16-4 shows the seaward (windward) and leeward reefs, the lagoon with patch reefs and a seaward island. The lagoon substratum is usually a thick layer of sand (~100 m) lying over a limestone base.

Indo-Pacific and Caribbean Reefs

Differences between Pacific and Caribbean coral reefs are mostly based on the species composition of the corals and algae and on the role calcareous algal play in reef formation. Adey (1978) showed that coral reefs subjected to high wind and wave action will have an outer ridge of coralline algae. The massive crustose coralline algae can withstand this heavy surf and will form the reef rim or ridge, slightly above or just below sea level. The impact of breaking waves is the greatest on the ridge; calmer conditions will then occur behind this seaward ridge (Figure 16-5). The most-massive algal ridges are seen in Pacific atolls. The lagoon of most atolls is the product of limestone-boring organisms that cause disintegration of calcareous material. The depth of an Indo-Pacific atoll lagoon is about 60–100 m based on the equilibrium between the boring organisms and sedimentation of calcified plant and animal material.

The reef margin extends seaward because of the growth and cementing action of coralline algae at the surf zone and the coral animals below. A gradual buildup of a talus slope below the reef edge forms the reef mount. The seaward advance of the reef edge results in a reef flat seaward and a slowly developing lagoon in the center. Storm-cast calcareous fragments, sand and coral rubble, accumulate on the reef flat, usually on the windward side. With fresh rainwater, the calcium carbonate in the sand becomes cemented together, forming the reef surface on which an island can develop from the cast-up sand and coral rubble.

Figure 16-5 The algal ridge of a Pacific atoll (Eniwetok) showing a surge channel and the covering by the coralline alga *Porolithon*.

Some atolls, such as Glover's Reef in the Caribbean, lack a well-developed algal ridge. Adey (1978) demonstrated that the occurrence of algal ridges in the Caribbean was related to the constancy, strength, and effect of the wind that resulted in a higher and more sustained wave action. The higher these factors were, the more pronounced the algal ridge. Since Glover's Reef lies in the northwest quadrant of the Caribbean, which is an area of weak winds, the dominant organisms are corals (Coelenterata), even at the surf zone. Instead of a solid reef ridge, at or above low-tide mark, the breaker zone on a windward reef is occupied by two corals, *Acropora* and *Millepora,* which are slightly below the low-tide mark (Figure 16-6). Coralline algae as well as other types of algae are present throughout the coral reef but are not the dominant organisms.

Origin of Atolls

Darwin saw some reefs and atolls during his voyage on the Beagle in 1831–1836. He suggested that a volcanic island initially provided a shallow-water base for the growth of a fringing reef (Figure 16-7*a* and *b*). With a slow subsidence of the volcanic island, the reef continued to grow. The fringing reef gradually became an atoll with the central portion of the volcano sinking beneath the ocean (Figure 16-7*d*). Through the action of boring organisms, the central portion became a lagoon. This theory is well supported by studies of Pacific Ocean atolls, where volcanic rock has been found by drilling over 1000 m through coral and coralline rubble in the lagoon.

In the Caribbean Sea, the atolls appear to have developed from fringing reefs that ocurred along fault lines, the surrounding substrata having subsided more rapidly than the atoll block (Stoddard, 1962). The development of the Caribbean atolls also is based on subsidence and reef formation. Recent drilling on Glover's

Figure 16-6 *Acropora palmata* (bramble coral) forms a zone in the surge and breaker region of a Caribbean reef such as Glover's Reef. The zone will occur at low tide to about 3 m deep depending on wave intensity. The branches are oriented offshore toward the waves.

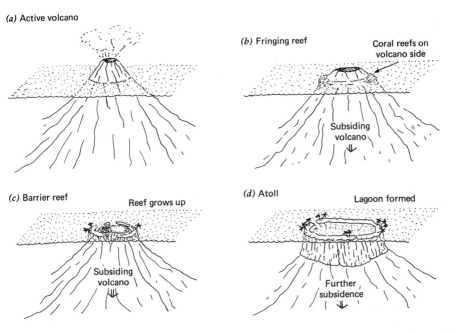

(a) Active volcano

(b) Fringing reef   Coral reefs on volcano side

Subsiding volcano

(c) Barrier reef   Reef grows up

Subsiding volcano

(d) Atoll   Lagoon formed

Further subsidence

Figure 16-7 Darwin's theory of atoll formation. (a)–(b) The formation of a fringing reef after subsidence of an active volcano, the transition from the fringing reef to a barrier reef (c), and finally the completion of an atoll (d). (After Ross, 1970.)

Reef showed a depth of reef coral of more than 1300 m. It is estimated that Glover's Reef is over 25 million years old.

An alternate theory to Darwin's one is that the oceans are rising, and reef formation is a reaction to that rise. In fact, there is a 200 m difference in oceanic levels during periods with and without glaciation. However, this difference would account only for the "fine tuning" of reef development, and subsidence does seem to be the best explanation.

ECOLOGY OF REEF ALGAE

The importance of the invertebrates, especially corals, in the ecology of coral reefs has been recognized since Darwin presented his hypothesis for the evolution of atolls. However, the role of macroalgae and microalgae in the ecology of coral reefs is only now becoming clear.

Zooxanthellae

If nutrients and plankton are in limited supply in the oceanic waters surrounding atolls, then how does such a high diversity and abundance of filter feeders occur in coral reefs? One answer is found in the symbiotic (mutualistic) relationship of the hermatypic corals that contain zooxanthellae. Yonge (1963) and Taylor (1973) have addressed this topic. The invertebrates that are taxonomically placed in the phylum Coelenterata, classes Anthozoa and Hydrozoa, have dinoflagellates (zooxanthellae) living within them that serve as primary producers. The zooxanthellae have been shown to be derived from two genera of free-swimming dinoflagallates, and these have been cultured free of the coral as well as found free-living in nature (see Chapter 9).

The algal symbiont gains protection and a wide variety of nutrients from the coral host. Zooxanthellae have been shown to intercept all the phosphate that would normally be excreted and will utilize a variety of nitrates, urea, uric acid, guanine and adenine, all produced by the host.

For the host coral, this symbiotic association contributes photosynthates and oxygen, and aids in calcification. The zooxanthellae supply the coral with fixed carbon, especially in the form of glycerol and glucose. This food source in hermatypic corals is so stable and abundant that less than 5% of the coral polyp's energy is derived from external (filter) feeding. The importance of this food source to the hermatypic coral can be seen in the rapid growth of such corals compared with the growth of those lacking zooxanthellae. Even so, the higher metabolic rates of tropical-water corals would require larger energy sources than can be supplied by the zooxanthellae (Yonge, 1963).

An important role of zooxanthellae involves their aid in the calcification of the coral skeleton. As diagrammed in Figure 16-8, calcium must pass from the seawater in the coral polyp coelenteron (body cavity) through living tissue before reaching the skeleton. The aragonite form of calcium carbonate may result from combination of calcium with bicarbonate to form first calcium bicarbonate (equation 1) and then calcium carbonate (equation 2). The bicarbonate ions can be generated by the enzyme carbonic anhydrase from carbonic acid (equation 3).

$$Ca^{2+} + 2HCO_3^- \rightleftharpoons Ca(HCO_3)_2 \tag{1}$$

$$Ca(HCO_3)_2 \rightleftharpoons CaCO_3 + H_2CO_3 \tag{2}$$

$$H_2CO_3 \overset{\text{carbonic anhydrase}}{\rightleftharpoons} H^+ + HCO_3^- \tag{3}$$

In addition, carbonic anhydrase can catalyze the production of carbon dioxide from carbonic acid (equation 4).

$$H_2CO_3 \overset{\text{carbonic anhydrase}}{\rightleftharpoons} CO_2 + H_2O \tag{4}$$

The zooxanthellae are involved in this process, since they utilize carbon dioxide in photosynthesis and thus can reduce the levels of carbonic acid and aid in the shift from calcium bicarbonate to calcium carbonate (equation 2; Figure 16-8).

Apparently, some filamentous green algae also occur in the coral skeletons and are symbiotic with the corals. It should be noted that the zooxanthellae and the green algal endosymbionts are not an insignificant portion of the living biomass of a coral, since they usually account for three fourths of the total organic matter and thus are three times the dry weight of the coral polyps. Womersley

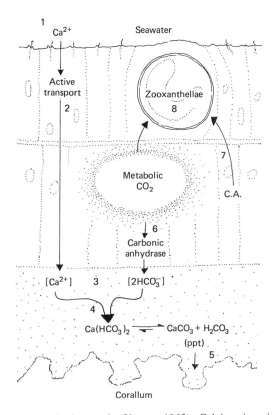

Figure 16-8 Calcium uptake by corals (Yonge, 1963). Calcium ions in the seawater (1) can be actively transported (2) through the gastrodermis of the coral animal and adsorbed on the mucopolysaccharide of the inner membrane (3) where it can be combined with bicarbonate to form calcium carbonate (4), which precipitates in the coral exoskeleton (corallum, 5). The bicarbonate in the inner membrane (3) is produced from metabolic carbon dioxide (process 6) with the aid of cabonic anhydrase. Carbon dioxide is also used in photosynthesis by zooxanthellae (8) found in the coelenteron and flagellated gastrodermis of the coral animal. Carbonic anhydrase (C.A.) will also produce carbon dioxide for the zooxanthellae directly (7). As depicted, the direction of the growth of the coral animal is upward, whereas the deposition of calcium is downward.

and Bailey (1969) have argued that the term *coral reef* is inaccurate and a better one would be "biotic reef" because of the importance of macroalgae. From a phycologist's viewpoint, it might be more reasonable to consider such reefs "algal reefs" because the algae support all the various animal systems.

Coral Reef Ecosystems

The dominant primary producers are the zooxanthellae and other mutualistic algae found in the corals as well as the free-living macrophytic algae. The calcified forms of algae contribute a significant amount of carbonate to the coral reef mass. The reef has been described as a giant filter, or trap, which is a closed system, removing the nutrients from the surrounding oceanic waters and utilizing these nutrients in the closed system of the reef. Although oceanic waters surrounding coral reefs are usually quite low in nutrients, with a constant flow of water over the reef organisms, sufficient nutrients can be obtained.

The abundance of zooplankton and phytoplankton is usually low in the surrounding oceanic waters but high (two to three times higher) in the atoll. The coral reef can be thought of as a complex forest, providing a wide variety of niches, tightly holding onto nutrients released through decomposition. However, van den Hoek et al., (1975) have suggested that this view may not be correct, that the reef is more open and that much is "lost" from the reef. Thus he compares the productivity of a reef to that of a typical benthic coastal marine community.

There is extreme competition for space and light in a coral reef, both by corals and their zooxanthellae symbionts as well as by free-living algae. The balance is delicate; epiphytic algae are removed or inhibited from growing over corals by toxic substances apparently released from the corals and by extensive grazing by fish and sea urchins. In studies in which coral heads were caged, preventing grazing by parrot fish (Scaridae) and sea urchins (echinoderms), the algae rapidly overgrew the corals, killing the polyps. An excellent paper on the importance of parrot fish in algal control was published by Brock (1979). In Guam, after corals were killed by the predatory starfish Acanthaster, blue-green algae colonies were initiated within 24 hr., and a completely filamentous community of algae was established afer 26 da (Belk and Belk, 1975). Similar effects by grazers are known in rocky shore communities (Chapter 15).

Ogden and Lobel (1978) have reviewed the role of herbivorous fishes and urchins in coral reef communities and Lobel (1980) has examined damselfish herbivory and its effects. They propose that herbivores and coral reef algae have a

Table 16-2   Benthic Plant Defense Mechanisms Against Herbivores

Defense	Examples
Escape in time (short-lived annuals)	Small filamentous forms (e.g., *Polysiphonia, Herposiphonia*)
Escape in space (cracks, surf)	Intertidal algal turfs, cryptic reef species, sea grass beds
Structural defense (hard, calcareous, lowered food quality)	*Turbinaria, Penicillus, Halimeda,* crustose corallines
Chemical defense (toxic, unpalatable)	*Sargassum* (tannins), *Asparagopsis* (ketones), *Caulerpa* (caulerpicin, caulerpin), *Laurencia* (halogenated compounds), *Dictyopteris* (dictyopterene), blue-green algae

Source: After Ogden and Lobel (1978.)

coevolved system of defense and counter-defense. They suggest that algal species have adapted toxic, structural, spatial, and temporal "defense" or "escape mechanisms" (see Table 16-2), whereas the herbivores employ strategies involving anatomical, physiological, and behavioral adaptations. They also noted that specific fish are highly selective in the algae they consume. In fact, Ogden and Lobel argue that selective feeding permits development of the coral animals, thus insuring the continuation of the coral reef.

A closed-system coral reef at the Smithsonian Institution has allowed Brawley and Adey (1981) to observe grazing efforts not only of larger animals but also of the smaller, amphipods. They found that algal diversity is decreased by intensive amphipod grazing, and filamentous algae (e.g., *Giffordia, Bryopsis, Ceramium*) quickly disappear. They also found that reduction in ampiphod populations resulted in a decrease in the larger algae populations due to increased diatom epiphytization and the reappearance of filamentous algae.

Fringing Reefs

The coralline algae of the Pacific Ocean fringing reefs of Waikiki in Hawaii have been thoroughly studied by Littler, both from the viewpoint of composition (1973a) and productivity (1973b). He found that crustose coralline algae covered 37% of the reef surface, exceeding all the other organisms as reef builders. Coelenterate corals covered less than 1% of the area (Figure 16-9). The major species of Hawaiian algal ridges were *Porolithon gardeneri* 6% cover) and *Lithophyllum kotschyanum* (7% cover). Another species, *Porolithon onkodes,* formed 41% of the cover on the heavily grazed (by parrot fish) seaward slopes of the algal ridge. This study demonstrates how significant coralline algae can be in a Pacific fringing reef.

In the Caribbean Sea, van den Hoek and his students have carried out detailed community studies of fringing reefs in Curacao (van den Hoek, *et al.,* 1975) as well as determining the productivity of the macroscopic algae (Wanders, 1976a, 1976b). An elegant set of zonation diagrams depict the reef transect from intertidal to 20.5 m in depth (van den Hoek *et al.,* 1975). Wanders was able to identify 142 algal specimens and seven zones through a mathematical analysis of the data. Although coral animals dominated this reef, fleshy and filamentous algae were also abundant and diverse (54 species/25 m^2), although heavily grazed. Van den Hoek *et al.* (1975) pointed out that the reef extended to only 30m, which is shallow when compared to the depths of hermatypic corals of atolls such as Glover's Reef. They believed that turbidity from the island runoff might be the major limiting factor in coral development at greater depths.

The seven zones included an upper intertidal blue-green alga community that for the most part was endolithic (Figure 16-10a) in the limestone. The second community was the lower intertidal zone, containing a variety of fleshy and filamentous algae (Figure 16-10a). The third community (Figure 16-10a) was a shallow-breaker zone dominated by small brain coral *(Diploria);* it had the highest diversity of algae (64 species/25 m^2). The fourth community was dominated by

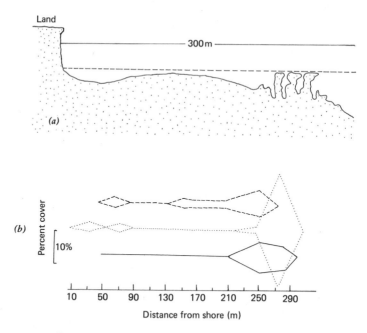

Figure 16-9 A fringing reef off the Waikīki Natatorium on the island of Oahu (Hawaii) has been studied by Mark Littler (Littler, 1973a). (*A*) The bottom profile. (*B*) Three nonarticulated corallines dominate the fringing reef, *Porolithon gardineri* (dashed line), *P. onkodes* (dotted line) and *Lithophyllum kotschyanum* (solid line). The percentage are cover of each species can be seen in (*B*). All three are most abundant in the area of the fringing reef, but each appears to have a specific niche.

surge-orientated corals such as *Acropora* and *Millepora* (Figure 16-10*b*) as well as a community of *Diadema,* the long-spined black sea urchin. There was an average of 60 algal species/25 m², including the highest percentage cover of coralline algae, especially *Porolithon*. The fifth community was essentially a transition with sandy substratum and scattered soft corals (Figure 16-10*c*), *Pseudopterogorogia* and *Plexaura*. The sixth community was dominated by the large boulder-like corals *(Montastrea),* and the seventh community occurred at the precipice of the submarine plateau (Figure 16-10*d*). The last area had shingle-type corals, flattened to receive the maximum amount of light because of the low levels of illumination.

Atolls

During an expedition to the Solomon Islands by the Royal Society of London, Womersley and Bailey (1969) were able to describe four types of reefs based on wave action. By the use of transect lines they described algal zonation for each of the reef types. Figure 16-11 is taken from their diagram of the reef flat and reef

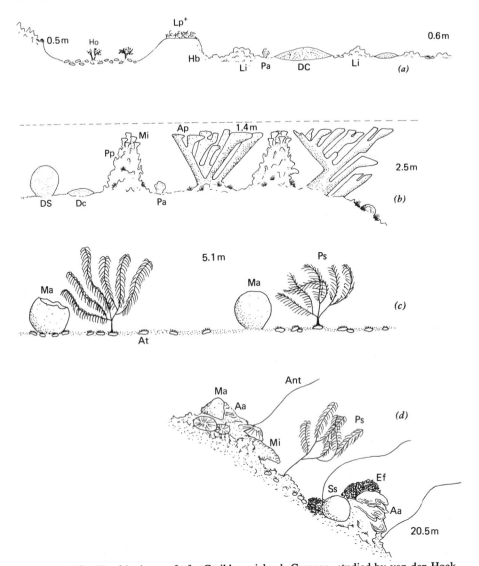

Figure 16-10 The fringing reef of a Caribbean island, Curacao, studied by van den Hoek *et al.* (1975). Four sections are drawn from their detailed zonal diagram. (*a*) The intertidal to 0.6 m region contains *Halimeda opuntia* (Ho), *Laurencia papillosa* (Lpt), *Hydrolithon boergesenii* (Hb), *Lithophyllum intermedium* (Li), *Porites astreoides** (Pa), and *Diploria clivosa** (Dc). (*b*) Another section, taken 1.2–2.5 m deep includes such species as *Diploria strigosa** (Ds), Dc*), *Porolithon pachydermun* (Pp), *Millepora sp.** (Mi), Pa*), and *Acropora palmata** (Ap). (*C*) At an intermediate depth (5–5.3 m) the following domi-nant species are found: *Montastrea annularis** (Ma), algal turf (At), and *Pseudopter-ogorgia acerosa* and *P. americana* (soft corals; Ps*). At the final drop off (*d*), where hermatypic corals terminate (17–20.5 m), a number of species are found including (Ma*), *Agaricia agaricites** (Aa), *Antipatharian* sp* (Ant), (Mi*), (Ps*), *Siderastrea siderea** (Ss), *Eusmilia fastigiata** (Ef), and (Aa*). (The asterisks denote coral animals.)

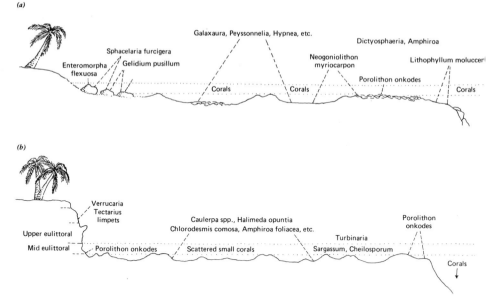

Figure 16-11 Transects of the seaward reefs off Mamara Island (moderate wave action) on Guadalcanal (*a*) and off Banika Island in the Russell Islands (*b*), made by Womersley and Bailey (1969). Both reefs were about 25–30 m wide. Both reefs had a *Porolithon* or *Lithophyllum* algal ridge and a wide variety of algae found on the reef flat. Mid to upper eulittoral zones had a vertical range of about 2/3 m on both islands.

edge of a "moderate wave energy" reef at Mamara Island on the atoll of Guadalcanal (Figure 16-11*a*) and a "high wave energy" reef at Banika Island in the Russell Islands (Figure 16-11*b*). The seaward rim consisted of crustose coralline algae (predominantly *Lithophyllum*) and corals. Behind the rim, the coralline alga *Porolithon onkodes* occurred as the dominant covering. A wide variety of non-calcified algae were found on both reef flats.

The reef flat gradually dropped toward the island, forming a "moat" in which a variety of fleshy and filamentous algae occurred. Moats are characteristic of the inner reef flat and apparently are formed by boring organisms and currents that run parallel to the island during high tide. Crustose and endolithic algae occurred in the intertidal rocks at the island edge. Van den Hoek *et al.* (1975) were able to denote distinct zonations within the intertidal zone where firm substrata occurred. The coralline alga not only dominated the reef edge (*Lithophyllum* and *Porolithon*) but also cemented the rubble behind the reef ridge (*Neogoniolithon*).

In the Caribbean, Tsuda and Dawes (1974) found a total of 104 species of macroscopic algae, consisting of 8 blue-green, 39 green, 19 brown, and 34 red algae on the atoll of Glover's Reef off Belize, Central America. The number of algae counted was estimated to be about 60–70% of the actual number present. Five zones were distinguished on the seaward side, off Long Cay (Figure 16-12).

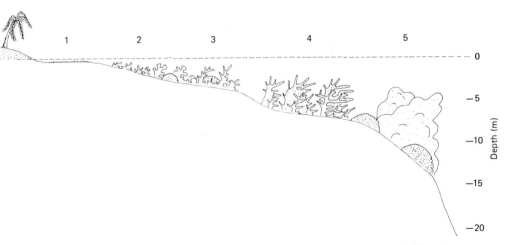

Figure 16-12 Transect of the seaward reef of Long Cay on the Caribbean atoll Glover's Reef. The reef width is about 20–25 m. Zone 1 is the reef flat proper with various coralline algae (*Lithothamnion, Goniolithon*), small *Thalassia* beds, and small corals such as *Millepora*. Zone 2 is the breaker zone, extending in depth from 0.1 to about 1 m and dominated by *Acropora palmata* and *Millepora alcicornis*, both corals. Zone 3 includes an upper region called the lower palmata zone (1–3 m deep) and a lower region called the buttress zone (3–5 m deep). The buttress zone is dominated by large corals, especially *Agaracia, Montastrea, Siderstrea, Porites*, and *Millepora*. Zone 4 is the *Acropora cervicornis* zone, in which large beds of the elkhorn coral occur; the zone extends from 5 to 15 m in depth. The final coral zone is the annularis zone (zone 5) that extends from 15 to 40 m in depth and is dominated by *Montastrea* (*M. annularis*) forming hugh multilobar heads, *Diploria, Siderastrea, Porites, Agaracia*, and tubular sponges. Below the drop-off, the shingle corals and leafy algae dominate.

Intertidal

This zone ranged from an upper region of coral rubble and bleached sand to a lower area of shell or "coquina substratum" (Figure 16-13). The algae were predominantly blue-green, endolithic forms at the upper and mid intertidal region, grading into filamentous or turfy forms (*Cladophorophsis*, other green alga) and larger brown algae (*Turbinaria, Colpomenia, Sargassum*) at the intertidal fringe. A number of other green (*Valonia, Dictyosphaeria*) and red algae (*Centeroceras, Pterocladia*) were evident in the lower intertidal region (Figure 16-13).

Breaker zone

The next major region ranged in depth from 0.1–3 m and was dominated by increasingly larger corals (see Figure 16-6), especially *Acropora palmata* (bramble coral) and *Millepora alcicornis* (fire coral). These corals were always found rising just to the lowest tide level and growing outward in the direction of the

Figure 16-13   The intertidel zone of the reef forming the Caribbean atoll Glover's Reef in Belize, Central America. The intertidal zone consists of rubble, beach rock and sand. Blue-green algae predominate here. See Figure 16-6 for the surge zone dominated by *Acropora plamata*, the bramble coral.

Figure 16-14   Underwater photo of the *Acropora cervicornis* zone at Glover's Reef, showing some of the buttress corals such as *Diploria* at depths of 4–7 m.

surge (Figure 16-6). The algae primarily consisted of small filamentous fuzzes, which were heavily grazed by parrot fish and the black sea urchin, *Diadema*. The dominant algae were *Giffordia, Gelidiella, Pterocladia Polysiphonia,* and *Sphaceleria* spp.

Lower palmata and buttress zone

The lower region of the reef had a depth range of 3–5 m and was so termed because it was dominated by the large buttress or boulder-type coral, *Montastrea*. Other corals included *Acropora palmata, Agaricia, Diploria, Porites* (as massive buttresses), and *Millepora*. The algae were mostly filamentous and were found growing on dead coral rubble. *Valonia, Sphacelaria, Gelidiopsis,* and *Pterocladia* as well as articulated coralline algae such as *Jania* and *Amphoria* were present.

Cervicornis zone

The upper region of the seaward slope was dominated by the elkhorn coral, *Acropora cervicornis,* which forms large thickets (Figure 16-14). The zone continues (5–15 m depth) on an increasingly steep slope, forming *spurs* (corals) and *grooves* (sand). The grooves then become quite evident in the final zone. The algae were divided between siphonaceous green *(Udotea, Penicillus, Caulerpa, Halimeda)* and filamentous forms *(Polysiphonia, Pterocladia, Ceramium, Heterosiphonia)* in the sand-covered groves, as well as larger, leafy brown algae *(Sargassum Zonaria, Spatoglossum)* on the dead coral rubble.

Montastrea zone

The remaining coral region of the seaward slope was a sharp drop from 15–40 m with an angle of about 60°. The spurs and grooves became very pronounced here. The zone was abruptly cut off at the drop off when the slope changes from 25–60°. It was dominated above by the large boulder corals such as *Montastrea annularis* and at the drop-off, the flattened, shingle forms of coral such as *Diploria, Sidesastrea,* and *Agarcia* (Figure 16-15) occurred.

The final region, below the drop-off, contained leafy forms of algae including such species as *Dictyopteris, Pocockiella* (= *Lobophora*), and *Zonaria* (brown algae), as well as long chains of *Halimeda discoidea* and *H. opuntia* (Figure. 16-16) (green algae), and crustose red algae *(Hildenbrandtia, Peyssonnelia)*. Articulated coralline red algae were present *(Amphoria, Jania)* as well as some crustose forms *(Lithophyllum* and *Neogoniolithon)* at the upper levels. The corals all have shingle-type morphologies and are much smaller because of the low light levels.

Figure 16-15 Underwater photo at Glover's Reef, showing the spur and grooves of the *annularis* zone at a depth of 20 m. Large (2–5 m high) heads of *Montastrea annularis* are visible here.

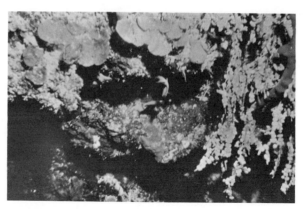

Figure 16-16 Underwater photo at Glover's Reef, showing some deep-water algae growing in about 50 m of water below the drop-off, along with leaf-like coral sheets. Two species of *Halimeda* as well as a leafy, crustose brown alga, *Lobophora*, are evident with flash illumination. (Underwater photos in Figures 16-14 through 16-16 courtesy of David Greenfield, Northern Illinois University, Dekalb.)

PRODUCTIVITY OF CORAL REEFS

The energetic role of algae in a reef ecosystem was considered in a review paper by Marsh (1976). Published net productivity values range from 0.18 g C/m$^2 \cdot$ hr for coralline algae of the Pacific atoll Rongelap to 0.72 g C/m^2/hr for algae studied on Eniwetok. The highest-producing component seems to be a population of *Sargassum* (0.057–1.95 g C/m$^2 \cdot$ hr). A comparison of the rates of carbon fixation per day indicates that the entire production of a reef ecosystem is high (Table 16-3).

Table 16-3   General Productivity Figures, a Comparison

Region	Productivity(g C/m$^2 \cdot$ da)
Open sea	0.1–0.35
Coral reef	5.0–10.0
Sugarcane field	19.0
Thalassia seagreass community	*11.0*
Wheat field	*5.0*

The productivity of crustose corallines on the fringing reefs of Hawaii has been studied by Littler, using various techniques (labeled carbon, oxygen probe, pH electrode). His studies (1973b) demonstrated an average value of 5.7 g C/m$^2 \cdot$ da, which was in agreement with the results of previous workers but higher than values reported by Marsh (1969) for the reef edge of Eniwetok Atoll in the Pacific. (1.5g C/m$^2 \cdot$ da).

Wanders (1976a, 1976b) studied a shallow, fringing reef in Curacao in the

Caribbean and compared the productivity of both calcareous and fleshy algae. He used the light/dark-bottle method and Winkler titration for oxygen and assumed 1 mol O_2 was released for each mol CO_2 consumed in photosynthesis. His findings were higher than those of previous studies, perhaps in part due to extrapolation from a few points in time. However, the comparative rates of photosynthesis of the different algal components of a fringing reef are of interest (Table 16-4).

Table 16-4   Productivity of Algae from a Fringing Reef in the Caribbean (Curacao)

Type of Algae	Productivity (g $C/m^2 \cdot da$)
Encrusting coralline algae	2.7
Fleshy and filamentous algae	3.3
Coral zooxanthellae	11.0
Total	17.0

Source: After Wanders (1976b).

As noted by Wanders (1976a), the shallow reef of Curacao has a gross productivity 50–100 times higher than the productivity of the open Caribbean. He considered water movement to be the major factor determining the high primary production by a coral reef. The continuous water flow would supply new nutrients, although in low concentrations, and the waves would enhance diffusion rates. Surprisingly, the fleshy algae were among the most important producers in the coral reef that fringes Curacao. Buesa (1977) also found that fleshy algae in tropical waters (off Cuba) showed high rates of productivity (7.2 mg O_2/g dry wt $\cdot$ hr). Similarly, Colinvaux (1974) demonstrated that tropical calcareous algae such as *Halimeda* can have high rates of carbon fixation (4 g $C/m^2 \cdot da$).

In a parallel study, Wanders (1976b) found that a combination of *Sargassum* and members of the brown algal order Dictyotales produced about 18.6 g $C/m_2 \cdot da$. Thus Wanders argued that coral reefs were not unique in being areas of high productivity in the tropics. Rather, he said any well-developed algal or sea grass population might be expected to show similar rates. Certainly all the present evidence thus far supports the concept that coral reefs are highly productive. However, the closed concept of a coral reef is probably not accurate. The most critical factor in an open system would be adequate water movement to insure efficient nutrient diffusion, although actual concentrations of the nutrients may be quite low.

MICROBIAL INFLUENCES IN A CORAL REEF

The microbial effects on coral reefs are poorly known, but some interesting features such as ciguatera, nitrogen fixation, and symbiotic bacteria and blue-green algae in sponges can be considered. Ciguatera is commonly associated with coral reefs throughout the tropical world (Banner, 1976). The illness, which may result in death, is brought on by eating fish that contain the toxin(s). The symptoms of the disease are mostly neurological, but may include gastrointestinal symptoms (nausea and diarrhea). Banner (1976) suggested that the sources of ciguatera are in the food chain, probably from the blue-green algae or bacteria of coral reefs on which fish feed. More-recent evidence implicates bacteria or yeasts (fungal, asco-mycetes) that may occur in the detritus and thus be fed upon by detrital feeders (fish, invertebrates). In addition, a ciguatera-type toxin has been found in a benthic, tropical dinoflagellate (see Chapter 9).

Wiebe (1976) presented data on the source of fixed nitrogen in a coral reef. It was evident that more nitrogenous compounds escaped the reef flat than were present at the surrounding edge of the reef, suggesting nitrogen fixation on the reef. He found that the increase in nitrogen compounds in reef flat water resulted from nitrogen fixation, mostly by blue-green algae on the reef flat. The average level of fixed nitrogen production was 985 kg/ha • yr, which is as high as that reported for an alfalfa field. The free-living blue-green alga *Calothrix crustacea* was the most important blue-green involved in nitrogen fixation on Enewitok At-oll. Thus Wiebe and his workers were able to explain how coral reefs maintain such high productivity rates in the face of extremely low levels of nitrogenous compounds in oceanic waters.

Finally, symbiotic blue-green algae have been found in a number of coral reef sponges (Wilkinson and Fay, 1979). Using electron microscopy, they found that the blue-green algae were located along with bacteria in specialized vacuolated cells called *bacteriocytes*. These cells were found throughout the tissues of the sponge *Siphonochalina* and in the brown-pigmented ectosome of *Theonella*. The mixed populations of bacteria and blue-green algae in the sponges were thought to be beneficial to both procaryotes. Baceteria probably enhance nitrogen fixation by creating a reducing atmosphere. The net result is a general leakage of fixed ni-trogenous compounds from the sponge to the surrounding reef organisms.

REFERENCES

Adey, W. H. 1978. Algal ridges of the Caribbean Sea and West Indies. *Phycologia* **17**: 361–367.

Banner, A. H. 1976. Ciguatera: a disease from coral reef fish. In Jones, O. A. and R. Endean (eds.). *Biology and Geology of Coral Reefs,* vol. 3 Part 2, Academic, New York pp. 117–214.

Belk, M. S. and D. Belk. 1975. An observation of algal colonization on *Acropora aspera* killed by *Acanthastera planci. Hydrobiologia* **46**: 29–32.

Brock, R. E. 1979. An experimental study on the effects of grazing by parrotfishes and role of refuges in benthic community structure. *Mar. Biol.* **51**: 381–388.

Brawley, S. H. and W. H. Adey. 1981. The effect of micrograzers on algal community structure in a coral reef microcosm. *Mar. Biol.* **61**: 167–177.

Buesa, R. J. 1977. Photosynthesis and respiration of some tropical marine plants. *Aq. Bot.* **3**: 203–216.

Colinvaux, L. H. 1974. Productivity of the coral reef alga *Halimeda* (Order Siphonales). *Proc. Second Intern. Coral Reef Symp.* **1**: 35–42.

Davis, R. A. 1977. *Principles of Oceanography.* Addison-Wesley, Reading, Mass.

Littler, M. M. 1973a. The population and community structure of Hawaiian fringing-reef crustose corallinaceae (Rhodophyta, Cryptonemiales). *J. Exp. Mar. Biol. Ecol.* **11**: 103–120.

Littler, M. M. 1973b. The productivity of Hawaiian fringing-reef crustose corallinaceae and an experimental evaluation of production methodology. *Limnol. Oceanogr.* **18**: 946–952.

Littler, M. M. 1976. Calcification and its role among macroalgae. *Micronesica* **12**: 27–41.

Lobel, P. S. 1980. Herbivory by damselfishes and their role in coral reef community ecology. *Bull. Mar. Sci.* **30**: 273–289.

Marsh, J. A., Jr. 1969. Primary productivity of reef-building calcareous red algae. *Ecology.* **51**: 255–263.

Marsh, J. A., Jr. 1976. Energetic role of algae in reef ecosystems. *Micronesica* **12**: 13–21.

Ogden, J. C. and P. S. Lobel. 1978. The role of herbivorous fishes and urchins in coral reef communities. *Env. Biol. Fish.* **3**: 49–63.

Ross, D. A. 1970. *Introduction to Oceanography* Appleton-Contrary-Croft, New York.

Stoddard, D. R. 1962. *Three Caribbean Atolls: Turneffe Islands, Lighthouse Reef, and Glover's Reef, British Honduras.* Atoll Res. Bull. No. 87. Smithsonian Inst., Washington, D. C.

Taylor, D. L. 1973. The cellular interactions of algal-invertebrate symbiosis. In Russell, F. S. (ed.). *Advances in Marine Biology.* Academic, New York. pp. 1–56.

Tsuda, R. T. and C. J. Dawes. 1974. *Preliminary Checklist of the Marine Benthic Plants from Glover's Reef, British Honduras.* Atoll Res. Bull. No. 173a.

Van den Hoek, C., A. M. Cortel-Breeman, and J. B. Wanders. 1975. Algal zonation in the fringing coral reef of Curacao, Netherlands Antilles, in relation to zonation of corals and gorgonians. *Aq. Bot.* **1**: 269–308.

Wanders, J. B. W. 1976a. The role of benthic algae in the shallow reef of Curacao (Netherlands, Antilles). I. Primary productivity in the coral reef. *Aq. Bot.* **2**: 235–270.

Wanders, J. B. W. 1976b. The role of benthic algae in the shallow reef of Curacao (Netherlands, Antilles). II. Primary productivity of the *Sargassum* beds on the northeast coast submarine plateau. *Aq. Bot.* **2**: 327–335.

Wiebe, W. J. 1976. Nitrogen cycle on a coral reef. *Micronesica* **12**: 23–26.

Wilkinson, C. R. and P. Fay. 1979. Nitrogen fixation in coral reef sponges with symbiotic cyanobacteria. *Nature* **279**: 527–529.

Womersley, H. B. S. and A. Bailey. 1969. The marine algae of the Solomon Islands and their place in biotic reefs. *Phil. Trans. Roy. Soc. London. Series B.* **255**: 433–442.

Wray, J. L. 1977. *Calcareous algae.* Elsevier, Amsterdam.

Yonge, C. M. 1963. The biology of coral reefs. In Russell, F. S. (ed.). *Advances in Marine Biology,* Academic, New York, pp. 209–260.

CHAPTER 17

Seagrass Communities

Seagrasses are marine monocots that have a number of important ecological roles in the shallow waters of tropical and temperate coasts, especially with regard to productivity, which is highest in the tropics (Figure 17-1). The coasts of North America have a number of species of seagrasses in both the temperate and tropical waters (seven genera); they are almost completely absent in South America. The most-abundant genera of North American marine phanerogams include *Zostera, Phyllospadix, Halodule,* and *Thalassia. Zostera* is the only seagrass of the European Atlantic coast; two other genera, *Posidonia* and *Cymodocea,* occur in the Mediterranean.

ADAPTATIONS TO THE MARINE ENVIRONMENT

The various genera of seagrasses are not closely related to each other and in fact are not even true grasses (family Poaceae). Rather, this is an ecological grouping of monocot genera that appears to be more closely related to members of the lily family. One important feature of seagrasses is their adaptation to submerged conditions (hydrophytes). In this first section, consideration will be given to the anatomical and morphological features that are common to these primarily subtidal marine monocots. The features of their environment that have influenced the evolution of these hydrophytic monocots will also be examined.

Morphological Adaptations

Arber (1920) and den Hartog (1970) list four properties that are indispensible for a marine angiosperm:

1 It must adapt to a saline medium.
2 It must grow completely submerged (hydrophytic adaptations).
3 It must be able to withstand wave action and tidal currents.
4 It must be able to carry out hydrophilous pollination and seed dispersal.

Thus it is evident that the marine environment of the seagrasses has to be considered in order to understand their hydrophytic adaptations.

Figure 17-1   Beds of the seagrass *Thalassia testudinum*. They are common in the shallow waters surrounding the Florida Keys. The broad (2–5 cm) blades are usually heavily epiphytized by algae.

A number of distinctive morphological adaptations to the marine environment have evolved in seagrasses. The plants have well-developed rhizomes (horizontal stems) that usually occur below the surface of the substratum and in close association with one another. Thus the seagrass beds are sites of sedimentation due to the trapping of debris and stabilization of the substratum. All of the seagrasses have alternating leaves occurring in two ranks that usually arise from small erect side stems called short shoots, or from the rhizomes (Tomlinson, 1974). Roots are common and develop from the rhizomes as well as from the basal portion of each short shoot. Rather than having fibrous roots such as those found on terrestrial grasses, seagrasses usually have roots that are thicker and more fleshy (but see *Zostera*). The leaves are flat, ribbon-shaped, or cylindrical in cross section and are supple; thus they can withstand water movement, yet remain erect. The flowers are small and pale white and occur at the base of the leaf clusters. The stamens (anthers) and pistils (style and stigma) extend above the petals. Usually the pollen is released in gelatinous strands that are carried by water currents. The pollen grains are elongate (family Potamogetonaceae) or spherical (Hydrocharitaceae), and are arranged in coherent moniliform chains. *Enhalus*, a genus of the Southern Hemisphere, has a flower that can project above the water; globular pollen is released from the free-floating spathe after it is broken off the plant. Most seagrasses are dioecious and the few monoecious genera exhibit early development of the ovary (proterogyny: *Zostera Heterozostera, Halophila*). An excellent comparison of shoot organization for all 12 seagrass genera is presented by Tomlinson (1974). Flowering in cultured plants has been observed in

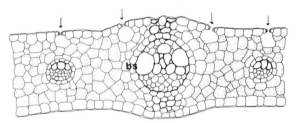

Figure 17-2 A cross section of the wheat leaf (*Triticum aestivum*), showing the typical mesophytic leaf anatomy of a monocot. A main vein and two secondary, parallel veins are visible. The upper side of the leaf contains most of the stomates (four are visible here, arrows) since the leaf curls upward with wilting, thus confining the stomates. The primary photosynthetic tissue is a ground mesophyll (stippled lightly) in which the chloroplasts are found. No chloroplasts occur in the epidermal cells. Fiber bundles with heavily lignified fiber cells occur on both sides of the main vein and throughout the leaf. The bundle sheath (*bs*) cell walls of all the veins contains suberin, a waxy substance, and controls water movement from the vein into the ground mesophyll. Each vein has a number of vessels making up the xylem (upper portion of vascular bundle, large cells) as well as some phloem sieve elements (lower portion of vascular bundle). A well-developed cuticle is found on both leaf surfaces.

Cymodocea serrulata, Halophila stipulacea, Syringodium isoetifolium, Zostera capensis, and *Thalassia hemprichii,* all species from Kenya (McMillan, 1980b).

Anatomical Adaptations

Internal structure is quite uniform among species of seagrasses and is typical of a hydrophyte (submerged plant). One of the most characteristic anatomical features of the blades, short shoots, roots, and rhizomes of all seagrasses is an aerenchyma common to hydrophytes (Haberlandt, 1914). This specialized parenchymatous tissue consists of a regular arrangement of air spaces, or lacunae, that aid in flotation of the leaves and allow gas exchange throughout the plant. In general the anatomical structure in hydrophytes is highly reduced.

A cross section of a blade of a mesophytic monocot such as *Triticum* (wheat) (Figure 17-2) can be compared with the blade of a hydrophyte such as *Thalassia* (Figure 17-3*A*) or *Halodule* (Figure 17-3*B*) to determine what adaptations the hydrophytes might have. Table 17-1 summarizes comparative differences between the leaves of *Triticum* and *Thalassia*. The epidermis of the hydrophyte has an almost-invisible, thin cuticle. The outer cell wall of the epidermal cells is thick, and both the wall and the cuticle can be seen in electron micrographs of epidermal cells (Figure 17-4). The epidermal cells of seagrasses contain chloroplasts, whereas in land plants the epidermal cells usually lack them. Evidence for osmoregulation by specialized epidermal cells has been suggested for *Thalassia testudinum* (Jagels, 1973). It is based on the highly invaginated plasmalemma and numerous mitochondria found in some of the epidermal cells (Jagels, 1973) and by the presence of pores in the cuticle (Gessner, 1971). Stomates and associated

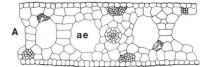

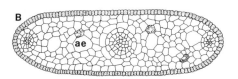

Figure 17-3 Cross sections of two seagrass leaves showing typical hydrophytic reduction of leaf anatomy. (*A*) *Thalassia testudinum* has leaves that are 2–3 mm in thickness. The leaf contains large air spaces (*ae*) due to development of aerenchyma instead of ground mesophyll as found in the wheat plant. (*B*) *Halodule wrightii* is a smaller, thinner (1 mm) leaf but has the same air system (*ae*). In all seagrass species, the chloroplasts are limited to the epidermal layer, and the cuticle, although present, is thin (see also Figure 17-4). The veins are small and the xylem is usually unlignified (see also Figure 17-5). Bundles of small but thin-walled cells are evident in *Thalassia* leaves, but no lignin is present. No stomates are evident in the epidermis of either leaf.

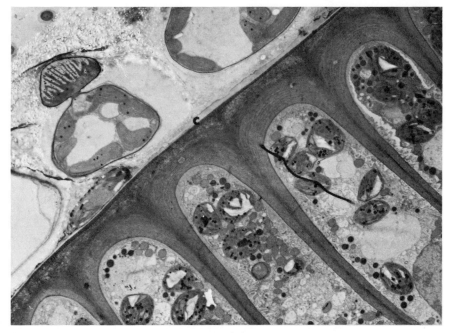

Figure 17-4 A transmission electron micrograph of the epidermal cells in the manatee grass, *Syringodium filiforme*, showing a thin cuticle on the cell wall (*c*). The chloroplasts, found only in the epidermis, contain numerous starch grains. The algal epiphyte seen in section on the grass blade is the calcified, nonarticulated coralline *Fosliella* sp. A pit connection is visible between the basal cell and the specialized upper *star cell*. Holes in the red algal wall are due to CaCO₃ crystals that have dissolved. × 3400.

guard cells are lacking in the epidermis of *Thalassia* and *Halodule* in contrast to the epidermis of terrestrial monocots such as *Triticum*. *Triticum* is filled with a general chlorenchyma tissue called ground mesophyll that contains plastids; in the seagrasses the aerenchyma has well-developed lacunae, or air spaces, and no chloroplasts are present. Vascular bundles of seagrasses are highly reduced, and little, if any, lignification is evident. Small fiber bundles, running parallel to the vascular bundles, are evident in the larger blades of *Thalassia,* but they lack lignification. In contrast, lignin is an evident cell wall component in the fiber bundles and vascular bundles of *Tricitum*. Because of the lack of lignin and simplicity of the conducting cells in vascular bundles of seagrasses, it is difficult to interpret which cells are phloem and which are xylem. Even under the electron microscope (Figure 17-5) the phloem and xylem cells are relatively indistinct in the vascular bundle of *Halodule*. A recent comparative study of the anatomy of vegetative organs of the family Hydrocharitaceae (*Halophila, Thalassia, Enhalus*) by Ancibar (1979) describes anatomical features with relation to the adaptations of the species. Although the xylem is considered relatively primitive in structure, the phloem shows a number of evolutionary developments.

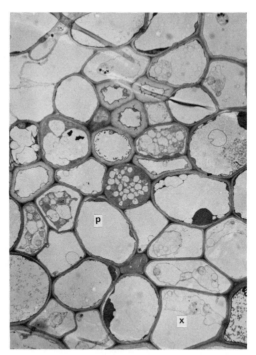

Figure 17-5 A transmission electron micrograph of the main vascular bundle of *Halodule wrightii*, showing the thin-walled xylem (*x*) and phloem (*p*) cells and the presence of cytoplasm in many of the cells. The thicker-walled cells near the phloem are fiber cells. Compared with a mesophytic terrestrial monocot, little differentiation is evident in this vascular bundle. × 1700.

Table 17-1   Thalassia Testudinum Leaf and Sheath Structures Compared with a Leaf of a Typical Monocotyledon Grass (Triticum)

Structure	Typical Monocotyledon (*Triticum*)	Seagrass (*Thalassia testudinum*)
Leaf blade shape	Linear, 1–2 mm	Ribbonlike, broad, 5–10 mm
Veination	Parallel	Parallel
Meristem tissue	Intercalary	Intercalary
Cuticle	Present, 1–2 μm thick	Thin—2.3 μm thick
Epidermis	Few to no chloroplasts; bulliform cells often present	Many chloroplasts; no bulliform cells
Stomata	Most on lower epidermis, often associated with trichomes, cork, or silica cells	No stomata
Ground tissue		
Palisade layer and mesophyll layer	No distinct differentiation, with mesophyll cells surrounding vascular bundles in orderly manner; many chloroplasts	No palisade layer, aerenchyma tissue with regularly spaced air spaces and vascular bundles; no chloroplasts
Vascular tissue		
Primary xylem	lignification obvious oriented toward upper epidermis;	In center of bundle surrounded by phloem, also toward upper epidermis; lignification absent or limited
	Two large vessels with tracheids between	One large vessel, tracheids obscure
Primary phloem Sieve tubes Companion cells	Toward lower epidermis	Surrounding xylem and toward lower epidermis
Bundle sheath	One thin-walled sheath	No sheath in lamina, sometimes present in leaf sheath, especially around midvein
Bundle sheath extension	Present in most vascular bundles, in form of one to many sclerenchyma cells	Noted in leaf sheath above and below midvein only
Fiber bundles	At margins, lignified	At margins, nonlignified

TAXONOMIC AND ECOLOGICAL CONSIDERATIONS

The marine flowering plants are not true grasses of the family Poaceae but belong in two related families of aquatic monocots; Hydrocharitaceae and the Potamogetonaceae.

Taxonomic Considerations

There are about 50 presently recognized species placed in 12 genera in the families Hydrocharitaceae *(Halophila, Thalassia, Enhalus)* and Potamogetonaceae *(Zostera, Phyllospadix, Posidonia, Halodule, Cymodocea, Syringodium, Amphibolus, Heterozostera,* and *Thalassodendron).* The question of seagrass evolution can be viewed from two rather-different perspectives. Arber (1920) proposed that seagrasses evolved through a gradual transition from freshwater and brackish-water hydrophytic species. This has been the prevailing view until recently. In contrast, den Hartog (1970) has considered some evidence from the fossil record that seagrasses may have evolved before the Tertiary period (in the

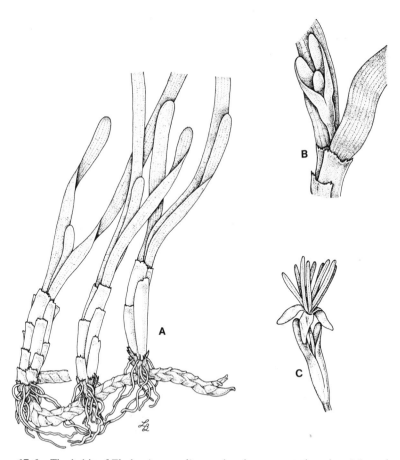

Figure 17-6   The habit of *Thalassia testudinum*, showing a vegetative plant (*A*), and male (*C*) and female (*B*) flowers. The flowers are found at the base of the leaf clusters and therefore are near the substrate level (top of leaf sheaths in *A*). (*A*) × 0.40; (*B*) and (*C*) × 2.

Cretaceous period) from xerophytic, salt marsh forms. On the basis of geographical distribution, McCoy and Heck (1976) argued for a pantropical seagrass evolution.

Geographical Distribution

The largest abundance of seagrasses occurs in the tropics (seven genera vs. five from temperate waters) and seagrasses are concentrated in two major areas of the Indo-Pacific and the Caribbean and Pacific coasts of Central America (den Hartog, 1970). Seagrasses are common in lagoons of coral reefs as well as in the shallower waters of continental shelves. Although quite common throughout the world, seagrasses have been poorly understood mainly because most higher plant taxonomy studies did not include the marine environment, and most oceanographic studies were deep-water projects.

Distribution is disjunct, with about three quarters of the species occurring in the Old World, having a center of distribution in the Indian and western Pacific Oceans. The remaining species occur in the Caribbean. It appears that seagrasses are relatively old types of marine plants, originating in the early Tertiary period (or perhaps earlier) when parts of Central America were submerged. It is believed that the distribution between the Caribbean and Pacific coasts took place before the rising of the Panamanian land barrier. Today these are a number of closely related Caribbean and Pacific species: *Thalassia testudinum (Figure 17-6) and T. hemprichii; Syringodium filiforme* and *S. isoetifolium;* and *Halodule wrightii* (Figure 17-7) and *H. uninervis*. Based on these species pairs, McCoy and Heck (1976) argued that the present global patterns are due to the existence of worldwide ancestral biotas, which developed in the Cretaceous, before the breakup of Gondwanaland. They said the separation of continents and resulting land and water barriers resulted in parallel speciation of the seagrasses. How important land barriers are was pointed out by den Hartog (1970) when he showed that *Halophilia stipulaceae* did not occur in the Mediterranean until after the opening of the Suez Canal. Now this species has become established even along the northern coasts of the Mediterranean.

Selected Species

In this section a brief discussion of four species of seagrasses will be presented. In each case the plant will first be described based on its taxonomic and morphological features, and then various aspects of its growth, ecology, and distribution will be given. The reader is referred to the text by den Hartog (1970) and the paper by Tomlinson (1974) for other genera. A key to the 12 genera, modified from Tomlinson (1974), is presented in Table 17-2.

Zostera marina

Eelgrass is common throughout the North Atlantic and Eastern Pacific oceans. It has a distinctive short shoot, which bears the flat, ribbon-shaped leaves and a

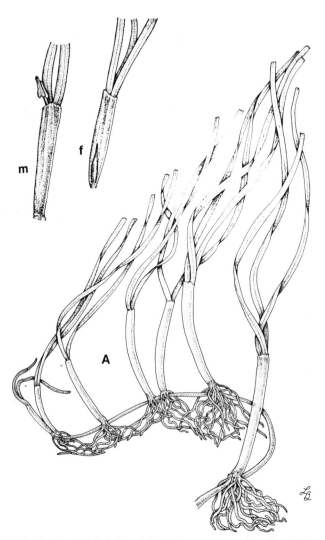

Figure 17-7 The habit of *Halodule wrightii* shows a vegetative plant (*A*), male (*m*), and female (*f*) flowers. The blades will reach 2–3 mm in width but are usually 1.5 mm wide. (*A*) × 1; (*m*) and (*f*) × 1.50.

horizontal, indeterminate rhizome. The roots occur in two clusters at each node (Figure 17-8). The leaves have a basal meristem. The flowers are unisexual and there are about 10 flowers on a spadix; the ovary develops before the stamens (proterogyny). A single plant will produce about 200 seeds in a season (up to 500 for *Zostera* from the east coast of the United States). The seeds are cylindrical, 1 × 2 mm in size, and marked with about 20 ribs on the coat. Germination and formation of a mature plant usually take 1–2 yr.

  Zostera marina is a widely distributed seagrass, found in temperate waters of the North Atlantic and Pacific oceans and even extending into the Artic Circle. It

Table 17-2 A Key to the Genera of Seagrasses[a]

1 All axes alike, that is producing only foliage leaves[b]................................ **2**

1 Axes of two types, either bearing only scale leaves or only foliage leaves **8**

   2 No specialized erect shoots (short shoots[c]), stems all of one type, flowers grouped on specialized lateral shoots.. **3**

   2 Stems differentiated into main horizontal rhizomes and erect short shoots and both bearing foliage leaves **4**

3 Inflorescences in every alternate leaf axis, roots unbranched.................. **Enhalus**

3 Inflorescences irregularly produced, roots branched ....**Posidonia**

   4 Short shoots (erect stems) annual, well differentiated, elongated when compared to rhizome (horizontal stems), roots unbranched; flower clusters obvious, spikelike **5**

   4 Short shoots (erect stems) not well differentiated or distinct from main, horizontal rhizomes, flowers inconspicuous and in terminal pairs **7**

5 Rhizomes congested and monopodial ....**Phyllospadix**

5 Rhizomes elongated, not congested **6**

   6 Rhizomes elongated, branching monopodial ....**Zostera**

   6 Rhizomes elongated, branching synpodial ....**Heterozostera**

7 Rhizomes and short shoots clearly differentiated, internodes easily seen, roots branched .... **Cymodocea**

7 Rhizomes and short shoots and regions of nodes and internodes not easily differentiated, roots unbranched ....**Halodule**

   8 Scale-bearing rhizome sympodial; erect shoots long (to 20 cm or more) and distinct from rhizomes. **9**

   8 Scale-bearing rhizome monopodial, erect shoots long or short................ **10**

9 Branching of scale-bearing rhizome diffuse, with 2–7 or more interveining branch-free nodes...**Amphilbolis**

9 Branching of scale-bearing rhizome periodic, from every fourth node ....
.... **Thalassodendron**

   10 Branching continuous **11**

   10 Branching periodic, interderminante short shoots produced at regular intervals and separated by 9–13 nodes, flowers in lateral groups .... **Thalassia**

11 Internodes of rhizomes uniformly long, short shoots variable and usually determinate, leaf terete ....**Syringodium**

11 Internodes of rhizomes alternate, long and short, short shoots determinate, leaves flat
....**Halophila**

[a]This key utilized primarily vegetative features only and therefore is not considered to be a definitive key.
[b]The first leaves produced are always scale-like leaves.
[c]The term "short shoot" is used in this text to include all erect axes that are determinate, producing foliage leaves. Tomlinson (1974) distinguished between erect axes and short shoots.
Source: Modified from Tomlinson (1974).

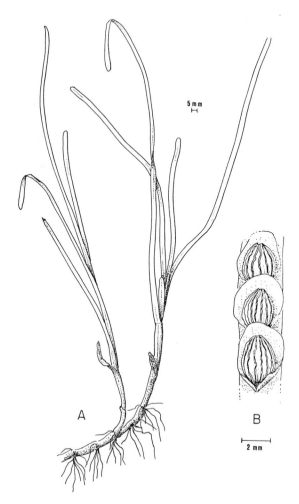

Figure 17-8 The habit of *Zostera marina*, showing a vegetative plant (*A*), and a portion of the spadix with mature seeds (*B*). (Courtesy of Ronald Phillips, Seattle Pacific University, Seattle, Washington.)

occurs sparsely in the Mediterranean and Black seas. It is a subtidal species but can occur in the lower intertidal regions of shallow lagoons on sand, mud, or a mixture of gravel, sand, and mud. The competition between *Z. americana* (introduced) and *Z. marina* (natural populations) has been studied in Boundary Bay, British Columbia, by Harrison (1979). He found that *Z. americana* is an upper intertidal plant that grows rapidly; it is an annual that uses up to 25% of its reserves in flowering (an *r* strategist). *Zostera marina* is a *k* strategist; it grows in the lower intertidal regions, where it is a perennial, and uses a large proportion of its resources for maintenance of rhizomes and roots.

Thalassia testudinum

The common name is *turtle grass,* and the plant is characteristic of the Caribbean and tropical western Atlantic Ocean. The plant has erect short shoots that produce a cluster of three to seven broad (to 2 cm wide) leaves, all developing from a basal meristem in the short shoots (Figure 17-6). The leaves have sheaths that envelop the upper portion of the short shoots. The short shoots arise from a rhizome, usually buried from 3 to 15 cm in the substratum. The rhizome grows by means of an apical meristem and branches left, then right. Roots develop on the rhizome close to the short shoots, as well as from the base of the short shoots. The plant is dioecious, with staminate flowers having a long base (pedicel), whereas the pistillate flowers are sessile. Flowering is common (Loraamm, 1980), occurring throughout the year but not at the some time in all plants. In Tampa Bay, Florida, fruits were evident in January and early inflorescences in February (Moffler and Durako, 1980). Seed production appears to be more seasonal in the northern part of the Caribbean, mostly occurring in May to July and extending into October for the central Caribbean. The fruit is beaked and contains four to five seeds.

Thalassia testudinum is the most common and abundant seagrass in the Caribbean, occurring from the northern Gulf of Mexico to the northern portion of South America. The plant is well adapted for soft sediments and is found in areas of relatively calm water to depths up to 25. However, the lush seagrass beds of *Thalassia* occur in shallow water (up to 10 m) and in areas of normal salinity (25–40 ppt). Vegetative propagation appears to be the most-important method of plant distribution, although flowering is common and the species appears to be a *k* strategist (Dawes and Lawrence, 1980).

Syringodium filiforme

The synonym is *Cymodocea manatorum,* and the plant is commonly called *manatee grass.* The species is second in abundance and importance in the Caribbean waters. The leaves arise in clusters of two or three, ensheathing the short shoots; they are cylindrical in cross section. Scale leaves are present on the rhizome and on the short shoots. The rhizome is cylindrical, and proliferation is by damage to the existing meristem or by proliferation of the short shoots. Roots occur at rhizome nodes; usually three are produced within the terminal (tip) meristem of the rhizome. Flowers occur in cymose clusters (Phillips, 1960). The pistillate flowers are subtended by hyaline bracts with a short style and two stigmas. The fruit is 3 mm long and is beaked by the persistent style.

Syringodium occurs throughout the Caribbean; it grows in pure or mixed stands with *Thalassia* and *Halodule.* Phillips (1967) reported that *Syringodium* can tolerate salinities as low as 20 ppt. However, it is usually found in areas of higher salinities, along with *Thalassia.* The plant can be found in deeper waters (e.g., 18 m off St. Croix) but it is usually found at 10 m or less. McMillan (1980a) found that its reproduction is controlled by temperature. Texas coastal plants are induced to flower at temperatures of 20–24°C, whereas Southern Gulf of Mexico and Caribbean plants are induced at 23–24°C. Plants from St. Croix require temperatures above 25°C for flowering.

Phyllospadix spp

Three species of this genus occur in the northeast Pacific (Figure 17-9). *Phyllospadix serrulatus* is found from 1.5 m to mean-low low water; *P. scouleri* is intermediate (MLLW to -1.5 m); and *P. torreyi* occurs in deeper water (Phillips, 1979). These two named species are commonly called *surf grass* and are typical intertidal plants in areas of high surf. Phillips (1979) compared the three species and presented the following key using vegetative characters:

1   Rhizome nodes with 6–10 roots; leaf tip somewhat apiculate with inconspicuous "fin cells"; leaves with three vascular bundles **2**

1   Rhizome nodes with two roots, leaf tips truncate with prominent "fin cells"; leaves with five or seven vascular bundles .... **P. serrulatus**

  2   Leaves 0.5–3.0 mm wide .... **P. torreyi**
  2   Leaves 2.0–5.0 mm wide .... **P. scouleri**

Another difference between *P. torreyi* and *P. scouleri* is that the veins of the former are practically invisible, whereas the latter has three distinct ones.

The genus is common in areas of surf, and the species grow in tight clusters; the rhizomes are tightly packed in clusters or mats and are usually attached to flat rock surfaces. Sand accumulates around the plants because of the packed conditions and short internodes as well as the dense root clusters. The roots have

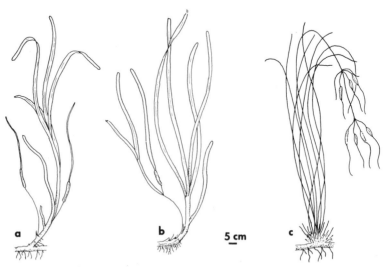

Figure 17-9   Habits of the three species of *Phyllospadix* found on the west coast of North America (after Phillips, 1979). (*a*) *P. serrulatus*. (*b*) *P. scouleri*. (*c*) *P. torreyi*. Note organization of the flowering shoots in *P. torreyi*.

branched root hairs that form dense packets and act as tiny "holdfasts" on the rocks, functioning similar to holdfasts of kelps. Thus the species are highly adapted to surf conditions and compete well with macroalgae. The flowers occur in two ranks along a flattened spadix, and the entire inflorescence is enclosed by a spathe. The male plants are less numerous than females, with a ratio of about 1:10. *Phyllospadix torreyi* has flowering stems 20–30 mm long, and *P. scouleri* has a spadix 1–6 cm long. Fruits have a pulpy exterior that is worn away by the surf. The mesocarp is horny and has two projecting, incurved arms that bear bristles. The bristly arms attach the fruit to any algae (coralline or otherwise) or rocky outcrops.

ECOLOGICAL ROLES

Seagrasses are adapted to shallow coastal waters and estuarine environments but are completely absent in fresh water (den Hartog, 1967). The plants have a high salt tolerance, are able to grow fully submerged, have well-developed rhizomes, and employ hydrophilous pollination (Arber, 1920). Furthermore, it is apparent that seagrasses are highly successful in their invasion of shallow waters of the continental shelves (den Hartog, 1967).

The complexity of seagrass communities, especially their influence on the physical environment, has been recognized by a number of authors. Six growth forms of seagrasses were recognized by den Hartog (1967), and he could link these forms to environmental conditions. Simple structured forms (e.g., *Halodule*) are early stages in succession, allowing for stabilization of the substratum. More complex forms (e.g., *Thalassia, Zostera*) follow, causing stratification and permitting epiphytes to develop because of blade width and the perennial nature of growth. *Zostera* can either be a colonizer or a pioneer seagrass, although there is some question as to whether the pioneer plants and climax plants are the same "biotypes."

Because seagrasses are important in bottom stabilization, there has been a recent interest in transplantation of seagrasses. *Zostera* (eelgrass) has been transplanted in studies in San Diego Bay (Robilliard and Porter, 1976), in regions of Great Britian (Ranwell *et al.*, 1974) and in Alaska and Puget Sound (Phillips, 1974). *Thalassia* (turtle grass) has been transplanted in Florida (Kelley *et al.*, 1971; van Breedveld, 1975). The use of *Thalassia* seeds as a means of spreading the plant has been promoted by Thorhaug (1974). She followed the sequence from seed germination to flowering in the field (Thorhaug, 1979). Studies in laboratory-grown seagrasses have been summarized by McRoy and McMillan (1977) and are presented here in a modified table (Table 17-3).

Six roles can be assigned to seagrass communities regarding ecological importance in the marine environment.

1 Seagrasses serve as sediment traps, stabilizing bottom sediments (den Hartog, 1967), and improving water clarity (see Chapters 2 and 13).

Table 17-3 Culturing of Seagrasses

	McMahan (1968)	Marmelstein et al. (1968)	Fuss and Kelley (1969)	McMillan (1974 & unpublished)	Phillips (1972)
Location of study	Point Isabel, Texas	Panama City, Florida	St. Petersburg, Florida	Austin, Texas	Puget Sound, Washington
Seagrasses cultured	Syringodium Halodule	Thalassia	Thalassia Halodule	Thalassia Halodule Syringodium	Zostera
Collection site	Lower Laguna Madre	St. Andrews Bay, Biscayne Bay	Boca Ciega Bay	Redfish Bay	Puget Sound
Duration of culture studies (da)	42	14–28	Up to 365	Up to 365	29
Type of study	Salinity tolerance	Photoperiodic response	Culture methods	Salinity tolerance, photoperiodic response, culture methods	Growth
Indoor	13 l glass jars	75 l polyethylene cans	Plexiglass aquaria, 125 l	5 l glass jars	Glass tubes
Substrate	Gravel or bay sediment	Beach sand	Builder's sand over gravel	Calcareous river sand or fine sandy loam	Water culture, no substrate
Water exchange	Seawater changes every 7 da	8 l changed every 7 da	Flowing seawater (140 l/hr)	Tapwater added	Enriched seawater, changed weekly
Temperature (°C)	25	27–30 (outdoors)	Not controlled 25 (indoors)	20–30	Controlled but not stated
Salinity (ppt)	3.5–87.5	28–34	32.36	37 and 0.95 seawater (Instant Ocean)	Not stated

Source: McRoy and McMillen (1977).

482

2 Seagrass communities are primary producers with high rates of production (Dawes *et al.*, 1979; Jones, 1968). This topic is discussed in detail in the next section, and only one example is presented here. The loss of fixed carbon from a seagrass bed of *Posidonia australis* in an estuary near Port Hacking, Australia, was found to be 2.6 mg C/g dry wt/da (Kirkman and Reid, 1979). About 48% of this loss was in the form of dissolved organic carbon, 3% by grazing, 12% by leaves floating away and 37% by sinking. Thus the *Posidonia* bed yielded a large amount of fixed carbon in a variety of ways (Kirkman and Reid, 1979). Similar findings were reported by Ott (1980) for *Posidonia oceanica* beds in the Bay of Naples.

3 Seagrasses are a direct food source (den Hartog, 1967) for many animals. In a study of seagrass beds in Texas bays, Fry and Parker (1979) found that juvenile shrimp and fish utilized seagrasses and associated algae as their primary sources of nutrients. Over 340 animals analyzed showed some level of seagrass consumption based on measurements of stable carbon isotope ratios.

4 Seagrass communities provide important habitats and shelters for a number of animal species (den Hartog, 1967). Kikuchi (1974) found that any decline in eelgrass (*Zostera marina*) production resulted in a significant drop in the numbers of decapods and juvenile and young-adult stages of commercial fishes. Because of the dramatic faunal changes in both fish and epiphytic-invertebrate populations when the eelgrass beds declined, Kikuchi has argued for conservation of inshore grass beds.

5 Seagrasses are important substrata for attachment, as shown by the studies of seagrass epiphytes (Humm, 1964; Ballantine and Humm, 1975). McRoy and McMillian (1977) and Harlin (1980) have reviewed the literature on seagrass epiphytes and report wide variations in epiphyte populations over a year's time and from one study to another. Harlin (1980) listed over 450 macroalgal epiphytes, over 150 microalgal epiphytes (mostly diatoms), and over 180 invertebrate epiphytes found on seagrass blades.

Epiphyte productivity can accout for 20 g C/m$^2 \cdot$yr in Massachusetts beds of *Zostera* and 200 g C/m$^2 \cdot$yr in Florida beds of *Thalassia*. However, the effect of shading, damage, and nutrient filtering by the epiflora and epifauna can limit the photosynthetically useful life of the seagrass leaves. Johnstone (1979), in a study of the Papua New Guinea seagrass *Enhalus acoroides*, found that the leaf segment has a 25 da photosynthetically useful life due to heavy epiphytism and epifaunalism.

It should also be pointed out that these epiphytic populations provide important food sources for small fish and invertebrates. The formation of such epiphytic populations, typically under nutrient-deficient conditions, may be due to direct transfer of carbon, nitrogen, and phosphorus from the seagrass substratum. However, electron micrographs (Figure 17-4) do not show any apparent connections between the epiphytes and epidermal cells, although pores in the cuticle have been reported (Gessner, 1971). It is more probable that the algae obtain nutrients from the seawater after excretion by the seagrass (see Kirkman and Reid, 1979).

6   Seagrasses are highly efficient in removing nutrients (nutrient stripping) from
    marine waters and surface sediments. Thus these may be critical plants in the
    control of water quality of shallow waters (Patriquin, 1972). Seagrasses can
    grow in waters that are low in dissolved nutrients. Patriquin (1972) has shown
    that the high productivity of seagrass communities in such waters is su̱ ͗ined
    by nutrients withdrawn in solution from the sediments in which they grow.
    Bell (1979) found that the morphology of *Thalassia* plants changed with rela-
    tion to the percentage of ash-free dry weight of the sediment. She showed that
    as the percentage of organic content increased in the sediments, the *leaf area
    indices* (LAI) increased, and the ratio of leaf-plus-epiphyte biomass to the
    biomass of the remaining plant increased. Nitrogen fixation carried out on
    Thalassia leaves by associated epiphytes was in the range of 20–110
    $\mu l/m^2 \cdot hr$ and on the roots in the range of 2–6 $\mu l/m^2 \cdot hr$. Patriquin (1972)
    showed that fixation of nitrogen occurred in the sediments and on the rhi-
    zomes of *Thalassia*. Oxygen can also be transported from the leaves to the
    rhizomes and roots of eelgrass and then released into the sediments (Iizumi *et
    al.*, 1980). The oxygen is used for nitrification in the rhizosphere in the anox-
    ic sediments, and the nitrogen is absorbed by the eelgrass. Nitrogen fixation
    has also been demonstrated by seagrass epiphytes (Goering and Parker, 1972)
    so that is appears fixed nitrogen is available from both the rhizosphere and the
    hydrosphere of a seagrass bed.

ORGANIC PRODUCTION

McRoy and McMillan (1977) introduced their chapter on production ecology and
physiology of seagrasses with the following statement: ''A seagrass meadow is a
highly productive and dynamic ecosystem; it ranks among the most productive in
the ocean.'' The various papers dealing with standing stock (biomass) and productivity (rate of carbon fixation) were summarized in a table that has been modified for this text (Table 17-4).

Productivity

Productivity measurements are few, since earlier studies used dissolved oxygen
measurements to determine productivity. Because the oxygen released in photosynthesis is internally recycled in the lacunar (aerenchyma) spaces of the leaves,
such determinations are questionable (Zieman, 1974). As much as 60% of the leaf
volume is essentially a gas sac (due to aerenchyma) in such genera as *Thalassia*
and *Zostera*. During periods of rapid photosynthesis, blades of such species can
double in volume with stored gasses, which burst from the edges of the leaves
and stream to the surface, causing a hissing sound. Because of this gas storage
and apparent recycling of CO_2 and O_2, measurements of dissolved oxygen or use
of C^{14} in productivity studies are not accurate (but see Clough and Attiwill,
1980).

Table 17-4 Biomass and Productivity of Seagrasses

Species	Locality	Biomass (g dry/m²) Range	Biomass (g dry/m²) Mean	Productivity (g C/m²·da)	References
Zostera marina	Denmark		443	7.9	Sand-Jenson (1975)
	Canary Islands	70–89	89	0.35	Johnston (1969)
	Nova Scotia	785–1251	1018	—	Mann (1972)
	Massachusetts	15–29		0.04	Conover (1958)
	Rhode Island	—	100	2.6	Nixon and Oviatt (1972)
	New York	247–2062	762	—	Burkholder and Doheny (1968)
	North Carolina	82–156		0.2–1.2	Williams (1966)
	Washington	116–231	117	0.7–4	Phillips (1972)
	Alaska	62–1840	1000	3.3–3.8	McRoy (1970a, 1970b)
	Japan	70–235		—	Kita and Harada (1962)
Cymodocea nodosa	Mediterranean	13–342		5.4–18.7	Gessner and Hammer (1960)
Halodule wrightii	North Carolina	105–200	200	0.48–2.0	Dillon (1971)
Posidonia oceanica	Malta	543–1072	742	2–5	Drew (1971)
	Bay of Naples	—	1296	1.52–4.51	Ott (1980)
Thalassia testudinum	Cuba	20–800	340	9.3–12.5	Buesa (1972)
	Puerto Rico	538–7376		—	Burkholder et al. (1959)
	Florida	20–400		—	Odum (1963)
		74–89	81	0.35–1.14	Pomeroy (1960)
		300–1800	835	0.9–2.5	Jones (1968)
	Bahamas	—		0.5–3	Patriquin (1972)

Source: Modified from McRoy and McMillan (1977).

McRoy and McMillian (1977) and Zieman and Wetzel (1980) have reviewed productivity in seagrasses. Rates of carbon fixation for *Zostera marina* varied from 0.9 g $C/m^2 \cdot da$ in North Carolina (Penhale, 1977) to 4.8 g $C/m^2 \cdot da$ in Alaska (McRoy, 1974). A rate of 15 mg $C/m^2 \cdot da$ was obtained for *Thalassia testudinum* in the northern Gulf of Mexico by Bittaker and Iverson (1976) and 13 mg C/g dry wt $\cdot hr$ for *Phyllospadix scouleri* (Barbour and Radosevich, 1979) in California. In an earlier study, Benedict and Scott (1976) found that *Thalassia testudinum* may have a C_4 type of carbon metabolism (Hatch-Slack CO_2 fixation pathway in photosynthesis) involving the epidermal and mesophyllic cells. However, more recently Andrews and Abel (1979), working with *Thalassia hemprichii* and *Halophila spinulosa*, showed that although C^{14}/C^{12} ratios (isotopes of carbon) were similar to those found in C_4 plants, the C^{14} incorporation patterns were definitely those of a C_3 system (Calvin CO_2 fixation pathway in photosynthesis).

Zieman (1974) found a simple technique to measure seagrass growth and productivity. Bittaker and Iverson (1976) confirmed that the Zieman procedure was as good as the more-complicated C^{14} measurement in yielding accurate estimates of net particulate carbon production. Because the technique is so simple and ingenious, it is presented here for possible consideration.

Zieman used a 20 × 20 cm permanent quadrat set in a *Thalassia* bed that was divided into four 10 × 10 cm sections. The square was randomly placed and was monitored every 2–3 wk. Growth and production were measured by stapling the leaves at the level of the frame. An inexpensive ($1) Swingline stapler was used and was discarded when too rusty, and the staple packs were coated with lacquer, (fingernail polish) to prevent separation. A non-water-soluble marker pen can also be used and may be more useful for narrow-bladed species of seagrasses (Zieman and Wetzel, 1980). After 2–4 wk, the blades were cut at the reference level of the frame. Because of growth, new blades lacking staples and older blades with staples, in varying distances from the cut end, were collected. The distance from the base (cut end) of the blade to the staple was measured on old blades. The section of new growth was cut off the blade, and the collection was washed, dried, and weighed. From these weights the standing crop of leaves (g/m^2) could be calculated by summing the weight of new blades, new growth of cut blades, and the remaining older-blade portion. Leaf production $(g/m^2 \cdot da)$ was determined by the weight of the new leaves plus the weight of the new growth of old leaves (area below the staple). Community replacement, or turnover, was calculated by dividing the standing crop by the productivity value.

Biomass

Information on biomass, or standing crop (Table 17-4), is also scanty, and unfortunately the methods used vary so that comparisons are difficult to make. From the information available, the species producing high biomass levels are those described by den Hartog (1967) as the larger forms of a seagrass meadows (e.g., *Thalassia*, *Phyllospadix*, *Syringodium*, *Zostera*). McRoy and McMillan (1977) believed that the factor controlling standing stock was related to leaf size. The control is by light and the effect of leaves shading one another. Increased self-shading will ultimately determine the standing crop. Irradiance has been found by Backman and Barilotti (1976) to be a factor in the size of leaf clusters of *Zostera marina* in California. McRoy and McMillan (1977) compared the leaf area indices (LAI) of four major seagrass species as used in studies of terrestrial crops (Table 17-5). In the latter study, it was found that seagrass blades had higher LAIs than terrestrial plants. The higher LAIs were attributed to the lower light conditions in submarine meadows when compared to terrestrial light levels.

Table 17-5 Estimates of Leaf Area Index in Seagrasses

Species	Region	LAI[a] (M^2/m^2)	Reference
Posidonia oceanica	Mediterranean	6.9–8.1	Drew (1971)
	Mediterranean	1.1–2.18	Ott (1980)
Cymodocea nodosa	Mediterranean	8–11	Gessner and Hammer (1960)
Thalassia testudinum	Columbia	18.6	Gessner (1971)
Zostera marina	Puget Sound	1–4	Phillips (1972)
	Alaska	12–21	McRoy (1970b)

Source: McRoy and McMillen (1977); Ott (1980)
[a]Evans (1972)

Organic Composition

Because of their conspicuous role in productivity, sea grasses have been analyzed for organic composition and caloric levels (Table 17-6). Few of these studies, however, have considered seasonal or individual variations (statistical) in composition or have investigated all organs of the seagrass, as pointed out by Walsh and Grow (1972). Hocking *et al.* (1980) have shown that the nutrient accumulation in the fruits of *Posidonia australis* and *P. sinuosa* parallels that found in herbaceous terrestrial plants. Accumulation will occur against large gradients and is concentrated in the seeds. Dawes *et al.* (1979) found that the algal component of a seagrass bed on the west coast of Florida was energetically equal in importance to *Thalassia* (4.2–4.3 kcal/g organic weight) for all seasons. The importance of various organic components also varied; in the spring proteins were highest, whereas in the winter carbohydrates were highest. In a comparison of the three dominant species in the Caribbean, Dawes and Lawrence (1980) found that levels of soluble carbohydrate in the rhizomes of *Thalassia testudinum, Syringodium filiforme,* and *Halodule wrightii* were highest in the fall and lowest in the spring (Figure 17-10). This information, coupled with a study on the effects of blade removal on the composition of the rhizome of *Thalassia testudinum* (Dawes and Lawrence, 1979), suggested that soluble carbohydrate is a nutrient reserve used to sustain the plants during periods of decreased productivity. Carbon transport from leaves to

Table 17-6 A Comparison of Published Data on Seagrass Constituents

Genus and Area	Seasonal	Kcal	Percent					Washing in Fresh Water
			Protein[a]	Lipid	Soluble Carbohydrate	Insoluble Carbohydrate	Ash	
Thalassia testudinum leaves								
Puerto Rico[c]	−	2.0	13.1	0.5	35.6	16.4	24.8	+
Florida[d]	−	—	9–12[a]	0.5	38.0	10.0	46.0	−
Florida[e]	+	4.5[b]	13–37[b]	—	18–36[b]	—	25[b]	+
Florida[f]	+	4.2	3–12	—	3–12	—	30–40	−
Thalassia testudinum rhizome								
Puerto Rico[c]	−	—	9.6[a]	—	—	—	50.0	+
Florida[e]	+	4.8[b]	8–15[b]	—	55–80[b]	—	24	+
Zostera marina leaves								
Alaska[g]	−	4.2	—	—	—	—	—	+
Nova Scotia[h]	+	—	10.6[a]	—	42 (total)	—	—	−
Korea[i]	−	—	10	0.8	—	5.5	10.6	−
Zostera marina rhizome								
Alaska[j]	0	3.6	—	—	—	—	—	+
Syringodium isoetifium leaves								
Australia[j]	−	2.2	10[a]	—	—	—	—	+
Syringodium isoetifium rhizome								
Australia[j]	−	3.0	3.1[a]	—	—	—	—	+

Source: After Dawes and Lawrence (1980).

[a]Protein analysis carried out by nitrogen determination and then calculated by multiplying N by 6.25.

[b]Expressed only as ash-free dry weight

[c]Burkholder et al. (1959)

[d]Bauersfeld et al. (1969)

[e]Walsh and Grow (1972)

[f]Dawes et al., (1979)

[g]McRoy (1970b)

[h]Harrison and Mann (1975)

[i]Park (1969)

[j]Birch (1975)

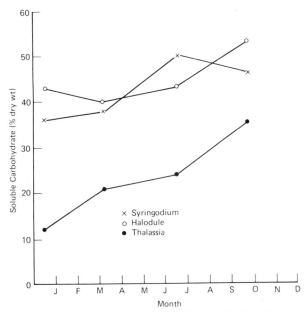

Figure 17-10 Seasonal changes in soluble carbohydrate in the rhizomes of *Thalassia testudinum*, *Syringodium filiforme*, and *Halodule wrightii*. This has been cited as evidence for mobilization of reserve foods in the rhizome during blade growth (Dawes and Lawrence, 1980).

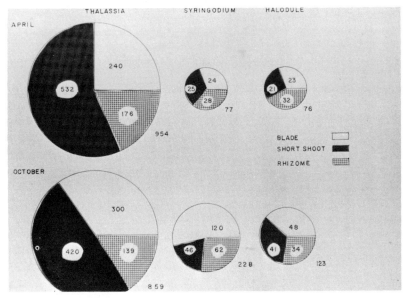

Figure 17-11 Comparison of the allocation of organic matter (mg) in three Caribbean seagrasses, *Thalassia testudinum*, *Syringodium filiforme*, and *Halodule wrightii*, in two seasons: spring (April) and fall (October). The areas of the circles and sectors are proportional to the amounts of organic matter found in each species. Thus *Thalassia*, the largest seagrass, is shown with the largest circles. Biomass was substituted for area and the radius was determined from the formula in which area = pi r² (Dawes and Lawrence, 1980).

the rhizome has been shown in *Zostera* and *Thalassia* by Wetzel and Penhale (1979). When the amount of organic biomass in the three species was compared (Dawes and Lawrence, 1980), it was apparent that *Thalassia* had 8–12 times more than *Syringodium* and *Halodule* (Figure 17-11). Allocation of organic matter also varied according to species (Figure. 17-11) and season. Thus *Thalassia* had the most organic matter in the short shoot throughout the year, whereas *Syringodium* and *Halodule* had the most in the blade in the fall. However, in the spring the latter two species had a greater proportion of organic matter allocated to their rhizomes. Such information can be used to interpret the strategy of each genus with regard to seasonal growth. In the case of *Thalassia* it appears the plants are designed to continue growth throughout the year, whereas in the other two species it appears they are more opportunistic and show seasonal responses.

REFERENCES

Ancibor, E. 1979. Systematic anatomy of vegetative organs of the Hydrocharitaceae. *Bot. J. Linn. Soc.* **78**: 237–266.

Andrews, T. J. and K. M. Abel. 1979. Photosynthetic carbon metabolism in seagrasses. C-labeling evidence for the C_3 pathway. *Plant Physiol.* **63**: 650–656.

Arber, A. 1920. *Water Plants, a Study of Aquatic Angiosperms.* Cambridge University Press, London.

Backman, T. W. and D. C. Barilotti. 1976. Irradiance reduction: effects on standing crops of the eelgrass *Zostera marina* in a coastal lagoon. *Mar. Biol.* **34**: 33–40.

Ballantine, D. and H. J. Humm. 1975. Benthic algae of the Anclote Estuary. 1. Epiphytes of seagrass leaves. *Fla. Scientist* **38**: 150–162.

Barbour, M. G. and S. R. Radosevich. 1979. [14]C uptake by the marine angiosperm *Phyllospadix scouleri*. *Amer. J. Bot.* **66**: 301–306.

Bauersfeld, P., Kifer, R. R., Durrant, N.W., and J. E. Sykes. 1969. Nutrient content of turtle grass (*Thalassia testudinum*). *Proc. Int. Seaweed Symp.* **6**: 637–645.

Bell, C. K. 1979. Nitrogen fixation (acetylene reduction) associated with seagrasses along the northern Florida Gulf Coast. Masters thesis. Florida State University, Tallahassee.

Benedict, C. R. and J. R. Scott. 1976. Photosynthetic carbon metabolism of a marine grass. *Plant Physiol.* **57**: 876–880.

Birch, W. R. 1975. Some chemical and calorific properties of tropical marine angiosperms compared with those of other plants. *J. Appl. Ecol.* **12**: 201–212.

Bittaker, H. F. and R. L. Iverson. 1976. *Thalassia testudinum* productivity: A field comparison of measurement methods. *Mar. Biol.* **37**: 36–46.

Buesa, R. J. 1972. Production primaria de las praderas de *Thalassia testudinum* de la platdorma noroccidental de Cuba. *INP, Cuba Cent. Inv. Pesqueras Reva. Bal. Trab. CIP* **3**: 101–143.

Burkholder, P. R., Burkholder, L. M., and J. A. Rivero. 1959. Some chemical constituents of turtle grass, *Thalassia testudinum*. *Bull. Torrey Bot. Club* **86**: 88–93.

Burkholder, P. R. and T. E. Doheny. 1968. *The Biology of Eelgrass.* Dept. Conserv. Waterways, Hempstead, N.Y. (unpublished).

Clough, B. F. and P. M. Attiwill. 1980. Primary productivity of *Zostera muelleri* Irmisch *ex* Aschers in Westernport Bay (Victoria, Australia) *Aq. Bot.* **9**: 1–13.

Conover, J. T. 1958. Seasonal growth of benthic marine plants as related to environmental factors in an estuary. *Publ. Inst. Mar. Sci. Univ. Texas* **5**: 97–147.

Dawes, C. J., K. Bird, M. Durako, R. Goddard, W. Hoffman, and R. McIntosh. 1979. Chemical fluctuations due to seasonal and cropping effects on an algal-seagrass community. *Aq. Bot.* **6**: 79–86.

Dawes, C. J. and J. M. Lawrence. 1979. The effects of blade removal on the proximate composition of the rhizome of the seagrass *Thalassia testudinum* Banks *ex* Konig. *Aq. Bot.* **7**: 255–266.

Dawes, C. J. and J. M. Lawrence. 1980. Seasonal changes in the proximate constituents of the seagrass *Thalassia testudinum, Halodule wrightii,* and *Syringodium filiforme. Aq. Bot.* **8**: 371–380.

den Hartog, C. 1967. The structural aspect in the ecology of sea-grass communities. *Helgo. Wiss. Meeresunt.* **15**: 648–659.

den Hartog, C. 1970. *The Seagrasses of the World.* North Holland, Amsterdam.

Dillon, C. R. 1971. A comparative study of the primary productivity of estuarine phytoplankton and macrobenthic plants. Ph.D. dissertation. University of North Carolina, Chapel Hill.

Drew, E. A. 1971. Botany. In Woods, J. D. and J. N. Lythgoe (eds.). *Underwater Science,* Oxford University Press, London, pp. 92–112.

Evans, G. C. 1972. *The Quantitative Analysis of Plant Growth.* Blackwell, Oxford.

Fry, B. and P. L. Parker. 1979. Animal diet in Texas seagrass meadows: δ ^{13}C evidence for the importance of benthic plants. *Estuar. Coast. Mar. Sci.* **8**: 499–509.

Fuss, C. M. and J. A. Kelly. 1969. Survival and growth of sea grasses transplanted under artificial conditions. *Bull. Mar. Sci.* **19**: 351–365.

Gessner, F. 1971. The water economy of the sea grass *Thalassia testudinum. Mar. Biol.* **10**: 258–260.

Gessner, F. and L. Hammer. 1960. Die primärproduktion in Mediterranean Caulerpa-Cymodocea Wiesen. *Bot. Mar.* **2**: 157–163.

Goering, J. J. and P. L. Parker. 1972. Nitrogen fixation by epiphytes on seagrasses. *Limnol. Oceanogr.* **17**: 320–323.

Haberlandt, G. 1914. *Physiological Plant Anatomy.* Macmillan, London.

Harlin, M. M. 1980. Seagrass epiphytes. In Phillips, R. C. and C. P. McRoy. (eds.). *Handbook of Seagrass Biology. An Ecosystem Perspective,* Garland STPM Press, New York, pp. 117–152.

Harrison, P. G. 1979. Reproductive strategies in intertidal populations of two co-occurring seagrasses (*Zostera* spp.). *Can. J. Bot.* **57**: 2635–2638.

Harrison, P. G. and K. H. Mann. 1975. Chemical changes during the seasonal cycle of growth and decay in eelgrass *(Zostera marina)* on the Atlantic coast of Canada. *J. Fish. Res. Bd. Canada* **32**: 615–621.

Hocking, P. J., M. L. Cambridge, and A. J. McComb. 1980. Nutrient accumulation in the fruits of two species of seagrasses, *Posidonia australis* and *P. sinuosa. Ann. Bot.* **45**: 149–161.

Humm, H. J. 1964. Epiphytes of the sea grass *Thalassia testudinum,* in Florida. *Bull. Mar. Sci. Gulf Carib.* **14**: 306–341.

Iizumi, H., A. Hattori, and C. P. McRoy. 1980. Nitrate and nitrite in interstitial waters of eelgrass beds in relation to the rhizosphere. *J. Exp. Mar. Biol. Ecol.* **47**: 191–201.

Jagels, R. 1973. Studies on a marine grass, *Thalassia testudinum.* I. Ultrastructure of the osmoregulatory leaf cells. *Amer. J. Bot.* **60**: 1003–1009.

Johnston, C. S. 1969. The ecological distribution and primary production of macrophytic marine algae in the Eastern Canaries. *Int. Rev. Ges. Hydrobiol.* **54**: 473–490.

Johnstone, I. M. 1979. Papua New Guinea seagrasses and aspects of the biology and growth of *Enhalus acoroides* (L.) Royle. *Aq. Bot.* **7**: 197–208.

Jones, J. A. 1968. Primary productivity by the tropical marine turtle grass, *Thalassia testudinum* Konig, and its epiphytes. Ph.D. dissertation. University of Miami, Coral Gables.

Kelley, J. A., C.M., Fuss, and J. R. Hall. 1971. The transplanting and survival of turtle grass, *Thalassia testudinum,* in Boca Ciega Bay, Florida. *Fish. Bull. Fish Wildl. Serv. U.S.* **69**: 273–280.

Kikuchi, T. 1974. Japanese contributions on consumer ecology in eelgrass (*Zostera marina* L.) beds, with special reference to trophic relationships and resources in inshore fisheries. *Aquaculture* **4**: 145–160.

Kirkman, H. and D. D. Reid. 1979. A study of the role of the seagrass *Posidonia australis* in the carbon budget of an estuary. *Aq. Bot.* **7**: 173–183.

Kita, T. and E. Harada. 1962. Studies on the epiphytic communities. I. Abundance and distribution of microalgae and small animals on the *Zostera* blades. *Publ. Seto Mar. Biol. Lab.* **10**: 245–257.

Loraamm, L. P. 1980. Flowers that took to the sea. *Sea Frontiers* **26**: 84–90.

Mann, K. H. 1972. Ecological energetics of the seaweed zone in a marine bay on the Atlantic coast of Canada. I. Zonation and biomass of seaweeds. *Mar. Biol.* **12**: 1–10.

Marmelstein, A. D., P. W. Morgan, and W. E. Pequegnat. 1968. Photoperiodism and related ecology in *Thalassia testudinum*. *Bot. Gaz.* **129**: 63–67.

McCoy, E. D. and K. L. Heck, Jr. 1976. Biogeography of corals, seagrasses, and mangroves. An alternative to the center of origin concept. *System. Zool.* **25**: 201–210.

McMahan, C. A. 1968. Biomass and salinity tolerance of shoalgrass and manateegrass in Lower Laguna Madre, Texas. *J. Wildl. Manag.* **32**: 501–506.

McMillan, C. 1974. Salt tolerance of mangroves and submerged aquatic plants. In Reinold, R. J. and W. H. Queen (eds.). *Ecology of Halophytes,* Academic, New York, pp. 379–390.

McMillan, C. 1980a. Reproductive physiology in the seagrass, *Syringodium filiforme,* from the Gulf of Mexico and the Caribbean. *Amer. J. Bot.* **67**: 104–110.

McMillan, C. 1980b. Flowering under controlled conditions by *Cymodocea serrulata, Halophila stipulacea, Syringodium isoetifolium, Zostera capensis* and *Thalassia hemprichii* from Kenya. *Aq. Bot.* **8**: 323–336.

McRoy, C. P. 1970a. On the biology of eelgrass in Alaska. Ph.D. dissertation. University of Alaska, Fairbanks.

McRoy, C. P. 1970b. Standing stocks and related features of eelgrass populations in Alaska. *J. Fish. Res. Bd. Canada* **27**: 1811–1821.

McRoy, C. P. 1974. Seagrass productivity: Carbon uptake experiments in eelgrass, *Zostera marina*. *Aquaculture* **4**: 131–137.

McRoy, C. P. and C. McMillan. 1977. Production ecology and physiology of seagrasses. In McRoy, C. P. and C. Helfferich (eds.). *Seagrass Ecosystems. A Scientific Perspective,* Marcel Dekker, New York, pp. 53–81.

Moffler, M. D. and M. J. Durako. 1980. Observations on the reproductive ecology of *Thalassia testudinum* (Hydrocharitaceae) in Tampa Bay, Florida. *Fla. Scientist* **43**: (Suppl. 1) 8.

Nixon, S. W. and C. A. Oviatt. 1972. Preliminary measurements of midsummer metabolism in beds of eelgrass, *Zostera marina*. *Ecology* **53**: 150–153.

Odum, H. T. 1963. Productivity measurements in Texas turtle grass and the effects of dredging an intracostal channel. *Publ. Inst. Mar. Sci. Texas* **9**: 45–58.

Ott, J. A. 1980. Growth and production in *Posidonia oceania* (L.) Delile. *P.S.Z.N.I: Mar. Ecol.* **1**: 47–64.

Park, M. S. 1969. Studies on the chemical composition of *Zostera marina*. *Korean J. Bot.* **12**: 1–6.

Patriquin, D. G. 1972. The origin of nitrogen and phosphorous for growth of the marine angiosperm *Thalassia testudinum*. *Mar. Biol.* **15**: 35–46.

Penhale, P. A. 1977. Macrophyte-epiphyte biomass and productivity in an eelgrass (*Zostera marina* L.) community. *J. Exp. Mar. Biol. Ecol.* **26**: 211–224.

Phillips, R. C. 1960. *Observations on the Ecology and Distribution of the Florida Seagrasses*. Prof. Papers Series No. 2, Fla. State Bd. Conser.

Phillips, R. C. 1967. On species of the seagrass, *Halodule,* in Florida. *Bull. Mar. Sci.* **17**: 672–676.

Phillips, R. C. 1972. Ecological life history of *Zostera marina* L. (eelgrass) in Puget Sound, Washington. Ph.D. dissertation. University of Washington, Seattle.

Phillips, R. C. 1974. Transplantation of seagrasses, with special emphasis on eelgrass, *Zostera marina* L. *Aquaculture* **4**: 161–176.

Phillips, R. C. 1979. Ecological notes on *Phyllospadix* (Potamogetonaceae) in the Northeast Pacific. *Aq. Bot.* **6**: 159–170.

Pomeroy, L. R. 1960. Primary productivity of Boca Ciega Bay, Florida. *Bull. Mar. Sci. Gulf Carib.* **10**: 1–10.

Ranwell, D. S., D. W. Wyler, L. A. Boorman, J. M. Pizzey, and R. J. Waters. 1974. *Zostera* transplants in Norfolk and Suffolk, Great Britain. *Aquaculture* **4**: 185–198.

Robilliard, G. A. and P. E. Porter. 1976. *Transplantation of Eelgrass (Zostera marina) in San Diego Bay*. Naval Undersea Center, San Diego, Calif. NUC TN 1701.

Sand-Jensen, K. 1975. Biomass, net production and growth dynamics in an eelgrass (*Zostera marina* L.) populations in Vellerup Vig, Denmark. *Ophelia* **14**: 185–201.

Thorhaug, A. 1974. Transplantation of the seagrass *Thallasia testudinum* Konig. *Aquaculture* **4**: 177–183.

Thorhaug, A. 1979. The flowering and fruiting of restored *Thalassia* beds: a preliminary note. *Aq. Bot.* **6**: 189–192.

Tomlinson, P. B. 1974. Vegetative morphology and meristem dependence. The foundation of productivity in seagrasses. *Aquaculture* **4**: 107–130.

van Breedveld, J. R. 1975. *Transplanting of Seagrasses with Emphasis on the Importance of Substrate*. Fla. Mar. Res. Publ. Fla. Dept. of Nat. Res. Number 17.

Walsh, G. E. and T. E. Grow. 1972. Composition of *Thalassia testudinum* and *Ruppia maritima*. *Quart. J. Fla. Acad. Sci.* **35**: 97–108.

Wetzel, R. G. and P. A. Penhale. 1979. Transport of carbon and excretion of dissolved organic carbon by leaves and roots/rhizomes in seagrasses and their epiphytes. *Aq. Bot.* **6**: 149–158.

Williams, R. B. 1966. *Annual Production of Marsh Grass and Eelgrass at Beaufort, N.C.* Ann. Rept. USBCF Biol. Lab., Beaufort, N.C.

Zieman, J. C. 1974. Methods for the study of the growth and production of turtle grass, *Thalassia testudinum* Konig. *Aquaculture* **4**: 139–143.

Zieman, J. C. and R. G. Wetzel. 1980. Productivity in seagrasses: methods and rates. In Phillips, R. C. and C. P. McRoy (eds.). *Handbook of Seagrass Biology. An Ecosystem Perspective*, Garland STPM Press, New York, pp. 87–116.

CHAPTER 18

Salt Marsh Communities

Tidal marshes are found throughout the world in temperate and warm temperate latitudes (Figure 18-1). Salt marshes have considerable economic potential and serve as nursery grounds for numerous invertebrates and fish. Odum (1974) estimated that salt marshes return $2,600/acre · yr. in Georgia. The salt marshes, consisting of halophytic grasses, rushes, and dicot bushes (Figure 18-2A), replace the tropical tidal swamps of mangroves in shallow coastal regions where frost or freezing temperatures occur. Some general texts and articles on salt marsh communities include Chapman (1964), Ranwell (1972), Redfield (1972), and Nixon and Oviatt (1973).

Figure 18-1 The black rush *Juncus roemerianus* dominates salt marshes such as that found at the Weeki Wachee River on the west coast of Florida. The mud banks in front of the marsh, the lower intertidal region, contain oysters and can become covered by *Vaucheria* spp. as well as various blue-green algae. (Courtesy of Mary Ann Davis, University of South Florida, Tampa.)

ADAPTATIONS TO THE MARINE ENVIRONMENT

Environmental factors supporting the development of a salt marsh are protection from wave action, influence of fresh water, sediment deposition, and shallow intertidal slopes. At least some freezing temperatures will occur, limiting or excluding mangrove development. The plants of a salt marsh occur in the mid to upper intertidal zones, where tidal currents and wave action do not cause erosion, and where plants are not permanently submerged. Estuaries are common areas for salt marsh development (see Chapter 10 regarding estuaries) since most of the necessary factors are present. In a typical salt marsh, the elevation will increase only gradually from the seaward to the landward side, with submergence persisting longest at the seaward side. Because of the differences in submersion, drainage, and associated sediment types, the vegetation is stratified (Figure 18-2a), especially along tidal channels.

Salt marshes occur in estuaries, where sedimentation rates are typically high; thus the substratum is usually a mixture of sand, silt, and mud. The substratum contains sea salts, and may even have crystals of sodium chloride due to excessive evaporation. Salt flats can develop in the upper zones of salt marshes.

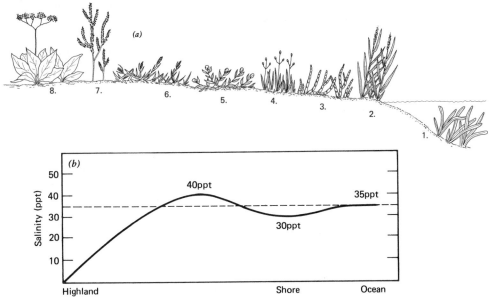

Figure 18-2 (*a*) Sketch of the dominant plants found in a California salt marsh, showing the zonation of various species: (*1*) The subtidal seagrass *Zostera marina*; (*2*) *Spartina foliosa* (cordgrass); (*3*) *Salicornia virginica* (pickle weed) and an algal mat of *Cladophora* sp. and *Vaucheria* sp.; (*4*) *Triglochin maritimum* (arrow grass); (*5*) *Jaumea carnosa* (jaumea); (*6*) *Distichlis spicata* (salt grass); (*7*) *Suaeda californica* (sea bite); and (*8*) *Limonium californicum* (sea lavender). (*b*) Salinity, or salt distribution, does not show a gradual decrease from coastal water to terrestrial substrates. There is a concentration of salt in the substrates of the lower intertidal zone of salt marshes.

Soil structure is also affected by the high concentrations of the positively charged sodium ion. Because of repulsion by the positive charge, the soil particles holding the ions are evenly distributed and won't clump. Without clumping of the soil particles, a reduction of soil air spaces results. With a decrease in soil air spaces, air and water movements are reduced, resulting in partial anaerobiosis. Both sulfur bacteria involved in the sulfur cycle (Appendix B) and nitrogen-fixing bacteria are present in these sediments. Salt distribution in the substratum of a salt marsh increases from seaward to landward areas and then decreases in typical nondesert environments as shown in Figure 18-2b.

Morphological Adaptations

Most plants of a salt marsh have xerophytic adaptations and are facultative halophytes because of the high water stress and concentration of salt in the substratum. A facultative halophyte is one that can tolerate salt conditions but can also grow in more-mesophytic, freshwater environments. The characteristic morphological features are demonstrated by two dominant monocots of salt marshes, *Spartina* and *Juncus*.

Salt marsh angiosperms are usually shallow rooted with rhizomes (horizontal stems) just below the substratum surface and extending down to about 100 cm (Reimold, 1972). As will be emphasized in the section on ecological roles, these rhizomaceous plants stabilize the soft sediments and reduce erosion. In addition, rhizomes of *Spartina* act as phosphorus pumps, transferring phosphates produced in the phosphorus cycle in the substratum (Appendix B) into the plant's leaves and stems (Reimold, 1972). The roots are usually adventitious, developing from the rhizome rather than from another root. There are usually two types of roots. The first type functions in anchoring, lacks root hairs and has a corky covering. The second type is thin, much branched, and has numerous root hairs. The second, fibrous type of root is short-lived and functions as an absorbing root. The erect stems (short shoots, culms) develop from the rhizome, usually on a regular basis, and these produce the leaves.

Anatomical Adaptations

Salt marsh plants show xerophytic adaptations including increased complexity and elaboration of the organs. Specifically, there is (1) increased lignification, (2) a high degree of epidermal development, (3) well-developed bundle sheaths and an endodermal layer with an extensive Casparian strip of suberin, and (4) a modification of the leaf structure to either a thin "dry" type or a thick "succulent" type. Since both monocots and dicots are common in salt marshes, the reader is referred to the mesophytic leaf anatomy of the monocot *Triticum* (Figure 17-2) and the dicot *Ligustrum* (Figure 18-3), for comparative purposes.

The xerophytic adaptations of the two dominant monocots, *Spartina* and *Juncus*, include a thin and a thick type of leaf anatomy. Both monocots have a thick and continuous cuticle resulting in a grayish, glaucous appearance. The upper

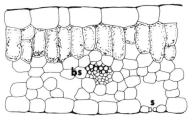

Figure 18-3 A section through a secondary vein of a leaf of the mesophytic dicot *Ligustrum*, showing the simple leaf structure. The cuticle is thin and not easily seen. Stomates (*s*) are present in the lower epidermis and the palisade layer is only one cell thick. Some lignified xylem cells are evident in the upper portion of the vein. The bundle sheath (*bs*), although present, is not extensively developed.

epidermis of *Spartina* leaves (Figure 18-4) contains large cells with a thinner cuticle and outer wall. The specialized cells are called hydathodes; they occur at the base of longitudinal, parallel grooves. The hydathodes collapse when they lose water, and this results in the lengthwise curling (rolling) of the leaf so that the lower epidermis and cuticle are on the outside of the leaf. A majority of the stomates occur along the sides of the grooves of the leaf; thus upon curling they are covered, and water loss is minimized. A dry type of xerophytic leaf, the *Spartina* leaf has a very limited system of air spaces. A general chlorenchyma (ground mesophyll) fills the upper side of the leaf in each ridge. A vein occurs in the middle of each ridge, and an air, or narrow lacunar, space occurs between each ridge, below the groove. As the leaf matures, the lower epidermal cells become lignified, as do the fiber bundles under each vein and at the top of each ridge. The bundle sheath is endodermal, the inner wall has a continuously suberized layer (Casparian strip). The "Krantz type" anatomy of the *Spartina* leaf supports the more recent findings that *Spartina* is a C_4 plant (Kuramoto and

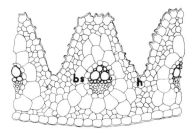

Figure 18-4 A cross section of the leaf of *Spartina alterniflora*, showing the anatomy of a "thin" type of xerophytic leaf. The cuticle is thick on both sides of the leaf, the upper side having deep groves, the bases of which have thin-walled hydathodes (*h*). Stomates also occur in the grove region but are not shown in the Figure. A ground mesophyll contains chloroplasts, and each vein has two tiers of cells forming the bundle sheath (*bs*). This leaf has Krantz type anatomy and has C_4 carbon fixation. Xylem cells are found in the upper portion of the veins. Little lignification is evident in the fiber bundles since this was a young, immature leaf. Compare with Figure 17-2.

Brest, 1979), along with *Distichlis,* another grass of salt marshes. The bundle sheath cells are large, highly specialized endodermal cells that contain chloroplasts with numerous starch grains but poor granal structure. The mesophyll cells surrounding the bundle sheath have numerous, large chloroplasts with well-developed grana and no starch grains; they carry out C_4 carbon fixation (Hatch-Slack pathway). The bundle sheath cells carry out C_3 (Calvin cycle) carbon fixation.

The tubular leaf of *Juncus* (Figure 18-5) is somewhat succulent in nature because the central region has a well-developed spongy parenchyma with numerous longitudinal air channels, or lacunae (aerenchyma). The vascular bundles are found mostly around the periphery of the leaf, but a few are scattered throughout the leaf. A large number of sclerenchyma strands, or fiber bundles, are characteristic of the outer-leaf region in the zone of the palisade mesophyll. The stomates are aligned in rows and are sunken along the epidermis.

The two dicots *Salicornia* and *Sesuvium,* are presented as examples of leaf anatomy because they are probably the most-common genera found in salt marshes. Both exhibit a thick, succulent type of leaf anatomy. *Salicornia* is rather unique in that a modified fleshy leaf-petiole is actually wrapped around the stem (Figure 18-9). The leaf has a thick cuticle and the epidermal cells have a thick outer wall (Figure 18-6). Stomates are sunken in the epidermis. The outer portion of the leaf, the palisade mesophyll, contain columnar cells that contain most of the chloroplasts. The inner region contains spongy mesophyll cells, which are larger, irregular, and contain high salt concentrations. Some small fiber bundles and vascular bundles are also found in the central region. The inner (lower) epidermis consists of smaller, elongated, and irregularly thin walled cells that essentially lack a cuticle or stomates. The outermost cells of the stem (epidermal layer), which is found inside the leaf, are thin walled, large, and lacking a well-developed cuticle (Figure 18-6). Immediately beneath the stem epidermal layer, developing cork cells are found; they develop after abscission of the leaf. The succulent nature of the *Salicornia* leaf is supported by the large water-containing cells of the spongy mesophyll.

Sesuvium has leaves that are succulent and elongated (Figure 18-8). In cross section (Figure 18-7) its leaf has a spongy mesophyll, consisting primarily of

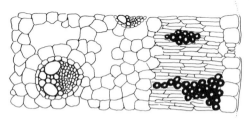

Figure 18-5 Cross section through the tubular leaf of *Juncus roemerianus*, the black rush. Clusters of fiber bundles and a three-to-four-tiered palisade mesophyll are evident in the outer portion of the leaf. The inner portion consists of a spongy parenchyma (aerenchyma) and veins that are primarily found in the outer region. Well-developed two-layered bundle sheaths are found around each vein, and the xylem is on the inside of the veins.

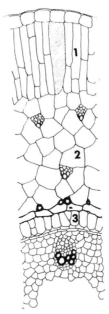

Figure 18-6 Cross section through the "stem" of a young *Salicornia virginica* plant, showing the modified leaf petiole wrapped around the stem. In the leaf portion of the cross section, an outer palisade with a salt-gland cell (*1*) and inner-ground mesophyll (*2*) is evident. Small veins are seen in the inner mesophyll. The stem proper has an outer layer of thin-walled cells (*3*) that will suberize, forming cork cells after the leaf base falls off. Scattered fiber cells are seen in the inner part of the leaf base, and xylem vessels are visible in the stem vascular bundle.

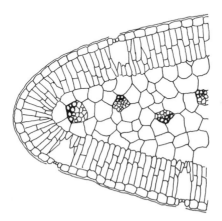

Figure 18-7 Cross section of an edge of the succulent leaf of *Sesuvium portulacastrum*, showing an outer palisade layer that is continuous around the leaf. The inner portion is a modified spongy, or ground, mesophyll that functions in water retention. Some veins are evident in the inner tissue.

water-containing cells and vascular bundles. The outer mesophyll is a palisade mesophyll with numerous chloroplasts. Stomates are sunken and are found on both sides of the leaf. The epidermis is covered by a thick cuticle. The two common salt marsh succulent dicots *Batis* and *Salicornia* do not show the modified C_4 pathway, crassulacean acid metabolism (modified Hatch-Slack pathway for carbon fixation in photosynthesis; Kuramoto and Brest, 1979). In a yearlong study of three succulents, *Borrichia frutescens, Batis maritima,* and *Salicornia virginica,* Antlfinger and Dunn (1979) came to the same conclusion using gas-exchange techniques.

TAXONOMIC AND ECOLOGICAL CONSIDERATIONS

The dominant angiosperms of salt marshes are monocots *Spartina* and *Juncus.* However, a number of other genera are also common, such as the monocot *Distichlis* and the dicots *Batis, Borrichia, Salicornia, Sesuvium,* and *Suaeda.*

Taxonomic Considerations

More than 18 angiosperm families (monocots and dicots) have halophytic representatives and occur in salt marshes. Salt marsh plants are believed to have evolved from freshwater forms that grew along streams that emptied into the ocean. By gradual adaptation to brackish and salt water through development of xerophytic mechanisms, some of these species became the xerophytic salt marsh plants of today. Forms that could withstand saltwater intrusion into the streams were probably the original types able to spread gradually into the mud flats and sand barriers formed in the new estuaries.

Geographical Distribution

Salt marshes occur throughout the temperate regions of the world, wherever the conditions listed at the beginning of this chapter are found. Although genera and species may differ, the xerophytic adaptations of salt marsh plants and the environmental features common to all estuaries result in similar marshes in Oregon and New England. Because mangroves cannot tolerate prolonged (4 hr) frosts, salt marshes replace them in cooler temperate climates. Thus on the west coast of Florida there are nearly continuous mangrove swamps south of Tampa Bay whereas salt marshes occur to the north.

Selected Species

The two dominant monocots, *Spartina* and *Juncus,* and two dicots, *Sesuvium* and *Salicornia,* will be used as examples of morphological and taxonomic features. All of the plants exhibit xerophytic adaptations as do all emergent angiosperms found in salt marshes. *Salicornia* seems to require a high saline condition for

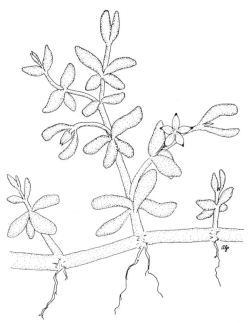

Figure 18-8 Habit view of *Sesuvium portulacastrum*. The flower is about 1 cm in diameter. The plant has a rhizome that is succulent and helps the plant adapt to the upper intertidal region.

growth (obligate halophyte) whereas most other genera do not (facultative halophytes).

Sessuvium portulacastrum (L.) L

This plant is called *sea purslane* and is placed in the family Aizoaceae (capretweed family). It is a perennial, creeping shrub with succulent leaves that are opposite in arrangement, 1–5 cm long, and obovate in cross section (Figure 18-8). The flowers are bisexual, solitary, and pink to white. The fruit is a capsule that splits open along five sutures.

   Sesuvium portulacastrum is a common seaside plant of tropical and subtropical coasts, and is especially abundant in the upper zones of tidal marshes (salt marsh and mangrove swamps). The family Aizoaceae is large, having over 22 genera with about 1100 species. It has a temperate to tropical distribution. Most of the species are succulent low shrubs or creeping herbs. A number of species are halophytic and can be found in tidal marsh environments.

Salicornia virginica L

The common name is *perennial glasswort* and the genus is placed in the Chenopodiaceae (goosefoot family). The plant is a succulent, perennial herb with decumbent stems that form mats over the substratum (Figure 18-9). The leaves are

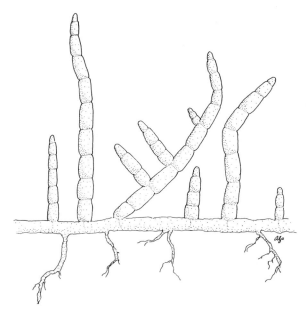

Figure 18-9 Habit view of a vegetative portion of *Salicornia virginica*. The erect branches, arising from a rhizome, are bulbous because of the bases of leaves that are wrapped around the stems.

opposite and reduced to scales, and the leaf petiole is modified to wrap about the stem, producing a bulbous, segmented appearance. Old stems lose these petioles and appear segmented and woody.

Salicornia is an excellent example of a succulent xerophyte and shows a major modification through the loss of the leaf blade and the wrapping of the petiole around the stem. The genus is found as an outer fringe, beyond the main salt marsh (lower intertidal zone) in England, whereas in tidal marshes of the United States, it is typically found above the *Spartina* and *Juncus* zone (upper border zone). The family Chenopodiaceae has a number of genera that are common in salt marshes, including *Salicornia, Suaeda, Glaux,* and *Atriplex*. These plants are all herbs or low shrubby flowering annual or perennial plants. The flowers are small, inconspicuous, often have green, persistent sepals (1–5), and lack petals. In the past many of these halophytes were harvested and burned for potash (see Chapter 2).

Spartina alterniflora Loisel

The common name is *smooth cordgrass* and the genus is placed in the Gramineae (Poaceae) or true grass family. The erect stems (culms) of the tall form is usually about 2 m long. The stems are soft and spongy at the base where the intercalary meristem occurs (Figure 18-10). The leaves are smooth and blades flat, furrowed and involute at the tips. The lower epidermal cells of the leaves have thick cuticles, whereas the upper sides are thinner. The leaf curls upward because of

Figure 18-10 Habit sketch showing the lower (*B*) and upper (*U*) portion of the salt grass, *Spartina alterniflora*. The rhizome is not visible but adventitious roots arising from the stem bases are evident. The floral heads are present (U). × 0.40.

thin-walled hydathodes at the bases of the furrows on the upper leaf side. The rhizomaceous stem grows just under the substratum and down to 100 cm. It produces stubby, anchoring roots that have no epidermis or cortex, but a cork covering instead. Thin, ephemeral absorbing roots are also produced. Maximum uptake of phosphate occurs in these roots and rhizomes at depths of 50–100 cm (Reimold, 1972).

Spartina patens (Ait) Muhl is the other common species and is called *salt meadow cordgrass*. This species usually occurs at a higher elevation than *S. alterniflora*. Because of the extensive meadows sometimes formed by *S. patens*, cattle are allowed to feed on it, and it is cut for hay. The two species of *Spartina* extend from Newfoundland to Texas amd Florida on the east coast of the United States. About 15 species of *Spartina* are found in salt marshes throughout the world. The grass family, Poaceae, has members that are annual or perennial herbs having culms with hollow internodes and solid nodes. The roots are fibrous, and horizontal stems (stolons, rhizomes) are common in the various genera found in salt marches *(Distichlis, Monanthochloe, Sporobolus, Spartina, Paspalum)*.

Juncus roemerianus Scheele

The species is commonly called *black rush* and is placed in the rush family, Juncaceae. The rush family is small, having about eight genera and occurring predominantly in the Southern Hemisphere. The species has stiff, 2 m tall blades, which are needlelike, tubular, and have spongy parenchymatous centers (Figure 18-11). The epidermis produces a thick cuticle, and numerous bundles occur as strands in the palisade mesophyll. The stem is rhizomaceous; it has a cork layer and scale leaves.

Although present all along the Atlantic and gulf coasts of the United States, the black rush becomes the dominant plant of the northern Florida salt marshes, resulting in a brown to black appearance of the swamps. The species is usually found near mean high tide level (0.94 ft) in Florida. It appears grasslike, with flat or terete blades. Flowers are in dense cymes. Some species of the genus are found in moist to aquatic habitats, and many are halophytic.

Figure 18-11   Habit view of the black rush *Juncus roemerianus* showing the basal portion with the rhizome and anchoring roots (with some absorbing roots) and the top portion with a floral raceme. × 0.40.

ECOLOGICAL ROLES AND ZONATION

The steady, gentle increase in elevation in a salt marsh results in a series of zones that correspond to the frequency of tidal flooding. Tidal inundation and exposure have frequently been cited as the controlling factors in plant zonation, although Eleuterius and Eleuterius (1979) could not substantiate this relationship.

Ecological views regarding the status of salt marshes differ. Chapman (1964) and Emory *et al.* (1957) described this community as a pioneer or serial stage of succession toward a terrestrial climax community. These authors hold a similar view with regard to the other tidal marsh, the mangrove swamp. Others, including this author, consider these tidal marshes to be stable communities that gradually progress seaward during periods of declining sea levels and landward during periods of rising sea levels. Either view includes the concept that the salt marsh is a continuum for a subtidal seagrass or shallow benthic community to a freshwater flat woods community.

Ecological Roles

Four major ecological roles have been assigned to the salt marsh, as follows:

1 The production of a large quantity of organic matter (Eilers, 1979; Heald, 1971; Teal, 1962), the release of dissolved organic carbon from salt marsh leaves (Gallagher *et al.*, 1976), and the production of peat and detritus (Squires and Good, 1974; de la Cruz and Gabriel, 1974).

2 The development of a habitat for numerous animals. In fact, Odum (1961) suggested that 95% of all sport and commercial neritic fish caught are nurtured by salt marsh productivity.

3 The stabilization of coastal substratum through the massive surface root systems.

4 The filtration of coastal runoff and removal of organic waste. Teal and Valiela (1973) found that salt marshes are highly efficient in the removal of organic wastes, acting as tertiary sewage treatment systems. Salt marsh plants also concentrate heavy metals. In a yearlong study, Chalmers (1979) found plant biomass increased substantially with sewage sludge application in a Georgia salt marsh. However, she also found that about half of the added nitrogen probably was removed during tidal exchange and pointed out the dangers of overloading a salt marsh with organic wastes. The finding needs further support but suggests problems with dumping wastes in tidal marshes.

Zonation

The zonation of salt marsh vegetation is dependent on a number of ecological factors to which the various salt marsh species are adapted. Many of these factors are interrelated, including the degree of submersion (flooding), topographical features of the marsh, textural variations of the soil, exposure to wave and wind

Table 18-1   Soil Profile in a Northern Florida Salt Marsh[a]

Characteristic (with Soil Depth in cm)	Lower Border	Lower Slope	Upper Slope	Upper Border	Upland Forest
Chlorinity (ppt)					
7.5	1.51	1.48	3.75	0.13	0.12
22.5	1.16	1.37	2.85	0.07	0.22
75.0	0.95	1.19	6.16	0.36	0.07
Percentage water					
7.5	66.8	46.8	18.0	18.8	10.2
22.5	72.2	27.7	21.1	17.6	8.2
75.0	31.4	40.0	23.3	17.0	18.2
pH					
7.8	5.6	5.3	8.6	5.7	4.3
22.5	4.8	5.3	7.0	6.0	5.8
75.0	4.6	4.8	6.1	5.6	5.6

Source: Modified from Kurtz and Wagner (1957).
[a] See also Figure 18-13.

action as well as current patterns (erosion) and soil chemistry (chlorinity, soil water, organic matter content, and pH; see Table 18-1 for an example of changes in salt, pH, and water in the substratum of a salt marsh.) In addition, biological factors such as competition (Pielou and Routledge, 1976) play a role in plant zonation.

Zedler (1977) found elevation to be a critical factor controlling zonation of a salt marsh in a California estuary. Similarly, Mendelssohn and Seneca (1980) and Linthurst and Seneca (1980) showed that soil drainage (hence elevation) influenced the distribution of *Spartina alterniflora* in North Carolina salt marshes. Soil chemistry is strongly influenced by tidal flushing, and thus tidal fluctuation is a critical factor, as well as elevation. Mahall and Park (1976a, 1976b, 1976c) carried out a series of studies and found that the rate of sedimentation, soil salinity, and tidal fluctuation were the prime factors in determining an ecotone (ecological separation) between *Spartina* and *Salicornia* populations in a salt marsh of San Francisco Bay. A combination of factors listed above was found by Pemadasa *et al*. (1979) to be statistically important in determining the distribution of salt marsh species in Sri Lanka.

Parrondo *et al*. (1978) found that substratum chemistry, especially salinity, was important in controlling zonation of *Spartina alterniflora*. Salinity increases of the root medium reduced shoot growth more than root growth. Smart and Barko (1978) also found distinctive tolerances to sediment salinity by various salt marsh plants. Linthurst (1979) found that redox potential and/or pH were the two main physical variables regulating *S. alterniflora* growth in natural environments, since these regulated the availability of nutritional elements. As shown in Table 18-1, the levels of chlorinity, water and pH change through the various zones in a northern Florida salt marsh. The pH rises from a low of about 5.3 in the lower slope to a high of 8.6 in the upper slope primarily because of salt increases in the substratum caused by evaporation.

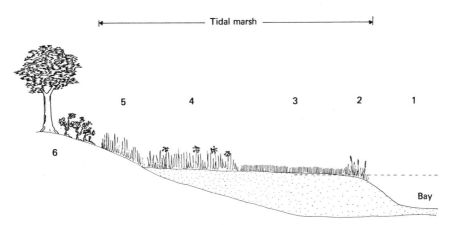

Figure 18-12 Profile of a New England salt marsh as depicted by Miller and Egler (1950). The zones are as follows: (*1*) lower intertidal (mud, blue-green algae, and green algae), (*2*) lower border (*Spartina alterniflora*), (*3*) lower slope (*S. patens*), (*4*) upper slope (*Juncus roemerianus*), (*5*) upper border (*Salicornia* sp., *Aster*, *Limonium*), and (*6*) upland shrub and forest.

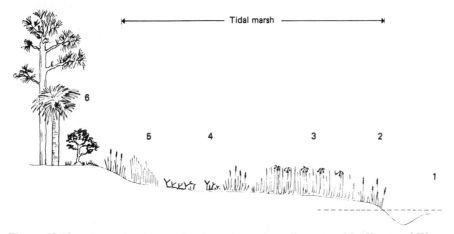

Figure 18-13 Zones of a northern Florida salt marsh as diagrammed by Kurtz and Wagner (1957). The zones are as follows: (*1*) and (*2*) as in Figure 18-12, (*3*) lower slope (*Juncus roemerianus*), (*4*) upper slope (*Distichlis spicata*, *Spartina patens*, *Batis maritima*), (*5*) upper border (*Salicornia*, sp. *Iva frutescens*, *Fimbristylus castana*), and (*6*) upland shrub and forest (pine and palmetto).

Two examples of plant distribution in salt marshes are presented in order to demonstrate latitudinal differences. A salt marsh in New England would have the plants shown in Figure 18-12, whereas one in northern Florida would have the plants shown in Figure 18-13 (see also Table 18-1). Both salt marshes have similar zones, a low intertidal band, a low border, a lower and upper slope, and an upper border followed by the upland shrub and forest. Detailed studies of a northern Florida salt marsh (Kurtz and Wagner, 1957) and New England salt marshes (Nixon and Oviatt, 1973; Redfield, 1972) are available.

Both salt marshes (Figures 18-12 and 18-13) have lower intertidal bands free of angiosperms, being mud or sand that can be covered by blue-green algae, benthic diatoms and some forms of green algae such as *Enteromorpha*. In both salt marshes, *Spartina alterniflora* is the major angiosperm forming the lower border, its zone apparently controlled by tidal planes (Provost, 1976). The large tidal fluctuation (2–3 m) on the east coast of North America probably allows a broad lower border and lower slope that is dominated by *Spartina alterniflora* (Nixon and Oviatt, 1973; Redfield, 1972). On the other hand, the smaller tidal range (0.5–1 m) in the eastern Gulf of Mexico allows only a narrow lower border, thus restricting *Spartina* to a small band, usually along tidal streams (Kurtz and Wagner, 1957; Dawes, 1974). Instead of *Spartina, Juncus romerianus* is the dominant meadow plant in the lower slope of a Florida, west coast salt marsh (Dawes, 1974). The upper slope, which is usually of much shorter expanse, is dominated in New England salt marshes by *Juncus* or *Spartina patens,* whereas in northern Florida a combination of *Distichlis spicata* (spike grass) and *Spartina patens* dominate. The upper borders of both salt marshes will usually have *Salicornia* sp. as well as other herbs or scrubby bushes such as *Iva imbricata* and *Fimbrystylus castana* (in Florida) or *Aster* and *Limonium* (in New England).

Algae in Salt Marshes

The algal flóra is significant from a productivity standpoint (see Zedler, 1980) and characteristic species are usually found in salt marshes. In the lower slopes of Floridian salt marshes the algal flora is similar to that found in the mangrove swamps. The algae include filamentous blue-green algae such as *Lyngbya, Oscillatoria,* and *Phormidium,* diatoms, green algae such as *Rhizoclonium, Cladophora, Chaetomorpha, Ulva,* and *Enteromorpha,* and red algae such as *Bostrychia, Polysiphonia, Catenella,* and *Centroceras.* A larger variety of algae is typical of tidal channels (Rhodophyta: *Gracilaria, Griffithsia, Spyridia:* Chlorophyta: *Ulva, Batophora*) as well as the estuarine seagrass, *Ruppia maritima.* Sullivan (1977a) found 91 species of diatoms inhabiting the sediments of a New Jersey salt marsh. He also found (1977b) a number of epiphytic species on the blades of the estuarine seagrass *Ruppia maritima.*

The salt marsh macroalgae of New England have been studied by Blum (1968), who found most algae on the upper slope. Larger fucoid marsh algae occur in the temperate and cold temperate marshes of the North Atlantic and Baltic Sea, such as *Pelvetia, Fucus,* and *Ascophyllum.* The rockweeds are usually attached to substrata and apparently reproduce by asexual fragmentation. *Ascophyllum nodosum* is common in the tidal marshes of New England; two ecological forms (ecads) are usually described; *A. nodosum* f. *scorpioides* and *A. nodosum* f. *mackaii.* The ecads have been the subject of detailed studies on growth and morphology (Brinkhuis, 1976; Brinkhuis and Jones, 1976) as well as physiological responses (Brinkhuis *et al.*, 1976; Chock and Mathieson, 1976). In another study, Mathieson and associates (Mathieson *et al.*, 1976) found that the reproduction and growth rates of *Ascophyllum nodosum* and *Fucus vesiculosus* var. *spiralis* were related to seasonal variations in temperature and light within the Great Bay Estuary system of New Hampshire. The maximum reproductive activity occurred in early spring (April and March). Niemeck and Mathieson (1976) studied the distribution of *Fucus spiralis* in the same region and found that the plant showed a discontinuous distribution due to its association with specific types of substrates; metasedimentary or metavolcanic rock outcrops were the primary site of attachment. Unlike *Fucus vesiculosus*, *F. spiralis* had maximum growth and reproduction during the summer. Such studies are of importance in understanding estuarine energy flows, since the algal components, especially diatoms (Sullivan, 1977a,b), serve as major sources of food for most estuarine herbivores. However, most of the fixed carbon ends up in the detrital, rather than herbivore, food chains.

Estuaries and Man

The impact of man on estuaries, especially over the last 25 years, has been enormous in many parts of the world. Estuaries are especially "sensitive" since they are areas of environmental stress because of the interaction of marine and fresh waters. One need only note the wide ranges in salinities and temperatures found in these shallow-water sites, and the high rates of sedimentation, high organic loading, and resultant anaerobic substrate or bottom-water zones. Tidal marsh plants must be able to grow and reproduce under a constant flux of conditions in order to survive. Most marine botanists would consider these plants to be under "stress," as they utilize a large percentage of their organic production to survive in such a harsh environment. With the expansion of cities, man has turned to estuaries as sites of farms (because of the organically rich soil) and settlement, especially for higher-cost developments because of the scenic beauty and access to the ocean. Thus dredge-and-fill operations channelize the meandering tidal streams and cover the salt marshes. Industrial and human wastes are dumped into the rivers that support salt marshes. The initial result of organic waste dumping is an increase in primary production and catabolism (respiration) due to increased nutrients. Subsequently there is a drop in the already-low level of dissolved oxygen, as anaerobic conditions simply worsen. Many toxic industrial wastes kill or damage plant and animal life. In many cases these toxic substances remain in the substrata. Furthermore, power plants utilize estuarine waters to cool atomic reactors or fossil fuel systems and then return the heated waters to the estuary (thermopollution). Although cooling-water temperatures may be only a few degrees centrigrade higher then normal, some organisms may die because of increased respiration. Other species will disappear because they cannot reproduce. Rapid recovery of the algal and angiosperm components of a thermally stressed salt marsh has been documented (Vadas *et al.*, 1978; Keser *et al.*, 1978). In both cases, after shutdown or relocation of thermal discharges, highly damaged salt marsh floras recovered to near-original states of productivity.

As pointed out in Chapter 13, such problems are now being recognized, as man begins to better understand the value of salt marshes. For example, the importance of meandering tidal channels, as opposed to straight, deeper boating channels, has become obvious. The result of channelization has been rapid erosion of sediment, which damages salt marsh plants because of too rapid drainage and subsequent desiccation. Channelization will usually result in increased turbidity and pollution of the open coastal region adjacent to salt marshes.

ORGANIC PRODUCTION

Probably the most significant contribution of salt marshes is their role as primary producing systems, as shown in the final section of this chapter. A *Spartina*-dominated salt marsh is one of the most-productive systems on earth, equal to the most fertile agricultural lands. Whereas agricultural productivity is maintained through fertilizers, pesticides, and a strong subsidy from fossil-fuel energy, a salt marsh is naturally fertilized twice a day by the incoming tide. There is a north-south gradient in macrophyte production in salt marshes of the North Atlantic that parallels the solar energy input (Turner, 1976). In a review of previous studies, Turner showed that turnover of organic matter increased with decreasing latitudes, and in some warm temperate marshes the organic turnover almost equaled the annual changes in standing biomass. Although there was considerable variation between studies of marshes in similar latitudes, *Spartina alterniflora* was the most common dominant marsh plant.

The net production by *Spartina* finds its way into the food chain in several ways. A small amount of organic matter, about 10%, is lost by leaching from the leaves when *Spartina* is covered by tides. The leached organic matter is uthlized by the various aquatic heterotrophs. Although there are no lrge herbivores such as cattle in the salt marsh (but see Chapter 2 for a discussion of harvesting for fodder), grasshoppers and sucking insects utilize a significant quantity of the *Spartina* production. Most of the plant material dies, decomposes, and becomes part of the detrital food chain. The bacteria and fungi that colonize the small pieces of each *Spartina* fragment (detritus) are the actual sources of nutrition for the detritus feeders (protozoa, meiofauna, snails, worms, crabs, clams, insect larvae). The more-resistant cellulose and lignin provide little nutrition to the animals, and these compounds break down slowly. A study of decomposition of the black rush, Juncus, was conducted by de la Cruz and Gabriel (1974). Their careful analyses showed a major increase in available nitrogen from the detritus due to microbial activity. A decay rate of 40% was determined by litter-bag studies; this was higher than expected. The rapid decay of *Spartina* allowed nutrient recycling at a higher rate as well.

Tenore and Hanson (1980) found that seaweeds may account for much of the assimilated food in detrital food chains compared to *Spartina* detritus. They found that even through all of the decay-resistant *Spartina* detritus was eventually oxidized, less was used in macroconsumer growth compared to seaweed detritus. In addition, there was a strong correlation between the lower production of macroconsumers associated with salt marshes and the essentially unavailable *Spartina* detritus. Tenore and Hanson suggested that ancillary sources of food such as benthic algae and terrestrial runoff may play a major role in detritrial food chains of salt marsh ecosystems.

In a study of *Spartina alterniflora* production above and below ground, Smith et al. (1979) found the highest levels of biomass below ground (Figure 18–14). The rhizome and root portion had a biomass of almost 3 kg/m² in the top 30 cm of substratum. Thus the New Jersey *Spartina* community provided a large flow of

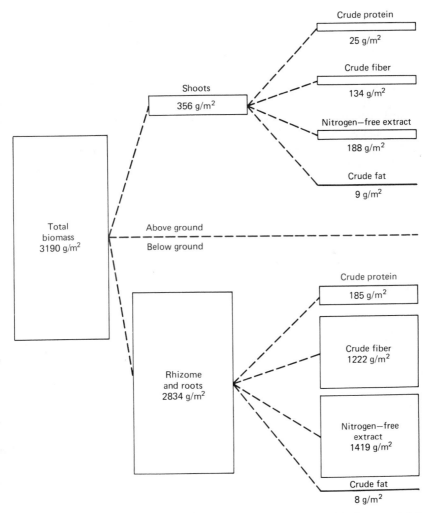

Figure 18-14 The accumulation of photosynthetically fixed material above and below ground in *Spartina alterniflora* shows the dominance of rhizome and root biomass (after Smith *et al.*, 1979).

organic compounds within the estuarine ecosystem, with the below-ground component and its decomposition acting as a sink.

Algae, filamentous forms and diatoms, are also abundant on the mud surface and on the bases of the salt marsh macrophytes. It is now becoming evident that these smaller plants may be important direct food sources not only to herbivores but also to detritus feeders (Tenore and Hanson, 1980). The smaller herbivores and detritus feeders are eaten by carnivores of various sizes that enter the salt marsh from land at low tide or from the channels at high tide. Large amounts of organic matter, including detritus, are also washed off the marsh and enrich the surrounding water. Haines (1977), however, found that in Georgia marshes the particulate detritus was not formed from the dominant plant, *Spartina alterniflora*. Perhaps much of the detrital material is of terrestrial orgin from stream flow, algal macrophytes, and phytoplankton (Haines, 1977).

In Brunswick County, North Carolina, *Spartina* has an average net primary production of 474–1751 g $C/m^2 \cdot yr$, whereas *Juncus* has a primary production of 758 g $C/m^2 \cdot yr$. Stiven and Plotecia (1976) found that the production range was dependent on the size of the plant. In New Jersey, Squires and Good (1974) estimated that *Spartina alterniflora* had a maximum standing crop of 1592 g $C/m^2 \cdot yr$ and a summer productivity of about 26 g $C/m^2 \cdot da$. Eilers (1979) found a strong correlation between net production, elevation, and length of submergence in a tidal marsh in Oregon. Above-ground net production ranged from 230 to 2800 g $C/m^2 \cdot yr$. The highest rates were found only at the higher elevations, where submergence was brief.

Pomeroy (1959) demonstrated that algae also can play an important role in the Georgia salt marshes, with annual primary production of about 200 g $C/m^2 \cdot yr$. *Ascophyllum nodosum* of northern Atlantic salt marshes showed highest photosynthetic rates under reduced light while slightly desiccated (Brinkhuis *et al.*, 1976) and the largest biomass (1.93 kg/m^2) in November at the end of the growing season (Chock and Mathieson, 1976). The same type of biomass data was obtained for *Fucus spiralis* (Niemeck and Mathieson, 1976) and *Fucus vesiculosus* var. *spiralis* (Mathieson *et al.*, 1976).

In southern California, Zedler (1980) found that the productivity of salt marsh algal mats was highest under a canopy of *Jaumea carnosa* (341g $C/m^2 \cdot$ yr). Because of the arid environment of southern California and hypersaline soils, Zedler suggested that the vascular-plant cover allows this high algal productivity. In north Wales, Oliveira and Fletcher (1977) studied the open coast rockweed *Pelvetia canaliculata* and the salt marsh ecad *P. canaliculata* ecad *libera*.

The rates of water loss were similar for both forms under laboratory conditions, but the ecad lost water more slowly because of the environmental features of a salt marsh. The salt marsh ecad could photosynthesize for much longer periods than the rocky shore species, even with a 20% water loss. Finally, they demonstrated that continuous submergence was deleterious to both forms.

From these studies it does appear that continuous or prolonged submergence of intertidal algae will be damaging (See Chapter 14). In a comparative study of algae from a salt marsh and a mangrove swamp, Dawes *et al.* (1978) showed similar net photosynthetic and respiratory rates for intertidal algae (*Bostrychia, Catenella, Cladophora*) of both tidal marshes. However, subtidal (channel) algae (*Gracilaria, Polysiphonia, Acanthophora*) collected in the mangrove swamp showed much higher rates of photosynthesis. One explanation was that the subtidal site is a more-protected site, which results in healthier plants. Intertidal species showed their highest rates of photosynthesis when monitored in the air. Thus it was concluded that these species were adapted to exposure and were more productive when exposed than when submerged. The concept that algae are under stress because of exposure should probably be reevaluated, and studies should include plants in air as well as those submerged.

REFERENCES

Antlfinger, A. E. and E. L. Dunn. 1979. Seasonal patterns of CO_2 and water vapor exchange of three salt marsh succulents. *Oecologia* **43**: 249–260.

Blum, J. L. 1968. Salt marsh Spartinas and associated algae. *Ecol. Monog.* **38**: 199–221.

Brinkhuis, B. H. 1976. The ecology of temperate salt-marsh fucoids. I. Occurrence and distribution of *Ascophyllum nodosum* ecads. *Mar. Biol.* **34**: 325–338.

Brinkhuis, B. H. and R. F. Jones. 1976. The ecology of temperate salt-marsh fucoids. II. *In situ* growth of transplanted *Ascophyllum nodosum* ecads. *Mar. Biol.* **34**: 339–348.

Brinkhuis, B. H., N. R. Tempel, and R. F. Jones. 1976. Photosynthesis and respiration of exposed salt-marsh fucoids. *Mar. Biol.* **34**: 349–359.

Chalmers, A. G. 1979. The effects of fertilization on nitrogen distribution in a *Spartina alterniflora* salt marsh. *Estuar. Coast. Mar. Sci.* **8**: 327–337.

Chapman, V. J. 1964. *Coastal Vegetation*. Pergamon, London.

Chock, J. S. and A. C. Mathieson. 1976. Ecological studies of the salt marsh ecad *scorpiodes* (Hornemann) Hauck of *Ascophyllum nodosum* (L) Le Jolis. *J. Exp. Mar. Biol. Ecol.* **23**: 171–190.

Dawes, C. J. 1974. *Marine Algae of the West Coast of Florida*. University of Miami Press, Coral Gables.

Dawes, C. J., Moon, R. E. and M. A. Davis. 1978. The photosynthetic and respiratory rates and tolerances of benthic algae from a mangrove and salt marsh estuary: A comparative study. *Estuar. Coast. Mar. Sci.* **6**: 175–185.

de la Cruz, A. A. and B. C. Gabriel. 1974. Caloric, elemental, and nutritive changes in decomposing *Juncus roemerianus* leaves. *Ecology* **55**: 882–886.

Eilers, H. P. 1979. Production ecology in an Oregon coastal salt marsh. *Estuar. Coast. Mar. Sci.* **8**: 399–401.

Eleuterius, L. N. and C. K. Eleuterius. 1979. Tide levels and salt marsh zonation. *Bull. Mar. Sci.* **29**: 394–400.

Emery, K. O., R. E. Stevenson, and J. W. Hedgpeth. 1957 Estuaries and lagoons. In Hedgpeth, J. W. (ed.). *Treatise on Marine Ecology and Paleoecology*. Geol. Soc. Amer. Memoir 67, New York, pp. 673–750.

Gallagher, J. L., W. J. Pfeiffer, and L. R. Pomeroy. 1976. Leaching and microbial utilization of dissolved organic carbon from leaves of *Spartina alterniflora*. *Estuar. Coast. Mar. Sci.* **4**: 467–471.

Haines, E. B. 1977. The origins of detritus in Georgia salt marsh estuaries. *Oikos* **29**: 254–260.

Heald, E. J. 1971. *The Production of Organic Detritus in a South Florida Estuary*. Tech. Bull. 6, Univ. Miami Sea Grant Program, Miami.

Keser, M., B. R. Larson, R. L. Vadas, and W. McCarthy. 1978. Growth and ecology of *Spartina alterniflora* in Maine after a reduction in thermal stress. In Thorp, J. H. and J. W. Gibbons (eds.). *Energy and Environmental Stress in Aquatic Systems*. DOE Symp. Ser. (Conf-771114), Nat. Tech. Info. Serv., Springfield, Va., pp. 420–433.

Kuramoto, R. T. and D. E. Brest. 1979. Physiological response to salinity by four salt marsh plants. *Bot. Gaz.* **140**: 295–298.

Kurtz, H. and K. Wagner. 1957. Tidal marshes of the Gulf and Atlantic coasts of northern Florida and Charleston, South Carolina. *Fla. State Univ. Studies* **24**: 1–168.

Linthurst, R. A. 1979. The effect of aeration on the growth of *Spartina alterniflora* Loisel. *Amer. J. Bot.* **66**: 685–691.

Linthurst, R. A. and E. D. Seneca. 1980. The effects of standing water and drainage potential on the *Spartina alterniflora*-substrate complex in a North Carolina salt marsh. *Estuar. Coast. Mar. Sci.* **11**: 41–52.

Mahall, B. E. and R. B. Park. 1976a. The ecotone between *Spartina foliosa* Trin. and *Salicornia virginica* L. in salt marshes of northern San Francisco Bay. I. Biomass and production. *J. Ecol.* **64**: 421–433.

Mahall, B. E. and R. B. Park. 1976b. The ecotone between *Spartina foliosa* Trin. and *Salicornia virginica* L. in salt marshes of northern San Francisco Bay. II. Soil water and salinity. *J. Ecol.* **64**: 793–809.

Mahall, B. E. and R. B. Park. 1976c. The ecotone between *Spartina foliosa* Trin. and *Salicornia virginica* L. in salt marshes of northern San Francisco Bay. III. Soil aeration and tidal immersion. *J. Ecol.* **64**: 811–819.

Mathieson, A. C., J. W. Shipman, J. R. O'Shea, and R. C. Hasevlat. 1976. Seasonal growth and reproduction of estuarine fucoid algae in New England. *J. Exp. Mar. Biol. Ecol.* **25**: 273–284.

Mendelssohn, I. A. and E. D. Seneca. 1980. The influence of soil drainage on the growth of salt marsh cordgrass *Spartina alterniflora* in North Carolina. *Estuar. Coast. Mar. Sci.* **11**: 27–40.

Miller, W. R. and E. E. Egler. 1950. Vegetation of the Wequetequock-Pawcatcck tidal marshes, Connecticut. *Ecol. Monog.* **20**: 141–172.

Niemeck, R. A. and A. C. Mathieson. 1976. An ecological study of *Fucus spiralis* L. *J. Exp. Mar. Biol. Ecol.* **24**: 33–49.

Nixon, S. W. and C. A. Oviatt. 1973. Ecology of a New England salt marsh. *Ecol. Monog.* **43**: 463–498.

Odum, E. P. 1961. *The Role of Tidal Marshes in Estuarine Production*. New York State Conservat. **16**: 12–15.

Odum, E. P. 1974. Halophytes, energetics and ecosystems. In Reimold, R. J. and W. H. Queen (eds.). *Ecology of Halophytes*. Academic Press, New York. pp. 599–602.

Oliveira, F.°, E. C. de, and A. Fletcher. 1977. Comparative observations on some physiological aspects of rocky-shore and salt marsh populations of *Pelvetia canaliculata* (Phaeophyta). *Bo. Botanica. Univ. S. Paulo.* **5**: 1–12.

Parrondo, R. T., G. G. Gosselink, and C. S. Hopkinson. 1978. Effects of salinity and drainage on the growth of three salt marsh grasses. *Bot. Gaz.* **139**: 102–107.

Pielou, E. C. and R. D. Routledge. 1976. Salt marsh vegetation: latitudinal gradients in the zonation patterns. *Oceologia* **24**: 311–321.

Pemadasa, M. A., S. Balasubramaniam, H. G. Wijewansa, and L. Amarasinghe. 1979. The ecology of a salt marsh in Sri Lanka. *J. Ecol.* **67**: 41–63.

Pomeroy, L. R. 1959. Algal productivity in salt marshes of Georgia. *Limnol. Oceanogr.* **4**: 386–397.

Provost, M. W. 1976. Tidal datum planes circumscribing salt marshes. *Bull. Mar. Sci.* **26**: 558–563.

Ranwell, D. O. 1972. *Ecology of Salt Marshes and Sand Dunes. Chapman and Hall, London.*

Redfield, A. C. 1972. Development of a New England salt marsh. *Ecol. Monog.* **42**: 201–238.

Reimold, R. J. 1972. The movement of phosphorous through the salt marsh cord grass, *Spartina alterniflora* Loisel. *Limnol. Oceanogr.* **17**: 606–611.

Smart, R. M. and J. W. Barko. 1978. Influence of sediment, salinity, and nutrients on the physiological ecology of selected salt marsh plants. *Estuar. Coast. Mar. Sci.* **7**: 487–495.

Smith, K. K., R. E. Good, and N. F. Good. 1979. Production dynamics for above and belowground components of a New Jersey *Spartina alterniflora* tidal marsh. *Estuar. Coast. Mar. Sci.* **9**: 189–201.

Squires, E. R. and R. E. Good. 1974. Seasonal changes in the productivity, caloric content, and chemical composition of a population of salt-marsh cord-grass (*Spartina alterniflora*). *Chesapeak. Sci.* **15**: 63–71.

Stiven, A. E. and R. K. Plotecia. 1976. *Salt Marsh Primary Productivity Estimates for North Carolina Coastal Counties: Projections from a Regression Model.* Univ. N. Carolina Sea Grant Publ. UNC 76–06. Chapel Hill.

Sullivan, M. J. 1977a. Edaphic diatom communities associated with *Spartina alterniflora* and *S. patens* in New Jersey. *Hydrobiologia* **52**: 207–211.

Sullivan, M. J. 1977b. Structural characteristics of a diatom community epiphytic on *Ruppia maritima.* *Hydrobiologica* **53**: 81–86.

Teal, J. M. 1962. Energy flow in the salt marsh ecosystem of Georgia. *Ecology* **43**: 614–624.

Teal, J. M. and I. Valiela. 1973. The salt marsh as a living filter. *Mar. Tech. Soc. J.* **7**: 19–21.

Tenore, K. R. and R. B. Hanson. 1980. Availability of detritus of different types and ages to a polychaete macroconsumer, *Capitella capitata.* Limnol. *Oceanography* **25**: 553–558.

Turner, R. E. 1976. Geographic variations in salt marsh macrophyte production: A review. *Contr. Mar. Sci.* **20**: 47–68.

Vadas, R. L., M. Keser, and B. Larson. 1978. Effects of reduced temperatures on previously stressed populations of an intertidal alga. In Thorp, J. H. and J. W. Gibbons (eds.). *Energy and Environmental Stress in Aquatic Systems,* Doe Symp. Ser. (Conf-771114), Nat. Tech. Info. Serv., Springfield, Va., pp. 434–451.

Zedler, J. B. 1977. Salt marsh community structure in the Tijuana estuary, California. *Estuar. Coast. Mar. Sci.* **5**: 39–53.

Zedler, J. B. 1980. Algal mat productivity: comparisons in a salt marsh. *Estuaries* **3**: 122–131.

CHAPTER 19

Mangrove Communities

The word *mangrove* is derived from a combination of the Portuguese word for tree ("mangue") and the English word for a stand of trees ("grove"). The term is ecological and is used to include both shrubs and trees (dicots and monocots) that occur in the intertidal and shallow subtidal zones of tropical and subtropical tidal marshes (Figure 19-1). A mangrove forest is also called a *mangale* (Walsh, 1974). Two excellent reviews of mangrove ecology (Lugo and Snedaker, 1974; Walsh, 1974) are available.

Figure 19-1 A view of the edge of a mangrove swamp found on Joe Island in Tampa Bay, Florida, showing the drop and prop roots of the red mangrove (*Rhizophora mangle*, outer edge of swamp) and pneumatophores of the black mangrove (*Avicennia germinans*, inner region). Oysters are attached on the intertidal zone of the prop roots. The dark band on both the prop roots and pneumatophores is due to blue-green algae that develop along the high tide zone. (Courtesy of Lherif Loraamm, University of South Florida, Tampa.)

ADAPTATIONS TO THE MARINE ENVIRONMENT

Mangrove swamps are found in tropical areas of the world that are protected from wave action and usually have high sedimentation. The mangrove forest is a fringing community of shallow sandy or muddy areas, ranging from highest-tide mark to the intertidal fringe and subtidal regions.

Morphological Adaptations

Warming (1883) pointed out that most mangroves have adapted to their environment through (1) development of mechanical adaptations for attachment in soft or loose substrata, (2) formation of respiratory roots and aerating devices, (3) evolution of vivipary, (4) use of specialized means for seed dispersal, and (5) development of xerophytic structures. In other words, the plants are adapted to an environment in which water stress is high, salt must by removed, and water must be conserved. Although one finds mangrove trees surrounded by water, the water is saline, and the intertidal substratum may have even higher levels of salt. Thus the plants show typical morphological adaptations for water retention just as desert and salt marsh plants do.

Usually mangroves are shallow rooted and lack well-developed tap roots as a result of the high salt concentrations as well as the water-saturated, organically-rich anaerobic substratum. A number of adaptations in root morphology are typical of most mangroves. The roots may be *prop* (from lower part of stem) or *drop* (from branches and upper part of stem) types that terminate after a few centimeters in the ground (e.g., *Rhizophora mangle,* the red mangrove). In other cases, surface (1–5 cm deep) horizontal roots (*cable roots*) grow out from the stem base and produce negatively geotrophic erect aerial roots called *pneumatophores (Avicennia germinans).* Pneumatophores, prop roots, and drop roots all act as ventilation systems by having well-developed aerenchymas, or specialized air space systems, in the cortex. Arenchyma is an important tissue for gas exchange in roots, that usually develop in an anerobic substratum. The prop, drop, and cable roots all produce *anchoring* and *feeding* types of roots as well. The latter are small, temporary horizontal fibrous roots that have root hairs. They occur just below the surface and function in absorption. The anchoring root develops a thick protective cork layer and grows into the anerobic substratum.

Another relatively common feature of mangroves is vivipary, meaning the embryo initiates germination from the seed while still on the tree. It is considered of special importance since the developing radicle and hypocotyl can respond more rapidly by anchoring in a substratum when washed ashore. In many cases the radicle sinks directly into the mud after falling from the tree. The dispersal properties of mangrove propagules have been correlated with mangrove zonation (Rabinowitz, 1978). Species found at the higher elevations on the landward edge of an intertidal zone, usually produce small propagules. The propagules require a period of freedom from tidal inundation of approximately 5 da in order to establish firmly in the substratum (e.g., *Laguncularia*). Genera whose adults are found on the seaward edge of the swamp or in the lower intertidal zone have large heavy propagules (e.g., *Rhizophora*).

At one point there was a major argument between two German ecologists regarding the importance of vivipary in the red mangrove *Rhizophora mangle*. The arguments could be summed up in a "plop theory" and a "plunk theory." Proponents of the former theory argued that the seedlings plopped directly into the mud, whereas the latter argued that the seedlings plunked into the water and then established. The argument was unnecessary since both forms of establishment occur. Most commonly the seedlings fall in the water and float, the root slowly curves down, and the epicotyl becomes erect. Upon reaching the substratum, the roots attach.

Anatomical Adaptations

Mangroves exhibit xerophytic adaptations, and thus there is an increase in complexity of the leaf structure when compared to the leaf of a mesophytic dicot such as *Ligustrum* (Figure 18-3). Essentially all of the mangrove trees have leaves that have the nonsucculent "dry" type of anatomy, although they may be thick and leathery. The features of mangrove leaves have been listed by Stace (1966) and are summarized in Table 19-1.

Cross sections of red mangrove (*Rhizophora mangle*) and black mangrove (*Avicennia germinans*) leaves are given in Figures 19-2 and 19-3. The epidermis of these leaves has a thick cuticle, and epidermal hairs may be common on the

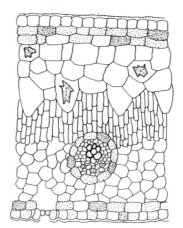

Figure 19-2   Cross section of a leaf of the red mangrove *Rhizophora mangle*. The epidermis is covered by a thick cuticle. Below the upper epidermis are tannin-containing cells (stippled), hydrocytes, and cystoliths (with calcium oxalate crystals), all forming the hypodermis. The palisade mesophyll is 2–4 cells thick, below it is the spongy mesophyll, a lower hypodermal layer of tannin cells, and the lower epidermis. A single veinlet is shown with a well-developed bundle sheath containing tannin. A few vessels are evident in the upper portion of the vascular bundle. The leaf is about 2 mm thick.

Table 19-1 *Characteristics of the Leaves of Mangroves*

	Rhizophora	Ceriops	Laguncularia	Conocarpus
Venous system on upper epidermis	Midrib only, very broad and conspicuous	Midrib only, narrow and inconspicuous	Midrib only, broad and conspicuous to very inconspicuous	At least midrib and lateral veins distinct
Venous system on lower epidermis	Midrib only, conspicuous, lateral veins discernible.	Midrib only, conspicuous	Midrib only, conspicuous	Midrib, lateral, secondary veins distinct
Epidermal cells of nonvenous areas	Straight- or curved-walled, not second divided	Straight- or curved-walled, not second divided	Straight- or slightly curved-walled, many second divided	Mostly curved or straight-walled, not second divided
Stomata	Sunken; outer stomatal ledge conspicuous	Sunken; outer stomatal ledge conspicuous	Scarcely sunken, not protected by hairs, usually more frequent on upper epidermis	Not sunken, usually protected by dense hairs, usually more frequent on lower epidermis
Water stomata, hydathodes and cork-warts	Large conspicuous cork-warts on lower epidermis; waterlike structures on both epidermides	Cork-warts more or less absent; frequent water stomatalike structures on both epidermides	Large water stomata present on both epidermides; hydathodelike areas present	Apparently absent, a few water stomata present
Hypodermis, including extra epidermal layers	Upper three to five layered, sparsely chloroplasted	Upper two-layered, sparsely chloroplasted	None	None
Mesophyll	One to three layers of palisade and ~8–10 layers of spongy below upper hypodermis	Usually one layer of palisade and ~8–10 layers of spongy below upper hypodermis	Two layers of palisade below upper epidermis; one to two layers of spongy	One or two layers of palisade below upper epidermis, one layer of palisade above lower epidermis
Water stage	Upper hypodermis	Upper hypodermis	6–12 layers of centrally placed ± isodiametric cells	4–6 layers of centrally placed ± vertically elongated cells

Source: Modified from Stace (1966).

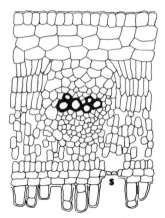

Figure 19-3 Cross section of the leaf of the black mangrove *Avicennia germinans* showing hairs that develop from the lower epidermis. Note the hypodermis and multiple-layered palisade mesophyll. No tannin-containing cells are present in the leaf, and the general anatomy is slightly more "mesophytic" than that of the red mangrove (compare with Figure 19-2). A stomate (*s*) is evident in the lower epidermis.

lower side of the leaf, as seen in the black mangrove leaf (Figure 19-3). The stomates are usually confined to the lower side of the leaf and are sunken. Salt glands, which are modified hydathodes (vein terminations in the epidermis), are common in the upper epidermis of the black mangrove leaf. These glands usually consist of thin-walled epidermal cells and a single xylary tracheid that terminates in a veinlet. The hydathodes function in the exudation of concentrated salt solutions from the leaves.

Below the epidermis, on both sides of the blades of red and black mangroves, there are usually one to three layers of chloroplast-free cells. Cells in this layer, called a hypodermis, may contain tannins, and typically contain a large amount of water. Sclereids (stone cells) occur in the hypodermis of *Rhizophora*. Chloroplast-containing palisade mesophyll occurs below the hypodermis, and in the red mangrove there are some crystoliths. The cystoliths are enlarged cells containing calcium oxalate crystals, the function of which is not known.

The spongy mesophyll is usually reduced in mangrove leaves, and air spaces are not as abundant as in mesophytic dicot leaves (Figure 18-3). The largest air chambers are usually found around the stomates, which are present in the lower epidermis. The veins have vascular bundle sheaths that exhibit a well-developed suberized band around each cell (Casparian strip). The function of the bundle sheath cells with the Casparian strip is to control lateral water movement into and out of the vein.

Aerenchyma is a common tissue in the pneumatophores, prop, and drop roots of mangroves. The tissue is a modification of the cortex of these specialized roots and consists of large, thin-walled parenchyma cells so organized as to produce well-developed air spaces throughout. An example of this special modification is shown in the cross section of the black mangrove pneumatophore (Figure 19-4).

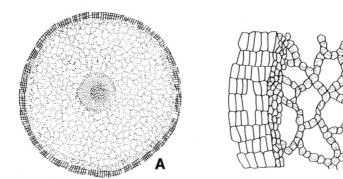

Figure 19-4 (*A*) Cross section of a young pneumatophore of a black mangrove, showing the general structure. (*B*) detail of the cork tissue and outer aerenchyma that forms the cortex. The central region in (*A*) is the vascular system with xylem to ward the inside and phloem to ward the outside, all enclosed by a well-developed endodermis.

TAXONOMIC AND ECOLOGICAL CONSIDERATIONS

There are approximately 80 species of mangroves distributed throughout the tropical and subtropical regions of the world (Table 19-2). The largest number of species occurs in southeast Asia (65 species), whereas about 11 species are found in the New World and the Caribbean.

Taxonomic Considerations

Waisel (1972) lists 12 genera of dicots that he includes under the ecological term *mangrove*, whereas Walsh (1974) lists 16 genera, including one monocot, *Nypha* (Table 19-2). To this list should be added another species, *Pelliciera rhizophorae* (family Theaceae), which is found along the western coast of Central and South America, from central Costa Rica to northern Ecuador. The genus is of special interest since a species of the *Azteca* ant inhabits the trees and defends them against phytophagous insects and other tree pests (Collins *et al.*, 1977). The monocot *Pandanus* and the dicot *Conocarpus* may also be included, although they typically grow in a more freshwater environment, behind the mangrove swamps.

In regions of the Caribbean, the fern *Acrosticum danaefolium* and the palm *Cocos nudifera* are found if fresh water flows slowly into the upper edge of the mangrove swamp.

It is apparent from Table 19-2 that mangroves are unrelated plants. The species, for the most part, probably evolved from mesophytic border forests, gradually being selected for xerophytic features and specialized root systems. Certainly all mangroves have adapted to high water stress, exhibit mechanisms that permit water uptake against a gradient, and have evolved xerophytic adaptations. Based on generic similarities in the Caribbean and the Indian Ocean and the West Pacif-

Table 19-2 A List of Families and Genera of Mangroves

Families and Genera	Number of Species	Indian Ocean West Pacific	Pacific America	Atlantic America	West Africa
Dicots					
Avicenniaceae					
Avicennia	11	6	3	2	1
Bombacaceae					
Camptostemon	2	2	0	0	0
Chenopodiaceae					
Suaeda[a]	2	0	0	1	1
Combretaceae					
Conocarpus[b]	1	0	1	1	1
Laguncularia	1	0	1	1	1
Lumnitzera	2	2	0	0	0
Euphorbiaceae					
Exoecaria[b]	1	1	0	0	0
Leguminosae					
Machaerium[b]	1	0	1	1	1
Meliaceae					
Xylocarpus	10	8	?	2	1
Myrsinaceae					
Aegiceras	2	2	0	0	0
Myrtaceae					
Osbornia	1	1	0	0	0
Plumbaginaceae					
Aegiatilis	2	2	0	0	0
Rhizophoraceae					
Bruguiera	6	6	0	0	0
Ceriops	2	2	0	0	0
Kandelia	1	1	0	0	0
Rhizophora	7	5	2	3	3
Rubiaceae					
Scyphiphora	1	1	0	0	0
Sonneratiaceae					
Sonneratia	5	5	0	0	0
Sterculiaceae					
Heritiera[b]	2	2	0	0	0
Theaceae					
Pellicera	1	0	1	0	0
Tiliaceae					
Brownlowia[b]	17	17	0	0	0
Monocots					
Arecaceae					
Nypha[a]	1	1	0	0	0
Pandanaceae					
Pandanus[b]	1	1	0	0	0
Total	80	65	9	11	9

Source: Waisel (1972) and Welsh (1974).

[a]*Suaeda* typically is a small to medium sized bush, but can become a small tree.

[b]At least some of the species in these genera are more typical of freshwater swamps behind the mangrove coastal swamp and are not considered to be true mangroves by some botanists.

ic, McCoy and Heck (1976) have argued that the origin of mangroves, as that of seagrasses, was during the early Cretaceous period. That is when Gondwanaland existed, and marine angiosperms were migrating into the shallow seas between the portions of the continent. With the separation of the various continents in the late Cretaceous period, both land and oceanic barriers developed, and speciation of the isolated mangrove populations then began (see McCoy and Heck, 1976).

Figure 19-5   Worldwide distribution of the red mangrove, *Rhizophora* spp.

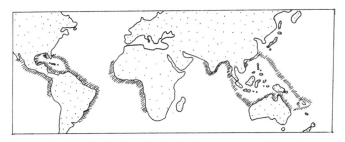

Figure 19-6   Worldwide distribution of the black mangrove, *Avicennia* spp.

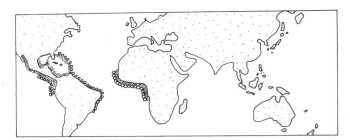

Figure 19-7   Worldwide distribution of both *Conocarpus erectus* (buttonwood; hashlines) and *Laguncularia racemosa* (white manrove; circles). Figures 19-5 through 19-7 are modified from Chapman. Note the pantropical distribution of all genera.

Geographic Distribution

The number and distribution of mangroves are somewhat ambiguous. About 11 species occur in the New World tropics, whereas about 65 species are reported for the Indian Ocean and the West Pacific (Table 19-2). However, many tropical tidal marshes have been poorly studied, and more species may be found. In addition to the trees themselves, a number of succulent perennial plants, mostly dicots, are usually associated with the forests. Distribution patterns for *Rhizophora, Avicennia, Conocarpus,* and *Lagunculara* are given in Figures 19-5, 19-6, and 19-7.

Mangroves are usually found in saline lagoons; they occur most commonly in estuaries such as those produced by tropical rivers. Some of these rivers include the Amazon and Orinoco in South America, the Congo and Zambesi in Africa, and the Ganges and Mekong in Asia. Mangrove swamps are not limited to estuarine conditions since they are also typical in high-stable-salinity areas such as lagoonal, protected sides of islands or atolls.

Selected Species

Four species of mangroves that are common to the Caribbean will be described. All four plants are dicots that are usually found in distinct zones in the intertidal and shallow subtidal coastal regions. A brief vegetative key to these four species is presented in Table 19-3, based on leaf features (Humm, 1973). Taxonomically these four trees are very diverse, being placed in two orders and three families.

Table 19-3   A Key to the Four Dominant Mangroves Found in the Northern Caribbean

---

1   Leaves alternate, simple, entire and ovate, fruit persistent woody aggregate ....
....**buttonwood**

2   Leaves opposite, fruit not a persistent woody aggregate **2**

    2   Leaves whitish on the underside, dark green and lustrous on the upper surface, broadly elliptical, 4–9 cm. long, 1.8–3.0 cm. wide; fruit egg-shaped, flattened, 2.5 cm. wide, 3.5 cm. long.. **black mangrove**

    2   Leaves green on both upper and lower surfaces................................... **3**

3   Leaves somewhat lighter green on the underside, ovate, 7–12 cm. long, 2–5 cm. wide; seeds germinating on the tree to form an embryo 15–30 cm. long.....**red mangrove**

3   Leaves almost identical on upper and lower surfaces, broadly oblong, 2–7 cm. long, 2.5–3.5 cm. wide, fruit about 1.5 cm. long with 10 ribs............. **white mangrove**

Some: Modified from Humm (1973).

Rhizophora mangle L.

Gill and Tomlinson (1969, 1971a, 1971b, 1977) have published a series of excellent papers on morphological and anatomical aspects of this species. The red mangrove is placed in the family Rhizophoraceae and its species have opposite, thick, leathery (coriaceous) leaves and a figlike bud at the apex of each branch (Figure 19-8). The flowers are small and bisexual, occurring throughout the summer. The fruit has a single seed and is viviparous. The radicle and hypocotyl (called "pencil seeds") of the germling extrude up to 20 cm from the fruit. The plant contains a large amount of tannin and the stem lacks distinctive growth rings. Vessels are evenly distributed throughout the xylem, and xylary parenchyma is common (Panshin, 1932). The root morphology is characteristic of the spe-

Figure 19-8 The habit of the red mangrove *Rhizophora mangle*. (*a*) A vegetative branch with flower buds. (*b*) A flower. (*c*) The pencil-like fruits that have germinated while on the tree.

cies, having both prop roots and drop roots for attachment. The roots are shallow, producing large, knoblike growths from which anchoring roots are produced. Lenticles and a cork layer (periderm) cover the surfaces of the anchoring roots. Absorbing, or feeding, roots are formed from the knoblike outgrowths of the prop or drop roots near the substratum surface. The family, Rhizophoraceae, is large, having 17 genera and about 70 species. Of these four genera and 17 species can be considered to be "true" mangroves.

Figure 19-9   The habit of the white mangrove *Laguncularia racemosa*. (*a*) A vegetative branch with flower buds. (*b*) The base of a leaf showing the two salt glands characteristic of the species. (*c*) A raceme of flowers. (*d*) The fruit. (*a*) ×0.60, (*b*) × 1.70 (*c*) × 3.40 (*d*) × 3.40.

Laguncularia racemosa Gaetner

The white mangrove is placed in the family Combretaceae or Terminaliaceae in the same order as the Rhizophoraceae (Myrtales). The family has 18 genera and 450 species. The leaves are opposite, thick, shiny, oblong, 2–7 cm. long and 2–3 cm. wide (Figure 19-9). A pair of salt glands is evident at the base of each blade. The blade is cordate. Although prop roots are reported, the species usually lacks specialized root systems, and it is usually found at the upper edges of mangrove swamps. The flower is perfect and the fruit viviparous. The leaves are thinner

Figure 19-10 The habit of the buttonwood *Conocarpus erectus*. (*a*) Vegetative branch and young flowers. (*b*) The base of the leaf showing salt glands. (*c*) An aggregate of flowers. (*d*) The developing woody fruits (aggregates) (*a*) × 0.50 (*b*) × 1.60 (*c*) × 3.25, (*d*) × 3.25.

than those of red and black mangroves, but a well-developed hypodermal layer and vascular bundle with a Casparian strip are evident. Tannin cells are also present.

Conocarpus erectus L.

The buttonwood tree has alternate leaves that are simple, entire, ovate, 4–9 cm. long, and wedge-shaped at the base (Figure 19-10). The genus, like *Laguncularia*, also belongs to the family Combretaceae. The flowers are perfect, and the fruit is a persistent woody aggregate. The buttonwood tree is usually found at the highest elevations of a mangrove swamp. Some botanists do not consider the plant a "true" mangrove.

Figure 19-11   The habit of the black mangrove *Avicennia germinans*. (*a*) A vegetative branch and young flowers. (*b*) A floral cluster. (*c*) A mature fruit. (*a*) × 0.60, (*b*) × 3.40, (*c*) × 1.70.

Avicennia germinans (L.) L.

The black mangrove tree is placed in its own family, Avicenniaceae and has 11 species. All of the species are trees or shrubs that occur in mangrove swamps. The leaves are opposite; they are dark green on top and whitish below (because of hairs), 4–9 cm. long, and have hydathodes (Figure 19-11). The flowers are perfect. The fruit is viviparous; it is a fleshy, compound capsule with one seed. The roots can be divided into five types: (1) primary adventitious roots that develop from the stem; (2) horizontal roots that grow just below the substratum and produce (3) erect pneumatophores and (4) descending anchoring roots; (5) absorbing roots that are usually produced just below the substratum surface on the pneumatophores. Chapman (1944) found that pneumatophores were only present in sites with low oxygen concentrations, suggesting their function in gas exchange. The stem has interxylary phloem arranged in concentric bands.

ECOLOGICAL ROLES AND ZONATION

Most studies have classified the mangrove communities (mangles) as either "pioneer" or seral succession stages. Such a viewpoint implies that the zonation within a mangrove forest is evolving toward a freshwater, terrestrial plant community through soil formation (Davis, 1940). Chapman (1944) presented a succession scheme for the mangrove swamps of Florida (Figure 19-12). Other investigators consider the mangrove forest a "steady state" community, which is in part changing the environmental conditions within it, and/or in part under the control of external environmental forces. The external forces might be changes in sea level, substratum composition, and wave action (Thom, 1967). Salinity appears to be most critical in eliminating competition by terrestrial plants. Although mangroves can grow in freshwater habitats, they grow slowly and do not compete well against typical freshwater plants. Based on the physiological studies of Bowman (1917) and Davis (1940), it is now evident that mangroves do not require saline conditions to grow; rather they tolerate these conditions. Thus, mangroves are facultative halophytes, taking advantage of the lack of competition in tropical tidal marshes.

In this text, mangrove populations are viewed as climax communities, having a distinct set of plants that interact with the marine environment. In addition, they are not directly involved in the succession of a terrestrial forest. Thus in periods of stable or slightly decreasing sea levels, mangrove forests will colonize more land, whereas in periods of rising sea levels the land will be lost.

Walsh (1974) described five basic requirements for extensive development of mangrove forests (mangales, manglares):

1 **Tropical temperatures.** The average temperatures of the coldest months should be higher than 20°C and the seasonal temperature range should not exceed 5°C, although some differences occur in Florida mangrove swamps.

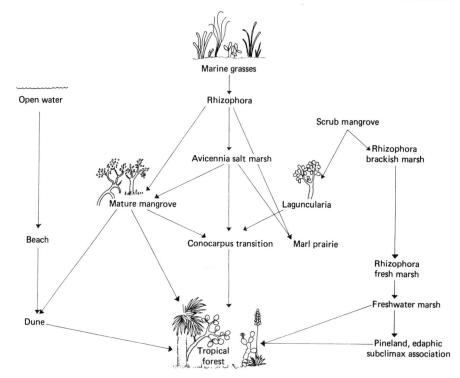

Figure 19-12 The proposed succession in the mangrove swamps of Florida as described by Chapman (1944). Chapman considered the red mangrove *Rhizophora* to be a pioneer species replacing marine grasses. He said the succession could take a number of pathways, ultimately ending up in a tropical forest.

2 **Fine-grained alluvium.** The most-extensive stands of mangroves occur along deltic coasts (mouths of rivers) or in estuaries where soft mud consisting of fine silt, clay, and organic matter is available for seedling development. Volcanic soils are also highly productive sites for mangrove development, whereas quartzitic and granitic alluvia are poor substrata.

3 **Shores free of strong wave and tidal actions.** The most-extensive developments occur in areas of protected shores within estuaries or behind islands, as wave action (even by boats in channels) will uproot seedlings and erode the soft sediments.

4 **Salt water.** As pointed out earlier, salt water is not a requirement since mangroves are facultative halophytes. However, the removal of competition by freshwater plants is critical. Mangrove swamps are found in estuarine as well as stable, high-salinity environments.

5 **Large tidal range.** The wide horizontal tidal range is considered important because with a gentle gradient the substratum will not erode during tidal changes. The shallow, extensive slope inland insures the settlement of sediment necessary for seedling development.

Ecological Roles

Because of the importance of mangroves in subtropical and tropical estuarine systems as well as in shoreline protection, major efforts are under way to reforest disturbed coastlines, especially in Florida (Teas, 1977). Planting costs were estimated at $1,140/ha) if unrooted red mangrove propagules were spaced 91 cm apart (see Chapter 2).

The unique nature of mangroves was noted in earlier writings, but as late as the 1960s a number of ecologists considered mangrove forests to have low productivity and to be nonessential transitional communities. Curtiss (1888) and Bowmann (1917) postulated a number of broad ecological roles for mangroves, especially in the formation of islands and extension of shorelines. Today most ecologists recognize four major roles of mangrove swamps.

1 Mangroves aid soil formation by trapping debris. Davis (1940) demonstrated that prop roots and pneumatophores accumulate sediments in protected sites and form mangrove peats. The process occurs with a stable sea level or one that is slowly dropping. The filamentous algae typical of Florida mangrove swamps (Cladophora, Rhizoclonium, Vaucheria, Centroceras) also help to stabilize the fine sediments trapped by mangroves, usually forming a green-to-red felty mass over the substratum.

2 Mangroves filter land runoff as well as removing terrestrial organic matter.

3 Mangroves serve as habitats for many species of small fish, invertebrates, and various epiflora and epifauna as well as larger birds. Heald (1971) and Odum (1971) both demonstrated the existence of food webs that are dependent on the organic production of mangrove swamps.

4 Mangroves are major producers of detritus that will contribute to offshore productivity (Heald, 1971).

Mangrove communities tolerate saline conditions and show xerophytic adaptations to water "stress." The term *stress* is used here to mean a drain of potential energy that otherwise would be available for the plant. If the mangroves are under ecological stress, the effects by man (channelization, drainage, herbicides, thermal loading) or the environment (hurricanes, siltation) may overload the system and cause a breakdown. Lugo and Snedaker (1974) presented a simple energy model showing potential stresses on mangrove swamps along with major ecological features (Figure 19-13).

Their model has two components, aboveground components (leaves, stems, photosynthesis, and aerobic processes) and belowground components (mud, roots, aerobic, and anaerobic processes). Five mangrove processes were included: primary productivity, aboveground respiration, belowground respiration, mineral and nutrient recycling, and organic matter export to other systems. Six potential stress factors were included: channelization, drainage, siltation, hurricanes, herbicides, and thermal loading. If one examines this energy model, it is obvious that remov-

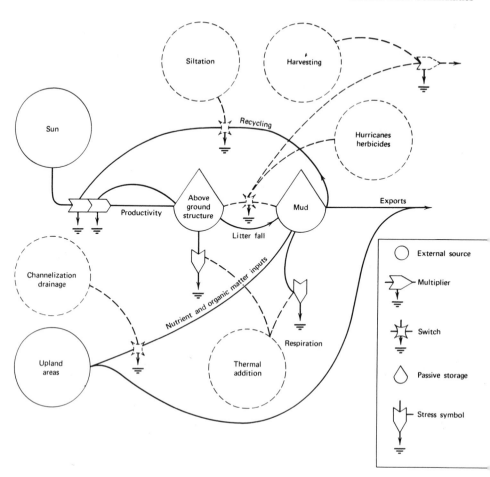

Figure 19-13 Model of a mangrove community as modified from Lugo and Snedaker (1974). The symbols used are identified in the box on the right. Potential stresses are represented by dashed lines; thus such factors as siltation, hurricanes, herbicides, thermal pollution, and channelization are enclosed in dashed lines.

al of mangroves by harvesting will result in an export of the primary production as will other modifications such as hurricane damage, thermal loading, and channelization.

Zonation

In the Caribbean, the mangrove swamps typically have a zonation of trees with the red mangrove forming the outer fringe in the shallow subtidal region (Figure 19-14). Behind these red mangroves are the black mangroves, which are found from the lower to the mid or upper intertidal zones. The white mangroves then

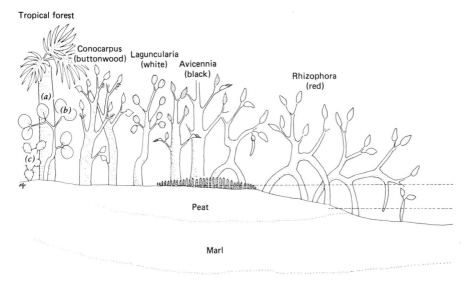

Figure 19-14 Typical zonal pattern of a mangrove swamp in southern Florida, modified from Davis (1940). Spring high and low tides are shown in dashed lines. The tropical forest consists of sabal palm (*a*) sea grape bushes (*Cocca loba*) (*b*) and *Opuntia* (cactus) (*c*). The leaves are oversized to indicate leaf features. The red mangrove, *Rhizophora*, is considered to be the species best adapted to submergence and thus forms the fringe, projecting into the shallow subtidal region. Tidal range is about 1 m.

occur from the middle intertidal region to the upper intertidal, leaf litter, or wave-deposit margin of the swamp. The buttonwood, *Conocarpus,* may occur in the upper intertidal region, but more typically it is found behind the white mangrove trees, in areas of sand/strand plants.

Black mangroves can also occur in the outer fringe, acting as pioneer mangroves in disturbed regions (i.e., because of heavy wave action or in areas of frost kills of red mangroves). Other zonation patterns are known. For example in Puerto Rico the manglares (mangroves) show a zonation pattern in a sea to land direction, with the red mangroves followed by the white, and then in turn the black mangroves. On the western side of Puerto Rico, *Avicennia* forms extensive thickets toward the dry land, and this genus can be replaced by *Conocarpus.*

If there is constant freshwater runoff or a freshwater swamp behind the mangrove swamps, the fern *Acrosticum danaefolium* and the palm *Cocos nudifera* as well as the Indo-pacific, *Nypha,* may occur. In such cases the white mangrove and buttonwood may be absent, since these two genera are not tolerant of long periods of submersion.

In the earlier discussion of mangrove communities (Figure 19-12; Chapman, 1944; Thom, 1967), it was noted that mangroves are facultative halophytes. According to Thom, salinity is simply an eliminator of competition and not the determining factor in zonation. He believes that substratum and water effects (submersion, wave energy) are the important factors controlling zonation and that each species has a given set of tolerances to these factors.

Based on this information it is not surprising that each species shows its greatest productivity when found in an appropriate zone (Lugo and Snedaker, 1974). The prop and drop roots of the red mangrove are especially well suited for the shallow, subtidal zone. Debris, trapped by these roots and by the pneumatophores of the black mangrove can allow the buildup of the substratum. As Davis (1940) and Lugo and Snedaker (1974) noted, as long as the sea level is stable or slightly decreasing, land formation may result because of debris trapping. During periods of decreasing tidal energy and lowering sea levels, mangroves can advance, but when the flushing action of water is high or the sea rises, the zones move landward.

Macnae (1968) has described the animal distribution in Indo-West Pacific mangrove swamps in terms of environmental gradients within habitats. He concluded that since animals show no marked zonation, but preferences for specific habitats, the fauna is not directly associated with the mangroves. He found that animal distributions were dependent upon their resistance to water loss and need for protection from the sun, the water table level, soil consolidation, and availability of food sources (microfauna, organic debris). Walsh (1967) also found that animal zonation of mangrove forests in Hawaii depended mostly on salinity differences, oxygen gradients, and substrata.

There are wide differences in algal components of mangrove swamps. The algal floras of the Caribbean, for example in Puerto Rico (Almodovar and Pagan, 1971; Burkholder and Almodovar, 1973) are highly developed, with an average of 5–10 algal species/root and a total of 22 species of macroscopic algae. Two major types of algal vegetation are found in a Caribbean community (Dawes, 1974). One consists of filamentous forms that produce felty mats or carpets on the substratum; the mats are involved in trapping fine sediments *(Cladophora, Boodleopsis, Vaucheria, Cladophorophsis)*. Members of another group, which mostly consists of leafy forms, are attached directly to the roots of mangroves (pneumatophores, drop and prop roots) and show microzonation *(Caloglossa, Bostrychia, Catenella)*. The alga found in the upper zone is the green seaweed *Rhizoclonium;* below it *Bostrychia* usually attaches itself by discoid haptera, forming soft, tiny fernlike clusters of deep-red to purple mats. The two red algal genera, *Caloglossa* and *Catenella,* usually occur below the *Bostrychia* zone. Both of these appear segmented. The former is a delicate, monostromatic series of blades, the latter a thicker, cactuslike segmented alga. All of these genera are found in the shaded, cool mangrove forest and do not dry out while exposed during low tide.

Mangrove forests that occur in arid regions, such as portions of the tropical Pacific, typically lack the algal flora described above. In such areas, salinities can be higher than normal sea water because of the high rate of evaporation at low tide.

ORGANIC PRODUCTION

Economically, mangroves have been harvested for tannins and lumber, especially in Thailand and Malaysia (Chapter 2). The lumber of several species is highly resistant to termite attacks; thus mangroves have been used extensively in ship construction. However, the major use seems to be as fuel, after conversion to charcoal.

Mangrove productivity seems to be regulated by two major factors, tidal fluctuation and water chemistry. Carter (1973) proposed seven critical factors affecting productivity. The first four are related to *tidal fluctuation:*

1 Oxygen transport to the root system.
2 Soil water and amount of water exchange in order to remove toxic sulfides.
3 Tidal flushing and its effect on deposition and erosion of the substratum.
4 Water table fluctuation and the resultant availability of nutrients.

The other three refer to *water chemistry:*

1 The saline content in the substratum and the ability of the leaves to transpire.
2 The levels of macronutrients in the soil. High levels permit high rates of productivity even though transpiration rates may be curtailed because of high salt concentrations in the substratum.
3 The amount of surface runoff and resulting availability of macronutrients from the land. This can dominate the nutrient budget of a mangrove swamp.

Detrital flow (leaf litter) from a mangrove swamp has been shown to be significant in determining estuarine biomass and productivity of associated waters (Heald, 1971). Heald determined that 60% of red mangrove leaf production was decomposed in 1 yr. Lugo and Snedaker (1974) have produced an energy model (Figure 19-13) that demonstrates the exportation of mangrove organic matter to the surrounding communities. Bunt *et al.* (1979) have devised a method to estimate potential levels of mangrove forest primary production through the correlation of light attenuation and leaf pigment assay. They obtained production estimates of 16–26 kg C/ha • da in a Queensland, Australia, mangrove swamp.

Lugo and Snedaker (1974) divided mangrove swamps into five physiognomic groups and presented biomass estimates for each (Table 9-4). Pool *et al.* (1977) did a similar but more detailed comparative study of the structure of Caribbean mangrove forests. The five physiognomic groups were similar in both studies. The fringe forest occurs along protected shorelines and islands with elevations extending above the highest tide mark with well-marked zonation. The *riverine* forest is the tall floodplain forest occupying river and creek drainages. The *overwash* forest occurs on low islands and mainland extensions into shallow bays that are flushed by normal tidal action. *Basin* forests occur inland along drainage depressions with reduced tidal changes. A *dwarf* forest is common in flat coastal fringes where external nutrient sources are limited. The plants in dwarf forests are small, usually about 1.5 m tall, but may be 40 yr old or older. The biomass estimates for the five forest types were given by Lugo and Snedaker (1974) and the four types common to Florida are reproduced in Table 19-4. The highest biomass occurs in overwash, riverine, and fringe forests; the amount of biomass is lower than in terrestrial tropical forests. A variety of physical factors were found to be important in biomass production (Pool *et al.,* 1977).

Table 19-4   Mangrove Biomass Estimates of Various Types of Florida Mangrove Forests[a]

Mangrove Source	Mangrove Type					
	Florida Overwash	Florida Riverine		Florida Fringe		Florida Dwarf
Leaves	7,263	3,810	9,510	5,843	7.037	712
Fruit and flowers	20	148	0.4	210	131	—
Stems and branches	—	16,770	27,670	19,120	18,550	3,950
Wood	70,380	62,850	133,660	65,150	109,960	—
Prop roots	51,980	14,640	3,060	27,200	17,190	3,197
Pneumatophores	—	—	—	—	—	—
Subsurface roots	—	—	—	—	—	—
Litter	17,310	42,950	33,930	60,250	98,410	1,140
Total ground biomass excluding litter	129,643	98,218	173,900	117,523	152,868	7,868

Source: After Lujo and Snedaker (1974).
[a]All figures are kilograms of weight per hectare.

Table 19-5   Primary Productivity of Mangrove Communities in Florida and Puerto Rico[a]

Location	Gross Primary Productivity	Net Primary Productivity	Types of Mangroves
Lower Fahka Union River, Florida	11.8	7.5	Red, black, white
La Parguera Puerto Rico	8.2	0.0	Red
Rookery Bay Florida	6.3	4.4	Red
Dwarf Forest Dade County, Florida	1.4	0.0	Red

Source: Lugo and Snedaken (1974).
[a]Figures are g C m/$^2 \cdot$da.

The net primary productivity based on respiration and photosynthesis data ranges from <1–7.5 g $C/m^2 \cdot da$ for mangrove forests in Florida (Table 19-5). For comparison, a highly fertilized sugarcane field has a maximum productivity rate of 10 g $C/m^2/da$. Table 19-5 gives several primary productivity measurements of mangrove ecosystems in Florida and Puerto Rico (Lugo and Snedaker, 1974). It is evident from this table that mangrove forests are highly productive.

REFERENCES

Almodovar, L. R. and F. A. Pagan. 1971. Notes on a mangrove lagoon and mangrove channel at La Parguera, Puerto Rico. *Nova Hedwigia* **21**: 241–259.

Bowman, H. H. M. 1917. Ecology and physiology of the red mangrove. *Proc. Am. Phil. Soc.* **61**: 589–672.

Bunt, J. S., K. G. Boto, and G. Boto. 1979. A survey method for estimating potential levels of mangrove forest primary production. *Mar. Biol.* **52**: 123–128.

Burkholder, P. R. and L. R. Almodovar. 1973. Studies on mangrove algal communities in Puerto Rico. *Fla. Scientist* **36**: 66–74.

Carter, M. R. 1973. *Ecosystems Analysis of the Big Cypress Swamp and Estuaries*, U.S. EPA, Athens, Ga.

Chapman, V. J. 1944. 1939 Cambridge University Expedition to Jamaica. I. A study of the environment of *Avicennia nitida* Jacq. in Jamaica. *J. Lin. Lond. Soc. Bot.* **52**: 448–486.

Collins, J. P., R. C. Berkelhamer, and M. Mesler. 1977. Notes on the natural history of the mangrove *Pelliciera rhizophorae* Tr. and Pl. (Theaceae). *Brenesia* **10/11**: 11–29.

Curtiss, A. H. 1888. How mangroves form islands. *Garden and Forest* **1**: 100.

Davis, J. H. Jr. 1940. The ecology and geological role of mangroves in Florida. Papers from Tortugas Lab. #32. *Carnegie Inst. Wash. Publ.* **517**: 305–412.

Dawes, C. J. 1974. *Marine Algae on the West Coast of Florida*. University of Miami Press, Miami.

Gill, A. M. and P. B. Tomlinson. 1969. Studies on the growth of red mangrove *(Rhizophora mangle* L) 1. Habit and general morphology. *Biotropica* **1**: 1–9.

Gill, A. M. and P. B. Tomlinson. 1971a. Studies on the growth of red mangrove *(Rhizophora mangle* L.) 2. Growth and differentiation of aerial roots. *Biotropica* **3**: 63–77.

Gill, A. M. and P. B. Tomlinson. 1971b. Studies on the growth of red mangrove *(Rhizophora mangle* L.) 3. Phenology of the shoot. *Biotropica* **3**: 199–224.

Gill, A. M. and P. B. Tomlinson. 1977. Studies on the growth of red mangrove *(Rhizophora mangle* L.) 4. The adult root system. *Biotropica* **9**: 145–155.

Heald, E. J. 1971. *The Production of Organic Detritus in a South Florida Estuary*. Univ. Miami Sea Grant Program. Tech Bull 6, Miami, Fla.

Humm, H. M. 1973. Mangroves. In *A summary of Knowledge of the Eastern Gulf of Mexico*, State Univ. Syst. Fla. Instit. Oceanog., Tallahassee, Fla., pp. IIID 1–6.

Lugo, A. E. and S. C. Snedaker. 1974. The ecology of mangroves. *Ann. Rev. Ecol. System.* **5**: 39–64.

Macnae, W. 1968. A general account of the fauna and flora of mangrove swamps and forests in the Indo-west-Pacific region. *Adv. Mar. Biol.* **6**: 73–270.

McCoy, E. D. and K. L. Heck Jr. 1976. Biogeography of corals, seagrasses, and mangroves: An alternative to the center of origin concept. *System. Zool.* **25**: 201–210.

Odum, W. E. 1971. *Pathways of Energy Flow in a South Florida Estuary*. Univ. Miami Sea Grant Program. Tech Bull 7. Miami, Fla.

Panshin, A. J. 1932. An anatomical study of the woods of the Philippine mangrove swamps. *Philipp. J. Sci.* **48**: 143–208.

Pool, D. J., S. C. Snedaker, and A. E. Lugo. 1977. Structure of mangrove forests in Florida, Puerto Rico, Mexico, and Costa Rica. *Biotropica* **9**: 195–212.

Rabinowitz, D. 1978. Dispersal properties of mangrove propagules. *Biotropica* **10**: 47–57.

Stace, C. A. 1966. The use of epidermal characters in phylogenetic considerations. *New Phytol.* **65**: 304–318.

Teas, H. J. 1977. Ecology and restoration of mangrove shorelines in Florida. *Environ. Conserv.* **4**: 51–58.

Thom, B. G. 1967. Mangrove ecology and deltaic geomorphology: Tabasco, Mexico. *J. Ecol.* **55**: 301–343.

Walsh, G. E. 1967. An ecological study of a Hawaiian mangrove swamp. In Lauff, G. H. (ed.). *Estuaries,* AAAS Publ. #83, Washington, D.C., pp. 420–431.

Walsh, G. E. 1974. Mangroves a review. In Reimold, R. J. and W. H. Queen (eds.). *Ecology of Halophytes.* Academic, New York, pp. 51–174.

Waisel, Y. 1972. *Biology of Halophytes.* Academic, New York.

Warming, E. 1883. Tropische Fragmente. II. *Rhizophora mangle* L. *Bot. Jab.* **4**: 519–548.

CHAPTER 20

Phytoplankton Communities

Plankton is a term that refers to free-floating and suspended plants and animals. The term was first used by Hensen in 1887 to describe microscopic organisms. Plankton can be separated according to their sizes: *nanoplankton* (small forms, $<$ 20 μm), *microplankton* (unicellular, filaments), and *macroplankton* (large, visible). Or they can be separated according to their life histories: *holoplankton* (entire life history spent as plankton), and *meroplankton* (portion of life history spent as plankton). Sieburth *et al.* (1978) have described a classification of plankton based on their trophic states and have correlated this with size classes.

Plankton can also be divided into zooplankton and phytoplankton. Most phytoplankton are holoplanktonic, and over half of these are nannoplankton. Many phytoplankton species may spend a part of their life histories in deep water or on the sea floor as resting cells. Although plankton may be flagellated and may have some independent locomotion, especially with relation to vertical migration (Staker and Bruno, 1980), they cannot migrate horizontally against the tides or currents. Thus phytoplankton are primarily carried by winds and currents, including horizontal and vertical currents, turbulence, eddies, and shifts in larger bodies of water. The reader is referred to Raymont's (1980) text on these organisms for a variety of details.

Phytoplankton include autotrophic, photosynthetic algae. They are primarily unicellular, although some of them are chain formers (e.g. filamentous blue-green algae such as *Trichodesmium* and colonial diatoms) or macroscopic multicellular plants (two species of *Sargassum* of the Sargasso Sea). The group is, therefore, heterogeneous and includes members of the Chrysophyta (silicoflagellates, coccolithophoids, diatoms), Pyrrhophyta (dinoflagellates), Cyanophyta (blue-green algae) as well as members of the Chlorophyta (green algae) and Euglenophyta. Phytoplankton are very important in food chains as they are primary producers, forming the "meadows" of the oceans. Since 72% of the earth is covered by ocean, phytoplankton are the most important group of primary producers on earth.

The nanophytoplankton are represented by a heterogeneous grouping of small phytoplankton from different divisions. There are a number of distinctly smaller species of coccolithophorids (*Crystallolithus*, 8–18 μm diameter), dinoflagellates (*Exuviella* and *Prorocentrum*), and cryptomonads (*Hemiselmis*). Nanophytoplankton have reproduction rates two to three times higher than larger-sized phytoplankton. In the sea, the actual numbers per sample can be quite high. At times the biomass may outweigh the mass of the more obvious microplankton and macroplankton. Nanoplanktonic algae are abundant in shallow coastal waters between depths of 20 and 80 m, and in warmer oceanic water such as the tropics. Although the abundance of nanoplanktonic algae is greatly outweighed by the larger diatom biomass in temperate and higher latitudes, the smaller-sized phytoplankton are of great importance as primary producers, especially in the tropics.

TYPES OF ORGANISMS

The taxonomy, cytology, and other features of the various phytoplankton were presented in Chapters 3–9. In this section only a few of the planktonic features are considered for each group.

Diatoms

(Division Chrysophyta, Class Bacillariophyceae). The diatoms are by far the most obvious and numerous phytoplankton. Diatoms are unicellular algae that often form chains. The marine planktonic forms primarily belong to the order Centrales, as they have radial symmetry. Since oil or fatty acids are the normal food storage compounds of the diatoms, this group is a high-energy food source for zooplankton grazers. In fact, as diatoms age the stored oil accumulates so the "oily patches" in the ocean may occur during the final portions of a diatom bloom.

Since all diatoms have siliceous cell walls (SiO_2 frustule) with two halves (thecae), their fossil record is excellent. In fact, marine diatomaceous deposits are one of the major constituents of deep-ocean sediments.

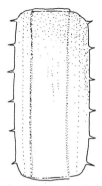
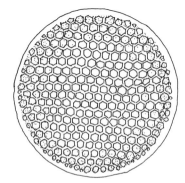

Figure 20-1   The centric diatom *Coscinodiscus excentricus*, an example of a bladder-type cell morphology. It can have a diameter of 70 μm.

Figure 20-2   A ''needle cell,'' *Rhizosolenia hebetata*. It is only 10 μm in diameter but $>$ 100 μm in length.

Although a number of marine diatoms are bottom dwellers (benthic), most are planktonic species with a number of modifications to insure their suspension and survival in the photic zone of oceanic waters. Smayda (1970) has suggested that morphological, physiological, and physical adaptations can aid in the suspension of diatoms and other phytoplankton. He includes cell shape, size, and colony formation under morphological adaptations. Three major modifications of frustules have been found to improve diatom flotation. (1) In the bladder-type cell, the cell is large, the frustule is thin, and the protoplast has a large, central vacuole (Figure 20-1). (2) The hair, or needle, cell is another type of adaptation that is evident in the pencil- or fusiform-shaped frustule of *Rhizosolenia* (Figure 20-2).

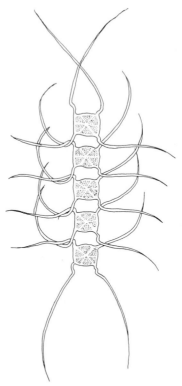

Figure 20-3 A colonial chain-forming centric diatom, *Chaetoceros laciniosus*. The cells are 30 μm in diameter. The diatom has extensive setae, processes that aid in flotation. See figures in Chapter 8 for other examples of morphologies that aid diatom flotation.

The needlelike extensions of the two valves and the cylindrical body of the cell covered by imbricating bands increases the frustule surface area and controls the orientation of a sinking valve, often introducing a spiralling motion with the broad side of the cell facing down. (3) The branching type of frustule with processes and projections called setae as seen in *Chaetoceros* (Figure 20-3) can increase frustule surface area. Summer populations of diatoms in northern latitudes as well as tropical forms tend to have thinner and more slender frustules with more extensions. All of these features appear to be an acclimation to warmer, lower-viscosity waters. Smayda (1970) reported that the larger colonial diatoms tended to sink more rapidly because of the increase in mass and reduced surface area.

Physiological mechanisms to increase flotation of nonmotile diatoms include ion regulation in the cytoplasm and vacuole, gas production due to photosynthetic activity, and secretion of thin threads of mucilage. Fat accumulation has been shown to be a symptom of senescence, with diatoms having large amounts of fat tending to sink rapidly. Thus fat accumulation is not considered to be effective in flotation (Smayda, 1970). Light, photoperiod, and nutrients also play a role in diatom suspension (Sournia, 1974). Cyclic or seasonal vertical migration of planktonic diatoms is also known (Raymont, 1980). Diatoms are able to sink and obtain nutrients at greater depths before vertical mixing again brings the cells nearer the surface. Smayda (1970) considers two physical factors in the suspension of phytoplankton: viscosity of the water and vertical circulation; the latter is probably very important in seasonal migrations of the cells.

The possibility that vertical migration is part of a circadian periodicity in phytoplankton was reviewed by Sournia (1974) This was demonstrated by Staker and Bruno (1980). Sournia suggests that the flagellated dinoflagellates and nonflagellated phytoplankton (diatoms as well) concentrated in the surface waters in the morning hours and sank as much as 200 m during the remaining part of the day (especially coccolithophorids.) Staker and Bruno (1980) found that dinoflagellates migrated about 4 m each day. Any number of physical factors could be the driving force for most types of phytoplankton *migration*. In addition, grazing pressures may mask the actual plankton populations at any one depth.

Dinoflagellates (division Pyrrhophyta, class Dinophyceae)

In contrast to the centric diatoms, which are nonmotile, a major constituent of the phytoplankton is flagellated, namely the dinoflagellates. Some dinoflagellates are large, unicellular algae; they are unique in shape with both transverse and trailing flagella (Figure 20-4). Both armored (with cell walls composed of cellulosic plates) and unarmored forms occur in the phytoplankton community. The exten-

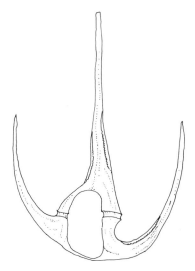

Figure 20-4 *Ceratium tripos*, an example of an armored dinoflagellate that can become quite large, reaching 270 μm or more in length. The drawing is a sulcal view, and the cingulum is also evident. Details of the dinoflagellate structure including other figures, are presented in Chapter 9.

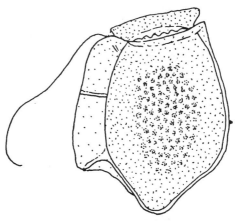

Figure 20-5 *Dinophysis* sp., another dinoflagellate that is armored and has a prominent cell wall "keel." The trailing flagellum is visible projecting from the cell.

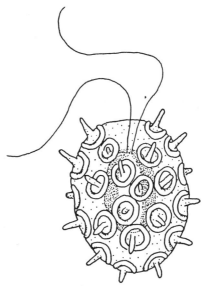

Figure 20-6 *Syracosphaera subsalsa*, a coccolithophorid (Chapter 8, Prymnesiophyceae; that has a pair of flagella and a haptonema (not visible) as well as the calcium carbonate coccoliths visible on the cell surface. The cells are about 3– 8 μm in diameter but can reach 10 μm in length.

sive development of the epithecal and hypothecal horns, as seen in *Ceratium* (Figure 20-4), as well as the various plate modifications as seen in *Dinophysis* (Figure 20-5) are thought to be flotation aids. Many dinoflagellates are heterotrophic, feeding on zooplankton, or they are even parasitic.

Because some dinoflagellates produce toxins, their blooms may result in large fish kills as well as high mortalities of crustacea and molluscs (Schmidt and

Loeblich, 1979). Such blooms are termed *red tides* because of the rust color of the water, in which 500,000 to 2 million dinoflagellate cells/liter may occur (see Chapter 9 for details). Two important red-tide-causing dinoflagellates are *Ptychodiscus (Gymnodinium)* and *Gonyaulax* (Figures 9-12 and 9-15).

Coccolithophorids and Silicoflagellates (division Chrysophyta)

The coccolithophorids are placed in the class Prymnesiophyceae and are so named because many of these unicellular, flagellated algae have small calcium carbonate plates (coccoliths) on the outside of their cells (see Black, 1965; Figure 8-16; Figure 20-6). Coccoliths were first described as organic concretions in a study of chalk and were called "crystalloids." During the laying of the first transatlantic cable from Ireland to Newfoundland in 1861, studies of the deep sea floor ooze again revealed these coccoliths. Based on the spherical arrangement of the coccoliths, it was finally proposed that they were produced organically. Shortly afterward coccolithophorids were discovered as part of the phytoplankton flora (Black, 1965). This group is particularly common in warm and temperate seas. The cells are small, 5–40 μm in diameter, with two equal or unequal flagella, as well as a modified structure called a haptonema on some motile stages.

The silicoflagellates are placed in the order Dictyochales of the class Chrysophyceae; they are characterized by the presence of an internal silicoskeleton (hollow tubes of SiO_2) that is covered by the protoplast (Figure 20-7). These cells are small, 10–30 μm in diameter, and usually have a single flagellum. The protoplast is naked, somewhat amoeboid, and may have elongated pseudopodia (Figure 20-7). Chrysophycean algae are most common in colder waters; they often become so abundant that they clog fishing nets, as does the genus *Phaeocystis* in New England (Raymont, 1980).

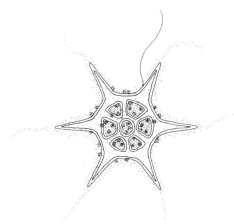

Figure 20-7 *Distephanus speculum*, a silicoflagellate (Chapter 8, Chroysohpyceae) that has long, thin pseudopodia and a single flagellum. The protoplast is "draped" over a central silica sklelton. The cells reach 15 μm in diameter.

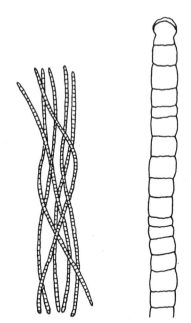

Figure 20-8 *Trichodesmium thiebautii*, a filamentous blue-green alga that lacks a sheath and has filaments woven into fascicles. The innermost filaments can fix nitrogen and the species can be abundant during red tides along the west coast of Florida.

Cyanophyta, Chlorophyta, and Cryptophyta

Blue-green algae (Cyanophyta) are more common in phytoplankton populations of neritic or shallow-water zones of warm waters. Only one species, a filamentous blue-green alga called *Trichodesmium thiebautii (Skujaella thiebautii)* (Figure 20-8) has ever been reported in abundance in open water and then only in subtropical and tropical waters, especially during red tides. In shallow tropical waters a number of filamentous blue-green alage *(Oscillatoria, Lyngbya)* may occur as mats that ultimately break free from the substratum and float on the surface. *Trichodesmium erythraeum* is in part responsible for the coloration of the Red Sea. The common species in the Gulf of Mexico is *T. thiebautii* which has been shown to fix nitrogen; thus it may be a very significant component of the phytoplankton community.

Few green algae (Chlorophyta) appear in oceanic waters, in sharp contrast to the number of species present as phytoplankton in fresh waters (Volvocales, Chlorococcales, Zygnematales). A few motile, unicellular forms are present in neritic waters; most of these such as *Bipedinomonas* (Figure 20-9), belong to the class Prasinophyceae.

Members of the Cryptophyta are more common in coastal than oceanic waters; some of these are nannoplankton. *Hemiselmis* (Figure 20-10) is one example of this small and poorly known division.

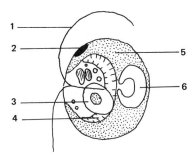

Figure 20-9 *Bipedinomonas rotunda* (*Heteromastix rotunda*), a biflagellated green alga (Class Prasinophyceae). It has an eyespot (*2*), a large chloroplast (*5*) with starch grains (*6*), a single large mitochondrion (*4*), and a single nucleus (*3*). The cells are about 14 μm in diameter (Manton *et al.*, 1965).

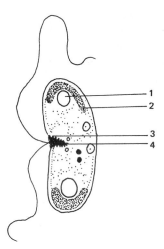

Figure 20-10 *Hemiselmis rufescens*, a member of the Cryptophyceae (Chapter 9). This biflagellated unicell has typical cryptophycean features: pyrenoids (*1*) in a bluish-green plastid (*2*), a furrow (*4*) lined with ejectile organelles (*3*) called trichocysts or ejectosomes. The cells are about 12 μm in width.

ECOLOGY: FACTORS AFFECTING PRIMARY PRODUCTION

In order to photosynthesize, the phytoplankton must remain in the photic zone; thus flotation devices and accessory pigments are critical. Nutrient levels vary not only seasonally but vertically. Typically higher nutrient levels are found at greater depths within the photic zone. Thus vertical migration, or sinking, is common among phytoplankton communities. Grazing can also profoundly affect the development of a phytoplankton community. Nutrients, light and temperature, grazing, seasonal cycles, and succession will be considered in this section.

Nutrients

In the chapter on chemical factors, both phosphorus (as orthophosphate) and nitrogen (as NH_4 or NO_3) were shown to be important elements for plant growth. Both are in relatively low concentrations, yet remarkable growth of phytoplankton occurs. In part, this growth can be explained by the rapid nutrient recycling among the unicellular phytoplankton. Seasonally, in temperate climates, there is a total depletion of phosphorous and nitrogen during the summer, following rapid phytoplankton growth in the spring. Nitrate and phosphate concentrations undergo marked seasonal cycles (Corner and Davies, 1971). The nutrient concentrations drop rapidly in the spring, remain low in the summer, and rise in the late fall. The maximum levels of nutrients in oceanic water occur in the winter, when phytoplankton populations may be depleted, at least in higher latitudes.

In regions nearer the equator, persistently higher light energies and temperatures allow continuous phytoplankton production throughout the year, resulting in a faster cycling of nutrients. Thus the seasonal maxima and minima of nutrient levels are more difficult to assess, and in general the nutrient concentrations are low (Corner and Davies, 1971). However, in areas of equatorial countercurrents, divergences can result in vertical circulation above the thermocline. In areas such as the Central and South Atlantic, a large increase in nutrients, especially phosphorus, results in increased phytoplankton density. It is evident that in the tropics, nutrient replenishment is probably the single most important factor in controlling phytoplankton growth (Steemann Neilson, 1959).

Other important elements are silicon (for diatom frustules), copper (which has a low solubility in seawater), and iron. Iron is of special interest since it is part of

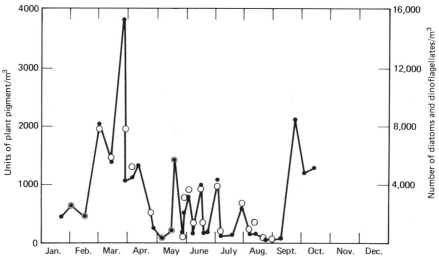

Figure 20-11 Yearly changes in the density of phytoplankton in coastal waters off Plymouth, England (Harvey *et al.*, 1935). The solid circles represent measurements as units of chlorophyll *a* per cubic meter. The open circles represent the total number of diatoms and dinoflagellates per cubic meter. Where the two overlap, a double circle results.

the cytochrome system. However, iron is in low concentration in oceanic water (< 1 mg/m^3) and is usually in nonusable forms in neritic waters. Thus iron can be a limiting factor for phytoplankton growth. Along the eastern Gulf of Mexico, chelated forms of iron may be carried into coastal waters by humic acids from the coastal tidal swamps and interior cypress heads. It is thought that this source of usable iron could, in some cases, support a red tide bloom. However, a number of other factors such as wind, concentration of the dinoflagellate cells, and breakdown of salinity barriers also play important roles in red tide development (see Chapter 9).

Light and Temperature

In temperate and high latitudes a rapid growth in marine phytoplankton (mostly diatoms) occurs in the spring and then drops off rapidly. During the summer, changes in phytoplankton density are erratic (Figure 20-11). Water temperatures and light (including photoperiod) show distinct seasonal cycles in higher latitudes, but such cycles are not evident in the tropics.

Temperature plays a role through its effect on enzymatic activities (respiration and photosynthesis) of phytoplankton and decomposers. The more rapidly cells are broken down, the more rapid nutrients become available, reducing the amount lost due to sinking to the deep, aphotic zone of the oceans. The spring growth of phytoplankton cannot easily be correlated with seasonal cycles in growth but can be shown to correlate well with the overturn of low-nutrient-containing surface water with deeper, high-nutrient-containing water. Increased water temperatures in the spring create a thermocline, thus isolating nutrient-rich waters in the photic zone to help initiate the spring bloom. Decreased temperatures in the winter cause a breakdown of the thermocline and subsequently a mixing of the nutrient-rich deeper waters with the nutrient-depleted surface waters (Corner and Davies, 1971).

Light is obviously an important factor, but again it cannot easily be correlated with rapid periods of phytoplankton growth. Although tropical oceans are subjected to high light intensities and have clear water, many regions show low productivity due to low nutrient concentrations as a result of a permanent thermocline. Light, along with increased water temperatures, initiates the spring bloom in temperate zones. In the spring, once the mixed layer is isolated in the photic zone by the establishment of a thermocline, the photosynthetic rate of the phytoplankton is greater than the respiratory rate, and spring growth can occur. The amount of suspended matter can significantly influence the quality and quantity of light that penetrates the water. Thus the coastal, or neritic, zones may have only half to a quarter the light levels when compared to open ocean water at similar depths.

Grazing

The sharp drop in phytoplankton in the summer in temperate zones may not be caused entirely by a depletion of nutrients. Instead, the decline in phytoplankton biomass may be due to a rapid rise in zooplankton, especially amphipods (Figure

20-12). Thus the determination of standing crop (biomass) of a phytoplankton community may not reflect the actual production because the community may be under constant, intense grazing. As Dawson (1966) pointed out, an initial population of 100 cells would produce 6400 cells after only six cell divisions, but only 1692 cells if 20% were grazed by zooplankton. The point is that a phytoplankton community is a dynamic entity and cyclic changes in biomass may be caused by a wide variety of factors. In this regard, Jackson (1980) has found that the so-called biological desert of the Sargasso Sea and North Pacific gyre are actually highly productive phytoplankton communities. Fast phytoplankton growth rates are balanced by equally fast zooplankton-caused mortality.

Seasonal Cycles (Succession)

As in many communities of plants, seasonal successions in phytoplankton species can be expected, as well as increases and decreases in biomass. In temperate climates there is typically a spring bloom with the largest increase in biomass, followed by a series of smaller variations with a final rise in biomass in the fall, and then a decline. Arctic regions usually have a single rise in biomass and a subsequent decline in the midsummer months.

Succession of phytoplankton species is also evident. For example, the eastern Gulf of Mexico has a series of cosmopolitan species that constitute resident populations. The diatom *Skeletonema costatum* is a dominant species year-round, whereas the diatom *Rhizosolenia alata* has a summer peak. Occasionally a large segment of the Gulf Stream will break off the loop current as it passes through

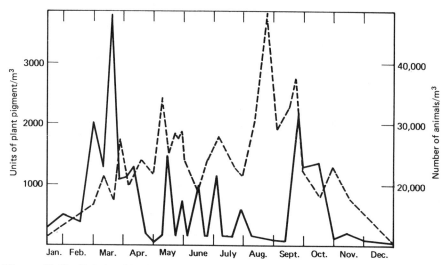

Figure 20-12 Relationship between density of phytoplankton (solid line, units of plant pigment per cubic meter) and density of zooplankton (dashed line) in oceanic waters off Plymouth, England. When zooplankton counts are high, the phytoplankton counts are usually lower because of grazing (Harvey *et al.*, 1935).

the Gulf of Mexico and these giant (5–10 km diameter) eddies will drift into the eastern side of the Gulf, bringing more species. Seasonally, dinoflagellates increase to a spring maximum, having the lowest numbers in the winter in certain areas of the Gulf of Mexico (Steidinger, 1973).

In the Gulf of Main, a seasonal change in diatom species has been shown to occur from April to August (Sverdrup *et al.*, 1964 as follows:)

April (3°C)	*Thalassiosira nordenskioldii*
	Chaetoceros diadema
May (6°C)	*Chaetoceros debilis*
June (9°C)	*Chaetoceros compressus*
August (19°C)	*Chaetoceros constrictus*
	Skeletonema costatum

Different forms of the same species occur in the warmer summer waters than in the colder winter waters.

ORGANIC PRODUCTION

Phytoplankton production varies with latitude, season, and the time of day. Generally phytoplankton production is lowest during winter months in middle and high latitudes because of the short days and lower temperatures. At lower latitudes the constant light and temperatures and low nutrient levels result in an essentially steady-state production. The nutrient level is probably the most important factor and is considered to be limiting for most phytoplankton communities. As noted in Chapter 14, available energy in a plant community can be measured either as the amount of organic matter present (biomass, standing crop) or as the rate of carbon fixation (productivity); both measures have been used extensively in phytoplankton studies.

Biomass

This term refers to the amount of organic matter present in a unit area per specified time. The *gross production* is the total amount produced. Some of this is utilized by plants through respiration or is lost through leakage, death, or grazing. The *standing crop* is the actual biomass present at the time of sampling and may be considered the *net production*. Thus the standing crop may be low and may not reflect the growth of a phytoplankton population because of high grazing pressures.

Productivity

In a review article on primary production of marine microphytes, Platt and Rao (1975) presented an eight-page table summarizing primary production studies throughout the world. The primary production of the ocean was estimated as 31

$\times 10^9$ metric tons C/yr. Furthermore, they compared the production of the three major oceans and found a relationship between primary production, nutrient content, and water exchange (upwelling). As shown in Table 20-1, the Pacific Ocean has a lower productivity based on its annual exchange through upwelling, when compared to the Indian and Atlantic oceans.

Table 20-1   Primary Production of the Three Major Oceans

Ocean	Area (10^6 km^2)	Primary Production (10^6 tons C/yr)	Amount of Exchange Divided by Volume of Ocean	Amount of Primary Production Divided by Ocean Area
Indian	73.82	6600	0.24	89.4
Atlantic	92.57	7760	0.19	105.0
Pacific	117.56	11400	0.008	64.2

Source: After Platt and Rao (1975).

Areas of high organic production may also have high standing crops for example 300 g C/m²·yr found in the coastal waters of the North Atlantic. Ryther (1963) demonstrated a wide range of production statistics based on chlorophyll levels and standing crops (Table 20-2). Regional production of major zones of the world decreases as one progresses toward the tropics. However, productivity in the warmer waters can continue for longer periods of the year, so the yearly production of fixed carbon per square meter can be higher in the tropics than in arctic waters. Also, the production levels of oceanic systems are considerably lower than those of terrestrial monocultures such as rice fields or natural pine forests (Table 20-3). Although rice fields and pine forests have production rates greater by a factor of 10 than those of the phytoplankton communities on a square meter basis, one must remember that the land masses are small. Thus Steeman Nielson (1959), comparing worldwide production, found a close comparison between the seas (1.5×10^{10} tons C/yr) and the land (2×10^{10} tons C/yr).

The rate of carbon fixation (organic synthesis) is measured by photosynthetic or respiratory rates. A number of methods are employed. Productivity is the best approach to use if one is interested in the ongoing organic synthesis and more than standing-crop information is needed. The rate is usually expressed as grams of carbon per period of time, in terms of dry weight or milligrams of chlorophyll.

Table 20-2   Primary Production of Phytoplankton from Various Areas of the World with Relation to an Average Square Meter

Area	g C/m²·yr	g C/m²·da
Open oceanic water	18–55	0.05–0.15
Equatorial Pacific Ocean	180	0.50
Equatorial Indian Ocean	73–90	0.20–0.25
Upwelling areas	180–360	0.50–1.00
Sargasso Sea	72	0.10–0.89
Arctic Ocean	1	0.005–0.024
Mean of all oceans	50	0.137

Source: After Ryther (1963).

Table 20-3 Regional Primary Production of Various World Oceanic Zones Compared with a Rice Field and a Pine Forest

Area	Production (g $C/m^2 \cdot da)^a$
Arctic	1.0^b
Antarctic	0.32–1.0
Temperate	0.05–0.4
Tropical (upwelling)	0.1
	0.3
Rice field	4.0
Pine forest	2.0–3.0

Source: Modified from Platt and Rao (1975).
[a]Seasonal data. Thus northern latitudes are limited to a shorter peak of production.
[b]Largely ice covered and so estimates are for only exposed portions.

Measurement of the evolution of oxygen (in photosynthesis), release of carbon dioxide (in respiration), or uptake of labeled carbon (as C^{14} in HCO_3^-) to supply carbon dioxide in photosynthesis) are the most commonly used methods. Also, pH changes (due to carbon dioxide release or uptake) can be measured.

The rates of carbon assimilation or oxygen evolution by marine microphytes was summarized by Platt and Rao (1975). They presented a six-page table of published data for all oceans; selected data are reproduced in Table 20-4. The problem of comparing various studies is apparent from the large variation in published rates. Platt and Rao attributed this to variation in experimental and analytical methodology, including physical factors such as light. They concluded that valid comparisons of worldwide rates of productivity were difficult to make, based on then available data.

The highest rates of carbon fixation occur during the spring months in the coastal waters of mid- to high latitudes (25 mg C/mg chl *a* • hr). Although tropical waters generally have lower rates, the rates can be high in limited areas such as equatorial current divergences (17 mg C/mg chl *a* • hr). The time of day the productivity measurements are made is also critical. Doty *et al.* (1967) demonstrated a diurnal cycle in photosynthetic rates whether under sunlight or under continuous, even illumination (Figure 20-13). Thus variable rates would be obtained if the phytoplankton were sampled at 0800, 1200, or 1600. The mean hourly rate a 24 hr period closely approximates the hourly rate about 1 hr after sunrise and 1 hr before sunset.

Methods of Sampling

Standard phytoplankton samples can be collected by a number of methods; all have good and bad points. The reader is referred to the UNESCO (1974) publication on phytoplankton methodology, as well as the handbooks of phycological methods edited by Stein (1973) and Hellebust and Craigie (1978), and the texts by Strickland and Parsons (1968) and Parsons and Takahashi (1973). Vollenweider (1974) has published a manual for the measurement of primary produc-

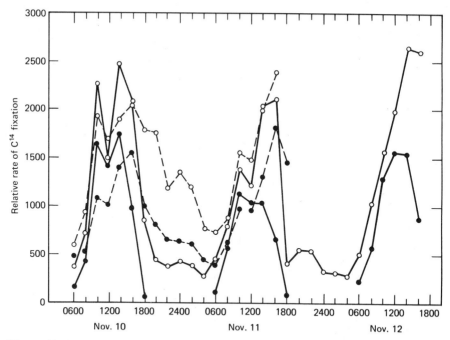

Figure 20-13 Diurnal fluctuations in primary production. This graph demonstrates the problem of measuring C^{14} uptake at only one time during a day (Doty *et al.*, 1967). Note the cyclic rise and fall in C^{14} fixation regardless of whether the sample was held in sunlight (solid circles) or under fluorescent lighting (open circles), and whether the plants were held under continuous illumination (open circles, dashed line) or on normal daylengths (solid circles, dashed line).

tion in aquatic environments and Raymont (1980) has published a general text on phytoplankton. Three types of sampling are presented here.

Plankton nets

Nets permit qualitative studies, since the mesh size will select the type of phytoplankton collected. Various sizes are available to allow the capture of at least the larger phytoplankton, but cell damage of fragile forms is common. Typical net sizes range from 20 to 60 μm mesh openings (a number 25 mesh has 60 μm openings). Nets are excellent for concentrating larger forms for detailed taxonomic studies. Nets are not appropriate for studies of primary production or quantitative estimates.

Water bottles

Closing samplers are basically water bottles that close automatically at prearranged or selected depths (Figure 20-14). Closure may be activated by a solenoid hooked to a pressure-sensing device or by a drop weight (messenger) that is released and slides down the line. There are a number of water bottles, including the nansen bottle (reversible bottle that tips upright and is sealed by jerking the wire), the van dorn sampler (a container with two spring-loaded covers that close when a messenger hits the spring) and the niskin sampler (a bottle that has a reversing thermometer to record temperature at sampling depths and has O-ring seals on the two closing caps).

Plankton pumps

These pumps are integrating samplers that pump a continuous stream of water to the surface and the phytoplankton can then be rapidly concentrated by continuous filtration. Because the pumps can collect continuously as the tube is lowered through the water column the samples are integrated from surface to desired depth. Problems include the turbulence from pumping, which can damage the more-delicate cells. The cells can also be contaminated with oil or metalic residues from the pump itself.

Table 20-4 Productivity of Phytoplankton[a]

Region	Latitude	Date	Maximum	Minimum
			(mg C/mg chl*a* · hr)	
Atlantic Ocean				
Woods Hole (coastal)	42 N	—	25.0	1.7
Sargasso Sea	28 N	January	2.0	0.0
Tropical Atlantic	1–15 S	October	17.0	2.2
Pacific Ocean				
Oregon (coastal)	44 N	—	8.9	4.5
Peru (coastal)	10–12 S	June	3.4	2.2
Hawaii	21 N	—	11.5	7.5
Antarctic Ocean				
Drake Passage	55–62 S	March	10.5	2.8
South Pole	50–64 S	Dec.–Jan.	2.0	0.8
Indian Ocean				
Erniokula	19 N	—	10.7	1.2
Arctic Ocean				
Davis Straits	64 N	July	3.3	—

Source: Modified from Platt and Rao (1975).
[a]All studies utilized C^{14} uptake methods.

If kept alive, the samples should be stored in an ice chest or a refrigerator and then only for a few hours. Usually the samples are preserved in a buffered fixative. A very popular fixative is Lugol's solution, which consists of 10 g iodine and 20 g potassium iodide dissolved in 200 ml of distilled water and 20 g of glacial acetic acid. The solution can be made up a few days ahead and stored in a dark bottle for convenience. Lugol's solution is added in a ratio of 1 part to 100 parts of the seawater sample. The fixation will cause sedimentation of nannoplankton in a liter in about 4 hr. The flagella and cilia are usually well preserved.

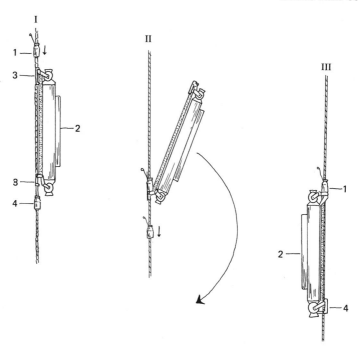

Figure 20-14 The nansen bottle. This one type of water sampler that will collect a sample at whatever depth the unit is triggered. The sample is lowered to the desired depth and a weight (*I-1*) is sent down the line triggering a release mechanism (*3*) and causing the bottle to pivot (*II*) on its lower axis (*3*). That releases the lower weight (*4*) allowing the caps on both ends to open. Once the Nansen bottle has reversed itself (*III*) and hoisting of the bottle begins, the two caps are reclosed, sealing the water sample. A reversing thermometer (*2*) is also attached and upon reversing of the sampler (*II*), the temperature recording is fixed.

Net Hauls (2% Final Concentration)		Water Samples (1% Final Concentration)	
Fixative (ml)	Sample (ml)	Fixative (ml)	Sample (ml)
25.0	500	12.5	500
12.5	250	6.25	250
5.0	100	2.5	100

Problems with Lugol's solution include its susceptibility to light, its relatively short life (~ 6 mo; iodine will oxidize), and the presence of glacial acetic acid, which dissolves coccoliths.

Another fixative recommended by Greta Fryxell (Texas A & M University) uses formalin neutralized with hexamine. To make this, 200 g of hexamine (hexamethylenetetramine) are added to 1 l of commerical 40% (37% formaldehyde) formalin. After approximately one week the solution is filtered. Since the 40% formalin-hexamine solution is not stable at low temperatures, equal volumes of the solution and distilled water are mixed before storage of the preservative, if the fixative will be stored at low temperatures.

When needed, the formalin-hexamine solution is poured into the collecting jar before the water sample is added. Addition of the fixative solution first insures complete mixing and prevents accidental omission of the preservative. Fryxell recommends the following volumes of the 40% fixative for various size samples:

Measurement of Biomass

After fixation, the biomass of the phytoplankton sample can be determined. Two basic procedures are given here, one utilizing cell counts, the other chlorophyll measurements.

Cell counts

The determination of cell numbers in a phytoplankton sample requires some form of concentration and a method for counting the cells. Concentration methods are abundant, and all have some problems. Settling, although slow, is usually the choice method for concentration since delicate cells are not damaged. Settling may be allowed to occur in a sample that has been fixed, and usually after a few hours the dish can be examined directly with an inverted microscope. Settling may also be carried out more rapidly with a Dodson concentration tube (Figure 20-15). The insert tube is placed in the sample container and allowed to sink. The water passes through a filter at the base of the insert tube and is then pumped out. By continuous water removal in the tube and the addition of more sample water to the outside, the phytoplankton can be concentrated without damage to their cells.

More rapid concentration methods include Millipore filtration and centrifugation. Both methods will damage the more delicate phytoplankton. Millipore filtra-

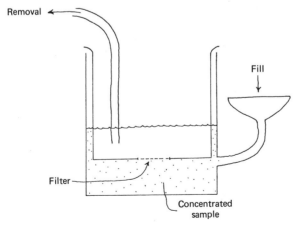

Figure 20-15 The Dodson settling tube. This is a simple device for concentration of phytoplankton. The water sample is poured in a side opening (*Fill*) and the water is removed (*Removal*) from the tube by pumping out an insert that has a filter in its bottom. This method is gentle, insuring little damage to delicate cells, and also allows for a calibrated sample since input and final volume can be determined.

tion is especially useful since a known volume of water can be pulled through a special filter that has a 100 μm^2 grid marked on it. The filter membrane can then be placed on a slide with immersion oil covering it. The filter is relatively transparent in the oil, having a similar refractive index, and the cells can be counted as they lie on the 100 μm^2 grid.

Probably the most accurate method for determination of cell numbers is to use an inverted microscope and a counting chamber with a sample obtained by settling (UNESCO, 1974). The cells from a known volume are counted directly on the bottom of the chamber after they have settled out. A more rapid method is to take a subsample of a concentrated (centrifuged or Millipore filtered) sample and count the cells, recording numbers by species; however, this is less accurate.

Also available are partical-size counters that will total the various volumes or size classes of phytoplankton. Cell volumes can also be calculated from direct measurements, and although this is quite time consuming, species counts and cell volume determinations yield the most-detailed information on phytoplankton samples.

Chlorophyll measurements and phytoplankton concentration

A rapid method for determining phytoplankton density in a sample involves the extraction and measurement of chlorophyll concentrations. The amount of organic matter associated with a given concentration of chlorophyll can then be determined from a standard curve in which a known amount of organic matter has been analyzed for its chlorophyll content. Organic levels can be determined by subtracting the ash wt from a known dry wt of a phytoplankton sample. Unfortunately, this is not always accurate, since the amount of plant pigment present can vary depending on the species present and their states of nutrition (Strickland and Parsons, 1968). However, the procedure is rapid and allows multiple comparisons, and so it is very useful in studies of phytoplankton densities.

After collection of the sample, it is filtered through a Millipore (HA, pore size 0.45 μm) or glass fiber (1 μm mesh) filter, and this is pumped to dryness. All steps should be carried out in the dark to avoid pigment breakdown. The filter containing the sample is placed in 90% spectroanalyzed acetone in a centrifuge tube and shaken vigorously to insure dissolving of the filter (Millipore) before storage in the refrigerator for 20–24 hr. A similar procedure is carried out with a clean filter, and this is used as the blank in the spectrophotometer. If the glass fiber filter is used, it is ground up at this step. Some recommend the addition of $MgCO_3$ to the filter during the concentration of the phytoplankton sample; this reduces acidity and therefore breakdown of the chlorophyll molecule. In this case, 1 ml of a 1% solution of $MgCO_3$ in distilled water is added to the phytoplankton sample during the final stages of filtration.

After 20–24 hr of extraction in the cold and dark, the centrifuge tube is brought to room temperature and the volume brought up to the original level by addition of 90% acetone. The sample is centrifuged (5–10 min at 3000–5000 rpm) and pigment determined by measuring absorbance at 630, 645, and 665 nm in a colorimeter. The appropriate formulas to use for the three main chlorophylls where D is the absorbance and C is the concentration of each chlorophyll type in mg/l are as follows:

$$C_a = 11.6 D_{665} - 1.31 D_{645} - 0.14 D_{630} \tag{1}$$

$$C_b = 20.7 D_{645} - 4.34 D_{665} - 4.42 D_{630} \tag{2}$$

$$C_c = 55.0 D_{630} - 4.6 D_{665} - 16.3 D_{645} \tag{3}$$

Equations 1, 2, and 3 allow for the determination of chlorophyll a, b, or c (mg/l) by distinguishing the contribution, in an extract mixture, of several pigments with overlapping absorption spectra (Vollenweider, 1974). Note that each optical density (OD) has its own extinction coefficient for that chlorophyll. The optical density (OD_{630}, OD_{645}, OD_{665}) results from a 1 cm pathway at the wavelength 630, 645, or 665, respectively.

Calculations can be carried out to convert the concentration of chlorophyll (mg/l) in the acetone extract sample tested to the amount present in a cubic meter of the original sample using the following equation:

$$\text{chlorophyll } a \text{ (or } b \text{ or } c) \text{ (mg/l)} = \frac{v}{V \times l} \tag{4}$$

where V = volume (in liters) of water filtered for extract
v = volume (in ml) of acetone (90%) used
l = pathlength (in cm) of cuvette

Note that equation 4 can be used for any of the chlorophylls. The concentration of chlorophyll per cubic meter can then be converted to grams of organic matter per cubic meter through the use of a standard curve as mentioned above, and this will yield an estimate of the standing crop of the phytoplankton population.

Measurement of Productivity

The basic concepts of primary productivity can be summed up in the equation for carbon fixation by autotropic aerobic algae:

$$6CO_2 + 6H_2O + \text{light} \rightarrow C_6H_{12}O_6 + 6O_2 \tag{5}$$

From this equation it is apparent that a number of methods can be employed to measure the rate of photosynthesis: carbon dioxide uptake, oxygen production, or the formation of carbon compounds. Littler (1973) compared three methods in his study of coral reef algae: labeled carbon (C^{14}) uptake, oxygen release (oxygen probe), and pH (CO_2 uptake). He found that all were effective, but the pH and oxygen-electrode methods were more useful and rapid, whereas the C^{14} procedure was the most precise. A similar approach was taken by Hoffman and Dawes (1980) in a study of benthic algae from a salt marsh and a mangrove swamp. Oxygen evolution measured by the Winkler titration procedure and manometric measurements were correlated so that field and laboratory studies could be compared. Also it was shown that C^{14} uptake was the most-effective method for measuring productivity. Although six methods are presented here, the first two are not used frequently.

Photosynthesis and pH

Changes in pH reflect both photosynthesis and respiration and are basically caused by a shift in the carbonic acid equilibrium, being part of the CO_2 system:

$$CO_2 + H_2O \rightleftharpoons H_2CO_3 \rightleftharpoons HCO_3^- + H^+ \rightleftharpoons H^+ + CO_3^- \qquad (6)$$

Photosynthesis will remove carbon dioxide, causing a shift to the left, whereas respiration will release carbon dioxide and cause a shift to the right. A rise in pH indicates an increase in photosynthesis. The problem with pH determination is that in seawater the shifts are small and the medium is basic. However, Littler (1973) has found this method to be both rapid and reliable with benthic algae, allowing for a large number of samplings over a short period of time.

Growth and biogenic salts

Seasonal and daily shifts in phosphates and nitrates are also a means for determining primary production. The analysis of seawater for levels of nitrate and phosphate over a growth season can demonstrate when plankton populations are high (low nutrient levels) or low (high nutrient levels). Since seawater analyses are usually carried out in most oceanographic studies, the nutrient information, if gathered on a seasonal basis and compared with known phytoplankton concentrations, can be used to estimate plankton abundance.

This method is useful in long-term studies, especially in cooler latitudes where seasonal changes in nutrients are apparent. However, the method is not specific (to organisms) and a patchiness or uneven distribution of nutrients in oceanic waters caused by factors unrelated to plankton populations may result in false information when levels of nutrients are determined. Factors causing nutrient patchiness include vertical currents, wind, and temperature.

Photosynthesis and oxygen evolution

Three primary methods are used to measure oxygen evolution: titration (Winkler), electronic (oxygen probe), and pressure changes (manometer). The first two methods can be used in the field.

Winkler (titration) method

The chemistry of this procedure was presented in Chapter 12 and the biological application was given in Chapter 14. There is no difference in the procedure when used with macroscopic algae or with phytoplankton, except that the plankton may have to be concentrated. The rate of oxygen evolution can be expressed either in milligrams of chlorophyll a (for more accurate comparisons between algae) or in grams dry weight. Both light and dark bottles should be used.

Oxygen probe

Both the chemistry of the process (Chapter 12) and the method (Chapter 14) have been presented. Czaplewski and Parker (1973) have presented a study showing the number of light and dark bottle replicates needed to obtain high accuracy in estimating photosynthetic rates with a (biological oxygen demand) oxygen probe and BOD bottles.

Manometric measurement

The entire procedure for measuring changes in oxygen evolution by measuring pressure changes is given in Chapter 14. As pointed out by Hoffman and Dawes (1980), this procedure, although a laboratory technique, can be correlated with field techniques.

Labeled carbon technique

The procedure was first outlined by Steeman Nielsen (1959). Doty and Oguri (1958) have presented a number of improvements. The plankton sample is placed in 300 ml BOD bottles in a standard light box under controlled temperature; a specified amount of sodium bicarbonate ($NaHC^{14}O_3$) is added to the bottle (~ 0.2 $\mu Ci/ml$ seawater; Hoffman and Dawes, 1980). A curie is a unit for measuring radioactivity; 1 Ci is that quantity of any radioactive isotope undergoing 3.7×10^{10} disintegrations per second.

After a given period of time (~ 1 hr) the light and dark bottles are fixed with 1 ml of neutral formaldehyde and filtered through a Millipore ($0.45\mu m$) filters and the filters washed free of carbonate with $0.001\ N$ HCl in 35 ppt NaCl. The filters are dried and stored until counting. Counting is carried out either on a gas or a scintillation counter. The former counter is a Geiger-Müller type of detector. The latter utilizes a fluorescent substance (e.g., Omnifluor) that emits light when bombarded with radiation. The emitted light is then measured. Standard blanks (lacking C^{14}) and samples of the seawater containing labeled C^{14} (background) are counted as well. Rates of mg C/hr per volume can be converted to mg C/mg Chl $a \cdot$hr by determination of chlorophyll a levels. Details and specific C^{14} methods are given by Doty and Oguri (1958), Vollenweider (1974) and Hoffman and Dawes (1980).

Labeled carbon is probably the most extensively used procedure for oceanic studies of productivity. It is especially advantageous because of the relatively safe, weak β-emission (0.15 Mev) as well as its long half-life (4700 yr) so that storage offers no major problems. However, Platt and Rao (1975) have found that the comparative values of most studies is low because of inconsistent methodology. For this reason, Vollenweider (1974) has proposed a standardized procedure; the procedure presented above is modeled on his method.

REFERENCES

Black, M. 1965. Coccoliths. *Endeavour* **24**: 131–137.

Corner, E. D. S. and A. G. Davies. 1971. Plankton as a factor in the nitrogen and phosphorous cycles in the sea. *Adv. Mar. Biol.* **9**: 102–335.

Czaplewski, R. L. and M. Parker. 1973. Use of a BOD oxygen probe for estimating primary productivity. *Limnol. Oceanogr.* **18**: 152–154.

Dawson, E. Y. 1966. *Marine Botany. An Introduction*. Holt, Rinehart and Winston, New York.

Doty, M. S., J. Newhouse, and R. T. Tsuda. 1967. Daily phytoplankton primary productivity relative to hourly rates. *Arch. Oceanogr. Limnol.* **15**: 1–9.

Doty, M. S. and M. Oguri. 1958. Selected features of the isotopic carbon primary productivity technique. *Rapp. et Proc. Verb.* **144**: 47–55.

Harvey, H. W., L. H. N. Cooper, M. V. Lebour, and F. S. Russell. 1935. Plankton production and its control. *J. Mar. Biol. Assoc. U.K.* **20**: 407–441.

Steidinger, K. A. 1973. Phytoplankton ecology: a conceptual review based on eastern Gulf of Mexico research. CRC Crit. Rev. Microbiol. **3**: 49–68.

Stein, J. R. (ed.). 1973 *Handbook of Phycological Methods. Culture Methods and Growth Measurements*. Cambridge University Press, London.

Strickland, J. D. H. and T. R. Parsons. 1968. *A Practical Handbook of Seawater Analysis*. Fisheries Res. Bd. Canada Bulletin 167, Ottawa.

Sverdrup, H. U., M. W. Johnson, and R. H. Fleming. 1964. *The Oceans. Their Physics, Chemistry, and General Biology*. Prentice-Hall, Englewood Cliffs, N.J.

UNESCO. 1974. *A review of Methods Used for Quantitative Phytoplankton Studies*. UNESCO Tech. Pap. in Mar. Sci. 18. New York.

Vollenweider, R. A. 1974. *A Manual on Methods for Measuring Primary Production in Aquatic Environments*. IBP No. 12 Blackwell Sc. Publ., Oxford.

Hellebust, J. A. and J. S. Craigie. 1978. *Handbook on Phycological Methods. Physiological and Biochemical Methods*. Cambridge University Press, London.

Hoffman, W. E. and C. J. Dawes. 1980. Photosynthetic rates and primary production by two Florida benthic red algal species from a salt marsh and a mangrove community. *Bull. Mar. Sci.* **30**: 358–364.

Jackson, G. A. 1980. Phytoplankton growth and zooplankton grazing in oliogrophic oceans. *Nature* **284**: 439–440.

Littler, M. M. 1973. The productivity of Hawaiian fringing-reef crustose corallinaceae and an experimental evaluation of production methodology. *Limnol. Oceanogr.* **18**: 946–952.

Parsons, T. R. and M. Takahashi. 1973. *Biological Oceanographic Processes*. Pergamon, New York.

Platt, T. and D. V. S. Rao. 1975. Primary production of marine microphytes. In Cooper, J. P. (ed.). *Photosynthesis and Productivity in Different Environments,* Cambridge University Press. IBP No. 13., pp. 249–275.

Raymont, J. E. G. 1980. *Plankton and Productivity in the Oceans,* 2nd ed., Vol. 1. Pergamon, New York.

Ryther, J. H. 1963. Geographic variations in productivity. In Hill, M. N. (ed.). *The Sea: Ideas and Observations in the Study of the Seas*. Interscience, New York.

Schmidt, R. J. and A. R. Loeblich III. 1979. Distribution and paralytic shellfish poison among Pyrrhophyta. *J. Mar. Biol. Assoc. U.K.* **59**: 479–487.

Sieburth, J. McN., V. Smetacek, and J. Lenz. 1978. Pelagic ecosystem structure: Heterotrophic compartments of the plankton and their relationship to plankton size fractions. *Limnol. Oceanogr.* **23**: 1256–1263.

Smayda, T. J. 1970. The suspension and sinking of phytoplankton in the sea. *Oceanogr. Mar. Biol. Ann. Rev.* **8**: 353–414.

Sournia, A. 1974. Circadian periodicities in natural populations of marine phytoplankton. *Adv. Mar. Biol.* **12**: 325–389.

Staker, R. D. and S. F. Bruno. 1980. Diurnal vertical migration in marine phytoplankton. *Bot. Mar.* **23**: 167–172.

Steemann Nielsen, E. 1959. Primary production in tropical marine areas. *J. Mar. Biol Assoc. India* **1**: 7–12.

Marine Fungi
and Lichens

The classification in this text (see Chapter 1) recognizes the fungal kingdom, Mycota (Myceteae), as distinct from the photosynthetic plant kingdom. In the past, fungi have been included in the plant kingdom itself, and even today these organisms are discussed in general botany texts. A number of reasons can be given for this delineation. Many fungi have cell walls, and at least in one class (Oomycetes) these walls contain a structural carbohydrate containing D-glucose, similar to cellulose. Many fungi have modes of reproduction as diverse as those of the algae, including asexual spores and sexual motile gametes. Life histories in the fungi are as widely diverse as in the algae, ranging from haplontic, to diplontic and haplodiplontic life histories.

The marine fungi are described in this text because they are involved in a number of symbiotic relationships (parasitic, mutualistic). In addition, they decompose organic matter that many plants require (saprophytes). Fungi are also important in many of the nutrient and mineral cycles that are critical to plants. The first Appendix therefore, provides a brief introduction to marine fungi, especially as they relate to marine plants. The reader is referred to the texts of Johnson and Sparrow (1961), Jones (1976), and Kohlmeyer and Kohlmeyer (1979) for detailed discussions of marine fungi.

How does one distinguish marine fungi from terrestrial species or species that are just passing through the marine environment? The two criteria presented by Barghorn and Linder (1944) seem to satisfy this question best:

1 The growth and reproduction of a marine fungus will occur either exclusively or predominantly in the sea or intertidal substratum.

2 The highest growth and optimal reproduction rates will take place in normal (~ 30 ppt) salinity seawater. In fact it appears marine fungi have a specific requirement for sodium that is linked to protein synthesis (Amon and Arthur, 1980).

However, marine fungi also occur in estuaries, and there are no distinctive morphological features to distinguish marine and terrestrial fungi. Thus the two criteria of Barghorn and Linder require some physiological information for a determination. The basic question seems to be whether the fungus will function best in a marine or a freshwater environment. Here, the definition offered by Kohlmeyer and Kohlmeyer (1979) seems appropriate. "Obligate marine fungi are those that grow and sporulate exclusively in a marine or estuarine habitat; facultative marine fungi are those from freshwater or terrestrial milieus able to grow (and possibly sporulate) in the marine environment."

Information on marine fungi is available in the two texts by Johnson and Sparrow (1961) and Kohlmeyer and Kohlmeyer (1979). The pioneer monograph of marine water molds was written by the Danish mycologist Peterson (1905). Sparrow published a series of papers in 1934 and 1936 clarifying the morphological details of marine water molds; this resulted in a 1960 monograph. An important account of marine fungi that attack wood (lignicolous fungi) was published by Barghorn and Linder in 1944. In this paper, they presented the two previously noted criteria for delineating marine fungi.

DISTRIBUTION OF MARINE FUNGI

According to Kohlmeyer and Kohlmeyer (1979) and Kohlmeyer (1979) there are about 500 species of marine fungi in oceans and estuaries. Of these, 209 are filamentous, 177 are yeasts, and 100 are placed in the simpler (water molds) fungi. Of the filamentous forms, they list 149 in the class Ascomycetes, 56 in the Deuteromomycetes and 4 in the Basidiomycetes (see final section for taxonomy). Fifty-four filamentous species have been reported from tropical coasts (Kohlmeyer 1980). Marine fungi do not seem to form series of discrete floristic units, such as with algal communities. Rather, marine fungi appear to be more associated with a type of substratum or range of host plant or animal. No vertical zonation is evident when erect substrata such as wood pilings or prop roots of mangroves have been studied.

A number of species have been identified from deep-sea sites (below 500 m) where there is sufficient oxygen to support respiration. For example, every piece of wood collected from the deep sea floor where oxygen was sufficient had a mycoflora on it. (Johnson and Sparrow, 1961). Kohlmeyer and Kohlmeyer (1979) believe that there are barophilic (adapted to high pressures) and psychrophilic (adapted to low temperatures) fungi, just as marine bacteria are specially adapted (Appendix B). Fungi have been described from marine and estuarine waters, sediments, beach sand, sea foam, algae, vascular plants, and marine animals.

Fungi are also abundant in salt marshes and mangrove swamps. The nutritive value of detritus in tidal marshes is much higher than the nutritive value of the dead plant material, primarily because of infestation by marine fungi (e.g., 24% protein in detritus vs. 6% protein in dead plant leaves). Many marine fungi are host specific, and as Kohlmeyer and Kohlmeyer (1979) pointed out, the number of fungi varies from one salt marsh host species to another (*Juncus romerianus*, 2 spp.; *Salicornia virginica* 2 spp.; *Spartina alterniflora*, 29 spp.). The same can be said for species of mangroves (*Avicennia germinans*, 14 spp.; *Rhizophora mangle*, 31 spp.). Fungi in tidal marshes may be simple saprobes, or may have symbiotic relationships with the plants. Both parasitic and mutualistic (microrhizoidal) associations are known in salt marshes.

Bark-dwelling fungi are common in mangrove swamps and are also present on the red mangrove seedlings before they drop from the trees. Four stages in a fungal infection of mangrove seedlings have been described including: (1) a primary superficial stage during the early seedling development, (2) a secondary superficial and subepidermal infection of the establishing seedling (2–5 mo old), (3) infection by obligate fungi on dying or dead parts of the seedling and (4) finally an infection of the entire seedling. Some specific fungi appear to be involved in each stage of the infection.

FUNGAL ACTIVITIES IN THE MARINE ENVIRONMENT

Most marine fungi show the following distributional and tolerance characteristics:

1   As salinity decreases, the number of water molds increases (Höhnk, 1956).

2   As salinity decreases, the number of species from the class Ascomycetes also decreases (Höhnk, 1956).

3   There are more marine fungi in tropical waters than in cold, temperate waters (Meyers, 1957).

4   As the average temperature of the water increases, salinity tolerance of lignicolous fungi (wood-rotting fungi) decreases (Ritchie, 1957). The salinity-temperature relationship in fungi has been termed the *phoma pattern*.

Fungi and Marine Algae

A number of symbiotic relationships are known between fungi and algae, including mutualistic ones, in which both species benefit (e.g., lichens); commensalistic ones, in which only one member benefits and the "host" is not damaged; parasitic and saprophytic associations.

Mutualistic relationships will be covered in the final section on lichens. Commensalistic (composite) relationships include mycophycobiotic examples such as the hyphomycete *Blodgettia bornetii* (=*B. confervoide*) that forms an anastomosing hyphal web throughout the cell wall of the green algae *Cladophora caespitosa* and *C. catenata* (Figure A-1). It is rare to find the alga without its mycobiotic partner, although the fungus does not penetrate the algal protoplast. Another example of commensalism can be found in the relationships between the ascomycete *Mycosphaerella ascophylla* and the brown algae *Ascophyllum nodosum* and *Pelvetia canaliculata*. The fungus grows intercellularly in the cortex of these two intertidal species. So close is this relationship that the fruiting bodies (ascocarps) are formed in the receptacles of the host alga, where its own reproductive organs develop. The fungus *Turgidosculum ulvae* forms eruptions on the green alga *Blidingia minima*, whereas *T. complicatulum* causes separation of the

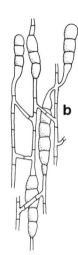

Figure A-1   The composite relationship between the fungus *Blodgettia bornetii* (=B *confervoides*) and the green alga *Cladophora catenata* (*C. fuliginosa*). The fungus grows within the cell wall of *Cladophora* (*a*) and forms chains of spores among its hyphal network (*b*). (*a*) × 19.50. (*b*) × 117.

green alga *Blidingia minima*, whereas *T. complicatulum* causes separation of the cells in the green alga *Prasiola tessellata*. Infected and uninfected plants of these two intertidal algae grow side by side and both seem to be "healthy." Such mycophycobiotic relationships are common in nature and do not appear to damage the hosts. In fact, the intertidal algal species may profit from such infections because the fungi probably supply water and prevent drying out of the algae. Such commensalistic, or mycophycobiotic, relationships have been termed *lichenoid* because of the similarity to lichen relationships.

Kohlmeyer and Kohlmeyer (1979) listed 32 species of higher fungi that are parasitic on marine algae. Johnson and Sparrow (1961) listed 6 species of slime molds and 50 species of water molds, almost all of them found on marine algae. Andrews (1976) reviewed fungal diseases of marine algae, and Kohlmeyer (1973, 1979) described a variety of pathogenic fungi found on marine algae. Kohlmeyer (1971) also described the types of fungi found on free-floating species of *Sargassum* from the Sargasso Sea. Kohlmeyer (1974) described three levels of fungal diseases on algae: (1) the outer appearance of the alga is not altered (2) discolorations may occur on the host, and (3) malformations such as galls are produced.

The present knowledge of algicolous fungi is weak; much more is known about lignicolous fungi (Kohlmeyer, 1979). He suggests eight areas that need further study in order to improve our understanding of fungal diseases of algae: (1) host physiology and life history, (2) mode of infection, (3) fungal growth in the host, (4) infection experiments, (5) susceptibility of an alga to infection, (6) life cycle of the fungus, (7) geographical distribution, and (8) fossil records.

Plasmodiophoromycetes and algal diseases

Probably the most dramatic example from the group of water mold fungi is the disease of the economically important red alga *Porphyra* spp. (Kazama, 1979). There are actually two serious fungal diseases, the *red rot* or *red wasting disease (akagusare)* caused by *Pythium porphyrae* and a *chytrid blight* disease caused by *Olpidiopsis* sp.

Pythium porphyrae infects the foliose (winter) phase of the two economically important *Porphyra* species in Japan, *P. tenera* and *P. yezoensis* (see Figure 2-11). Thalli of *Porphyra* infected by *Pythium* develop circular light green lesions. The fungus penetrates through the algal cell walls into the protoplasts, causing the cells to collapse (Kazama, 1979). In an ultrastructural study of another species of *Pythium*, *P. marinum*, which infects the North American species *Porphyra perforata*; Kazama and Fuller (1970) found that the fungus modified the host cell boundaries very little, and the fungal hyphae were also not modified. The starch and cellular structures of the host cells were simply dissolved in the region of the penetrating hyphae. Kazama (1979) isolated *Pythium marinum* from lesions of *Porphyra miniata* from the east coast of North America as well as from lesions of *P. perforata, P. lanceolata, and P. schizophilla* from the west coast. He suggested that the red rot disease of *Porphyra* spp. is probably caused by a single species of *Pythium*.

Another example, *Ectrogella perforans*, is a virulent pathogen of diatoms, infecting up to 99% of the genus *Licmophora* in nature (Figure A-2). Infection is

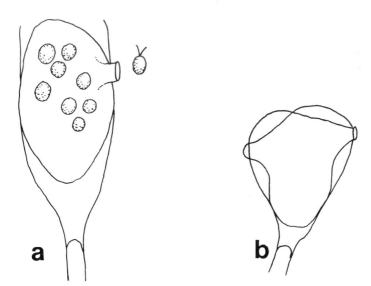

Figure A-2 *Ectrogella perforans*, a parasite of diatoms. It produces small (3–5) μm diameter) external saccate sporangia that release biflagellated zoospores (*a*). Empty sporangia (*b*) are found on the frustules of such diatoms as *Licmophora*.

by the adherance of fungal zoospores to the diatom frustule and their penetration by germination tubes. The pathogen is eventually situated in the diatom cell and forms an external saccate sporangium that releases zoospores. Other water mold parasites include *Eurychasmidium tumefaciens*, which causes localized hypertrophy in the nodal cells of the red alga *Ceramium*. *Eurychasma dicksonii* attacks brown algae such as *Ectocarpus* and *Striaria*.

Ascomycetes and algal diseases

Thirty-one ascomycetes have been identified as parasites of marine algae. Of these 16 occur on brown algae, 13 on red algae, and 2 on green algae (Kohlmeyer, 1979). Kohlmeyer suggested that the lower number of green algal pathogens is based on the ephemeral nature of most green algae. The Southern Hemisphere red alga *Ballia callitrichia* produces "a wild growth of hairs" in response to infection by the ascomycetes *Spathulospora clava, S. adelpha,* or *S. phycophila*. The growth response of the hairs is comparable to the *witches-broom*, a growth caused by insect or fungal infestations of vascular terrestrial plants (Kohlmeyer, 1979).

Discolorations of red algae by fungal parasites include *Chadefaudia marina* on *Palmaria palmata, Lulworthia kniepii* on calcified algae, and *Didymella gloiopeltidis* on *Gloiopeltis furcata*. The latter disease is called the *black dot disease* because the fungal ascocarps (perithecia) eventually develop in the infected region on *Gloiopeltis* (Kohlmeyer, 1979). This black dot disease is a major concern, since *Gloropeltis* is harvested for the phycocolloid funoran. *Chondrus crispus*, another phycocolloid-yielding (carrageenan) red alga, is frequently infected by the ascomycete *Didymosphaeria danica*, which causes blackening of the cystocarps. The infection will prevent reproduction but apparently does not stop algal growth.

Brown algae infected by ascomycetes usually show black discolorations or gall formation. The *raisin disease* the free-floating *Sargassum fluitans* (Fucales) in the Sargasso Sea is so called because its floats, or vesicles, become soft and wrinkled. The same fungus also attacks seagrasses and is named *Lindra thalassiae* (Figure A-3). A number of kelps, such as *Alaria* and *Laminaria*, show a *stipe splotch disease*, which appears as tar-stained areas on the stipes. The causal organism is *Phycomelaina* spp. *Phycomelaina laminariae* forms black circular or oblong patches on the stipes of *Laminaria*; these infection sites permit penetration by other fungi.

Brown algae may be attacked by gall-forming fungi such as *Haloguignardia* spp. (*Cystoseira, Halidrys, Sargassum*) and *Massarina cystophorae* (*Cystophora retroflexa*). These galls are subglobose to elongate outgrowths of the inner and outer cortical cells. Protuberances containing the fungal ascocarps will form on the galls.

Deuteromycetes and algal diseases

Although quite common as parasites of terrestrial plants, only one member of the Fungi Imperfecti (Deuteromycetes) has been described as an algicolous parasite

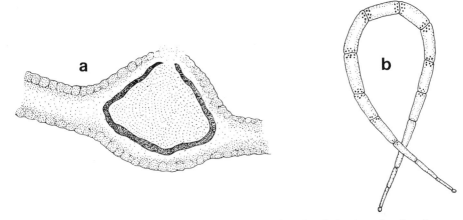

Figure A-3 (*a*) A section through the perithecium of *Lindra thalassiae*, showing the ascogenous layer (dense band). (*b*) The parasite of the seagrass, *Thalassia* produces elongated ascospores in the leaves that have spherical appendages and are adapted for water dispersal through flotation. The perithecia are about 2 mm in diameter; the ascospores are about 300 μm in length (Orpurt *et al.*, 1964).

(Kohlmeyer, 1979). *Sphaceloma cecidii* is a *hyperparasite,* infecting regions of another ascomycete (*Haloguignardia* spp.) in such brown algae as *Cystoseira,* *Halidrys,* and *Sargassum.* The secondary fungal infection causes the brown galls of the primary infection to turn black (Kohlmeyer, 1979).

Fungi and Marine Vascular Plants

Fungi are involved in a variety of ways with marine vascular plants. A sequence of fungal populations has been demonstrated in the decay of mangrove leaves (Fell and Master, 1980), and the mycoflora of turtle grass (*Thalassia testudinum*) has been described (Newell and Fell, 1980). The wasting disease of *Zostera marina* (eelgrass) was found to be caused by the slime mold *Labyrinthula* sp. (Figure A-4); the sequence of invasion of the sea grass cells was described by Renn (1936). Symptoms of infection are dark-streaking or splotching of the leaves, which is rapid, and the subsequent disintegration of the leaves. A catastrophic decline of *Zostera* along the southeastern seacoast of North America occured in the 1930s. The leaves were destroyed in a few days because of the removal of the photosynthetic tissue, and within two years entire populations of seagrass were gone. The biological and economic results were very significant since *Zostera* is a basic food for wild ducks and geese and an important food and habitat for young fish. The fishing industries of the region declined by over 50% and only began to recover after 5 yr, with the return of *Zostera*. An ascomycete, *Ophiobolus maritimus*, also appeared in the *Zostera* leaves a few days after the invasion by *Labryinthula*, but this fungus was considered to be a secondary invador, taking advantage of the already-damaged leaves.

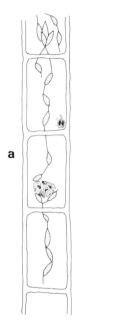

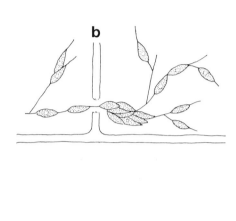

Figure A-4 An example of an important marine slime mold might be *Labyrinthula macrocystis*, shown here growing in cells of the green alga *Cladophora* sp. (*a*). The spindle-shaped slime mold cells are attached by cytoplasmic extensions (*b*) and can dissolve the cell wall, permitting growth from one cell to another in the algal filament. The cells measure about 3–4 μm in diameter.

Plasmodiophora diplantherae (ascomycete) is a pathogen of the seagrass *Halodule wrightii*, attacking the inner cortical cells of the internodes but not killing the plant. Another species of the same genus, *P. halophila*, has been found in the petioles of the seagrass *Halophila englemannia*, and a third species, *P. bicaudata* has been found in the cortical and epidermal cells of *Zostera nana* from West Africa. Two brackish-water grasses, *Ruppia* and *Zannichella*, can host a plasmodiophoromycete parasite called *Tetramyza parasitica*. *Lindra thalassia*, an ascomycete (subclass Plectomycetidae), will cause extensive damage to blades of the turtle grass, *Thalassia testudinum* (Figure A-3). Orpurt *et al.* (1964) isolated the species and demonstrated its distinctive role in these seagrass beds.

Graminicolous fungi are also parasitic on salt marsh plants, as shown by Gessner (1977) for *Spartina* (Figure A-10). Gessner (1978) also found that a number of parasitic fungi are associated with the seeds of *Spartina*, including ergot (*Claviceps purpurea*). His list of these predominantly terrestrial fungi included 8 ascomycetes, 1 basidiomycete, and 16 deuteromycetes. Gessner (1978) suggested that the *Spartina* seeds, being high-nutrient packets, can serve as a dispersal mechanism for many fungi. He divided seed colonization by fungi into two phases, an initial surface development and modification of the substratum, followed by invasion.

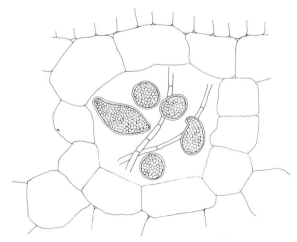

Figure A-5 *Melanotaenium ruppiae*, a smut (Basidiomycetes) that will infect the aerenchyma of the rhizome of the estuarine grass *Ruppia maritima* (Feldmann, 1959). A group of spores with hyphae is shown within a small lacuna of the rhizome. The spores are 12–25 μm in diameter.

The only parasitic basidiomycete reported from the marine environment is a smut, that attacks the rhizomes of the sea grass *Ruppia maritima* (Figure A-5). *Melanotaenium ruppiae* produces spores in the air spaces, or lacunae, of the rhizome but apparently does not kill the host.

Marine Animals and Fungal Diseases

Fungal parasites of marine animals have attracted more study than fungal parasites of marine plants; thus many examples are known. This is especially evident in cases of commercially important fish and shellfish such as herring and oysters.

One of the most-important fish pathogens is *Ichthysporidium hoferi*, a water mold of questionable taxonomy. The fungus attacks herring, flounder, plaice, mackerel, and other noncommercial fish. The hyphae invade the connective tissue (light infection) or occur throughout the internal cavity (heavy infection) and cause a yellow to brown coloration. The fungus is considered to be a natural parasite of herring. Infections will reach epidemic proportions under what are still poorly defined environmental conditions and then will decline to an endemic level.

Commercial sponges almost died out in the Bahamian and Floridian waters in 1938 and 1939 when a water mold fungus, *Spongiophagia communis*, invaded the large invertebrates (Figure A-6). The invader broke down the flagellated chambers and canal linings of the sponge, causing cytolysis of the mesoglea tissue. Hyphae were found attached to the sponge choanocytes.

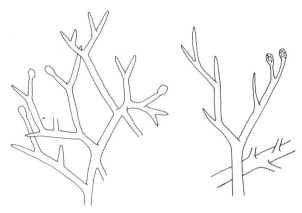

Figure A-6 A water mold, *Spongiosphagria communis*, that attacks sponges. Two hyphal branches, without cross walls, are shown; the branch at the right has sporangia. The main hyphae measure about 2–3 μm in diameter.

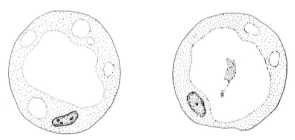

Figure A-7 A unicellular water mold, *Dermocystidium marinum*, a parasite of oysters. Two infecting cells, about 10 μm in diameter, are shown.

One of the best-studied fungal parasites of invertebrates is the oyster-infecting species *Dermocystidium marinum* (Figure A-7). The unicellular water mold invades all the animal's tissues, resulting in failure of the adductor muscles to close the shell. Such opened oysters are termed *gappers*. Before death the oyster stops growth and becomes emaciated. The fungus is more common in oysters of warm waters and older animals.

A number of species in the class Plasmodiophoromycetes have been shown to attack invertebrates. *Leptolegnia marina* is an important parasite of body organs of the pea crab, (*Pinnotheres* sp.), the gill tissue of *Barnea*, and the mantle of *Cardium*. *Plectospira dubia* (or *Atkinsiella dubia*) infects the ova of a number of marine invertebrates, such as *Pinnotheres, Gonoplax,* and *Typton*. Species of *Lagenidium* occur in the egg masses of the blue crab, *Callinectes* (Figure A-8*a*), the ova of the oyster crab, *Pinnotheres* sp., and the ova of barnacles (*Chelonibia, Chthamalus*).

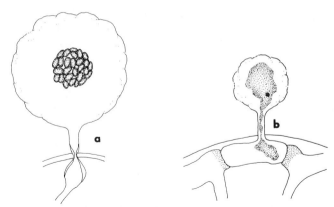

Figure A-8 Two species of *Lagenidium*, parasites of invertebrates, (*a*) *Lagenidium cal-linectes*, a parasite of the crab, is shown with a sporangial discharge tube and a cluster of spores (each 15 μm in diameter) within a gelatinous matrix. (*b*) The discharge tube and some of the hyphae (7 μm in diameter) of *L. chthamalophilium*, a parasite of eggs of barnacles.

Saprophytic Fungi

The largest number of marine fungi are saprophytes, utilizing dead organic matter such as plant detritus as their source of nutrients. Two categories of saprophytic fungi are presented here: lignicolous and sediment fungi.

Lignicolous fungi

Wood is a common substratum in coastal and deep-sea sites; thus lignicolous fungi are common in the marine environment. Islands of wood occur on the deep-sea floor, driftwood is trapped in the sand of beaches and between rocks in inter-tidal regions. In addition, man has installed pilings and seawalls, as well as using hemp rope for lines; all of these are excellent substrata for lignicolous and cel-lulolytic fungi. Wood-penetrating fungi were first documented by Barghorn and Linder (1944) in their classic paper on marine fungi. The marine species are members of the classes Ascomycetes (76 species), Deuteromycetes (29 species), and Basidiomycetes (3 species, Kohlmeyer and Kohlmeyer, 1979).

Wood-rotting fungi usually penetrate through transverse wood rays. The fungal hypha enters the cell lumen through pits and then grows into the secondary wall, enzymatically breaking down the low-lignin-containing S-2 layer of this wall. The hyphae then grow parallel to the cellulose microfibrils, which are arranged helically in the S-2 layer of its secondary cell wall of woodcells. Thus the tubes resulting from fungal penetration are helical and appear conical in a cross section of the wood. Through removal of the supporting cellulose microfibrils in the S-2 layer, the wood takes on a spongy texture. Hence the term *soft rot*. Fungal activity is restricted to the wood surface where oxygen levels are high.

Although cellulolytic bacteria are also important in wood rot, it appears they are usually more effective after the fungi have penetrated the wood. Brooks *et al.* (1972), using the scanning electron microscope, established that the initial colonization of wood panels in the sea was by fungi, bacteria being essentially absent. He also found that colonization by fungi was species-specific for each wood zone. The lignicolous hyphae were especially common in the intercellular pectic substances that along with cellulose form a good portion the middle lamellae of plant cell walls. Growth in the middle lamella results in the separation of the cells since the middle lamella is the cementing layer between cells. In studies of fresh wood blocks placed in the sea, it was evident that isolated marine lignicolous fungi have a number of active enzymes, including cellulases, hemicellulases, and pectinases. In addition, marine fungi have been isolated that contain enzymes such as amylases, laminarinases, proteases, and chitinases.

Fungi that infest wood structures such as pilings and ramps probably prepare them for subsequent attack by bacteria and other boring crustaceans and molluscs. A number of yeasts (Ascomycetes) may occur in the gut of marine boring invertebrates; the yeasts can digest cellulose and the animals utilize the released sugar glucose. By removing the cellulose component of the middle lamella, the fungi soften and break down the outer layers of a wood piling. With the softening of the surface, marine boring animals can penetrate and ultimately aid the fungus in its growth by the formation of bore holes into the wood. The two most important boring animals are the ship worm *Teredo* sp. and the boring gribble *Limnora* sp. These worms are present within the timber and are invisible until the wood

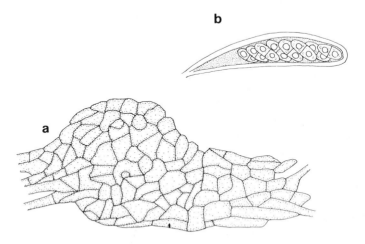

Figure A-9 (*a*) A spermogonial cluster of *Leptosphaeria discors* as found in the epidermal cells of *Spartina* detritus (Wagner, 1965). The spermogonium measures about 50 μm in length and produces small (1 μm) cells believed to be the "male" cells. (*b*) A thick-walled ascus with eight uniseptate ascospores ($\simeq$ 40 μm diameter). Both the inner and outer walls are still intact.

disintegrates. The attacks can be so severe that ships will fall apart quite suddenly, as did an entire set of ships used by Columbus on one of his missions. The cost of replacement of pilings, moorings, and so forth is enormous because of these fungi and wood-boring animals.

Sediment fungi

Sedimentary fungi found in detritus and various organic deposits include a number of poorly identified yeasts (Ascomycetes) and hyphal forms of the class Plasmodiophoromycetes. The importance of fungal activities in detritus breakdown is recognized, but few studies have been conducted. Wagner (1965) described an ascomycete growing on the dead culms of the cordgrass, *Spartina* sp. in salt marshes of the Atlantic. The fungus, which was identified as *Leptosphaeria discors*, penetrates the host through the cuticle and continues through the pits in the epidermal cell wall, forming ascocarps (Figure A-9) in the epidermal cells. A number of graminicolous fungi occur on salt marsh plants, as pointed out by Gessner (1977). He found a vertical distribution of 17 fungi growing on living *Spartina*, as well as two ascomycetes (including *Leptosphaeria*) on dead culms (Figure A-10).

Since detrital breakdown is an important feature of tidal marshes, subtidal sea grass beds, and algal communities, it would appear the graminicolous and lignicolous fungi should be common in neritic environments and marine sediments. Species of *Penicillium* have been found to be common in marine muds and beach sand and the genus *Metashnikowia* (yeast) is widely reported from marine sediments. A wide variety of fungal spores occur in sea foam (arenicolous fungi). Wagner-Merner (1972) suggested that this is an important dispersal mechanism. She found that the foam, blown from beaches in marsh areas, may contain spores from up to 20 different fungal species. The foam acts as an inoculum and a source of nutrition and moisture.

TAXONOMY OF MARINE FUNGI

A number of general mycological texts are available (Ross, 1979; Alexopoulos and Mims, 1979), as well as texts dealing with marine fungi (Johnson and Sparrow, 1961; Kohlmeyer and Kohlmeyer, 1979). The taxonomic approach in this Appendix follows Alexopoulos and Mims (1979); three divisions are identified in the kingdom Fungi: the Gymnomycota (slime molds), Mastigomycota (water mold and plantlike fungi), and the Amastigomycota (fungi lacking plantlike features, nonmotile).

The fungal kindom, Mycota or Myceteae, contains organisms that are achlorophyllous and are thus either saprophytic or parasitic. Thalli can be unicellular or filamentous and may be septate (uninucleate or multinucleate) or coenocytic. Sexual reproduction occurs by fusion of motile or nonmotile gametes or by gametic nuclei.

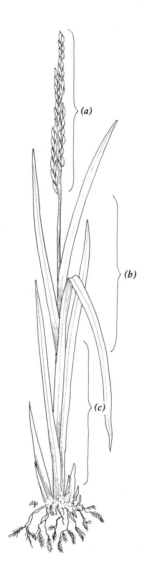

Figure A-10 Gessner (1977) demonstrated that a wide array of fungi can be found on salt marsh plants. Three major areas of a plant of *Spartina alterniflora* were examined and the fungi listed for each component. (*a*) Inflorescence and seeds: *Alternaria alternata*, *Claviceps purpurea*, *Stagonospora* sp., *Epicoccum nigrum*. (*b*) Leaves: *Phaeosphaeria typharum*, *Drechslera halodes*, *Pleospora vagans*, *Buergenerula spartinae*, *Puccinia sparganioides*, *Phoma* spp. (*c*) Culm: *Pleospora pleagica*, *Hallsphaeria hamata*, *Leptosphaeria albopunctata*, *L. obiones*, *L. pelagica*, *Lulworthia* sp., *Buergenerula spartinae*.

Division Gymnomycota

This division contains phagotrophic organisms that lack a cell wall and have a naked amoeboid assimilative stage. Thus these organisms are called *slime molds*. The mode of nutrition ranges from saprophytic to parasitic, many slime molds can ingest organic material such as bacteria. Three classes are recognized (Myxomycetes, Acrasiomycetes, and Labyrinthulomycetes); the third one contains a few marine species.

CLASS LABYRINTHULOMYCETES

The members of this class are mostly aquatic; their individual amoeboid cells are spindle shaped. When grouped together to form the plasmodial stage, the cells retain the spindle form. Thus the plasmodium appears netlike and is called a *net plasmodium*. About 12 species are placed in this class. The genus *Labyrinthula* (Figure A-4) contains about six marine species that infect seagrasses (*Zostera*, *Ruppia*) and algae (Chlorophyta: *Ulva*, *Bryopsis*; Phaeophyta: *Ectocarpus*, *Taeonia*, *Laminaria*; Rhodophyta: *Ceramium*, *Polysiphonia*). The "wasting disease" of *Zostera marina* is caused by a species of this genus and a number of major seagrass infections have been reported from the east coast of North America (1893–94; 1908; 1920) from the European and North American coasts of the North Atlantic (1930–31, 1931–32) and from the Pacific coast of North America (1937–39). The infection by this genus resulted in a 90% loss of *Zostera* on the eastern North American seaboard (1930–32).

Division Mastigomycota

The division contains fungi with a number of plantlike features, including centrioles that function during nuclear division, flagellated cells at some stage of their life cycles, and usually cellulose in their cell walls. A variety of primarily aquatic forms are placed in the four classes; most of the marine species are parasitic on algae and animals. Alexopoulos and Mims (1979) recognize two subdivisions and four classes.

CLASS 1. CHYTRIDIOMYCETES

Species placed in this class produce posteriorly uniflagellated motile cells, the flagella are acronematic. Four orders are recognized, one of which, the Chytridiales, contains marine species. Two examples can be cited, *Chytridium megastomum*, which grows on *Ceramium*, and *Coenomyces consuens*, which occurs in the sheaths and cells of blue-green algae (Figure A-11).

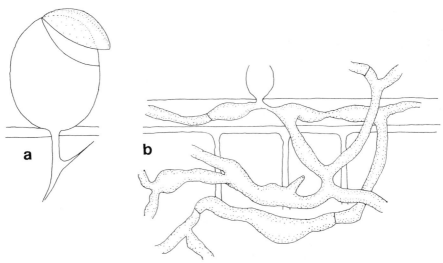

Figure A-11 Two examples of chytrids, common parasites of algae. *Chytridium megas-tomum* is an epibiotic fungus that penetrates the algal cell and produces a sporangium that opens by means of an operculum (*a*). to release the zoospores. (*b*) *Coenomyces consuens* has a hyphal system that can extend throughout the gelatinous sheath and the cells of the blue-green alga *Rivularia* sp., producing an external sporangium. The sporangia of both species are about 10 μm in diameter.

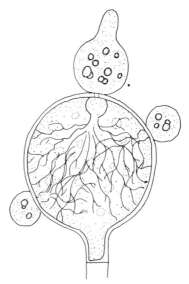

Figure A-12 A member of the Xanthophyceae, *Vaucheria* sp., is parasitized by the water mold *Rhizidiomyces apophysatus* (Karling, 1944). The oogonium of *Vaucheria* is shown parasitized by three monocentric cells, all of which are penetrating with haustorial hyphae. The external parasitic cells are about 5–8 μm in diameter.

CLASS 2. HYPOCHYTRIDIOMYCETES

This small class has an order that is characterized by its parasitic nature and uniflagellated zoospores. *Rhizidiomyces apophysatus* parasitizes the oogonia of the Xanthophycean alga *Vaucheria* (Figure A-12; Karling, 1944). *Anisopidium* spp. parasitize such brown algae as *Ectocarpus, Sphacelaria,* and *Cladostephus.*

CLASS 3. PLASMODIOPHOROMYCETES

This small class contains one order of obligate parasites on vascular plants, algae, and other fungi. The fungi are naked and form multinucleated thalli called *plas-modia.* Motile cells with two anteriorly placed acronematic flagella are produced. *Tetramyxa parasitica* is a small plasmodium that fills the epidermal cells of the estuarine seagrass *Ruppia maritima* and causes small whitish galls on its blades. *Plasmodiophora diplantherae* is so named because it parasitizes the cells of the rhizome of *Halodule wrightii (Diplanthera wrightii)*, again resulting in the production of galls at the internodes.

CLASS 4. OOMYCETES

This class is the largest group of water molds, with four orders, three of which have numerous marine representatives. The fungi are usually filamentous and coenocytic; their cell walls contain glucans and cellulose, and chitin is present in members of one order. The motile cells are biflagellated; one is acronematic the other pantonematic. The class name is derived from the oogamous type of sexual reproduction.

Lagenidiales

All three families placed in the Lagenidiales have marine representatives. *Peter-senia lobata* (Olsidiosidaceae) has been found parasitizing several members of the red algal family Ceramiaceae (*Ceramium, Callithamnion, Spermothamnion*). Members of the family Lagenidiaceae are parasites of crustaceans. *Sirolpidium bryopsidis* (family Sirolpidiaceae) parasitizes the green algae *Bryopsis* and *Cladophora* and forms unbranched chains of sporangia.

Peronosporales

The Peronosporales are important parasites of terrestrial and marine agricultural crops. Of the four families placed in this order, one has marine representatives. *Pythium porphyrae* and *P. marinum* parasitize the economically important *nori* crop (*Porphyra* sp.).

Saprolegniales

The order Saprolegniale contains five families, two have marine representatives. *Thraustochytrium proliferum* has been found on *Bryopsis plummosa* and in *Ceramium diaphanum* and produces an endobiotic rhizoidal system (haustoria mycelia). *Ectrogella perforans* is a pathogen of diatoms (Figure A-2), *Eurychasmidium tumefaciens* is a parasite of *Ceramium* sp., and *Eurychasma dicksonii* is a parasite of *Ectocarpus* sp.

Division Amastigomycota

This division contains the majority of fungi $\left($ 500,000 species); they lack centrioles and motile (flagellated) cells but spindle fibers are present during mitosis. Typically these "true fungi" are filamentous and usually septate. Asexual reproduction is common and occurs by budding, fragmentation, sporangiophores, and conidia. Sexual reproduction includes zygotic meiosis. Thus the life histories in many cases are haplontic, with a prolonged dikaryotic phase (delay between plasmogamy and karyogamy). Alexopoulos and Mims (1979) presented four subdivisions and five classes.

CLASS 1. ZYGOMYCETES

This class contains a variety of terrestrial saprobes and parasites, a number of which are present in oceanic and estuarine waters. According to Johnson and Sparrow (1961), "Not one member is known to be strictly marine." Sexual reproduction is a characteristic feature of the class; fusion is usually though equal (isomorphous) gametangial hyphae that result in a zygosporangium that contains the zygospore (zygote). After resting, the zygospore undergoes meiosis and usually produces a series of haploid spores. Depending on the taxonomist, three to seven orders are delineated. Tolerance to salinity has been studied in members of the genus *Mucor*. Those studied demonstrated decreasing growth rates with increasing salinity.

CLASS 2. TRICHOMYCETES

The members of this class are obligate parasites and commensals of arthropods; some species have been identified from marine invertebrates. Because the fungi are associated with animals, the mycelial development is limited. Asexual reproduction is by amoeboid cells or specialized spores. Sexual reproduction is poorly known. Alexopoulos and Mims (1979) recognized four orders in the class, whereas Johnson and Sparrow (1961) listed two orders with marine members. All species thus far identified have been isolated from invertebrate digestive tracts. *Arundinula incurvata* hyphae have been found in the stomach and intestine of a hermit crab (*Eupagurus prideauxi*). *Asellaria ligiae* occurs in the intestine of the isopod *Ligia mediterranea*.

CLASS 3. ASCOMYCETES

Five subclasses and a large number of orders are recognized in this class. The spores resulting from meiosis (meiospores) are distinctive and are called ascospores. The spores are produced in a saclike cell called the ascus (see Figure A-9*b*). The spores are nonmotile and usually eight in number. In the more advanced forms, the asci are found in fruiting bodies called ascocarps (Figure A-13). Of the five subclasses, two have marine representatives, namely the Hemiascomycetidae and the Plectomycetidae.

Members of the Hemiascomycetidae are unicellular or filamentous and lack reproductive mating hyphae and fruiting bodies. The entire cell may become the ascus, producing four or eight ascospores, or a cell will be cut off from the hyphae to form the ascus. Yeasts are found in this subclass; they have been reported from a variety of marine substrates and environments (Johnson and Sparrow, 1961). These authors listed nine species that have been identified from shrimp and algae. In addition they noted that numerous yeasts of uncertain designation are present in marine sediments and organic matter.

The subclass Plectomycetidae is based on the type of ascocarp present; this is also the basis for separation of the five orders. Three basic types of fruiting bodies are shown (Figure A-13). The *cleistothecium* is an enclosed ascocarp that lacks a natural opening for the release of ascospores; rather their release occurs by

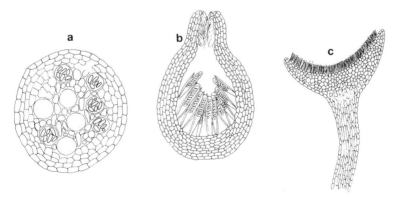

Figure A-13 The three basic types of ascocarps. (*a*) The Cleistothecium. (*b*) Perithecium. (*c*) Apothecium. Note that the cleistothecium has no opening, whereas the perithecium has an ostiole and the apothecium is a cup-shaped ascocarp. Asci with ascospores are shown in each type of fruiting body.

rupturing or breakdown of the wall. The *perithecium* is an oval-to-tear-shaped ascocarp that has a pore or an ostiole (with a neck). The third type of ascocarp, the *apothecium*, is an open, cup-shaped structure, with the asci found in the central cavity. Myers (1957) reported 30 genera and 70 species of marine origin in this subclass, whereas Johnson and Sparrow (1961) identified 41 genera and 99 species, and Kohlmeyer and Kohlmeyer (1979) listed 91 species. A variety of plectomycetidae are found as parasites of algae and sea grasses (see earlier sections). The class also has a number of common marine saprophytes, especially the lignicolous forms.

CLASS 4. BASIDIOMYCETES

The *club fungi*, which are placed in their own subdivision, are divided into three subclasses by Alexopoulos and Mims (1979). Members of the class, in addition to having septate hyphae, produce club-shaped basidia after fusion of the gametic hyphae. Nuclear fusion and meiosis occur in the basidia, and four exogenous meiospores (basidiospores) are budded off (Figure A-14). The class includes mushrooms and other terrestrial saprophytic fungi as well as a large number of parasites (rusts, smuts) that attack many terrestrial plants. Only four species in this class, three saprobes and one parasite, a smut, are known from the marine environment (Kohlmeyer and Kohlmeyer, 1979). *Melanotaenium ruppiae* (Figure A-5) is a smut that attacks the rhizomes and internodes of the estuarine seagrass *Ruppia maritima* (Feldman, 1959).

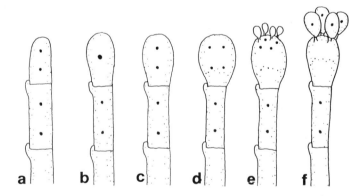

Figure A-14 A developmental sequence of the basidium. The two (dikaryotic) nuclei (*a*) fuse (*b*) and then undergo meiosis *c–d*. After this the four haploid nuclei migrate to the tip of the basidium, where small basidiospores are forming on stalks, or sterigmata (*e*). Ultimately the nuclei migrate into the basidiospores (*f*). Note the clamp connections by which the nuclei are dispersed in normal hyphal growth. (See Alexopoulos and Mims, 1979, for details).

CLASS 5. DEUTERMYCETES

The class is an artifical one that contains fungi lacking the sexual stage necessary for their placement in the Ascomycetes or Basidiomycetes. About 15,000 to 20,000 species are assigned to the "fungi imperfecti." All members have hyphae or are yeasts that produce nonmotile asexual spores. Two of the four subclasses have representatives in the marine environment.

Members of the Coelomycetidae subclass reproduce by asexual spores called conidia, which are produced in globose or other-shaped structures called pycnidia. Two families with five genera and five species are reported from the marine environment (Johnson and Sparrow, 1961).

Members of the Hypomycetidae subclass lack sack-shaped structures (pycnidia) in which the spores are produced. Instead, the conidia, or spores, are produced from vegetative hyphae on erect conidiophores. Three families with 19 genera and 21 species are known from the marine environment (Johnson and Sparrow, 1961).

MARINE LICHENS

A lichen is a mutualistic association of a fungus and an alga that results in a new, single thallus. The fungal component is called the mycobiont, and the algal component is the phycobiont. Ascomycetes are the most common mycobionts, although there are some basidiomycetes and deutermycetes as well. About 21 genera of blue-green algae and green algae have been identified as phycobionts, the most common being the green alga *Trebouxia*. In most cases, there is only a single algal phycobiont, although in one case the lower portion of the lichen contains the blue-green alga *Scytonema* and the upper portion the green alga *Chlorella* or a *Chlorella*-like alga.

The relationship is usually described as a mutualistic symbiosis (Alexopoulos and Mims, 1979), in which both the fungus and the alga receive some benefits from the association. However, none of the mycobionts can live without its phycobiont or at least some organic source. Furthermore, recent ultrastructural and physiological studies indicate that the mycobiont is the main benefactor in the association, parasitizing and even destroying algal cells. The phycobiont receives a protective covering, nutrients, and water from the mycobiont.

Four types of lichens can be described: foliose, crustose, squamulose, and fruticose. The foliose lichen is leaflike; the mycobiont forms an upper and lower cortex, and the phycobiont forms a central medulla. The crustose type is a thin continuous crust, and the squamulose type has a number of small, flattened lobes. The fruticose lichens are pendant or erect and branching.

Most marine lichens are crustose or squamulose, and they are found in the intertidal zone, especially the upper intertidal fringe. One species is said to occur in the subtidal waters off Antarctica. Marine lichens are more common along the North Atlantic coasts than along the coasts of the North Pacific in North America.

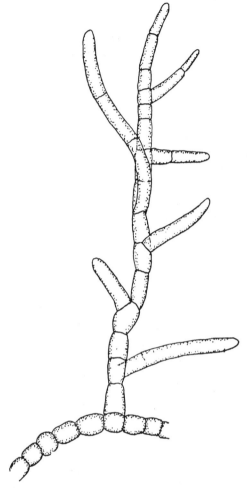

Figure A-15 The phycobiont oɪ the marine lichen *Verrucaria* sp. It is a green alga (*Dilabifilium arthopyreniae*) that can produce erect branches as well as more beadlike basal systems. (Tschermak-Woess, 1976). × 320.

About 21 species are obligate marine lichens (Kohlmeyer and Kohlmeyer, 1979); most genera are found in the upper intertidal and spray zones. Some of the more-common genera include *Verrucaria, Lichina, Caloplaca, Arthopyrenia,* and *Xanthorina.*

Verrucaria is a dark-brown to black crustose lichen that is common on the North Atlantic coasts. The species *V. adriatica* produces a distinct black belt in the upper intertidal zone. The phycobiont of this lichen is the green, filamentous alga *Dilabiflum arthopyreniae* (Figure A-15; Tschermak-Woess, 1976). The green alga is found living free next to the lichen, and when its quadriflagellated zoospores settle on the lichen they literally are drawn into the thallus.

Lichina confinis is another common species that occurs in the splash zone and resembles the brown alga *Pelvetia*. The intensity and magnitude of the spray determines the width of the lichen band.

Tidal amplitude and wave action are probably the two most critical factors controlling the vertical distribution of marine lichen belts. Seashore lichens tend to be distributed according to the amount of salt or fresh water in the environment (Fletcher, 1976). Tolerance to repeated wetting and drying seems to be most critical in controlling lichen zonation, whereas the actual effect of salinity is a secondary factor. Inorganic cations such as calcium are important in the growth of *Xanthoria parietina* (Fletcher, 1976). The presence of calcium decreases the potassium loss to fresh water. In another lichen, *Parmelia saxatilis*, no such relationship between calcium and potassium exists. On the whole lichens can be grouped according to the type of substratum they inhabit; this seems to be a worldwide correlation.

REFERENCES

Alexopoulos, C. J. and C. W. Mims. 1979. *Introductory Mycology*, 3rd ed. Wiley, New York.

Amon, J. P. and R. D. Arthur. 1980. The requirement for sodium in marine fungi: uptake and in corporation of amino acids. *Bot. Mar.* **23**: 639–644.

Andrews, J. H. 1976. The pathology of marine algae. *Biol. Rev.* **51**: 211–253.

Barghorn, E. S. and D. H. Linder. 1944. Marine fungi. Their taxonomy and biology. *Farlowia* **1**: 395–467.

Brooks, R. D., R. D. Goos, and J. McN. Sieburth. 1972. Fungal infestation of the surface and interior vessels of freshly collected driftwood. *Mar. Biol.* **16**: 274–278.

Feldmann, G. 1959. Une ustilaginale marine, parasite due *Ruppia maritima* L. *Rev. Gen. de Bot.* **66**: 35–39.

Fell, J. W. and I. M. Master. 1980. The association and potential role of fungi in mangrove detrital systems. *Bot. Mar.* **23**: 257–263.

Fletcher, A. 1976. Nutritional aspects of marine and maritime lichen ecology. In Brown, D. H., D. L. Hawksworth, and R. H. Bailey (eds.). *Lichenology: Progress and Problems*. Academic, New York, pp. 359–384.

Gessner, R. V. 1977. Seasonal occurrence and distribution of fungi associated with *Spartina alterniflora* from a Rhode Island estuary. *Mycologia* **69**: 477–491.

Gessner, R. V. 1978. *Spartina alterniflora* seed fungi *Can. J. Bot.* **56**: 2942–2947.

Höhnk, W. 1956. Studien zur Brack- und Seewassermykologie IV. Über die pilzliche Besiedlung verschieden salziger submerser. *Standorte. Veröff. Inst. Meeresforsch. Bremerhaven* **4**: 195–213.

Höhnk, W. 1958. Mykologische Notizen: I. Mikropilze im Eis. *Veröff. Inst. Meeresforsch. Bremerhaven.* **5**: 193–194.

Johnson Jr., T. W. and F. K. Sparrow Jr. 1961. *Fungi in Oceans and Estuaries*. J. Cramer, New York.

Jones, E. S. G. (ed.). 1976. *Recent Advances in Aquatic Mycology*. Wiley, New York.

Karling, J. S. 1944. Brazilian anisochytrids. *Amer. J. Bot.* **31**: 391–397.

Kaxama, F. Y. 1979. *Pythium* "red rot disease" of *Porphyra*. *Experimentia* **34**: 443–444.

Kazama, F. and M. S. Fuller. 1970. Ultrastructure of *Porphyra perforata* infected with *Pythium marinum*, a marine fungus. *Can. J. Bot.* **48**: 2103–2107.

Kohlmeyer, J. 1971. Fungi from the Sargasso Sea. *Mar. Biol.* **8**: 344–350.

Kohlmeyer, J. 1973. Fungi from marine algae. *Bot. Mar.* **26**: 201–215.

Kohlmeyer, J. 1974. Higher fungi as parasites and symbionts of algae. Ver*öff. Inst. Meeresforsch. Bremerhaven.* **Suppl. 5**: 339–356.

Kohlmeyer, J. 1979 Marine fungal pathogens among Ascomycetes and Deuteromycetes. *Experimentia* **35**: 437–439.

Kohlmeyer, J. 1980. Tropical and subtropical filamentous fungi of the western Atlantic Ocean. *Bot. Mar.* **23**: 529–540.

Kohlmeyer, J. and E. Kohlmeyer. 1979. *Marine Mycology. The Higher Fungi.* Academic Press, New York.

Meyers, S. P. 1957. Taxonomy of Marine Pyrenomycetes. *Mycologia* **49**: 475–528.

Newell, S. Y. and J. W. Fell. 1980. Mycoflora of turtlegrass (*Thalassia testudinum* Konig) as recorded after seawater incubation. *Bot. Mar.* **23**: 265–275.

Orpurt, P. A., S. P. Meyers, L. L. Boral, and J. Sims. 1964. Thalassiomycetes V. A new species of *Lindra* from turtle grass, *Thalassia testudinum* Konig. *Bull. Mar. Sci. Gulf Caribb.* **14**: 405–417.

Petersen, H. E. 1905. Contributions à la connaissance des phycomycetes marins (Chytridinae Fischer). *Oversigt Kgl. Danske Vidensk. Selskabs Forhandl.* **1905**: 439–488.

Renn, C. E. 1936. Persistence of the eel-grass disease and parasite on the American Atlantic coast. *Nature* **138**: 507–508.

Ritchie, D. 1957. Salinity optima for marine fungi affected by temperature. *Amer. J. Bot.* **44**: 870–874.

Ross, I. K. 1979. *Biology of the Fungi.* McGraw-Hill, New York.

Sparrow Jr., F. K. 1934. Observations on marine Phycomycetes collected in Denmark. *Dansk Bot. Arkiv.* **8**: 1–24.

Sparrow Jr., F. K. 1936. Biological observations on the marine fungi of Woods Hole waters. *Biol. Bull.* **70**: 236–263.

Sparrow Jr., F. K. 1960. *Aquatic Phycomycetes*, 2nd ed. University of Michigan Press, Ann Arbor.

Tschermak-Woess, E. 1976. Algal taxonomy and the taxonomy of lichens: the phycobiont of *Verrucaria adriatica*. In Brown, D. H., D. L. Hawksworth, and R. H. Bailey (eds.). *Lichenology: Progress and Problems*, Academic, New York.

Wagner, D. T. 1965. Development morphology of *Leptosphaeria discors* (Saccardo and Ellis) Saccardo and Ellis. *Nova Hedwiga* **9**: 45–61.

Wagner-Merner, D. T. 1972. Arenicolous fungi from the south and central gulf coast of Florida. *Nova Hedwiga* **23**: 915–922.

Marine Bacteria

INTRODUCTION

Bacteria carry out a number of significant biological functions in the marine environment and are especially evident in the anaerobic zone of sediments. Some of the bacterial activities in oceanic waters include converting and translocating minerals (elemental cycles, mineralization), producing and decomposing organic matter (food chain), and altering physical properties of marine communities (pH, oxygen level). A number of important interactions with marine plants are described in this Appendix.

Bacteria are found in all types of marine environments including the open-ocean water (planktonic bacteria or bacterioplankton; autotrophic and heterotrophic), the deep sea floor (epibacteria; heterotrophic), the neritic zone, and estuaries. The highest numbers of bacteria occur in subtropical to tropical water and along the coastal regions where organic matter and nutrients are high.

Marine Bacteria Defined

The question of whether marine bacteria really exist has been reviewed (Zobell, 1947; Brown, 1964; MacLeod, 1965; Wood, 1967) and a set of criteria has been established. The criteria are designed to separate indigenous marine forms, *autochthonous bacteria,* from terrestrial forms that are transient in the marine environment. Three basic criteria can be used:

1 A requirement for seawater media during at least the initial isolation and culture. However, many species can be acclimated by a series of gradual changes to freshwater media; the reverse is true of terrestrial bacteria acclimated toward seawater media.
2 An absolute requirement for chlorine and/or bromine.
3 An absolute requirement for a high level of magnesium (4–8 mM), whereas a terrestrial bacterium has a much lower requirement $\dashv$ 0.04–0.08 mM).

Organisms fitting these criteria are described in two recent texts, one by Sieburth (1979) and another edited by Colwell and Morita (1974). Another feature that is usually characteristic of marine bacteria is a tolerance to a narrow range of temperatures. Many marine bacteria are psychrophilic (growth is highest at temperatures below 20°C and will occur at 0°C) and have enzyme systems that are damaged or inactivated to temperatures above 28°C. In contrast, some marine species are barophiles, and require high pressures ($\sim$ 500 atm) for optimal growth. Psychrophily and barophily are typical of deep-sea bacteria.

General and Specific Characteristics of Marine Bacteria

All bacteria have prokaryotic cells (Figure B-1), as do the blue-green algae, or cyanobacteria (see Chapter 4). Thus these cells lack a membrane-bound nucleus and instead of chromosomes they have a circular strand of DNA found in the central region of the cell (nucleoid). The bacteria lack the typical organelles of eukaryotic cells (plastids, mitochondria, etc.). Photosynthetic forms have a series of vesiculate or lamellate cytoplasmic thylakoids in the peripheral portion of the cytoplasm. The bacterial flagellum, if present, is essentially a naked (outside the cell membrane) proteinaceous polymer consisting of the protein flagellin. One to many flagella are present; they may be polar or found all over the cell wall.

Bacteria can be grouped according to their morphology (Figure B-2). They have spherical (cocci), rod-shaped (bacilli), curved (vibrioid), or coiled (spiralloid) cells. Bacterial species can be separated into gram-positive and gram-negative cells, based on the staining reactions of the cell walls. The staining procedure is described in Chapter 4 (section on blue-green algal cell walls); basically it

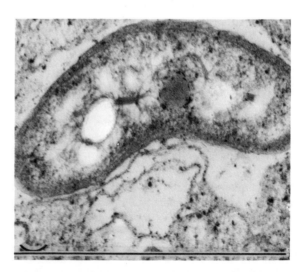

Figure B-1 A transmission electron micrograph of an endosymbiotic bacterium found in the green alga *Caulerpa prolifera*. The bacterial cell has a well-developed cell wall and some of the DNA fibrils are visible in the central region. The clear area is due to the removal of a lipid body during infiltration. The unit mark equals 1 μm (Dawes and Lohr, 1978).

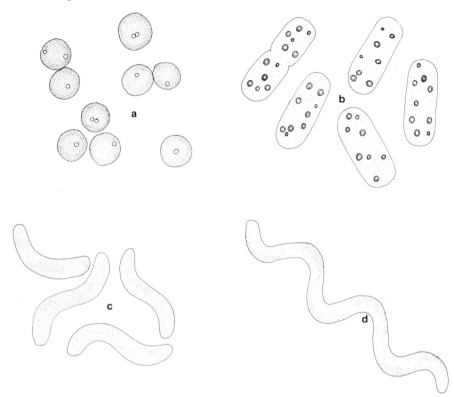

Figure B-2 Four types of bacterial cells, demonstrating the different morphologies. (*a*) The coccoid form is represented by *Thiocapsa roseopersicina*, common in muds of estuaries. The cells are 1.2–3 μm in diameter ($\times$ 8200). (*b*) The rod, or bacillus, form is represented by *Chromatium vinosum*; sulfur globules are visible in the cells. The cells measure 2 $\times$ 4–6 μm ($\times$ 8200). (*c*) A vibrioid form, *Rhodospirillum molischianum*, has cells 0.7–1 μm in diameter, each cell having about one twist. (*d*) A less-common form in the marine environment, the spirilloid form, is represented by *Cristispira* sp. isolated from the clam's crystalline style and found in many molluscan digestive tracts. The cells are 0.5 μm in diameter ($\times$ 9000).

separates those bacteria having an outer layer of lipopolysaccharide (gram-negative) from those with a cell wall that lacks the lipopolysaccharide layer (gram-positive).

The following generalizations can be made about marine bacteria:

1 Most marine species are single-celled, rodlike, and smaller than non-marine species.

2 About 95% of the marine bacteria are gram-negative, almost 70% are pigmented, and most of these pigmented cells fluoresce under ultraviolet light.

3 About 70% of the marine bacteria are mud-dwelling forms, most abundant at the surface. Most are able to carry on metabolic activities either in the pres-

ence or absence of oxygen (facultative aerobes). There are only a few ($\sim$ 10–15%) known marine forms that are either obligate anaerobes or obligate aerobes.

4 The highest diversity of species of marine bacteria occurs near shore (neritic zone) in depths of less than 50 m, and this parallels the distribution of phytoplankton.

5 Marine bacteria show optimal growth rates at lower temperatures (18–22°C), even if nonpsychrophylic, than their terrestrial or freshwater counterparts.

HABITATS OF MARINE BACTERIA

Microbial ecology encompasses the interrelationships between microorganisms and the environment, and in this section a variety of bacterial-environmental habitats will be considered. The role of bacteria in elemental cycles will be addressed in a separate section.

Sieburth (1979) divides microbial habitats into planktonic, epibiotic, benthic, and endobiotic, as well as others. The largest numbers (as high as 500,000 cells/ml) of marine bacteria occur in the neritic regions of the oceans. The majority of marine bacteria occur in the upper sediments (1–3 cm) of the seafloor, where there are sufficient nutrients to support growth of organisms dependent on organic material (heterotrophs). Cores taken from the ocean floor have demonstrated that living bacteria can be present in up to 5 m of mud, although the majority of species are found at the water/substratum interface.

Planktonic Bacteria

Bacterioplankton are suspended forms that live on soluble organic matter released from phytoplankton and zooplankton (Fuhrman *et al.*, 1980) as well as from the decomposition of planktonic material in the water column. In turn, the free-floating bacteria produce vitamins (e.g., B_{12}) essential to some phytoplankton species; they also release various nitrogenous and phosphorous compounds, and are eaten by unicellular organisms such as amoebas and dinoflagellates. Sieburth (1979) reported that bacterioplankton from 40–80 m in Narragansett Bay have doubling times of 4 hr. The growth rates showed a cyclic (diel) pattern, with rapid growth occurring in the afternoon through early evening and a lag phase lasting into the morning hours. Thus he points out that although autochthonous bacterioplankton may occur in low numbers in a given collection, it appears they are being heavily grazed on and may be important food sources.

Neustonic Habitat

The sea-air interface or the first 0.1 μm of seawater is a unique area of gas exchange, high light levels, and a wide variety of physical and chemical characteristics (slicks, debris, organic films). The level of dissolved organic carbon may be as high as 1400 mg C/l in neustonic waters, as opposed to 0.6 mg C/l in oceanic waters where phytoplankton occur (Sieburth, 1979). Because of these high levels of fixed carbon, nitrogen, and phosphorus, nutrients are not limiting and the bacterial populations found at this interface may be an order of magnitude higher than those found in planktonic waters. The significance of the *bacterial neuston* population lies in its importance in the food chain, serving as a food source for protozoans (tintinnids, radiolarians, amoebas) that are also most abundant in this interface region. Choanoflagellates (Chrysophyta) and dinoflagellates (Pyrrhophyta) are commonly found in this zone, and it is presumed these single cells, some of which are phagotrophic, feed in the neustonic layer.

Epibiotic Bacteria

The colonization of surfaces by bacteria is fairly well studied and has important ecological roles (Dempsey, 1981). Fouling of marine substrata results in unsightly appearances and causes major economic losses. From an ecological viewpoint, epibiotic bacteria are instrumental in the preparation of substrata for eukaryotic organisms. They also serve as a source of food for sessile and motile protozoans and are instrumental in the packaging of organic particles for assimilation into the food chain (e.g., detrital material).

Bacteria are usually the first detectable microorganisms to colonize new surfaces and are pioneer species (Figure B-3; Sieburth, 1979; Dempsey, 1981). The epibacteria adsorb to the surface, and through cell division (fission) they rapidly expand over the surface (Zobell, 1947). The result is a bacterial film, which can be covered by diatoms that slowly overtake the original bacterial population. Protozoa may become abundant enough to control the bacterial population by grazing; ultimately a balance between the two can result. The development of such

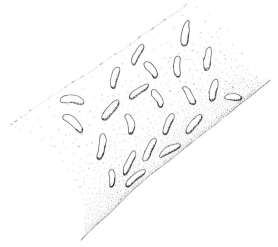

Figure B-3 Bacteria are common pioneer colonizers of new substrates such as the fibers of nylon nets, as shown in this sketch (Sieburth, 1979).

bacterial colonies on living seaweeds appears to be regulated by various antifouling substances (polyphenols, physodes), and by slimes that slough off the plant surface (carrageenan) or by cuticles produced by the algae.

More recently, marine bacteria have been found to package detrital organic matter that is then available for smaller organisms such as meiofauna and various ingesters in the food chain. The growth of bacterial populations on organic debris results in the production of slime, thus these colonies essentially cement together smaller bits of plant and animal matter; the sizes range from 1 to 10 μm in diameter, with the most common size about 2 μm.

Benthic Bacteria

Probably the best-studied microbial populations of the ocean are in the sediments. Marine bacteria are quite common there and a wide variety of metabolic types have been found in neritic and deep sea floor regions. The food source is primarily derived from the production of organics in the photic zone that sink and end up in the sediments of ocean basins. Much of the organic matter that sinks never reaches the seafloor however, because it is decomposed on the way down. Thus only about 10% of the organic matter produced in the photic zone reaches the seafloor. Chitin is aerobically broken down by bacteria such as *Beneakea* at the sediment surface if oxygen is available, or anaerobically decomposed in the sediments by *Desulfovibrio*.

Although only a small portion of the organic matter produced in the photic zone reaches the seafloor of ocean basins, an abundance of organic material is present because of the steady accumulation. The rate of bacterial decomposition of organic sediments is slow, in part because of the low temperatures (0–4°C) of the seafloor water. Most organic matter on the seafloor off Cape Cod takes 2 mo to decompose. Anaerobic decomposition is carried out below the aerobic surface of the sediments. About 50% of the mineralized carbon that occurs in the anaerobic sediments may be oxidized after sulfate reduction by such bacteria as *Desulfovibrio* (Sieburth, 1979). The bacterium obtains its oxygen by reduction of sulfate and releases hydrogen sulfide, which maintains anaerobic conditions.

The rate of anaerobic sulfate reduction can be used as a measure of organic productivity, as Howarth and Teal (1979) have shown for a New England salt marsh. They found that sulfate reduction in the substratum accounted for about 1800 g organic C/m^2 · yr; this approximately equaled the net primary productivity in the marsh. Three reasons for this high rate were noted: (1) belowground production by the salt marsh plant *Spartina alterniflora* provided a large amount of organic substrates, (2) sulfate was not limiting since it was rapidly resupplied to the peat by tidal exchange, and (3) sulfide concentrations remained below the toxic level because of the production of the stable pyrite, FeS$_2$ (Figure B-9).

A number of important bacteria that break down plant products have been identified from shallow aerobic sediment surfaces. Some of these decomposers possess enzymes specific for unique marine plant carbohydrates. Cellulose decomposers make up in abundance what they may lack in species diversity in the bottom sediments of the neritic region. Zobell (1947) reported about 1000 cellulose digesters, 10,000 starch digesters, and 100,000 glucose fermenters per gram of sea-bottom mud. Thus there is a miniature pyramid of digesters, one dependent on the other. The importance of digestion is evident when one considers the rapid deterioration of fishing nets and ship lines that are made of cellulose. As might be expected, bacterial enzymes have evolved that can digest the more-unique phycocolloids of marine algae such as agar (*Pseudomonas gelattica*, *Flavobacterium polysiphoniae*), alginic acid *Bacterium alginicum*), and carrageenan (*Pseudomonas* sp.; Weigel and Yaphe, 1966). Chitin digesters are also well known in marine sediments (*Bacillus chitonus*, *Desulfovibrio* sp.).

Endobiotic Habitats

Direct contact by marine bacteria with other organisms includes a variety of symbiotic or biotic relationships, ranging from mutualism to commensalism to parasitism. Mutualistic bacteria plant relationships are common, although endosymbiotic bacteria may fall into this group. Bacterial cells have been found in the vacuolar systems of a number of coenocytic green algae, including *Caulerpa* (Figure B-4; Dawes and Lohr, 1978), *Penicillus* (Turner and Friedmann, 1974), and *Bryopsis* (Burr and West, 1970). Leedale (1967) also observed endonuclear bacteria in euglenoid flagellates. The presence of such endosymbionts has been discussed by Dawes and Lohr (1978), and they concluded that the bacteria were not simply introduced during tissue preparation. Possible functions of endobacter-

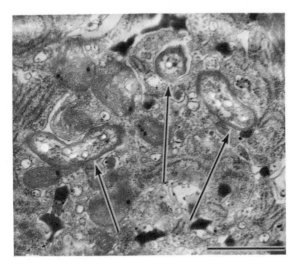

Figure B-4 Endosymbiotic bacteria are common in various green algal coencoytes (Dawes and Lohr, 1978). Three cells, one in cross section and two in longitudinal section (arrows), are shown in the cytoplasm of the green alga *Caulerpa prolifera* adjacent to typical eukaryotic organelles such as mitochondria and rough endoplasmic reticula. The unit mark is 1 μm.

ia may include nutrient cycling, especially elemental cycles within the vacuole. Commensalistic relationships might include the fungi/cyanobacteria relationship (see Appendix A) found in lichen associations. However, the best-known symbiotic relationships are the parasitic ones where bacterial pathogens have been identified as the disease-causing organisms of marine animals and plants. The best-known examples are the diseases of economically important marine organisms, especially those of commercial fishes (red disease of eels, fish lesions) and shellfish (oyster kills, red tail disease of lobster). The saprophytic actions of marine bacteria cause extensive economic losses through spoilage of fish and crustacean catches, as well as the rotting of cellulose fibers in nets and lines.

Andrews (1976) and Sieburth (1979) listed a number of examples of algal diseases. Bacteria are associated with algal galls, blights, and rots. Galls on *Chondrus, Cystoclonium,* and *Fucus* have been shown to contain bacteria; in the case of *Chondrus,* the bacterium is oval. The galls are proliferations of the inner (medullar) tissue of the alga; the bacterium is found in the intracellular spaces. It is thought that at least some of the so-called red algal parasites (see Chapter 7) may actually be "host" tumors resulting from bacterial infection.

The black rot of kelps (*Macrocystis, Pelagophycus, Egregia*) is visible as small dark lesions that enlarge at the blade tips. Infected blades or portions of the plants slough off, and the disease may cause a 90% drop in the harvesting of *Macrocystis* in the warmer summer months. The actual disease-causing bacterium has not been identified, and there is the possibility that the disease is caused by viruses since they are known to occur in algae (see LaClaire and West, 1977; Dodds, 1979; Hoffman and Stanker, 1976; and Sicko-Goad and Walker, 1979, for reviews). Dodds (1979) listed 19 algal species in which viruslike particles have been reported. As pointed out in Chapter 4, viral infections are well known in blue-green algae, and infection has been proposed as a method for controlling blooms of the more noxious species (Safferman and Morris, 1964). Daft and Stewart (1971) described four strains of bacterial pathogens of blue-green algae.

PHYSIOLOGICAL ASPECTS OF MARINE BACTERIA

Marine bacteria can be grouped according to their energy sources (Sieburth, 1979) and characterized by their growth requirements for sodium chloride, high salinities, and narrow tolerances to temperature and pressure (Colwell and Morita, 1974). Brock (1967) has shown that the rate of growth of a marine bacterium such as *Leucothrix mucor* will vary in culture (94 min doubling time) and the field (>600 min doubling time). Such differences are apparently dependent on the culture conditions and suggest that marine bacterial growth may be quite slow in nature.

Metabolic Types

Marine bacteria, like terrestrial ones, can be divided into heterotrophic and autotrophic forms. Autotrops include those obtaining energy through photosynthesis (phototrophs) or through oxidation of nitrogenous or sulfur compounds (chemolithotrophs). Heterotrophs derive energy from other organic compounds (saprophytes and parasites). Phototrophic bacteria can be divided into those that use reduced substances, such as sulfides, molecular hydrogen, or carbon compounds as external electron donors in photoassimilation of carbon dioxide, and those that use the hydrogen in water as the electron donor, releasing oxygen.

Anoxyphotobacteria

These bacteria possess only photosynthetic photosystem I*. The marine forms recognized in this category are found in the order Rhodospirillales (Pfennig and Truper, 1973) and can be divided into three families: Rhodospirillaceae (purple nonsulfur bacteria), Chromatiaceae (purple sulfur bacteria) and the Chlorobiaceae (green sulfur bacteria).

The purple nonsulfur bacteria are found in decomposing plant materials of inshore waters (Figure B-5). A number of these bacteria can oxidize sulfur to extracellular elemental sulfur (*Rhodospirillum rubrum, Rhodopseudomonas capsulata*; see Figure B-9) so that the name ''nonsulfur'' is no longer true for this family (Sieburth, 1979). The purple and green sulfur bacterria (Chlorobiaceae, Chromatiaceae) are limited to anaerobic environments where sulfides are present. Thus they are present in the upper layer of sulfur-containing sands and muds of tidal marshes (salt marshes, mangrove swamps) and in the anoxic, sulfide-containing waters of shallow basins in which water and gas exchange are poor. The purple and green bacteria act on sulfides produced by reduction of sulfate by the bacterium *Desulfovibrio* sp., which occurs deeper in the sediments. Such bacteria, being close to the sediment surface (for light), serve as an important food source and as decomposers of organic matter. The members of these three families are gram-negative and occur in a variety of shapes (cocci, rods, spirals).

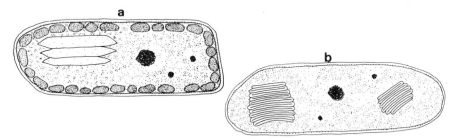

Figure B-5 A number of marine bacteria are photosynthetic species with bacteriochlorophyll found on photosynthetic membranes (thylakoids). (*a*) *Chlorobium* sp. is a green bacterium with intracytoplasmic thylakoid vesicles. (*b*) *Rhodospirilla* sp. is a purple bacteria with thylakoids organized in parallel lamellae. Both species are about 3–5 μm in diameter.

*Chlorophyll *a* bearing organisms (blue-green algae and eukaryotic plants) have two light-reaction systems, photosystem I and photosystem II. Each photosystem has its own unique form of chlorophyll *a*. Photosynthetic bacteria have only photosystem I in which neither water is used as an electron donor or oxygen released.

Most of the purple sulfur bacteria have vesicle-type photosynthetic membranes (thylakoids) that parallel the cell membranes (Figure B-5*a*). The green sulfur bacteria have specialized oblong vesicular thylakoids arranged parallel to the cytoplasmic membrane (Figure B-5*b*).

Oxyphotobacteria

The blue-green algae or cyanobacteria, are also prokaryotic, possess both photosystems I and II, use water as the electron donor, and produce oxygen in photosynthesis. All blue-green algae have chlorophyll *a*, as do all plants.

Because of their ability to evolve oxygen and possession of both photosystems, the cyanobacteria are included by most botanists under the general term algae. Typically the blue-green algae are covered in texts on algae as in Chapter 4 of this text. One unique blue-green alga, which is endozoic in sea squirts, has chlorophyll *b* as well. Lewin (1977) placed this unique green prokaryote (*Prochloron didemni*) in its own division, class, order, family, genus, and species because of the presence of both chlorophylls *a* and *b* (Figure B-6), the lack of bilisomes, and the appressed thylakoids. However, other bacteriologists such as Sieburth (1979) do not agree, and so it is retained here as a unique "blue-green" alga.

Chemolithotropic bacteria

These autotrophic bacteria derive their energy from the oxidation of ammonia, nitrite, and sulfur compounds, and they are important in various elemental cycles. The nitrifying bacteria oxidize ammonia through nitrite (equation 1) to nitrate (equation 2). Ammonia is a waste product in protein metabolism and is excreted by animals or released through decay of proteinaceous matter (ammonification).

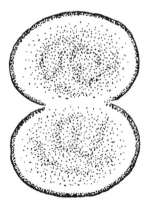

Figure B-6 Another photosynthetic prokaryotic organism is the unique coccoid genus *Prochloron* (Lewin, 1977). The cell is $10 \times 6.5 \, \mu$m and contains phycobilins, chlorophyll *a*, and chlorophyll *b*, making it the only known prokaryotic species with both chlorophylls. The phycobilins are not organized into bilisomes.

$$NH_3 \rightarrow NH_2OH \rightarrow 2(NOH) \rightarrow NO_2^-$$

(ammonia) (hydroxylamine) (?intermediate) (nitrite)   (1)

$$2NO_2^- + O_2 \rightarrow 2NO_3^-$$

(nitrite)   (nitrate)   (2)

All of the nitrifying bacteria are placed in the family Nitrobacteriaceae (Watson, 1965), which consists of aerobic species found on sediment surfaces where ammonia is being released. Nitrifying bacteria are especially common in fish mariculture operations; they are gram-negative and have a seven-layered cell envelope. Many genera have flattened cytomembranes (Figure B-7) where the oxidation is believed to take place. Details of the nitrogen cycle are presented under the section on ecological roles of marine bacteria in this Appendix and in Chapter 12.

Another group of chemolithotrophs comprises the colorless sulfur bacteria, and these can be divided into two subgroups, the sulfur-oxidizing bacteria (free sulfur to sulfate; sulfides to sulfate) and those containing sulfur granules. A number of species of the genus *Thiobacillus* occur in the marine environment and obtain their energy by the oxidation of sulfides and other reduced sulfur compounds to sulfate. One species of *Thiobacillus* is an obligate chemolithotroph that oxidizes elemental or free sulfur to sulfate. However, it can also oxidize hydrogen sulfide, metal sulfides, and sulfite as energy sources (Sieburth, 1979). Sulfur-oxidizing bacteria are widespread in marine muds and eutrophic systems and in the open sea; they are all gram-negative rods and many have polar flagella.

Bacteria that contain sulfur granules are also chemolithotrophic; they can form white to grey films near inshore sites where hydrogen sulfide is released from the decomposition of organic matter. These bacteria contain refractive sulfur granules. Marine forms include *Beggiatoa, Thioplaca,* and the large (up to 25 μm in diameter), motile genus *Thiovulum.* Both the sulfur-granule-containing bacteria and the colorless sulfur bacteria are involved in the sulfur cycle.

Heterotrophic bacteria

This is the largest group of marine bacteria and they exhibit a wide variety of metabolic features. Sieburth (1979) suggested that two general subgroups are

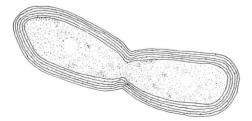

Figure B-7 The genus *Nitrosomonas* contains species that are chemolithotrophs and obtain their energy by oxidizing ammonia to nitrite. Note the parallel lamella in the rod-shaped cells. The cells are about 2.5 μm in diameter.

found in the oceans. One is a population of small bacterioplankton that do not grow well on nutrient-rich surfaces; the other is a population of larger epibacteria found in and on decaying matter and sediments of the shallow and deep sea floor. The former are autochthonous (indigenous), free-living forms that utilize dissolved organic matter found in low or limiting concentrations. The latter are autochthonous epibacteria that are usually larger, motile, have a high affinity for surfaces, and utilize solid organic substrates by means of exogenous enzymes (e.g., digesters).

Bacterioplankton are small-celled ($0.2-0.6$ μm diameter), truly planktonic species that have a sporadic growth pattern in the sea. The irregular growth pattern is due to their dependence on dissolved organic matter that is usually found in low amounts. Bacterioplankton are primarily gram-negative rods and can be collected in the same zones as the phytoplankton.

Epibacteria have been more intensively studied because they can be cultured on nutrient surfaces. The gram-negative forms are especially common perhaps because their cell wall retains various digestive enzymes that break down complex substrates exogenously. Thus gram-negative species can outcompete other forms for food. Gram-positive epibacteria appear to be limited primarily to protected

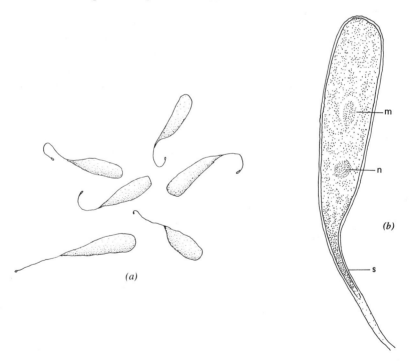

(a)

(b)

Figure B-8 *Caulobacterium*, a genus of stalked bacteria common in the marine environment. Cells are about 3.5 μm in diameter. (a) The narrow stalks are visible in entire cells. (b) The stalks contain cytoplasm (s) at least in the upper portion. The DNA is concentrated in a central body (n) called a nuceloplasm. A membraned body (m) called a mesosome is also evident.

habitats, such as sediments where there is a high concentration of organic matter (Sieburth, 1979). A number of the gram-negative epibacteria are distinguished by their ability to glide (Flexibacteria, Lewin, 1969), as well as by their unique stalks or hyphae (*Caulobacterium*, Figure B-8) and curved or spiral shapes. The heterotrophic epibacteria comprise the largest number of known marine bacteria and are involved in a variety of activities including diseases of animals and algae.

Physiological Requirements for Sodium

Most marine bacteria have an obligate adaptation to oceanic salinities, and this is best demonstrated by their sodium requirement (Colwell and Morita, 1974). It is not a simple osmotic need, since NaCl by itself cannot satisfy the requirement, although potassium may partially replace sodium. Marine genera, such as *Pseudomonas* spp., will lyse (break open) if grown in media with low concentrations of sodium. The addition of magnesium will reduce this lytic phenomenon; apparently magnesium functions in the maintenance of cell membranes.

Sensitivity to Temperature

Marine bacteria have limited tolerances to higher temperatures. For example, in one study it was found that about 25% of the bacteria died if seawater media or marine sediments were subjected to 30°C for 10 min, and 80% died if they were exposed to 40°C for 10 min (Morita, 1975). Obligate *psychrophilic* bacteria are common, especially in sediments from the deep sea floor. Usually they are rendered nonviable if exposed to 20°C for 10 min. Morita (1975) has defined psychrophilic bacteria as those "having an optimum temperature for growth at about 15°C or lower," whereas Inniss (1975) includes bacteria that can grow well at 0°C and have optimal growth below 20°C. Such organisms are well adapted to the low temperatures of the deep sea floor (~ 0–4°C) and play important roles in decomposition and detrital packaging in the cold deep-ocean waters; 90% of these waters are at 4°C. The sensitivity of marine bacteria to fairly "normal" terrestrial temperatures has been attributed to (1) disruption of intercellular organization and inactivation of enzymes or enzyme-forming systems; (2) accelerated use of the amino acid pool or an increase in lipid saturation; and (3) permeability loss or leakage of intracellular components (Colwell and Morita, 1974).

There is a relationship between temperature and salinity with regard to protein synthesis and bacterial growth. The tolerance to higher temperatures for *Vibrio marinus* increases with increasing salinity as seen in bacterial growth (Cooper and Morita, 1972). A similar pattern has been noted for marine fungi and has been termed the *phoma pattern*. With this pattern (Appendix A), fungi incubated at higher temperatures at various salinities exhibit increased tolerance to higher salinities. This may be due to changes in cell membrane properties.

The Adaptation to High Pressure

The average ocean depth (3800 m) exerts a pressure of 280 atm; at the greatest depths (12,000 m) there is a pressure of 1200 atm. Thus sediment bacteria of deep water are subjected to high hydrostatic pressures. Increased pressure results in a number of complex physicochemical changes, including changes in pH, water structure and gas solubility. Certain marine bacteria are called *barophiles* because they grow best, under high pressures, at which morphological changes (filament formation, pleomorphism and cell size) may occur. In many cases barophilic enzyme systems become inactive at normal pressures. The inactivity may be due to the type of enzymes a barophilic bacterium contains. Monomeric enzymes (simple forms) are stimulated by pressure and they may occur in barophilic bacteria. On the other hand, multimeric (complex forms) enzymes are inactivated at high pressures.

ECOLOGICAL ROLES OF MARINE BACTERIA

A number of bacterial activities in the marine environment were indicated in the section on habitats and the subsection on metabolic types. Three areas are singled out in this section, the food chain, elemental cycles, and uses by man. The conversion of organic forms of various elements to inorganic forms (mineralization) is not only important in various elemental cycles but is also significant from the viewpoint of mineral deposits such as iron ore (e.g., Minnesota Iron Ore Range), sulfur deposits (e.g., sulfur deposits in Louisiana and Texas), and petroleum deposits. Two other types of mineralization by marine bacteria should be considered as well: calcium carbonate deposition and manganese nodule formation.

Marine Bacteria and the Food Chain

Wood (1965) emphasized the importance of bacteria in making nutrients available for algae and marine angiosperms as well as their role in various elemental cycles. Marine bacteria not only function as decomposers of organic material, but they also package the detritus for primary consumers and serve as direct food for many zooplankton.

Colonization of detrital material was briefly discussed in the section on epibacteria; it is of major importance in the ultimate decomposition of detrital material. The associated microbial communities are complex (Sieburth, 1979), both qualitatively and quantitatively. Usually about 2–15 bacterial cells are found per 100 μm of surface area on detrital material such as eelgrass (*Zostera marina*) or macroalgal fragments. A number of larger marine animals feed on bacteria, including sponges, mussels, worms, oysters, and crayfish. Bacterial adsorption to new surfaces prepares the way for settlement of other organisms that feed on the epibacteria. Sorokin (1971) demonstrated that bacterial plankton were an important food source for various oceanic filter feeders even in the low, natural bacterial plankton concentrations of 0.3–0.7 g/m^3.

The Nitrogen Cycle

The role of marine bacteria in the oceanic nitrogen cycle is important. Bacteria are now known to be the sources of dissolved nitrate, nitrite, ammonia, and organically complexed forms of nitrogen. The cycle was presented in Chapter 12 and Figure 12-7; it can be divided into nitrogen fixation, nitrification, assimilation, ammonification, and denitrification. In this section some of the roles of bacteria are explored.

Since the nitrogen-fixing enzyme nitrogenase can reduce acetylene to ethylene and these two molecules are detectable at very low levels by gas chromatography, a rapid and simple method is now available to measure the activity of the enzyme. Using this acetylene-reduction method, a wide variety of marine environments have been examined, and nitrogen-fixing activity has been found in most. A large percentage ($\sim$ 80%) of all nitrogen fixation occurs in shallow bays, estuaries, tidal marshes, and shallow neritic sediments. Such habitats appear to be the primary sources of combined nitrogen in oceanic waters. Evidence for nitrogen fixation in mangrove swamps, salt marshes, seagrass beds and coral reefs has been published (see Chapters 16, 17, 18, 19), and a number of nitrogen-fixing bacteria have been isolated including *Azobacter, Clostridium,* and marine forms of *Desulfovibrio*. A wide variety of marine heterocyst-bearing and nonheterocystic blue-green algae have also been shown to fix nitrogen (see Chapter 4). The planktonic, filamentous, nonheterocystic blue-green alga *Trichodesmium* sp. (see Figure 20-8) is a particularly interesting case since it is found in open oceanic waters. Carpenter and Price (1977) found that this genus was an important nitrogen fixer in the Sargasso Sea. The genus is well known from the Red Sea (the name is derived from the reddish color of the filaments) and the Gulf of Mexico. In the Gulf of Mexico, *Trichodesmium* sp. is found in bloom proportions during some red tides (outbreaks or blooms of the dinoflagellate *Ptychodiscus brevis*) and may be important as a nitrogen fixer, thus supporting the red tide.

Marine nitrification involves the oxidation of ammonia to nitrite and then to nitrate (equations 1 and 2; see chemolithotrophic bacteria) by nitrifying bacteria found in the sediments and inshore waters. Watson (1965) and his co-workers (Watson and Waterbury, 1971) have identified a number of species of marine nitrifying bacteria including *Nitrocystis oceanus, Nitrospina gracilis,* and *Nitrococcus mobilis*. The first species oxidizes ammonia to nitrite and the other two oxidize nitrite to nitrate.

Phytoplankton are primarily involved in the assimilation, uptake, and incorporation of nitrates and ammonia into amino acids and proteins. In some studies it appears that ammonia can be directly incorporated, thus bypassing the nitrification process. Ammonification is the decomposition of proteins and amino acids and the subsequent release of ammonia; these processes are conducted by many heterotrophic epibacteria and fungi. A variety of marine bacteria can utilize peptone and protein nitrogen and some others (pseudomonads, corynebacterias) can utilize uric nitrogen.

Denitrification in seawater (conversion of nitrate and nitrite to molecular nitrogen) commonly occurs in regions of low oxygen (Goering and Dugdale, 1966).

A number of denitrifying strains of *Pseudomonas* spp. are known from shallow waters. For example *P. perfectomonas* contains denitrifying enzymes capable of anaerobically reducing nitrate, nitrite, or nitrous oxide. Thus marine bacteria can carry out all the major steps in the nitrogen cycle.

The Sulfur Cycle

The sulfur cycle, which is part biological and part chemical, has been well studied in the marine environment (Figure B-9). Sulfur is a common constituent of several amino acids (cysteine, cystine, methionine) and proteins, as well as a number of polysaccharides (carrageenan, agar, alginate). All organisms contain sulfur equal to about 1% of their dry weight. During anaerobic decomposition of organisms, the organic sulfur is released as free or methyl sulfates.

A variety of bacteria are involved in one or more parts of the sulfur cycle; these include chemolithotrophic sulfur-oxidizing bacteria (e.g., *Chromatium okenii*) and anoxyphototrophic purple nonsulfur bacteria (e.g., *Rhodospirillum rubrum*), as well as purple and green bacteria (e.g., *Chlorobium lunicola*). All of these bacteria can produce elemental sulfur by sulfide reduction. The photosynthetic forms that utilize H_2S and release elemental sulfur in photosynthesis may have a direct spatial relationship with the sulfide-producing bacteria. For example, the purple sulfur bacterium *Rhodothiobacterium* forms a pinkish to purple layer on intertidal sandy beaches and tidal marshes in tropical and subtropical areas. The main producers of H_2S in marine environments are anaerobic species of *Desulfovibrio*, which form a black anaerobic zone below the surface (Figure B-9). Many blue-green algae (cyanobacteria) will form crusts on the surface of high-organic sediments where H_2S is abundant. Cyanobacteria can tolerate the high reducing atmosphere and may directly utilize H_2S as a sulfur source. In most aquatic systems, H_2S can be chemically oxidized in the presence of O_2 to form SO_2, $S_2O_3^{2-}$, or SO_3^{2-}, depending on such factors as pH and rates of oxidation.

The sulfur now being mined in Texas and Louisiana probably resulted from the release of elemental sulfur by bacteria. Examples of chemolithotrophic bacteria that oxidize sulfur-containing compounds for energy include *Thiobacillus* (equation 3) and *Thiosulfus* (equation 4). Both bacteria can release extracellular sulfur granules as follows:

$$2S_2O_3^{2-}{}_3 + O_2 \rightarrow 2SO^{2-}{}_4 + 2S \tag{3}$$

$$S_4O^{2-}{}_6 + O_2 \rightarrow 2SO^{2-}{}_4 + 2S \tag{4}$$

Hydrogen sulfide concentrations from sulfate reduction could become quite high in substrata where anaerobic decomposition is occurring, such as tidal marshes. As noted earlier, Howarth and Teal (1979) showed that sulfate reduction can equal the rate of carbon fixation in a New England salt marsh. However, the sulfide levels can also be low because of pyrite (FeS_2) formation, thus preventing the occurrence of toxic levels of sulfide (Figure B-9). In the Black Sea, there is

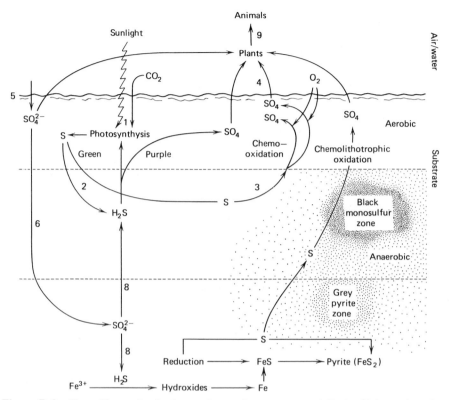

Figure B-9 The sulfur cycle. In the marine environment, especially in tidal marshes. It can be quite complicated, occurring in both aerobic and anaerobic regions of the substrate. Through photosynthesis (1) sulfur can be released from various sulfides, especially hydrogen sulfide (H_2S). Free sulfur can then be reduced to hydrogen sulfide (2) or oxidized to sulfates (3) by bacterial chemooxidation. The sulfates are incorporated into plants (4) and in turn eaten by animals (9). Sulfates can also be made available from land (5) and used directly by plants or reduced to H_2S (8) by bacteria after being transported to the anaerobic layer of the substrate (6). Within the pyrite region of the substrate of a tidal marsh, various iron and sulfur compounds can be formed that will result in pyrite (FeS_2), or through reduction will result in free sulfur that again can be chemolithotrophically used by bacteria to produce sulfate.

no available oxygen and only sulfur-reducing bacteria at 200 m. The name *Black Sea* is based in part on the sulfur activity of the ocean. On the west coast of South Africa there is a region where the Benguela and Guinea currents form a large eddy, which results in a major drop of organic matter to the seafloor. Anaerobic activity in this region can become sufficiently high to form large bubbles of hydrogen sulfide. The bubbles rise to the surface, killing the local plankton, fish, and even birds in the immediate region.

Other Elemental and Mineral Cycles

Phosphate can be a limiting compound for marine plant growth as it is usually found in low concentrations in marine waters. The enzyme phosphatase and phosphate-solubilizing bacteria have been reported from the sediments of neritic and estuarine systems (Ayyakkannu and Chandramohan, 1971). Such findings suggest that bacteria are important in releasing phosphates and increasing phosphate concentrations. This increase can affect the buffering capacities of sediments, and thereby support a greater microbial flora.

Bacteria have been found associated with the manganese nodules found on the deep sea floor (Greenslate, 1974). It is proposed that the bacteria growing on phytoplankton debris can change the oxidation states of iron and manganese, causing them to precipitate. The elements are then incorporated into the walls of foraminifera. It may be that the precipitation of iron by such bacterial activities (including chemolithotrophic activity) has been instrumental in the formation of the present iron ore deposits. For example, *Thiobacillus ferroxidans* can oxidize iron or reduced sulfur compounds, and *Ferrobacillus ferroxidans* obtains its energy from oxidation of iron as shown in equation 5.

$$4\ FeCO_3 + O_2 + 6H_2O \rightarrow 4\ Fe(OH)_3 + 4CO_2 \quad (\triangle F = -40\ kcal) \quad (5)$$

Calcium carbonate is one of the most important biological deposits. Carbonates are deposited by a wide variety of organisms (see Chapter 16), including red and green algae coccolithophorids, corals, bryozoans, molluscs, and a variety of bacteria. At least three types of calcium carbonate deposition have been identified as follows:

1 Precipitation of calcium carbonate may occur through the reaction of ammonium carbonate with calcium sulfate as shown in equation 6.

$$(NH_4)_2CO_3 + CaSO_4 \rightarrow CaCO_3 + (NH_4)_2SO_4 \quad (6)$$

The precipitation results from a pH change caused by an increase in CO_2 from microbial respiration during decomposition of nitrogenous organic substances.

2 Bacterial reduction of sulfates can also result in carbonate precipitation as shown in equation 7. Again, pH is critical.

$$CaSO_4 + CH_3COOH \rightarrow CaCO_3 + H_2S + H_2O + CO_2 \quad (7)$$

3 Denitrifying and ammonifying bacteria can also cause carbonate deposition as shown in equations 8 (denitrifying bacteria) and 9 (ammonifying bacteria).

$$Ca(NO_3)_2 + 3H_2{}^* + C^* \rightarrow CaCO_3 + 3H_2O + N_2 \quad (8)$$

*derived from the anaerobic oxidation of organic matter.

$$Ca(HCO_3)_2 + 2NH_4OH \rightarrow CaCO_3 + 2H_2O + (NH_4)_2CO_3 \quad (9)$$

Man and Marine Bacteria

Probably the most direct effect of marine bacteria on humans is the spoilage and disease of commercial marine organisms. The annual worldwide spoilage of fish and seaweed catches is approximately $1 billion, which is a loss of more than 15% of the total marine product. Bacterial diseases of fish and algae (Andrews, 1976; Sieburth, 1979) are known, but the importance of bacterial diseases in nature is not well documented.

Marine bacteria indirectly serve as important food sources for zooplankton, providing vitamins and nutrients for phytoplankton growth as well as supporting food chains and man's marine products. A direct use of marine bacteria for the biological breakdown of petroleum products that might be spilled or leaked into the oceans has been proposed (Gutnick and Rosenberg, 1977). Thus, they suggest that a screening process should be established to find microbes capable of decomposing and consuming specific types of oil. Three criteria would be used to select such bacteria: an efficient hydrocarbon uptake system, specific oxygenases to handle the hydrocarbons, and activation (growth) only in the presence of these hydrocarbons. The authors point out that such microbes should exist today, even though human-induced oil pollution is a twentieth century event.

TAXONOMY OF MARINE BACTERIA

There are a number of sources for taxonomic information on marine bacteria. Wood (1967) listed five orders (Pseudomonadales, Eubacteriales, Actinomycetales, Beggiatoales, Myxobacteriales), and about 20 families containing marine bacteria. Sieburth (1979) summarized a number of metabolic and physiological groups (phototrophic, chemolithotrophic, authochthonous planktobacteria, nondistinctive epibacteria, aerobic gram-positive epibacteria). Within these groupings he presented various orders, based on the eighth edition of *Bergey's Manual of Determinative Bacteriology* (Buchanan and Gibbons, 1974). Bergey's manual is the primary taxonomic account of bacteria, and it recognizes two divisions, the Photobacteria (photosynthetic) and Scotobacteria (bacteria indifferent to light). Three classes are recognized within the Photobacteria: blue-green, red, and green bacteria. Three classes are also recognized in the Scotobacteria: true bacteria, rickettsia (obligate intracellular forms), and mycoplasms (lacking cell walls). A number of excellent papers on the taxonomy of specific groups of marine bacteria have been published, including studies on aerobic eubacteria (Baumann et al., 1972), luminous bacteria (Reicholt and Baumann, 1973), Orndorff and Colwell, 1980) and the aerobic genus *Beneckea* (Baumann et al., 1971).

REFERENCES

Andrews, J. H. 1976. The pathology of marine algae. *Biol. Rev.* **51**: 211–253.

Ayyakkannu, K. and D. Chandramohan. 1971. Occurrence and distribution of phosphate and solubilizing bacteria and phosphatase in marine sediments at Porto Novo. *Mar. Biol.* **2**: 201–205.

Baumann, P., L. Baumann, and M. Mandel. 1971. Taxonomy of marine bacteria: the genus *Beneckea. J. Bact.* **107**: 268–294.

Baumann, L., P. Baumann, M. Mandel, and R. D. Allen. 1972. Taxonomy of aerobic marine *Eubacteria. J. Bact.* **110**: 402–429.

Brock, T. D. 1967. Bacterial growth rate in the sea: Direct analysis by thymidine autoradiography. *Science* **155**: 81–83.

Brown, A. D. 1964. Aspects of bacterial response to the ionic environment. *Bacteriol. Rev.* **28**: 296–329.

Buchannan, R. E. and N. E. Gibbons (eds.). 1974. *Bergey's Manual of Determinative Bacteriology.* 8th ed. Williams and Wilkins, Baltimore, Md.

Burr, F. A. and J. A. West. 1970. Light and electron microscope observations on the vegetative and reproductive structures of *Bryopsis hypnoides. Phycologia* **9**: 17–37.

Carpenter, E. J. and C. C. Price, IV. 1977. Nitrogen fixation, distribution, and production of *Oscillatoria (Trichodesmium)* spp. in the western Sargasso and Caribbean Seas. *Limnol. Oceanogr.* **22**: 60–72.

Colwell, R. R. and R. Y. Morita (eds.). 1974. *Effect of the ocean environment on microbial activities. Proc. Second U.S.-Japan Conf. on Mar. Microbiol.* University Park Press, Baltimore, Md.

Cooper, M. F. and R. Y. Morita. 1972. Interaction of salinity and temperature on net protein synthesis and viability of *Vibrio marinus. Limnol. Oceanogr.* **17**: 556–565.

Daft, M. J. and W. D. P. Stewart. 1971. Bacterial pathogens of freshwater blue-green algae. *New Phytol.* **70**: 819–829.

Dawes, C. J. and C. A. Lohr. 1978. Cytoplasmic organization and endosymbiotic bacteria in the growing points of *Caulerpa prolifera. Rev. Algol. N. S.* **8**: 309–314.

Dempsey, M. J. 1981. Marine bacterial fouling: A scanning electron microscope study. *Mar. Biol.* **61**: 305–315.

Dodds, J. A. 1979. Viruses of marine algae. *Experimentia* **35**: 437–439.

Fuhrman, J. A., J. W. Ammerman, and F. Azam. 1980. Bacterioplankton in the coastal euphotic zones. Distribution, activity, and possible relationships with phytoplankton. *Mar. Biol.* **60**: 201–207.

Goering, J. J. and R. C. Dugdale. 1966. Denitrification rates in an island bay in the equatorial Pacific Ocean. *Science* **154**: 505–506.

Greenslate, J. 1974. Microorganisms participate in the construction of manganese nodules. *Nature* **249**: 181–183.

Gutnick, D. L. and E. Rosenberg. 1977. Oil tankers and pollution: a microbiological approach. *Ann. Rev. Microbiol.* **31**: 379–396.

Hoffman, L. R. and L. H. Stanker. 1976. Virus-like particles in the green alga *Cylindrosapsa. Can. J. Bot.* **54**: 2827–2841.

Howarth, R. W. and J. M. Teal. 1979. Sulfate reduction in a New England salt marsh. *Limnol. Oceanogr.* **24**: 999–1013.

Innis, W. E. 1975. Interaction of temperature and psychrolphilic microorganisms. *Ann. Rev. Microbiol.* **29**: 445–464.

LaClaire, J. W. II, and J. A. West. 1977. Virus-like particles in the brown alga *Streblonema. Protoplasma* **93**: 127–130.

Leedlae, F. G. 1967. Observations on endonuclear bacteria in euglenoid flagellates.*Österr. Bot.* **116**: 279–294.

Lewin, R. A. 1969. A classification of flexibacteria. *J. Gen. Microbiol.* **58**: 189–206.

Lewin, R. A. 1977. *Prochloron,* type genus of the Prochlorophyta. *Phycologia* **14**: 149–152.

MacLeod, R. A. 1965. The question of the existence of specific marine bacteria. *Bacteriol. Rev.* **29**: 9–23.

Morita, R. Y. 1975. Psychrophilic bacteria. *Bacteriol. Rev.* **39**: 144–167.

Orndorff, S. A. and R. R. Colwell. 1980. Distribution and identification of luminous bacteria from the Sargasso Sea. *App. Environ. Microbiol.* **39**: 963–987.

Pfennig, N. and H. G. Truper. 1973. The Rhodospirialles (phototrophic or photosynthetic bacteria). In Laskin, A. E. and H. A. Lechevalier (eds.). *Handbook of Microbiology,* vol. 1. CRC Press, Cleveland, Ohio, pp. 17–27.

Reicholt, J. L. and P. Baumann. 1973. Taxonomy of the marine luminous bacteria. *Archiv Mikrobiol.* **94**: 283–330.

Safferman, R. S. and M. E. Morris. 1964. Control of algae with viruses. *J. Amer. Water Works Assoc.* **56**: 1217–1224.

Sicko-Goad, L. and G. Walker. 1979. Virplasm and large virus-like particles in the dinoflagellate *Gymnodinium uberrimum. Protoplasma* **99**: 203–210.

Sieburth, J. M. 1979. *Sea Microbes.* Oxford University Press, New York.

Sorokin, Y. I. 1971. Abundance and production of bacteria in the open water of the Central Pacific. *Oceanography* **11**: 85–94.

Turner, J. B. and E. I. Friedmann. 1974. Fine structure of capitular filaments in the coenocytic green alga *Penicillus. J. Phycol.* **10**: 125–134.

Watson, S. W. 1965. Characteristics of a marine nitrifying bacterium *Nitrococystis oceanus* sp. n. *Limnol. Oceanogr.* **10** (Suppl.): R274–R289.

Watson, S. W. and J. B. Waterbury. 1971. Characteristics of two marine nitrite oxidizing bacteria, *Nitrospina gracilis* nov. gen. nov. sp. and *Nitrococcus mobilis* nov. gen. nov. sp. *Archiv Mikrobiol.* **77**: 203–230.

Wood, E. J. F. 1965. *Marine Microbial Ecology.* Reinhold, New York.

Wood, E. J. F. 1967. *Microbiology of Oceans and Estuaries.* Elsevier, Amsterdam.

Weigl, J. and W. Yaphe. 1966. The enzymic hydrolysis of carrageenan by *Pseudomonas carrageenovora:* purification of a kappa carrageenase. *Can. J. Microbiol.* **12**: 939–947.

Zobell, C. E. 1947. *Marine Microbiology.* Chronica Botanica, Waltham, Mass.

Taxonomic Index

Page numbers in *italics* indicate illustrations.

Subject Index

Page numbers in *italics* indicate illustrations.

DATE DUE